TOPOS THEORY

TOPOS THEORY

P. T. JOHNSTONE

Dept. of Pure Mathematics and Mathematical Statistics
University of Cambridge
Cambridge, England

DOVER PUBLICATIONS, INC.
Mineola, New York

Bibliographical Note

This Dover edition, first published by Dover Publications, Inc., in 2014, is an unabridged republication of the work originally published by Academic Press, Inc., New York, in 1977.

International Standard Book Number

ISBN-13: 978-0-486-49336-7
ISBN-10: 0-486-49336-9

Manufactured in the United States by Courier Corporation
49336901 2014
www.doverpublications.com

Preface

The origins of this book can be traced back to a series of six seminars, which I gave in Cambridge in the winter of 1973/74, and which formed the nucleus of the present chapters 1–6. Further seminars in the same series, covering parts of chapters 0, 7 and 9, were given by Barry Tennison and Robert Seely. By popular request, the notes of these seminars were written up and enjoyed a limited circulation. In the summer of 1974, I began to revise and expand these notes, with the idea that they might some day form a book. During the winter and spring of 1975, whilst at the University of Liverpool, I was able to give a course of lectures covering the material of chapters 0–5 and 8 in some detail. By the end of this period, I had a fairly clear picture of the overall shape of the book; and (encouraged by Michael Butler) I began the actual writing of it in July 1975. From October 1975 to March 1976 I was at the University of Chicago, where there was a weekly seminar on topos theory organized by Saunders Mac Lane and myself; the material covered during this period was drawn mainly from chapters 2, 4, 5, 6 and 9, and the speakers (in addition to myself) were Kathy Edwards, Steve Harris and Steve Landsburg. Also during this period, I wrote the text of chapters 2–5 and most of chapter 6; the remainder of the text was completed during May–July 1976 after my return to Cambridge.

The lectures and seminars mentioned above had a very direct influence on the text of the book, and all those who attended them (in particular those whose names appear above) deserve my thanks for the part they have played in shaping it. But I have also benefited from more informal contacts with many mathematicians at conferences and elsewhere. Among those whose (largely unpublished) ideas I have gladly borrowed are Julian Cole, Radu Diaconescu, Mike Fourman, Peter Freyd, André Joyal and Chris Mulvey. John Gray gave me valuable advice on 2-categorical matters, and Jack Duskin and Barry Tennison helped to improve my understanding of cohomology. And I must thank Jean Bénabou for the many ideas I have consciously or inadvertently borrowed from him, and Tim Brook for his help in the compilation of the bibliography.

There remain four mathematicians to whom I owe a debt which must be acknowledged individually. Myles Tierney introduced me to topos theory through his lectures at Varenna in 1971; looking back on the published version [TV], I still find it incredible that he managed to teach me so much in eight short lectures. Gavin Wraith's help and encouragement have meant a great deal to me, and his Bangor lectures [WB] served as a model for some parts of this book. Like every other worker in topos theory, I owe Bill Lawvere an overwhelming debt in general terms, for his pioneering insights; but I have also benefited at a more personal level from his ideas and conversation. Above all, I have to express my indebtedness to Saunders Mac Lane: but for him I should never have become a topos-theorist in the first place; and the care with which he has read through the original typescript, and provided suggestions for improvement in almost every paragraph, has been altogether out of the ordinary. If there are any major errors or obscurities still remaining in the text, they are surely a testimony to my perversity rather than his lack of vigilance.

On a different, but no less significant, level, I must also thank the Universities of Liverpool and Chicago, and St John's College, Cambridge, for employing me during the writing of the book; Paul Cohn, for accepting it for publication in the L.M.S. Monographs series; and the staff of Academic Press for the efficiency with which they have transformed my amateurish typescript into the book which you see before you.

Cambridge, June 1977 P.T.J.

Contents

Introduction

Topos theory has its origins in two separate lines of mathematical development, which remained distinct for nearly ten years. In order to have a balanced appreciation of the significance of the subject, I believe it is necessary to consider the history of these two lines, and to understand why they came together when they did. I therefore begin this Introduction with a (personal, and doubtless strongly biased) historical survey.

The earlier of the two lines begins with the rise of *sheaf theory*, originated in 1945 by J. Leray, developed by H. Cartan and A. Weil among others, and culminating in the published work of J. P. Serre [107], A. Grothendieck [42] and R. Godement [TF]. Like a great deal of homological algebra, the theory of sheaves was originally conceived as a tool of algebraic topology, for axiomatizing the notion of "local coefficient system" which was essential for a good cohomology theory of non-simply-connected spaces; and the full title of Godement's book indicates that it was still viewed in this light in 1958. But well before this date, the power of sheaf theory had been recognized by algebraic and analytic geometers; and in more recent years, its influence has spread into many other areas of mathematics. (For two widely-differing examples, see [49] and [106].)

However, in algebraic geometry it was soon discovered that the topological notion of sheaf was not entirely adequate, in that the only topology available on an abstract algebraic variety or scheme, the Zariski topology, did not have "enough open sets" to provide a good geometric notion of localization. In his work on descent techniques [43] and the étale fundamental group [44], A. Grothendieck observed that to replace "Zariski-open inclusion" by "étale morphism" was a step in the right direction; but unfortunately the schemes which are étale over a given scheme do not in general form a partially ordered set. It was thus necessary to invent the notion of "Grothendieck topology" on an arbitrary category, and the generalized notion of sheaf for such a topology, in order to provide a framework for the development of étale cohomology.

This framework was built up during the "Seminaire de Géometrie Algébrique du Bois Marie" held during 1963–64 by Grothendieck with the assistance of M. Artin, J. Giraud, J. L. Verdier and others. (The proceedings of this seminar were published in a revised and greatly enlarged version [GV], including some notable additional results of P. Deligne, eight years later.) Among the most important results of the original seminar was the theorem of Giraud, which showed that the categories of generalized sheaves which arise in this way can be completely characterized by exactness properties and size conditions; in the light of this result, it quickly became apparent that these categories of sheaves were a more important subject of study than the sites (= categories + topologies) which gave rise to them. In view of this, and because a category with a topology was seen as a "generalized topological space", the (slightly unfortunate) name of *topos* was given to any category satisfying Giraud's axioms.

Nevertheless, toposes were still regarded primarily as vehicles for carrying cohomology theoreies; not only étale cohomology, but also the "fppf" and crystalline cohomologies, and others. The power of the machinery developed by Grothendieck was amply demonstrated by the substantial geometrical results obtained by using these cohomology theories in the succeeding years, culminating in P. Deligne's proof [159] of the famous "Weil conjectures"—the mod-p analogue of the Riemann hypothesis. And the machinery itself was further developed, for example in J. Giraud's work [38] on nonabelian cohomology. But the full import of the dictum that "the topos is more important than the site" seems never to have been appreciated by the Grothendieck school. For example, though they were aware of the cartesian closed structure of toposes ([GV], IV 10), they never exploited this idea to the full along the lines laid down by Eilenberg–Kelly [160]. It was, therefore, necessary that a second line of development should provide the impetus for the elementary theory of toposes.

The starting-point of this second line is generally taken to be F. W. Lawvere's pioneering 1964 paper on the elementary theory of the category of sets [71]. However, I believe that it is necessary to go back a little further, to the proof of the Lubkin–Heron–Freyd–Mitchell embedding theorem for abelian categories [AC]. It was this theorem which, by showing that there is an explicit set of elementary axioms which imply all the (finitary) exactness properties of module categories, paved the way for a truly autonomous development of category theory as a foundation for mathematics.

(Incidentally, the Freyd–Mitchell embedding theorem is frequently regarded as a culmination rather than a starting-point; this is because of what seems to me a misinterpretation (or at least an inversion) of its true significance. It is commonly thought of as saying "If you want to prove something about an abelian category, you might as well assume it is a category of modules"; whereas I believe its true import is "If you want to prove something about categories of modules, you might as well work in a general abelian category"—for the embedding theorem ensures that your result will be true in this generality, and by forgetting the explicit structure of module categories you will be forced to concentrate on the essential aspects of the problem. As an example, compare the module-theoretic proof of the Snake Lemma in [HA] with the abelian-category proof in [CW].)

This theorem was soon followed by Lawvere's paper [71], setting out a list of elementary axioms which, with the addition of the non-elementary axioms of completeness and local smallness, are sufficient to characterize the category of sets. (In a subsequent paper [72], Lawvere provided a similar axiomatization of the category of small categories, and D. Schlomiuk [105] did the same for the category of topological spaces.)

One may well ask why this paper was not immediately followed by the explosion of activity which greeted the introduction of elementary toposes six years later. In retrospect, the answer is that Lawvere's axioms were too specialized: the category of sets is an extremely useful object to have as a foundation for mathematics, but as a subject of axiomatic study it is not (*pace* the activities of Martin, Solovay *et al.*!) tremendously interesting—it is too "rigid" to have any internal structure. In a similar way, if the abelian-category axioms had applied only to the category of abelian groups, and not to categories of modules or of abelian sheaves, they too would have been neglected. So what was needed for the category of sets was an axiomatization which would also cover set-valued functor categories and categories of set-valued sheaves—i.e. the axioms of an elementary topos.

In his subsequent papers ([73], [75]), Lawvere began to investigate the idea that the two-element set $\{true, false\}$ can be regarded as an "object of truth-values" in the category of sets; in particular, he observed that the presence of such an object in an arbitrary category enables us to reduce the Comprehension Axiom to an elementary statement about adjoint functors. The same idea was at the heart of the work of H. Volger ([125], [126]) on logical and semantical categories.

Meanwhile, the embedding-theorem side of things was advanced by M. Barr [2], who formulated the notion of *exact category* and used it as the basis of a non-additive embedding theorem. The closely-related notion of *regular category* was formulated independently by P. A. Grillet [41] and D. H. Van

Osdol [122], who used it in their investigations of general sheaf theory; and Barr himself observed that Giraud's theorem could be regarded as little more than a special case of his embedding theorem. This perhaps represents (logically, if not chronologically) the first coming-together of the two lines of development mentioned earlier.

However, at about the same time Lawvere's attention also turned towards Grothendieck toposes; he observed that every Grothendieck topos has a truth-value object Ω, and that the notion of Grothendieck topology is closely connected with endomorphisms of Ω (see [LH]). During the year 1969–70, Lawvere and M. Tierney (who had earlier contributed to the theory of exact categories) began to investigate the consequences of taking "there exists an object of truth-values" as an axiom; the result was elementary topos theory. A remarkably large proportion of the basic theory was developed in that 12-month period, as will be apparent from the large number of theorems in chapters 1–4 of this book whose proof is credited to Lawvere and Tierney.

Once these theorems became known to mathematicians at large (i.e. after Lawvere's lectures at Zürich and Nice [LN] in the summer of 1970, and the Dalhousie conference [LH] in January 1971), they were immediately taken up and further developed by several people. One of the first and most important was P. Freyd, whose lectures at the University of New South Wales [FK] explored the embedding theory of toposes; in retrospect this seems to have been something of a blind alley, in that the inversion of the usual metatheorem, mentioned above in connection with abelian categories, applies with even more force to topos theory—since the great virtue of the topos axioms is their elementary character, one should not have to appeal to a non-elementary embedding theorem to prove elementary facts about toposes. (Freyd's embedding theorem will not be found in this book; but the most important (and elementary) part of it, which shows that any topos can be embedded in a Boolean topos, is proved in §7.5.) Nevertheless, Freyd's work contained a great many important technical results; in particular his characterization of natural number objects is a theorem of major importance.

Amongst other early workers on topos theory, one should mention J. Bénabou and his student J. Celeyrette in Paris [BC], and A. Kock and G. C. Wraith in Aarhus [KW]. C. J. Mikkelsen, a student of Kock, was the first to prove that one of the Lawvere–Tierney axioms, that of finite colimits, could be deduced from the others; his thesis [84] also contains many important contributions to lattice-theory in a topos.

In view of the Lawvere–Tierney proof of the independence of the continuum hypothesis [117], it became a matter of importance to determine the precise relationship between elementary topos theory and axiomatic set theory. The answer was found independently by J. C. Cole [18], W. Mitchell [85] and G. Osius [92]. W. Mitchell also introduced an idea which has since become central to the subject: namely that each topos gives rise to an internal language which can be used to make "quasi-set-theoretical" statements about objects and morphisms of the topos. Whilst the original idea is due to Mitchell, its most enthusiastic proponent has undoubtedly been J. Bénabou, and his students have used the internal language extensively in recent years.

The next major advance was made by R. Diaconescu, a student of Tierney whose thesis was completed in 1973. Diaconescu's theorem [30] was important not only for the insight it gave into the 2-categorical structure of $\mathfrak{Top}$, but also because it represented the first significant exploitation of the theory of internal categories. (This theory had developed over the years in a rather haphazard way, largely through unpublished work of J. Bénabou.) As an encore, Diaconescu proved the relative Giraud theorem; Giraud himself [39] had proved a relative version of his theorem (by non-elementary means) for Grothendieck toposes, and W. Mitchell had formulated the correct elementary form. But Mitchell was able to prove this only in the special case when the "object of generators" (see 4.43) is 1; it turned out that Diaconescu's theorem was the essential tool needed to prove the general case. At about the same time, P. T. Johnstone [52] also used internal categories in his proof that Grothendieck's construction of the associated sheaf functor could be carried over to the elementary setting.

The next development (which in fact overlapped the previous ones) was the rise of the notion of toposes as theories and the concept of classifying topos. In a sense, this goes right back to Lawvere's work [176] on algebraic theories, but its connection with topos theory began with the work of M. Hakim [45], a student of Grothendieck, on relative schemes, in the course of which she constructed the classifiers for rings and local rings, and established their fundamental properties. In 1972, A. Joyal and G. E. Reyes [RM] isolated the notion of "coherent theory" (=finitary geometric theory, in our terminology), and proved that every such theory has a classifying topos; their work was later extended by Reyes and M. Makkai [82] to cover infinitary geometric theories.

It was F. W. Lawvere [LB] who first observed that, in view of the work of Joyal and Reyes, the theorem of P. Deligne on points of coherent toposes was precisely equivalent to the Gödel–Henkin completeness theorem for finitary geometric theories; and Lawvere too conjectured the "Boolean-valued completeness theorem" for infinitary theories whose topos-theoretic equivalent was proved by M. Barr [4].

Once again, Diaconescu's theorem provided the key to the "relativization" of the Joyal–Reyes results; the decisive step was taken in 1973 by G. C. Wraith, who constructed an object classifier over an arbitrary topos with a natural number object. From there to the general existence theorem for classifying toposes was little more than a formality; it was achieved independently by A. Joyal, M. Tierney [119] and J. Bénabou [8].

This brings our historical survey up to date, at least where major results are concerned. Now let us consider the present position of topos theory, and its future prospects.

The first thing which must be said is that the basic theory of elementary toposes (i.e. the contents of chapters 1–5 of this book) seems to be almost completely worked out. Indeed, I am aware of only one substantial unanswered question arising from these five chapters (namely the existence of finite (pseudo-)colimits in $\mathfrak{Top}$, touched on in §4.2); doubtless there are many other minor points to be cleared up, and several theorems whose proofs will be improved and simplified in time, but the foundations of the subject do appear to be pretty stable. This is of course a bad thing: it is vital to the health of a subject as basic as topos theory that its fundamental tenets should be the subject of continual review and improvement, and I am uncomfortably aware that by writing this book I have contributed largely to the concreting-over of these foundations. My only defence against this charge is that it seemed to me that the solidification was taking place anyway, and it was better that it should happen in print than in an unpublished folklore accessible only to insiders.

The average mathematician, who regards category theory as "generalized abstract nonsense", tends to regard topos theory as generalized abstract category theory. (No doubt it has inherited this reputation from its parent, the Grothendieck approach to algebraic geometry.) And yet S. Mac Lane [179] regards the rise of topos theory as a symptom of the *decline* of abstraction in category theory, and of abstract algebra in general. I am convinced that Mac Lane is right, and that his insight points the way to the most probable future development of topos theory; for almost all the *recent* work of significance in topos theory has been concerned not with toposes as an abstract and isolated area of mathematics, but with toposes as an aid to understanding and clarifying concepts in other areas. (See, for example, [36], [57], [63], [79], [88], [90], [112], [130].)

To take a specific example, consider the general existence theorem for classifying toposes (6.56). One's first reaction on seeing this theorem is to admire its elegance and generality; the second reaction (which comes quite a long time later) is to realize its fundamental uselessness—a quality which, by the way, it shares with the General Adjoint Functor Theorem. For the only possible use of such a theorem is to reduce the study of a particular geometric theory to the study of its generic model (or conversely, to reduce the study of a particular topos to that of the theory whose generic model it contains), and the theorem as proved in §6.5 simply does not provide an effective means of passing from the one to the other. Thus the "syntactic" proof of the same theorem in §7.4, though appreciably messier, is much more valuable in practice—and it is this proof, not the later one given in the earlier chapter, which has inspired most subsequent work on the subject.

In saying that the future of topos theory lies in the clarification of other areas of mathematics through the application of topos-theoretic ideas, I do not wish to imply that, like Grothendieck, I view topos theory as a machine for the demolition of unsolved problems in algebraic geometry or anywhere else. On the contrary, I think it is unlikely that elementary topos theory itself will solve any major outstanding problems of mathematics; but I do believe that the spreading of the topos-theoretic outlook into many areas of mathematical activity will inevitably lead to the deeper understanding of the real features of a problem which is an essential prelude to its correct solution.

What, then, is the topos-theoretic outlook? Briefly, it consists in the rejection of the idea that there is a fixed universe of "constant" sets within which mathematics can and should be developed, and the recognition that the notion of "variable structure" may be more conveniently handled within a universe of *continuously variable* sets than by the method, traditional since the rise of abstract set theory, of considering separately a domain of variation (i.e. a topological space) and a succession of constant structures attached to the points of this domain. In the words of F. W. Lawvere [LB], "Every notion of constancy is relative, being derived perceptually or conceptually as a limiting case of variation, and the undisputed value of such notions in clarifying variation is always limited by that origin. This applies in particular to the notion of constant set, and explains why so much of naïve set theory carries over in some form into the theory of variable sets". It is this generalization of ideas from constant to variable sets which lies at the heart of topos theory; and the reader who keeps it in mind, as an ultimate objective, whilst reading this book, will gain a great deal of understanding thereby.

Next, a few words on some of the things which I have not done in this book.

(1) In the definition of a topos, I have taken cartesian closedness and the existence of Ω as two separate axioms, instead of combining them into a single axiom of power-objects as suggested by A. Kock [66]. (The equivalence of Kock's axiom is, however, covered in the Exercises to chapter 1.) At a practical level, I would defend this decision on two grounds: (a) that there are a number of results in the book (notably in chapter 2) which use only cartesian closedness and not the full topos axioms, and some (e.g. Theorem 1.47) where exponentials and Ω are used in essentially different ways in the same proof; and (b) that if one takes the power-object definition, one is obliged (as in [WB]) to follow it immediately with the rather technical proof that this definition implies cartesian closedness, and one is in danger of losing one's readers at this critical point. On a more philosophical level, I would add (c) that the definition via power-objects is really a set-theorist's rather than a category-theorist's definition of a topos, in that it subordinates the notion of "function" to that of "subset" by means of the set-theoretic device of identifying functions with their graphs. One of the principal features of category theory is that it takes "morphism" as a primitive notion, on a level with (*not*, incidentally, superior to) that of "object"; it is therefore right that the definition of a topos should include its closed structure.

(2) I have not introduced the Mitchell–Bénabou language until rather late in the book, at the end of chapter 5. I know that there are some people whose ideal textbook on topos theory would begin with the definition and just enough development of exactness properties to introduce the language and prove the soundness of its interpretation; thereafter all proofs would be conducted within the formal language. I do not agree with this approach; I believe that it is impossible to appreciate the full power of the Mitchell–Bénabou language until you have had some experience of proving things without it (indeed, this is almost the only place in the book where I have consciously followed a particular ordering of material for pedagogical rather than logical reasons). There is also the point that the formal-language approach breaks down when confronted with the relative Giraud theorem (4.46); whilst the Mitchell–Bénabou language is a very powerful tool in proofs within a single topos, it is not well adapted to proofs in which we have to pass back and forth between two toposes by a geometric morphism. (It is possible that the proof of 4.46 could be shortened by using the language of locally internal categories, but that is a different matter.)

(3) I have already mentioned that Freyd's embedding theorem [FK] will not be found in this book. In consequence, Freyd's concept of well-pointed topos plays a relatively minor role; it is not introduced until §9.3.

(4) I have not included any reference to Freyd's more recent development (unpublished as yet) of the theory of *allegories*. This theory sets out to do for the category of sets and relations what topos theory does for sets and functions; Freyd has been known to maintain that it provides a simpler and more natural basis than topos theory for many of the ideas developed in this book, but I personally remain unconvinced of this.

(5) I have not mentioned the work being done by D. Bourn [13], R. Street [113], [114] and others, on the development of a 2-categorical analogue of topos theory. It appears to me, however, that the fundamentals of this theory have not yet reached a sufficiently definitive state for treatment in book form.

(6) One generalization of topos theory whose omission I do slightly regreat is J. Penon's notion of *quasitopos* [99]. However, I feel that to introduce it early in the book would simply have introduced extra complications in the proofs without any benefits in the form of additional well-known examples; and to introduce it later on would have involved a good deal of duplication. I hope, nevertheless, that O. Wyler's forthcoming notes on quasitoposes (promised in [130]) will help to fill this gap.

(7) The phrase "Grothendieck universe" does not appear anywhere in the book. This is intentional; I have deliberately been as vague as possible (except in §9.3) about the features of the set theory which I am using, since it really doesn't matter. Topos theory is an elementary theory, and its main theorems are not—or ought not to be—dependent on recondite axioms of set theory. (In fact I am a fully paid-up member of the Mathematicians' Liberation Movement founded by J. H. Conway [157].) If pressed, however, I would admit to using a Gödel–Bernays-type set theory having a distinction between small categories (sets) and large categories (proper classes); but I also wish to consider certain "very large" 2-categories (notably $\mathfrak{Cat}$ and $\mathfrak{Top}$) whose objects are themselves large categories. If I wished to be strictly formal about this, I should need to introduce at least one Grothendieck universe; but since all the statements I wish to make about $\mathfrak{Cat}$ and $\mathfrak{Top}$ are (equivalent to) elementary ones, there is no *real* need to do so. In order to retain some set-theoretic respectability, I have limited myself to considering sheaves only on small sites; this has the disadvantage that we cannot state Giraud's theorem in its slickest form (a category is a Grothendieck topos iff it is equivalent to the category of canonical sheaves on itself), but is otherwise not as irksome as the authors of [GV] would have us believe.

Finally, I have to state my position on the most controversial question in the whole of topos theory: how to spell the plural of topos. The reader will already have observed that I use the English plural; I do so because (in its mathematical sense) the word topos is not a direct derivative of its Greek root, but a back-formation from topology. I have nothing further to say on the matter, except to ask those toposophers† who persist in talking about topoi whether, when they go out for a ramble on a cold day, they carry supplies of hot tea with them in thermoi.

† I am indebted to Miles Reid for suggesting the terms "toposopher" and "toposophy": and I urge my fellow-toposophers to adopt them.

Notes for the Reader

Throughout the book, a single numbering system is used for definitions, lemmas, theorems, remarks, etc.; the number $n.pq$ normally denotes the qth numbered reference in the pth paragraph of chapter n, except that in certain paragraphs which have more than nine numbered references, the numbers run on consecutively from $n.p9$. (Thus 8.20 is the tenth numbered reference in § 8.1, and 5.51 the eleventh in § 5.4.) Fortunately, it has been possible to do this without overlapping. While this system may not have logic on its side, it does combine the advantage of simple and easily-remembered numbers (no profusion of decimal points!) with that of references which are easy to locate. A.n denotes the nth numbered reference in the Appendix.

At the end of each chapter will be found a number of exercises: about ten on each of the earlier chapters, more on the later ones. They vary considerably in difficulty, some being completely routine, whereas others are quite substantial. I have not given any indication of which exercises I consider to be easier (the order is that of the material in the preceding chapter to which they refer), but I have given fairly copious hints for most of the harder ones. In quite a number of cases, the result of an exercise is used in either the exercises or the text of a subsequent chapter; these exercises are distinguished by an obelus (†).

The following summary of the logical interdependence of the various chapters may be useful to the reader who is interested in one particular topic. Chapter 0 contains a summary of certain background material which is required either to motivate the definition of a topos, or to provide a source of examples. Chapters 1–5 form the core of the book; of these chapters 1–4 follow a more or less geodesic path (with a few digressions such as §4.2) from the definition of a topos (1.11) to the relative Giraud theorem (4.46) and the existence of pullbacks in $\mathfrak{BTop}/\mathscr{E}$ (4.48). The logical dependence relation in these four chapters is thus fairly close to being a total order.

However, the majority of the material in chapter 2 (on internal categories) is fairly technical, and some readers may find it fairly difficult at a first reading. I would advise such readers to omit the whole of chapter 2, except for Theorem 2.32 (which is important, and has applications in other areas than internal category theory), and go on to chapter 3. (There are some references to chapter 2 in §3.3, but you can refer back for these as necessary.) You may then go on to the first paragraph (only) of chapter 4, the whole of chapter 5 except for some parts of §5.3, and even the first two paragraphs of chapter 6, before returning to tackle chapter 2.

Chapter 5 introduces a number of concepts which, although part of the mainstream of topos theory, are not involved in the proof of the relative Giraud theorem. In particular, it contains a description of the internal language of a topos, which is used freely in the second half of the book.

The last four chapters present various extensions and applications of the basic theory; I originally hoped to make them logically independent, so that they could be read in any order, but inevitably some cross-connections have established themselves. The following table summarizes the important ones:

Before reading	*it is advisable to read*
7.4	6.3 and 6.5
8.1	7.5
8.4	6.2
9.1	6.2 and 6.4

There are some further cross-connections between the exercises on these four chapters (see, for example, exercises 6.11, 8.7, 9.6 and 9.14).

The Appendix is a presentation of material originally intended for inclusion in chapter 2; it was removed from there because, whilst it seems certain to become part of the mainstream of topos theory before very long, it is possible that the basic definition of "locally internal category" has not yet reached its final form. It may be read at any time after chapter 2, although it does make a number of references to later chapters.

Throughout the book, references to the bibliography are enclosed in square brackets. The bibliography itself is divided into four sections: section A consists of "standard references" to other areas of mathematics (e.g. lattice theory, algebraic topology) which are used when a theorem or definition from one of these areas is quoted in the text. Section B contains works of a general nature on topos theory, and some introductory articles written for non-specialist readers (e.g. [MB], [WI]). Section C contains the remaining references on topos theory, and a number of closely-related papers on category theory, sheaf theory, etc. Sections B and C together aim to present a complete list of articles so far published on topos theory; however, I do not include short abstracts of talks, nor Ph.D. theses unless they contain important results not published elsewhere. Section D contains the remaining papers which are referred to in the text. References in sections A and B are indicated by two-letter codes; those in sections C and D are consecutively numbered. In all four sections, I have indicated the *Mathematical Reviews* review numbers, where they exist.

TOPOS
THEORY

Chapter 0

Preliminaries

0.1. CATEGORY THEORY

This paragraph is not intended to give a detailed introduction to the basic ideas of category theory; our intention is rather to indicate some of the concepts and theorems which we shall be assuming as familiar, and to set out some standard notations. The reader who considers himself unfamiliar with these basic concepts would be well advised to refer to the excellent book by Mac Lane [CW], or to any other standard text on category theory, before proceeding further with this book.

We shall normally use script capitals ($\mathscr{C}$, $\mathscr{D}$, $\mathscr{E}, \ldots$) to denote "large" categories. When we make the assertion "$\mathscr{C}$ is a category", without further qualification, we mean that $\mathscr{C}$ is a model of the "elementary theory of categories" [72], i.e. $\mathscr{C}$ is a *metacategory* in the sense of [CW], chapter I. This means that we do not regard $\mathscr{C}$ as being formally defined within a particular model of set theory; in particular, if X and Y are objects of $\mathscr{C}$, we do not require that the morphisms of $\mathscr{C}$ from X to Y shall form a set.

However, except in chapter 9, we shall normally assume that we are given a (fixed) model of some suitable set theory (including the axiom of choice when necessary); and we shall use the letter $\mathscr{S}$ to denote the *category of sets and functions* which we obtain from it. We use the term *small category* for a category whose morphisms form a set. If $\mathbf{C}$ is a small category, we write $\mathscr{S}^{\mathbf{C}^{\mathrm{op}}}$ for the category of *presheaves* on $\mathbf{C}$, i.e. contravariant functors from $\mathbf{C}$ to $\mathscr{S}$; amongst the objects of $\mathscr{S}^{\mathbf{C}^{\mathrm{op}}}$, we have the *representable* functors h_U, where U is an object of $\mathbf{C}$, defined by $h_U(V) = \hom_{\mathbf{C}}(V, U)$. (For typographical reasons, we shall sometimes write $h(U)$ for h_U; we shall also write h^U for the covariant representable functor $\hom_{\mathbf{C}}(U, -)$.) We shall make frequent use of the following two results:

0.11 LEMMA (Yoneda [187]). *For objects U, X of* $\mathbf{C}$ *and* $\mathscr{S}^{\mathbf{C}^{op}}$ *respectively, there is a bijection (natural in both variables) between morphisms* $h_U \longrightarrow X$ *in* $\mathscr{S}^{\mathbf{C}^{op}}$ *and elements of the set* $X(U)$. ■

0.12 LEMMA. *Any object of* $\mathscr{S}^{\mathbf{C}^{op}}$ *can be expressed as a colimit of a diagram whose vertices are representable functors.*

Proof. Let X be an object of $\mathscr{S}^{\mathbf{C}^{op}}$, and let $(\mathbf{C}\downarrow X)$ denote the (small) *comma category* whose objects are pairs (U, α), U an object of C and $\alpha: h_U \longrightarrow X$ in $\mathscr{S}^{\mathbf{C}^{op}}$, and whose morphism are commutative triangles

$$\begin{array}{ccc} h_U & \longrightarrow & h_V \\ & \searrow \quad \swarrow & \\ & X & \end{array}$$

in $\mathscr{S}^{\mathbf{C}^{op}}$.

Then we have an obvious "forgetful" diagram $(\mathbf{C}\downarrow X) \longrightarrow \mathscr{S}^{\mathbf{C}^{op}}$ given by $(U, \alpha) \longmapsto h_U$; and the colimit of this diagram is readily seen to be X. ■

We assume that the reader is familiar with the notions of *limit* and *colimit*, and of *adjoint functors*. (We use the notation $T \dashv G$) for "T is left adjoint to G".) A word of warning here: when we say that a category *has limits* of a particular type, we mean that there is given, for each diagram of the appropriate type, a *canonical* choice of limit. Thus for example, when we say that $\mathscr{S}$ has binary products we mean that, given sets X and Y, there exists not merely a set whose elements are in 1–1 correspondence with ordered pairs $\langle x, y\rangle$, but a canonical such set, namely the set of all ordered pairs. itself. However, when we say that a functor *preserves limits*, we do not imply that it preserves the canonical choice of limits; and the statement that a given object is *a* limit of a certain diagram does not imply that it is the canonical one.

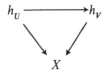

If a category $\mathscr{C}$ has a terminal object (i.e. a limit for the empty diagram), we denote it by 1; and if X is any object of $\mathscr{C}$, we use X also to denote the unique morphism $X \longrightarrow 1$, and 1_X (or simply 1) for the identity morphism on X. (The confusion inherent in this notation may be loosely justified by the fact that 1_X is the terminal object of the category $\mathscr{C}/X$ of *objects over* X, and that if $Y \xrightarrow{f} X$ is any object of this category, f is also the unique morphism of $\mathscr{C}/X$ from f to 1_X.) If $\mathscr{C}$ has products and/or pullbacks, we use the letter π to denote the canonical projection of a product or pullback onto one of its factors, with suffix 1, 2, 3, ... to denote the first, second, third, ... factor. Similarly, we shall usually (but not exclusively) use the letter ν, with appropriate suffix, to denote the inclusion of one factor in a coproduct.

We also assume familiarity with the notions of *monad* (or triple) and *comonad*, and of *algebras* over a monad. We shall make use of the "crude tripleability theorem" of Beck [153] in its "reflexive-coequalizer" form; recall that a parallel pair $X \underset{g}{\overset{f}{\rightrightarrows}} Y$ in a category $\mathscr{C}$ is said to be *reflexive* if there exists $Y \xrightarrow{h} X$ such that $fh = gh = 1_Y$. [In the case $\mathscr{C} = \mathscr{S}$, this is equivalent to saying that the image of $X \xrightarrow{(f,g)} Y \times Y$ is a reflexive binary relation on Y.]

0.13 THEOREM. *Let* $\mathscr{C} \underset{U}{\overset{F}{\rightleftarrows}} \mathscr{A}$ *be a pair of functors with* $F \dashv U$, *and let* $\mathbb{H}$ *be the monad on* $\mathscr{C}$ *induced by this adjunction. Suppose that* $\mathscr{A}$ *has coequalizers of reflexive pairs, that U preserves them and that U reflects isomorphisms. Then U is* monadic; *i.e. the comparison function* $K: \mathscr{A} \longrightarrow \mathscr{C}^{\mathbb{H}}$ *is an equivalence of categories, where* $C^{\mathbb{H}}$ *denotes the category of* $\mathbb{H}$*-algebras.* ■

We shall also make use of the following theorems about categories of algebras:

0.14 THEOREM (Eilenberg–Moore [162]). *Let* $\mathbb{H} = (H, \eta, \mu)$ *be a monad on* $\mathscr{C}$, *and suppose H has a right adjoint G. Then there is a unique comonad structure* $\mathbb{G} = (G, \varepsilon, \delta)$ *on G, such that the category* $\mathscr{C}_{\mathbb{G}}$ *of* $\mathbb{G}$*-coalgebras is isomorphic to* $\mathscr{C}^{\mathbb{H}}$, *by an isomorphism which identifies the two forgetful functors.* ■

0.15 THEOREM ("Adjoint lifting theorem"; see [54]). *Let* $\mathbb{H}$, $\mathbb{K}$ *be monads on categories* $\mathscr{C}$, $\mathscr{D}$ *respectively,* $T: \mathscr{C} \longrightarrow \mathscr{D}$ *a functor, and* $\bar{T}: \mathscr{C}^{\mathbb{H}} \longrightarrow \mathscr{D}^{\mathbb{K}}$ *a functor which is a "lifting" of T in the sense that the square*

$$\begin{array}{ccc} \mathscr{C}^{\mathbb{H}} & \xrightarrow{\bar{T}} & \mathscr{D}^{\mathbb{K}} \\ \downarrow{\scriptstyle U} & & \downarrow{\scriptstyle U} \\ \mathscr{C} & \xrightarrow{T} & \mathscr{D} \end{array}$$

commutes, where the U's denote forgetful functors. Suppose also that $\mathscr{C}^{\mathbb{H}}$ *has coequalizers of reflexive pairs. Then if T has a left adjoint, so has* $\bar{T}$. ■

0.16 THEOREM (Linton [178]). *Let* $\mathbb{H}$ *be a monad on a category* $\mathscr{C}$, *and suppose* *that $\mathscr{C}^{\mathbb{H}}$ has coequalizers of reflexive pairs. Then if $\mathscr{C}$ has all finite (resp. all $\mathscr{S}$-indexed) coproducts, $\mathscr{C}^{\mathbb{H}}$ has all finite (resp. all $\mathscr{S}$-indexed) colimits.* ■

It is clear from 0.13, 0.15 and 0.16 that coequalizers of reflexive pairs ("reflexive coequalizers") play an important rôle in the theory of monads. It is therefore appropriate to insert at this point a lemma which, although capable of very wide application (as we shall see in chapter 6), has not found its way into the standard texts on category theory.

0.17 LEMMA. *Let*

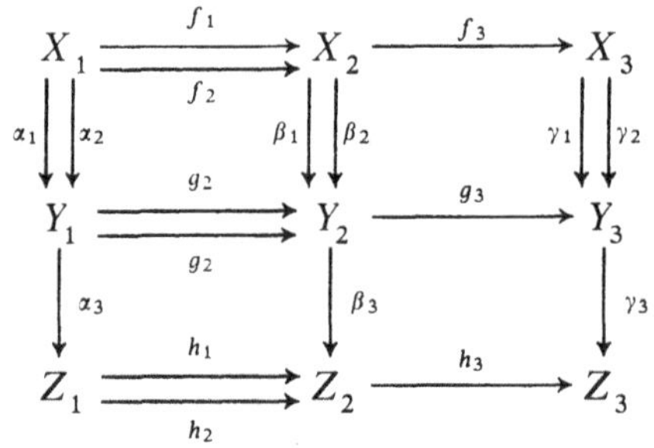

be a diagram in any category satisfying the "obvious" commutativity conditions (i.e. $\beta_i f_j = g_j \alpha_i$ for $i = 1, 2$, $j = 1, 2$, etc.), in which the rows and columns are coequalizers and the pairs (f_1, f_2) and (α_1, α_2) are reflexive. Then the diagonal $X_1 \underset{\beta_2 f_2}{\overset{\beta_1 f_1}{\rightrightarrows}} Y_2 \xrightarrow{\gamma_3 g_3} Z_3$ is a coequalizer.

Proof. First we note that

$$\begin{aligned}\gamma_3 &= \text{coeq}(\gamma_1, \gamma_2)\\ &= \text{coeq}(\gamma_1 f_3, \gamma_2 f_3) \text{ since } f_3 \text{ is epi}\\ &= \text{coeq}(g_3\beta_1, g_3\beta_2).\end{aligned}$$

So the lower right-hand square is a pushout; i.e. a morphism $Y_2 \xrightarrow{\theta} T$ factors through $Y_2 \xrightarrow{\gamma_3 g_3} Z_3$ iff it coequalizes both (g_1, g_2) and (β_1, β_2). But if this condition is satisfied, then $\theta\beta_1 f_1 = \theta\beta_2 f_1 = \theta g_1 \alpha_2 = \theta g_2 \alpha_2 = \theta\beta_2 f_2$. Conversely, if $\theta\beta_1 f_1 = \theta\beta_2 f_2$ and $X_2 \xrightarrow{s} X_1$ is a common splitting for f_1 and f_2, then $\theta\beta_1 = \theta\beta_1 f_1 s = \theta\beta_2 f_2 s = \theta\beta_2$; and similarly $\theta g_1 = \theta g_2$. So $Y_2 \longrightarrow Z_3$ is a coequalizer of $\beta_1 f_1$ and $\beta_2 f_2$. ■

In conclusion, we must mention two areas of category theory not covered in

[CW]. One is the theory of 2-categories: a *2-category* $\mathfrak{C}$ is a category whose "hom-sets" have the structure of (not necessarily small) categories. That is to say for each pair of objects (X, Y) of $\mathfrak{C}$, we have a category $\mathfrak{C}(X, Y)$ whose objects are the morphisms (or *1-arrows*) of $\mathfrak{C}$ from X to Y, and whose morphisms are called *2-arrows* of $\mathfrak{C}$; and the operation of composing 1-arrows of $\mathfrak{C}$ is functorial in both variables. Introductions to the theory of 2-categories will be found in [155], [167] and [172].

The reader is warned that, when discussing 2-categories, we shall normally use terms such as "functor" and "limit", and notations such as $\mathfrak{C}/X$, in what the Australian school [172] would call the "pseudo" sense, in which diagrams which commute "on the nose" are replaced by diagrams commuting up to a (specified) 2-isomorphism. For example, $\mathfrak{C}/X$ is the 2-category whose objects are 1-arrows $Y \xrightarrow{f} X$ of $\mathfrak{C}$, and whose 1-arrows are triangles

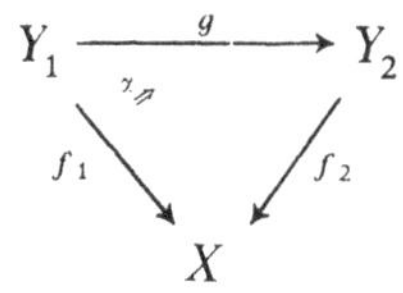

commuting up to a specified 2-isomorphism α; a 2-arrow of $\mathfrak{C}/X$ from (g_1, α_1) to (g_2, α_2) is a 2-arrow $g_1 \xrightarrow{\beta} g_2$ of $\mathfrak{C}$ such that $(f_2 * \beta).\alpha_1 = \alpha_2$. Similarly, when we speak of a functor $T: \mathfrak{C} \longrightarrow \mathfrak{D}$ between 2-categories, we do not imply that T commutes exactly with composition of 1-arrows, but only up to coherent natural 2-isomorphism. (It is understood that if $\mathscr{C}$ is an ordinary category, we identify it with the *locally discrete* 2-category whose objects and 1-arrows are the objects and morphisms of $\mathscr{C}$, and whose only 2-arrows are identities.)

Our reason for departing from the Australian convention on nomenclature is that it is these "pseudo" concepts which arise most frequently in practice. If we wish to emphasize the fact that a particular functor commutes exactly with composition of 1-arrows (i.e. is a functor in the Australian sense), we shall call it a *strict functor*. In one paragraph (4.2), we shall have occasion to consider the still weaker notion of *lax functor*, in which the 2-arrows up to which diagrams commute need not even be invertible; we shall define these lax concepts explicitly when we need them. And in §2.4 we shall encounter an example of a *bicategory*, which is a "pseudo-2-category" in the Australian sense; i.e. it is defined by the same data as a 2-category, but the unitary and associative laws for composition of 1-arrows hold only up to coherent natural 2-isomorphism.

Finally, we must introduce the notion of category of fractions. If $\mathscr{C}$ is a category and Σ a class of morphisms of $\mathscr{C}$, then the *category of fractions* $\mathscr{C}\Sigma^{-1}$ is defined (up to equivalence) by requiring that there exist a functor $P_\Sigma: \mathscr{C} \longrightarrow \mathscr{C}\Sigma^{-1}$ which is universal among functors $\mathscr{C} \longrightarrow \mathscr{D}$ sending all the morphisms in Σ to isomorphisms.

0.18 Definition. A class Σ of morphisms of $\mathscr{C}$ is said to admit a *calculus of right fractions* if:

(i) Σ is closed under composition, and contains all identity morphisms of $\mathscr{C}$.
(ii) Given a diagram

$$X \xrightarrow{\ f\ } Z \xleftarrow{\ \sigma\ } Y \quad \text{with } \sigma \in \Sigma \text{, there exists} \quad X \xleftarrow{\ \tau\ } T \xrightarrow{\ g\ } Y$$

with $\tau \in \Sigma$, such that $f\tau = \sigma g$. (Note: T need not be a pullback.)
(iii) Given a diagram $X \overset{f}{\underset{g}{\rightrightarrows}} Y \xrightarrow{\sigma} Z$ with $\sigma \in \Sigma$, $\sigma f = \sigma g$, there exists $T \xrightarrow{\tau} X$ in Σ with $f\tau = g\tau$. ■

0.19 Theorem (Gabriel–Zisman [CF]). *Let $\mathscr{C}$ be a category with finite limits, and Σ a class of morphisms of $\mathscr{C}$ admitting a calculus of right fractions. Then the category of fractions $\mathscr{C}\Sigma^{-1}$ has finite limits, and P_Σ preserves them.*

Proof. We show that $\mathscr{C}\Sigma^{-1}$ can be described as follows: its objects are those of $\mathscr{C}$, and morphisms $X \longrightarrow Y$ in $\mathscr{C}\Sigma^{-1}$ are equivalence classes of diagrams

$$\begin{array}{ccc} Z & \xrightarrow{\ f\ } & Y \\ \downarrow{\scriptstyle \sigma} & & \\ X & & \end{array}$$

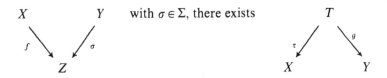

in $\mathscr{C}$ ($\sigma \in \Sigma$), under the relation that (σ_1, f_1) is equivalent to (σ_2, f_2) iff there exists

$$Z_1 \xleftarrow{\ \tau_1\ } T \xrightarrow{\ \tau_2\ } Z_2$$

such that $\sigma_1\tau_2 = \alpha_2\tau_2 \in \Sigma$ and $f_1\tau_2 = f_2\tau_2$. (It is not hard to verify that this does define an equivalence relation.) To compose two morphisms

$$X \xleftarrow{\ \sigma_1\ } T \xrightarrow{\ f_3\ } Y \xleftarrow{\ \sigma_2\ } U \xrightarrow{\ f_2\ } Z$$

in $\mathscr{C}\Sigma^{-1}$, we form a commutative square

$$\begin{array}{ccc} V & \xrightarrow{\ g\ } & U \\ {\scriptstyle\tau}\downarrow & & \downarrow{\scriptstyle\sigma_2} \\ T & \xrightarrow{\ f_1\ } & Y \end{array}$$

with $\tau \in \Sigma$ (using 0·18(ii)), and then define $(\sigma_2, f_2)(\sigma_1, f_1) = (\sigma_1\tau, f_2 g)$. Again it is easy to check that this definition is independent of the choice of representatives (σ_i, f_i) and of the choice of V; hence also the composition it defines is associative. The functor P_Σ is defined by

$$P_\Sigma(X) = X,\ P_\Sigma(X \xrightarrow{\ f\ } Y) = \begin{array}{ccc} X & \xrightarrow{\ f\ } & Y \\ {\scriptstyle 1}\downarrow & & \\ X & & \end{array};$$

its universal property is immediate. We now show that $\mathscr{C}\Sigma^{-1}$ has equalizers and P_Σ preserves them. (The argument for finite products is similar but easier.) Let

$$\begin{array}{ccc} Z_i & \xrightarrow{\ f_i\ } & Y \quad (i = 1, 2) \\ {\scriptstyle\sigma_i}\downarrow & & \\ X & & \end{array}$$

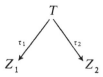

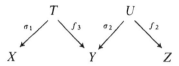

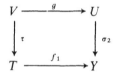

be a parallel pair of morphisms of $\mathscr{C}\Sigma^{-1}$; then by a suitable choice of representatives we may ensure that $Z_1 = Z_2$ and $\sigma_1 = \sigma_2$. Let $E \overset{q}{\rightarrowtail} Z$ be the equalizer of f_1 and f_2 in C. Now let

$$\begin{array}{ccc} U & \xrightarrow{g} & X \\ {\scriptstyle\tau}\big\downarrow & & \\ T & & \end{array}$$

be a morphism of $\mathscr{C}\Sigma^{-1}$ equalizing (σ, f_1) and (σ, f_2); then for a suitable choice of commutative square

$$\begin{array}{ccc} V & \xrightarrow{h} & Z \\ {\scriptstyle\rho}\big\downarrow & & \big\downarrow{\scriptstyle\sigma} \\ U & \xrightarrow{g} & X \end{array}$$

with $\tau\rho \in \Sigma$ we can ensure that $f_1 h = f_2 h$ in C, and so there exists a factorization $V \xrightarrow{k} E$ of h through q. Now

$$\begin{array}{ccc} V & \xrightarrow{k} & E \\ {\scriptstyle\tau\rho}\big\downarrow & & \\ T & & \end{array}$$

is a factorization of (τ, g) through $P_\Sigma(\sigma q)$ in $\mathscr{C}\Sigma^{-1}$; and it is not hard to verify that it is unique. So $P_\Sigma(\sigma q)$ is an equalizer of (σ, f_1) and (σ, f_2) in $\mathscr{C}\Sigma^{-1}$; hence also P_Σ preserves equalizers. ■

0.2. SHEAF THEORY

As in the previous paragraph, we do not intend to give a detailed account of the "classical" theory of sheaves on a topological space; for this the reader is referred to one of the standard texts on the subject, for example [ST] or [TF]. However, a brief review of the classical theory is indispensable as an aid to understanding the motivation for the more general theory of Grothendieck topologies which follows in the next paragraph.

Let $(X, \mathbf{T})$ be a topological space. The set $\mathbf{T}$ of open subsets of X is partially ordered by inclusion, so we can regard it as a small category in the usual way; i.e. the objects of $\mathbf{T}$ are the open sets, and $\mathbf{T}$ has one morphism from U to V iff $U \subseteq V$.

0.21 DEFINITION. A *presheaf* (of sets) on X is a presheaf on the category $\mathbf{T}$, i.e. a contravariant functor from $\mathbf{T}$ to $\mathscr{S}$. A presheaf P is thus specified by giving a set $P(U)$ for each open $U \subseteq X$, and a *restriction map*

$$\rho_V^U \colon P(U) \longrightarrow P(V)$$

for each $V \subseteq U$, subject to the obvious compatibility conditions. A *morphism of presheaves* is just a natural transformation of functors. ■

The notion of presheaf includes many examples which are familiar concepts in elementary topology.

0.22 EXAMPLES. (i) For any set A, we have the *constant presheaf* $\overline{A}$ defined by $\overline{A}(U) = A$, $\rho_V^U = 1_A$ for all $V \subseteq U \subseteq X$.

(ii) For any open $U \subseteq X$, we have the representable presheaf h_U, defined by

$$h_U(V) = \text{singleton if } V \subseteq U,$$
$$= \text{empty set otherwise.}$$

The restriction maps are obvious.

(iii) For any space Y, we have the presheaf C_Y of continuous Y-valued functions, defined by $C_Y(U) =$ set of continuous functions $U \longrightarrow Y$, $\rho_V^U(f) = f|_V$.

(iv) We have a presheaf Ω defined by $\Omega(U) =$ set of open subsets of U, $\rho_V^U(W) = W \cap V$.

(v) Similarly, we have a presheaf J defined by $J(U) =$ set of open coverings of U, $\rho_V^U(\{U_\alpha | \alpha \in A\}) = \{U_\alpha \cap V | \alpha \in A\}$. ■

However, many of these examples satisfy an additional "exactness condition" which says that elements of $P(U)$ may be built up by "glueing together" elements of $P(U_\alpha)$, where $\{U_\alpha\}$ is an open cover of U. This leads to the definition of a sheaf.

0.23 DEFINITION. Let P be a presheaf on X. We say P is a *sheaf* if, given any open covering $\{U_\alpha | \alpha \in A\}$ of an open set U and any family $\{\sigma_\alpha | \alpha \in A\}$ of elements of $P(U_\alpha)$, such that for all pairs (α, β) we have $\sigma_\alpha = \sigma_\beta$ when both are restricted to $P(U_\alpha \cap U_\beta)$, there exists exactly one $\sigma \in P(U)$ such that σ restricts to σ_α in $P(U_\alpha)$ for each α. P is said to be a *separated presheaf* if it satisfies the above condition with "exactly one" replaced by "at most one".

We write $\mathrm{Shv}(X, \mathbf{T})$ (or simply $\mathrm{Shv}(X)$) for the full subcategory of $\mathcal{S}^{\mathbf{T}^{\mathrm{op}}}$ whose objects are the sheaves. ■

Diagrammatically, we can say P is a sheaf iff the diagram

$$P(U) \longrightarrow \prod_{\alpha} P(U_\alpha) \rightrightarrows \prod_{\alpha, \beta} P(U_\alpha \cap U_\beta)$$

is an equalizer, where the maps are induced by restrictions in the obvious way. It is easily verified that the presheaves of 0.22(ii), (iii) and (iv) are in fact sheaves.

0.24 THEOREM. *Let* **esp** *denote the category of topological spaces and continuous maps. Then we have a pair of adjoint functors* $\mathcal{S}^{\mathbf{T}^{\mathrm{op}}} \underset{\Gamma}{\overset{L}{\rightleftarrows}} \mathbf{esp}/(X, \mathbf{T})$, *which restrict to an equivalence of categories between* $\mathrm{Shv}(X, \mathbf{T})$ *and the full subcategory of* $\mathbf{esp}/(X, \mathbf{T})$ *whose objects are* local homeomorphisms $E \xrightarrow{p} X$.

Proof. Let P be a presheaf on X, $x \in X$. We define the *stalk* of P at x to be the colimit

$$P_x = \varinjlim_{U \ni x} P(U)\,;$$

and if $\sigma \in P(U)$ for some $U \ni x$, we write σ_x for the image of σ in P_x. Now define $L(P)$ to be set of all pairs (x, t) with $x \in X$, $t \in P_x$, topologized by saying that $V \subseteq L(P)$ is open iff, for all open $U \subseteq X$ and all $\sigma \in P(U)$, the set $\{x \in U \mid (x, \sigma_x) \in V\}$ is open in X. We have a projection $L(P) \xrightarrow{p} X$ given by $(x, t) \longmapsto x$; and it is readily checked that a morphism of presheaves $P \longrightarrow Q$ induces a continuous map $L(P) \longrightarrow L(Q)$ over X. Moreover, the projection $L(P) \xrightarrow{p} X$ is a local homeomorphism; for given $(x, t) \in L(P)$, we can find a neighbourhood U of x and $\sigma \in P(U)$ such that $t = \sigma_x$, and then the set $V = \{(y, \sigma_y) \mid y \in U\}$ is open in $L(P)$, and is mapped homeomorphically onto U by p.

Conversely, let $E \xrightarrow{p} X$ be a space over X. We define a presheaf $\Gamma(E, p)$ by $\Gamma(E, p)(U) =$ set of continuous maps $U \xrightarrow{s} E$ which are sections of p, i.e.

such that ps is the inclusion map $U \longrightarrow X$, and $\rho_V^U(s) = s|_V$. Then it is not hard to check that $\Gamma(E, p)$ is a sheaf (the *sheaf of sections* of p). Now we have a natural transformation $\eta: 1 \longrightarrow \Gamma L$, given by defining η_P to be the morphism of presheaves which sends $\sigma \in P(U)$ to the section

$$\tilde{\sigma}: U \longrightarrow L(P); \ x \longmapsto (x, \sigma_x).$$

And we have a natural transformation $\varepsilon: L\Gamma \longrightarrow 1$, such that $\varepsilon_{(E,p)}$ sends a point (x, t) of $L\Gamma(E, p)$ to $s(x)$, where $s \in \Gamma(E, p)(U)$ is a section of p over a neighbourhood U of x such that $s_x = t$. (It is easily checked that this definition is independent of the choice of U and of s, and that $\varepsilon_{(E,p)}$ is a continuous map.) Moreover, η and ε satisfy the "triangular identities", and so define an adjunction $(L \dashv \Gamma)$.

But now if P is a sheaf and $s \in \Gamma LP(U)$, then there is a unique $\sigma \in P(U)$ such that $\tilde{\sigma} = s$, since we can construct σ locally at every point of U, and glue it together using the sheaf axiom. So η_P is an isomorphism. Similarly, if p is a local homeomorphism, then $\varepsilon_{(E,p)}$ is a homeomorphism. So the adjunction restricts to an equivalence of categories, as required. ■

0.25 COROLLARY. *The inclusion functor* Shv$(X) \longrightarrow \mathscr{S}^{\mathbf{T}^{\text{op}}}$ *has a left adjoint, namely the composite* ΓL *regarded as a functor* $\mathscr{S}^{\mathbf{T}^{\text{op}}} \longrightarrow$ Shv(X). *This functor is known as the* associated sheaf functor. ■

It follows from 0.24 that we can regard sheaves on X alternatively as presheaves satisfying an exactness condition (as in 0.23), or as spaces equipped with local homeomorphisms into X. Much of the power of sheaf theory comes from the fruitful interplay of these two ideas, for example:

0.26 PROPOSITION. *Let* $(X, \mathbf{T}) \xrightarrow{f} (Y, \mathbf{U})$ *be a continuous map. Then we have a pair of functors* $\text{Shv}(X, \mathbf{T}) \underset{f^*}{\overset{f_*}{\rightleftarrows}} \text{Shv}(Y, \mathbf{U})$, *such that* $f^* \dashv f_*$ *and* f^* *is left exact.*

Proof. f_* is defined using the presheaf representation. Observe that since f is continuous, it induces a functor (= order-preserving map) $f^{-1}: \mathbf{U} \longrightarrow \mathbf{T}$. Composition with f^{-1} then gives us a functor $f_*: \mathscr{S}^{\mathbf{T}^{\text{op}}} \longrightarrow \mathscr{S}^{\mathbf{U}^{\text{op}}}$, which is easily seen to carry sheaves to sheaves.

To define $f^*(E)$, where E is a sheaf on Y, we regard E as a local homeomorphism over Y and form the pullback

$$\begin{array}{ccc} f^*(E) & \longrightarrow & E \\ \downarrow & & \downarrow \\ X & \xrightarrow{f} & Y \end{array}$$

in **esp**. It is again easy to check that the projection $f^*(E) \longrightarrow X$ is a local homeomorphism. The same argument shows that the inclusion of (local homeomorphisms over X) in **esp**/X preserves pullbacks and hence all finite limits; from this it follows easily that f^* is left exact. The adjunction $f^* \dashv f_*$ is established by arguments similar to those of 0.24. ∎

0.3. GROTHENDIECK TOPOLOGIES

The notion of Grothendieck topology arose out of the desire of algebraic geometers to study objects with "sheaf-like" properties which were defined on categories more general than the lattice of open sets of a topological space. As we saw in §0.1, the definition of presheaf generalizes without difficulty to any small category **C**, but in order to talk about sheaves we must have a notion of "covering family" within **C**, which we can use as in 0.23.

0.31 Definition. Let **C** be a small category with pullbacks. A *Grothendieck pretopology* on **C** is defined by specifying, for each object U of **C**, a set $P(U)$ of families of morphisms of the form $\{U_i \xrightarrow{\alpha_i} U \mid i \in I\}$, called *covering families* of the pretopology, such that

(i) For any U, the family whose only member if $U \xrightarrow{1} U$ is in $P(U)$.
(ii) If $V \longrightarrow U$ is a morphism of **C** and $\{U_i \longrightarrow U \mid i \in I\}$ is in $P(U)$, then $\{V \times_U U_i \xrightarrow{\pi_1} V \mid i \in I\}$ is in $P(V)$.
(iii) If $\{U_i \xrightarrow{\alpha_i} U \mid i \in I\} \in P(U)$ and $\{V_{ij} \xrightarrow{\beta_{ij}} U_i \mid j \in J_i\} \in P(U_i)$ for each i, then $\{V_{ij} \xrightarrow{\alpha_i \beta_{ij}} U \mid i \in I, j \in J_i\} \in P(U)$. ∎

We could now define a *sheaf* for the pretopology P to be a presheaf F such that the diagram

$$F(U) \longrightarrow \prod_{i \in I} F(U_i) \rightrightarrows \prod_{i,j} F(U_i \times_U U_j)$$

is an equalizer for every covering family $\{U_i \to U \mid i \in I\}$. However, if F satisfies this condition for the family $R = \{\alpha_i \mid i \in I\}$, it will clearly also satisfy it for any family S which contains R, and indeed the converse is true if every morphism in S factors through one of the α_i. There is therefore a certain imprecision in the definition of a Grothendieck pretopology, in that two different pretopologies may give exactly the same sheaves. To remove this, we restrict our attention to those families R which are "saturated" in the sense that $(V \xrightarrow{\alpha} U) \in R$ implies $(W \xrightarrow{\alpha\beta} U) \in R$ for any $W \xrightarrow{\beta} V$. Such a family is called a *sieve* on the object U; we now rewrite 0.31 in terms of sieves.

0.32 DEFINITION. Let $\mathbf{C}$ be a small category. A *Grothendieck topology* on $\mathbf{C}$ is defined by specifying, for each object U of $\mathbf{C}$, a set $J(U)$ of sieves on U, called *covering sieves* of the topology, such that

(i) For any U, the maximal sieve $\{\alpha \mid \text{codomain}(\alpha) = U\}$ is in $J(U)$.
(ii) If $R \in J(U)$ and $V \xrightarrow{f} U$ is a morphism of $\mathbf{C}$, then the sieve

$$f^*(R) = \{W \xrightarrow{\alpha} V \mid f\alpha \in R\}$$

is in $J(V)$.
(iii) If $R \in J(U)$ and S is a sieve on U such that, for each $V \xrightarrow{f} U$ in R, we have $f^*(S) \in J(V)$, then $S \in J(U)$.

A small category equipped with a Grothendieck topology is called a *site*. ∎

Note that 0.32(i) and (iii) together imply that if $R \in J(U)$ and S is a sieve on U containing R, then $S \in J(U)$. Clearly, if we are given a pretopology P on $\mathbf{C}$, we can replace it by a topology having the same sheaves, by defining a sieve to be J-covering iff it contains a P-covering family.

But the change from 0.31 to 0.32 has another advantage besides removing the ambiguity mentioned earlier. The reader will note that we have been able to dispense with the assumption that $\mathbf{C}$ has pullbacks, since a sieve can always be pulled back along a morphism of $\mathbf{C}$ (as in 0.32(ii)), even if the individual morphisms in it cannot. The reason for this is that each sieve R on U can be identified with a sub-presheaf of the representable functor h_U, namely the presheaf $V \longmapsto \{\alpha \in R \mid \text{domain}(\alpha) = V\}$; and the presheaf category $\mathcal{S}^{\mathbf{C}^{\text{op}}}$ does of course have pullbacks. We make use of this identification of sieves with sub-presheaves of h_U in defining sheaves for a topology.

0.33 DEFINITION. Let $(\mathbf{C}, J)$ be a site, F a presheaf on $\mathbf{C}$. We say F is a *sheaf* (for the topology J) if, for every object U of $\mathbf{C}$ and every $R \in J(U)$, each morphism $R \longrightarrow F$ in $\mathscr{S}^{\mathbf{C}^{\mathrm{op}}}$ has exactly one extension to a morphism $h_U \longrightarrow F$; and F is a *separated presheaf* if it satisfies the above condition with "exactly one" replaced by "at most one".

We denote the full subcategory of $\mathscr{S}^{\mathbf{C}^{\mathrm{op}}}$ whose objects are J-sheaves by $\mathrm{Shv}(\mathbf{C}, J)$ (or simply $\mathrm{Shv}(\mathbf{C})$, if J is obvious from the context). ■

It is not hard to see that 0.33 is equivalent to our provisional definition of sheaves for a pretopology; for by 0.11 morphisms $h_U \longrightarrow F$ correspond to elements of $F(U)$, and it is easily seen that specifying a moprhism $R \xrightarrow{f} F$ amounts to specifying a "compatible" family of elements $f(\alpha) \in F(V)$, for each $V \xrightarrow{\alpha} U$ in R.

In general, there will be many different Grothendieck topologies which we can impose on a particular category. Two in particular deserve mention: the *maximal topology*, for which every sieve is covering, and the *minimal topology*, for which the only covering sieves are those of 0.32(i). For the latter topology, every presheaf is a sheaf; for the former, the only sheaf is the terminal object of $\mathscr{S}^{\mathbf{C}^{\mathrm{op}}}$. The topologies on $\mathbf{C}$ are partially ordered by inclusion, and we have

0.34 LEMMA. *Let $\{J_\alpha | \alpha \in A\}$ be a set of Grothendieck topologies on* $\mathbf{C}$. *Then*

(i) $\bigcap_{\alpha \in A} J_\alpha$ *is a topology, and is the greatest lower bound of the* J_α.

(ii) *There exists a topology* $\sum_{\alpha \in A} J_\alpha$ *which is the least upper bound of the* J_α.

Proof. (i) is immediate from the definition.

(ii) Apply (i) to the set $\{K | K \text{ is an upper bound for the } J_\alpha\}$. ■

0.35 LEMMA. *For any presheaf F, there exists a unique largest topology J_F for which F is a sheaf.*

Proof. Define J_F by $R \in J_F(U) \Leftrightarrow$ for every $V \xrightarrow{f} U$, each morphism $f^*(R) \longrightarrow F$ has a unique extension to a morphism $h_V \longrightarrow F$. It is then easily checked that J_F is a topology, and that F is a sheaf for it; and if J is any topology for which F is a sheaf, then each J-covering sieve must satisfy the above condition. ■

By combining 0.35 and 0.34(i), we deduce that there is a unique largest topology for which each of a given class of presheaves is a sheaf. In particular, we define the *canonical topology* on **C** to be the largest topology for which all the representable functors are sheaves; and we say that a topology J is *sub-canonical* if it is smaller than the canonical topology, i.e. if all the representables are J-sheaves. It should be noted that the topology which we imposed on the lattice **T** of open sets of X in §0.2 was in fact the canonical topology.

One aspect of sheaves on a topological space, which does not generalize to sheaves on a site, is their alternative representation (0.24) as local homeomorphisms. Nevertheless, the equivalent of 0.25 remains true; i.e. we have an *associated sheaf functor* $L: \mathcal{S}^{\mathbf{C}^{\mathrm{op}}} \longrightarrow \mathrm{Shv}(\mathbf{C}, J)$ which is left adjoint to the inclusion functor. The detailed construction of L in this case is deferred until chapter 3 (where we shall in fact consider a still more general situation); for the present, we merely note that it exists, and also the importance fact that it is a left exact funtor (3.39).

0.4. GIRAUD'S THEOREM

0.41 Definition. A category $\mathcal{E}$ is called a *Grothendieck topos* if there exists a site $(\mathbf{C}, J)$ such that $\mathcal{E}$ is equivalent to Shv $(\mathbf{C}, J)$. ∎

We note that the definition of Grothendieck topos includes any category of the form $\mathcal{S}^{\mathbf{C}^{\mathrm{op}}}$ (take J to be the minimal topology); hence in particular it includes $\mathcal{S}$ (take **C** to be the trivial category **1**). We conclude this chapter by sketching the proof of the remarkable theorem of J. Giraud ([GV], IV 1.2), which shows that Grothendieck toposes may be completely characterized by a combination of "exactness conditions' concerning existence and properties of limits, and "size conditions". Later on (in § 4.4) we shall be in a position to give a more detailed proof of this theorem; but its significance is such that it is essential to give some indication of how it works straight away.

Before we can state the theorem, however, we need a number of definitions.

0.42 Definition. Let $\mathcal{E}$ be a category with pullbacks.

(i) If X is a coproduct of a family of objects $(X_\alpha \mid \alpha \in A)$ of $\mathcal{E}$, we say it is *disjoint* provided

(a) each coproduct inclusion $\nu_\alpha : X_\alpha \longrightarrow X$ is a monomorphism, and

(b) for each pair of distinct indices (α, β), the pullback $X_\alpha \times_X X_\beta$ is an initial object of $\mathscr{E}$.

(ii) If $(X_\alpha \longrightarrow X \mid \alpha \in A)$ is a cone under a diagram in $\mathscr{E}$ with vertices X_α, we say X is a *universal colimit* of the diagram provided, for each morphism $Y \longrightarrow X$, the cone $(Y \times_X X_\alpha \longrightarrow Y \mid \alpha \in A)$ is a colimit for the "pulled-back" diagram with vertices $Y \times_X X_\alpha$. ■

0.43 Definition. Let $\mathscr{E}$ be a category with finite limits, $R \underset{b}{\overset{a}{\rightrightarrows}} X$ a parallel pair of morphisms of $\mathscr{E}$. We say (a, b) is an *equivalence relation* on X if

(i) $R \xrightarrow{(a,b)} X \times X$ is mono.
(ii) The diagonal subobject $X \overset{\Delta}{\rightarrowtail} X \times X$ factors through (a, b) [i.e. (a, b) is reflexive, cf. 0.13].
(iii) There exists a morphism $\tau\colon R \longrightarrow R$ such that $b\tau = a$ and $a\tau = b$ [i.e. (a, b) is *symmetric*].
(iv) If

$$\begin{array}{ccc} T & \xrightarrow{q} & R \\ {\scriptstyle p}\downarrow & & \downarrow{\scriptstyle a} \\ R & \xrightarrow{b} & X \end{array}$$

is a pullback, then $T \xrightarrow{(ap,\, bq)} X \times X$ factors through (a, b) [i.e. (a, b) is *transitive*]. ■

Note that if $X \xrightarrow{f} Y$ is any morphism, then the kernel-pair of f (i.e. the pullback of f against itself) is always an equivalence relation on X. An equivalence relation which is a kernel-pair of some morphism is said to be *effective*.

0.44 Definition. Let $\mathscr{E}$ be a category, $\mathscr{G}$ a class of objects of $\mathscr{E}$. We say $\mathscr{G}$ is a *class of generators* for $\mathscr{E}$ ([CW], p. 123) if, whenever we have $X \underset{g}{\overset{f}{\rightrightarrows}} Y$ with $f \neq g$, there exists $G \in \mathscr{G}$ and $G \xrightarrow{h} X$ such that $fh \neq gh$. ■

0.45 Theorem (J. Giraud). *Let $\mathscr{E}$ be a category. The following conditions are equivalent:*

(i) *$\mathscr{E}$ is a Grothendieck topos.*

(ii) *$\mathscr{E}$ satisfies the following conditions:*

(a) *$\mathscr{E}$ has finite limits.*

(b) *$\mathscr{E}$ has all set-indexed coproducts, and they are disjoint and universal.*

(c) *Equivalence relations in $\mathscr{E}$ have universal coequalizers.*

(d) *Every equivalence relation in $\mathscr{E}$ is effective, and every epimorphism in $\mathscr{E}$ is a coequalizer.*

(e) *$\mathscr{E}$ has "small hom-sets" (i.e. for any two objects X, Y, the morphisms of $\mathscr{E}$ from X to Y are parametrized by a set).*

(f) *$\mathscr{E}$ has a set of generators.*

Proof. (i) ⇒ (ii): The exactness conditions (a)–(d) are all trivial for $\mathscr{S}$; for $\mathscr{S}^{\mathbf{C}^{op}}$ they follow from the fact that limits and colimits in $\mathscr{S}^{\mathbf{C}^{op}}$ are constructed "pointwise" ([CW], p. 112). In a general topos, (a) follows from the fact that the sheaf axiom (0.33) involves only morphisms with codomain F; hence the limit in $\mathscr{S}^{\mathbf{C}^{op}}$ of any diagram of sheaves is itself a sheaf. To construct coproducts in Shv($\mathbf{C}$, J), we construct them first in $\mathscr{S}^{\mathbf{C}^{op}}$ and then apply the associated sheaf functor; their disjointness and universality follow from the fact that the associated sheaf functor preserves pullbacks and monos.

Existence and universality of coequalizers, and effectiveness of equivalence relations, are similarly proved. And if $X \xrightarrow{f} Y$ is an epimorphism in Shv($\mathbf{C}$, J), it is not hard to show that Y must be the associated sheaf of the image I of f in $\mathscr{S}^{\mathbf{C}^{op}}$; then since $X \twoheadrightarrow I$ is a coequalizer in $\mathscr{S}^{\mathbf{C}^{op}}$, f is a coequalizer in Shv($\mathbf{C}$, J).

The size condition (e) is trivial for any Grothendieck topos. For the generators in (f), we take the associated sheaves of the representable presheaves; 0.12 implies that the representables form a set of generators for $\mathscr{S}^{\mathbf{C}^{op}}$, and the result for Shv($\mathbf{C}$, J) follows from the adjunction between the inclusion and the associated sheaf functor.

(ii) ⇒ (i): Let $\mathbf{C}$ be a small full subcategory of $\mathscr{E}$ whose objects generate $\mathscr{E}$. (Such a subcategory exists, by (e) and (f).) Consider the functor $T: \mathscr{E} \longrightarrow \mathscr{S}^{\mathbf{C}^{op}}$ which sends the object X to the presheaf $U \longmapsto \hom_{\mathscr{E}}(U, X)$; since the objects of $\mathbf{C}$ generate $\mathscr{E}$, this functor is clearly faithful, but in fact it is also full since every object of $\mathscr{E}$ can be written as a colimit of objects in $\mathbf{C}$. (This uses the fact that epimorphisms in $\mathscr{E}$ are coequalizers.)

Now equip $\mathbf{C}$ with the largest Grothendieck topology J for which all the presheaves in the image of T are sheaves. (Observe that J is sub-canonical, since the image of T contains the representables.) It remains to show that every J-sheaf on $\mathbf{C}$ is isomorphic to a sheaf in the image of T; this will prove that $\mathscr{E} \simeq$ Shv($\mathbf{C}$, J). But it can be shown using the exactness conditions (b)–(d) that $T: \mathscr{E} \longrightarrow \mathrm{Shv}(\mathbf{C}, J)$ preserves set-indexed coproducts and coequalizers of equivalence relations; and since the representables generate $\mathrm{Shv}(\mathbf{C}, J)$, every J-sheaf can be obtained from the representables by means of these two constructions (cf. 7.40 below). So the result is established. ■

0.46 COROLLARY. *Let $\mathscr{E}$ be a Grothendieck topos. Then there exists a "site of definition" $(\mathbf{C}, J)$ for $\mathscr{E}$ (i.e. a site such that $\mathscr{E} \simeq \mathrm{Shv}(\mathbf{C}, J)$) such that $\mathbf{C}$ has finite limits and J is sub-canonical.*

Proof. We have to show that the category $\mathbf{C}$ used in the proof of (ii) ⇒ (i) above can be chosen to have finite limits. But if $\mathbf{C}_0$ is any small full subcategory of $\mathscr{E}$ whose objects are a set of generators, define $\mathbf{C}_n$ $(n \geqslant 1)$ to be the full subcategory whose objects are all (canonical) limits in $\mathscr{E}$ of finite diagrams in $\mathbf{C}_{n-1}$, and define $\mathbf{C}_\infty = \bigcup_{n=0}^{\infty} \mathbf{C}_n$. Then it is clear from the construction that $\mathbf{C}_\infty$ has finite limits; but by condition (e) there is only a set of finite diagrams in each $\mathbf{C}_n$, and so $\mathbf{C}_\infty$ is a small category. ■

In fact the result of the Corollary can be improved slightly; M. Barr ([2], p. 116) has shown that if the category $\mathbf{C}$ is closed under the formation of subobjects in $\mathscr{E}$, then the topology J is actually canonical, and not just sub-canonical. Since the category of sheaves on a site is clearly always well-powered ([CW], p. 126), it is always possible to choose $\mathbf{C}$ so that this condition is satisfied. However, we shall not need this sharper result except in one special case (Theorem 5.37 below), in which we shall prove it explicitly.

EXERCISES 0

†1. Let $\mathscr{C}$ be a category with finite products, such that the functor $(-)\times X: \mathscr{C} \longrightarrow \mathscr{C}$ preserves coequalizers for every object X of $\mathscr{C}$. If $X_1 \underset{g_1}{\overset{f_1}{\rightrightarrows}} Y_1 \xrightarrow{h_1} Z_1$ and $X_2 \underset{g_2}{\overset{f_2}{\rightrightarrows}} Y_2 \xrightarrow{h_2} Z_2$ are two reflexive coequalizer diagrams in $\mathscr{C}$, prove that their product

$$X_1 \times X_2 \underset{g_1 \times g_2}{\overset{f_1 \times f_2}{\rightrightarrows}} Y_1 \times Y_2 \xrightarrow{h_1 \times h_2} Z_1 \times Z_2 \text{ is a coequalizer.}$$

2. Let $\mathscr{C}$ be a category, Σ a class of morphisms of $\mathscr{C}$ and X an object of $\mathscr{C}$. Define $(\Sigma \downarrow X)$ to be the full subcategory of $\mathscr{C}/X$ whose objects are those

morphisms $Y \xrightarrow{\sigma} X$ which are in Σ. If Σ admits a calculus of right fractions on $\mathscr{C}$, prove that $(\Sigma \downarrow X)^{\mathrm{op}}$ is filtered for every X. [For the definition of filteredness, see 2.51 below, or alternatively [CW], p. 207.]

3. (i) Why is the presheaf $\bar{A}$ of 0.22(i) not, in general, a sheaf?
(ii) Describe the associated sheaf of $\bar{A}$.

4. Prove that the sheaf Ω defined in 0.22(iv) is isomorphic to the sheaf C_S, where S is the *Sierpinski space*, i.e. the two-point space with just one closed point.

5. Let $\mathbb{R}$ denote the real line with its usual topology, and define a space $E \xrightarrow{p} \mathbb{R}$ over $\mathbb{R}$ as follows: as a set, E is $\mathbb{R} \times \{t, f, r, l, b\}$ (where t, f, r, l, b are five abstract symbols) and p is projection on the first factor. E is topologized so that the basic open neighbourhoods of each point have the following form:

$$\begin{array}{lll}
\langle x, t\rangle & \text{has} & (x-\delta, x+\delta) \times \{t\} \qquad (\delta > 0) \\
\langle x, f\rangle & \text{has} & (x-\delta, x+\delta) \times \{f\} \\
\langle x, r\rangle & \text{has} & (x-\delta, x) \times \{t\} \cup \{\langle x, r\rangle\} \cup (x, x+\delta) \times \{f\} \\
\langle x, l\rangle & \text{has} & (x-\delta, x) \times \{f\} \cup \{\langle x, l\rangle\} \cup (x, x+\delta) \times \{t\} \\
\langle x, h\rangle & \text{has} & (x-\delta, x) \times \{t\} \cup \{\langle x, b\rangle\} \cup (x, x+\delta) \times \{t\}.
\end{array}$$

Prove that p is a local homeomorphism, and that the sheaf $\Gamma(E, p)$ may be naturally identified with a subsheaf of the sheaf Ω of 0.22(iv). [Hint: identify $s \in \Gamma(E, p)(U)$ with the set $\{x \in U \mid s(x) = \langle x, t\rangle\}$.] Show that an open subset V of U is in $\Gamma(E, p)(U)$ iff the frontier of V has no points of accumulation in U.

6. Let $F \rightarrowtail G$ be a monomorphism in $\mathbf{Shv}(X)$, and σ an element of $G(U)$ for some open $U \subseteq X$. Prove that there is a unique largest open set $V \subseteq U$ such that $\rho^U_V(\sigma)$ is in (the image of) $F(V)$. Hence prove that the sheaf Ω of 0.22(iv) is *injective* in $\mathbf{Shv}(X)$; i.e. given a diagram

$$\begin{array}{ccc}
F & \overset{s}{\rightarrowtail} & G \\
{\scriptstyle f}\downarrow & & \\
\Omega & &
\end{array}$$

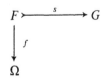

in Shv(X), there exists $G \xrightarrow{g} \Omega$ (not necessarily unique) such that $gs = f$.

†7. Let X be a topological space. Prove that the following conditions are equivalent:

(i) X is locally connected.
(ii) For every local homeomorphism $E \xrightarrow{p} X$, E is locally connected.
(iii) For every local homeomorphism $E \xrightarrow{p} X$, the space $\Pi_0(E)$ of components of E is discrete.
(iv) For every open $U \subseteq X$, $\Pi_0(U)$ is discrete.

Deduce that X is locally connected iff the functor $\mathcal{S} \xrightarrow{\Delta} \text{Shv}(X)$ which assigns to a set S the *constant sheaf* $\Delta(S)$ (i.e. the associated sheaf of the presheaf $\bar{S}$, or equivalently the sheaf of sections of $X \times S \xrightarrow{\pi_1} X$, where S is given the discrete topology) has a left adjoint Π_0.

8. Let **C** be a small category, $R \rightarrowtail h_U$ a sieve on an object U of **C**. Define R to be *epimorphic* if, for every $U \underset{\beta}{\overset{\alpha}{\rightrightarrows}} W$ in **C** such that $\alpha \neq \beta$, there exists $V \xrightarrow{\gamma} U$ in R such that $\alpha\gamma \neq \beta\gamma$; and say that R is *universally epimorphic* if f^*R is an epimorphic sieve on V, for every $V \xrightarrow{f} U$ in **C**. Show that the universally epimorphic sieves form a topology on **C**, and that it is the largest topology for which each representable presheaf is separated.

†9. Let G be a group, and let G-**esp** denote the category of topological spaces equipped with a G-action and G-equivariant continuous maps. If X is an object of G-**esp**, define the category $\text{Shv}_G(X)$ of *G-equivariant sheaves on X* to be the full subcategory of G-**esp**/X whose objects are local homeomorphisms. Show that $\text{Shv}_G(X)$ is a Grothendieck topos. [Hint: construct a site of definition $(\mathbf{U}_G, J_G)$ for it, where the objects of $\mathbf{U}_G$ are the open subsets of X, and the morphisms $V \longrightarrow U$ in $\mathbf{U}_G$ correspond to elements $g \in G$ such that $g(V) \subseteq U$.] If X/G denotes the quotient space of X under the action of G, show that pulling back local homeomorphisms along the quotient map $X \longrightarrow X/G$ defines a functor $\text{Shv}(X/G) \xrightarrow{u} \text{Shv}_G(X)$, and show that u is an equivalence if G acts effectively on X (i.e. if the stabilizer of any point of X is the trivial subgroup of G). Show also that the converse of this last statement is true provided G acts properly on X (i.e. the G-orbit of any point of X is a closed subset of X).

†10. (P. T. Johnstone [57]) Let $\mathbf{C}$ be the full subcategory of **esp** whose objects are the one-point compactification $\mathbb{N}^+$ of the discrete space $\mathbb{N}$ of natural numbers, and the one-point space 1. Show that the canonical topology J on $\mathbf{C}$ can be described as follows: 1 is covered only by the maximal sieve, and a sieve R covers $\mathbb{N}^+$ iff (a) $1 \xrightarrow{x} \mathbb{N}^+ \in R$ for every $x \in \mathbb{N}^+$, and (b) for each infinite $T \subseteq \mathbb{N}$, there is a monomorphism $\mathbb{N}^+ \xrightarrow{f} \mathbb{N}^+$ in R such that $\operatorname{im}(f) \subseteq T \cup \{\infty\}$. [Hint: observe that every morphism $\mathbb{N}^+ \longrightarrow \mathbb{N}^+$ in $\mathbf{C}$ with infinite image factors as a split epi followed by a split mono.] Show also that if X is any space, then the presheaf $\hom_{\mathbf{esp}}(-, X)\colon \mathbf{C}^{\mathrm{op}} \longrightarrow \mathscr{S}$ is a sheaf for this topology, and hence define a functor $F\colon \mathbf{esp} \longrightarrow \mathrm{Shv}(\mathbf{C}, J)$. Show that F is faithful, and that its restriction to the subcategory $\mathscr{F}$ of sequential spaces is full. (A space X is said to be *sequential* if its topology is determined by convergence of sequences, i.e. if every non-closed $A \subseteq X$ contains a sequence converging to a point of $X - A$; cf. [GT], p. 71. Observe that sequential spaces include first countable spaces (and hence all metric spaces); they also include all CW-complexes ([AT], §7.6).)

Finally, show that if $\{C_1, \ldots, C_n\}$ is a finite closed cover of a space X, then the canonical map $\coprod_{i=1}^{n} F(C_i) \longrightarrow F(X)$ is epi in $\mathrm{Shv}(\mathbf{C}, J)$. [Hint: let $I \rightarrowtail F(X)$ be the image of $\coprod F(C_i) \longrightarrow F(X)$ in $\mathscr{S}^{\mathbf{C}^{\mathrm{op}}}$; now observe that for every element α of $F(X)(\mathbb{N}^+)$ we can find a J-covering sieve R on $\mathbb{N}^+$ such that

$$\begin{array}{ccc} R & \longrightarrow & I \\ \downarrow & & \downarrow \\ h(\mathbb{N}^+) & \xrightarrow{\alpha} & F(X) \end{array}$$

commutes, and deduce that $F(X)$ is the associated sheaf of I.]

†11. (In this question, "ring" means "commutative ring with 1". **ann** denotes the category of rings.)

A ring homomorphism $A \xrightarrow{f} B$ is said to be *étale* if (a) B is finitely-presented as an A-algebra, and (b) for any diagram

$$\begin{array}{ccc} A & \xrightarrow{f} & B \\ {\scriptstyle g}\downarrow & & \downarrow{\scriptstyle h} \\ C & \xrightarrow{q} & C/N \end{array}$$

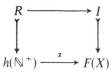

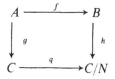

where N is a nilpotent ideal of C and q is the canonical projection, there is a unique $B \xrightarrow{k} C$ such that $qk = h$ and $kf = g$. (We also say B is an *étale A-algebra*.) Prove that

(i) if we are given $A \xrightarrow{f} B \xrightarrow{g} C$ with f étale, then g is étale iff gf is.
(ii) a pushout of an étale morphism is étale.
(iii) if S is a finitely-generated multiplicative submonoid of A, then the canonical map $A \longrightarrow A[S^{-1}]$ is étale. ($A[S^{-1}]$ denotes the usual ring of fractions.)
(iv) if K is a field and p is a non-constant polynomial in one variable t over K, then the inclusion $K \longrightarrow K[t]/(p)$ is étale iff p is *separable* (i.e. has distinct roots in some extension field of K). [Hint: consider diagrams of the form

$$\begin{array}{ccc} K & \longrightarrow & K[t]/(p), \\ \downarrow & & \downarrow \\ L[\varepsilon]/(\varepsilon^2) & \longrightarrow & L \end{array}$$

where L is an extension field of K. In general, a K-algebra $A = K[t_1, \ldots, t_n]/I$ is étale over K iff I can be generated by n polynomials $p_1, \ldots, p_n$ such that the Jacobian $\det(\partial p_i/\partial t_j)$ is invertible in A.]

Now fix a ring A, and let $\mathbf{C}$ denote the *opposite* of the category of étale A-algebras. [Observe that we may regard $\mathbf{C}$ as a small category, by choosing one representative for each finite presentation of an étale A-algebra.] Using (i) and (ii) above, prove that $\mathbf{C}$ has finite limits; and show that if we define a family of ring homomorphisms $(C \longrightarrow D_i \mid i \in I)$ to be *covering* provided every prime ideal of C is the inverse image of a prime ideal of some D_i, then the covering families form a pretopology on $\mathbf{C}$. Show also that the corresponding Grothendieck topology on $\mathbf{C}$ is sub-canonical. [Hint: if $(C \longrightarrow D_i \mid i \in I)$ is covering, then the kernel of $C \longrightarrow \prod D_i$ is a nilpotent ideal of C. In fact it is true more generally that the presheaf $\hom_{(A/\mathbf{ann})}(B, -)$ is a sheaf on $\mathbf{C}$ for any A-algebra B (not necessarily étale) [43]. The topos $\mathbf{Shv}(\mathbf{C})$ is called the *étale topos* of A, and denoted A_{et}.]

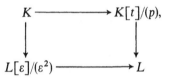

Chapter 1

Elementary Toposes

1.1. DEFINITION AND EXAMPLES

We saw in § 0.4 how the definition of a Grothendieck topos may be reduced to a set of axioms which refer directly to exactness properties of the topos itself, and not to any site of definition for it. This is an important advance, since it is clear that the same topos (at least up to equivalence) can be defined by many different sites. However, the definition of a Grothendieck topos is still explicitly dependent on the model of set theory which we suppose we are given; if we change our model of set theory, we will also change our notion of Grothendieck topos.

The definition of an *elementary* topos was introduced by F. W. Lawvere and M. Tierney [LN], [TV] mainly in response to the feeling that it should be possible to characterize a class of categories, which behave "internally" in the way in which we expect Grothendieck toposes to behave, but which are defined by "elementary" axioms which are independent of set theory.

1.11 DEFINITION. A category $\mathscr{E}$ is called an (elementary) *topos* if

(i) $\mathscr{E}$ has all finite limits. (Equivalently, $\mathscr{E}$ has pullbacks and a terminal object.)

(ii) $\mathscr{E}$ is *cartesian closed*, i.e. for each object X we have an *exponential functor* $(-)^X\colon \mathscr{E} \longrightarrow \mathscr{E}$ which is right adjoint to the functor $(-) \times X$.

(iii) $\mathscr{E}$ has a *subobject classifier*, i.e. an object Ω and a morphism $1 \xrightarrow{t} \Omega$ (called "true") such that, for each monomorphism $Y \overset{\sigma}{\rightarrowtail} X$ in $\mathscr{E}$, there is a unique $\phi_\sigma\colon X \longrightarrow \Omega$ (the *classifying map* of σ) making

$$\begin{array}{ccc} Y & \longrightarrow & 1 \\ {\scriptstyle \sigma}\big\downarrow & & \big\downarrow{\scriptstyle t} \\ X & \xrightarrow{\ \phi_\sigma\ } & \Omega \end{array}$$

a pullback diagram. ■

Informally, 1.11(iii) says that $\mathscr{E}$ contains a *generic subobject*, i.e. a monomorphism $S \rightarrowtail^{t} \Omega$ such that every mono in $\mathscr{E}$ is expressible as a pullback of t in a unique way. It is an easy exercise to prove that the domain of a generic subobject is a terminal object of $\mathscr{E}$, and object of $\mathscr{E}$, and conversely any morphism with domain 1 is mono, so this is equivalent to the definition as given in 1.11. Although the importance of cartesian closed categories—i.e. categories satisfying 1.11(i) and (ii)—had been realized for some time before the joint work of Lawvere and Tierney (see [71], [160]), their introduction of the notion of subobject classifier was the vital step forward which made the elementary development of topos theory possible.

1.12 PROPOSITION. *Any Grothendieck topos is a topos in the sense of* 1.11.

Proof. (i) Existence of finite limits has been established in the proof of 0.45.

(ii) For $\mathscr{S}$, we define Y^X = set of functions $X \longrightarrow Y$, and the adjunction is immediate.

In $\mathscr{S}^{\mathbf{C}^{op}}$, Y^X is determined by 0.11; for we have

$$Y^X(U) \cong \hom(h_U, Y^X) \cong \hom(h_U \times X, Y)$$

for every object U of $\mathbf{C}$. Taking the right-hand side as a definition, we have to verify that the adjunction $\hom(Z, Y^X) \cong \hom(Z \times X, Y)$ is valid for all presheaves Z, not just for representable ones; but by 0.12 we can express Z as a colimit $\varinjlim(h_\alpha)$ of representable functors h_α, and argue as follows:

$$\begin{aligned} \hom(Z, Y^X) &\cong \hom(\varinjlim(h_\alpha), Y^X) \\ &\cong \varprojlim(\hom(h_\alpha, Y^X)) \\ &\cong \varprojlim(\hom(h_\alpha \times X, Y)) \quad \text{by the definition of } Y^X \\ &\cong \hom(\varinjlim(h_\alpha \times X), Y) \\ &\cong \hom((\varinjlim(h_\alpha)) \times X, Y) \quad \text{since } (-)\times X \text{ preserves colimits} \\ &\quad\ \text{in } \mathscr{S} \text{ and hence in } \mathscr{S}^{\mathbf{C}^{op}} \\ &\cong \hom(Z \times X, Y). \end{aligned}$$

Finally, we can show that if Y is a sheaf for a topology J on $\mathbf{C}$, then so is the presheaf Y^X for any X (cf. 3.24 below); so exponential functors exist in Shv($\mathbf{C}$, J).

(iii) In $\mathscr{S}, \Omega$ is the two-element set $\{t, f\}$, and ϕ_σ is the characteristic function of the subset Y, i.e. $\phi_\sigma(x) = t \ (x \in Y)$
$= f \ (x \notin Y)$.

In $\mathscr{S}^{\mathbf{C}^{\mathrm{op}}}$, Ω is again determined by 0.11:

$$\Omega(U) \cong \hom(h_U, \Omega) \cong \{\text{sub-presheaves of } h_U\} \cong \{\text{sieves on } U\}.$$

And it is again straightforward from 0.12 to show that this definition has the required property for subobjects of any presheaf X.

Similarly for Shv($\mathbf{C}$, J), we find that $\Omega(U)$ is the set of those subobjects of the associated sheaf $L(h_U)$ which are sheaves. (It is readily checked that this does define a sheaf.) In particular for Shv(X), X a topological space, we find that Ω is the sheaf defined in 0.22(iv). ■

1.12 ensures that we have a plentiful supply of examples of toposes, on which we shall draw freely in order to illustrate points in the development of the subject. However, not every topos in the sense of 1.11 is a Grothendieck topos; we give here some examples which will be of importance later on.

1.13 Example. The category $\mathscr{S}_f$ of finite sets and functions satisfies 1.11 in the same way as $\mathscr{S}$; but it is not a Grothendieck topos since it does not have the infinite colimits required for 0.45. Similarly if $\mathbf{C}$ is a finite category, then the category $\mathscr{S}_f^{\mathbf{C}^{\mathrm{op}}}$ of presheaves of finite sets on $\mathbf{C}$ is a topos. ■

1.14 Example. Let G be a group; and regard it in the usual way as a small category $\mathbf{G}$ with one object, whose morphisms (the group elements) are all invertible. Then a presheaf on $\mathbf{G}$ is just a (left) G-*set*, i.e. a set on which G acts by permutations. Applying the prescriptions of 1.12, we find that in the topos $\mathscr{S}^{\mathbf{G}^{\mathrm{op}}}$ we have Ω = two-element set with trivial G-action, and Y^X = set of all (not necessarily G-equivariant) maps $X \longrightarrow Y$, with G acting by conjugation. It follows easily that the category $(\mathscr{S}_f)^{\mathbf{G}^{\mathrm{op}}}$ of finite G-sets is a topos, even if G itself is infinite. ■

Still further examples can be constructed using

1.15 Lemma. Let $\mathscr{E}_1$, $\mathscr{E}_2$ be toposes. Then the Cartesian product $\mathscr{E}_1 \times \mathscr{E}_2$ is a topos.

Proof. It is easily verified that finite limits in $\mathcal{E}_1 \times \mathcal{E}_2$ can be computed separately in each factor, that $(Y_1, Y_2)^{(X_1, X_2)} = (Y_1^{X_1}, Y_2^{X_2})$, and that the subobject classifier for $\mathcal{E}_1 \times \mathcal{E}_2$ is (Ω_1, Ω_2). ■

We must now define an appropriate concept of "morphism of toposes". It might be thought that a morphism of toposes would be simply a functor preserving finite limits, exponentials and the subobject classifier. This in fact defines the notion of *logical functor*; and examples of logical functors play an important rôle in topos theory. (For example, the inclusion functor $\mathcal{S}_f \longrightarrow \mathcal{S}$ is logical, as are the projections $\mathcal{E}_1 \times \mathcal{E}_2 \xrightarrow{\pi_i} \mathcal{E}_i$). However, there is another concept, namely that of geometric morphism, which is in many ways even more important.

1.16 Definition. Let $\mathcal{E}$, $\mathcal{F}$ be toposes. A *geometric morphism* $\mathcal{F} \xrightarrow{f} \mathcal{E}$ consists of a pair of functors $f_*: \mathcal{F} \longrightarrow \mathcal{E}$, $f^*: \mathcal{E} \longrightarrow \mathcal{F}$ (called the *direct* and *inverse images* of f) such that $f^* \dashv f_*$ and f^* is left exact. If $\mathcal{F} \overset{f}{\underset{g}{\rightrightarrows}} \mathcal{E}$ are two geometric morphisms, a *natural transformation* $\eta: f \longrightarrow g$ means a natural transformation of functors $f^* \xrightarrow{\eta} g^*$ (which of course induces, by adjunction, a unique $g_* \xrightarrow{\bar{\eta}} f_*$). Toposes, geometric morphisms and natural transformations clearly form a 2-category, which we denote by $\mathfrak{Top}$.

A geometric morphism $\mathcal{F} \xrightarrow{f} \mathcal{E}$ is said to be *essential* if f^* has a left adjoint $f_!$ as well as a right adjoint f_*. ■

1.17 Examples. (i) Let X, Y be topological spaces. Then we have already seen in 0.26 that each continuous map $X \longrightarrow Y$ gives rise to a geometric morphism $\mathrm{Shv}(X) \longrightarrow \mathrm{Shv}(Y)$; so we can regard the operation of taking sheaf categories as a functor **esp** $\longrightarrow \mathfrak{Top}$.

(ii) Let $(\mathbf{C}, J)$ be a site. Then it follows from the remarks at the end of §0.3 that the inclusion functor $\mathrm{Shv}(\mathbf{C}, J) \longrightarrow \mathcal{S}^{\mathbf{C}^{\mathrm{op}}}$ is the direct image of a geometric morphism, whose inverse image is the associated sheaf functor.

(iii) Let $\mathcal{E}_1$, $\mathcal{E}_2$ be toposes. Then the projection $\mathcal{E}_1 \times \mathcal{E}_2 \xrightarrow{\pi_1} \mathcal{E}_1$ is the inverse image of a geometric morphism, whose direct image is the functor $X \longmapsto (X, 1)$. If $\mathcal{E}_2$ has an initial object 0 (which is in fact always true, as we shall see in §1.3), then this morphism is essential, the left adjoint of π_1 being given by $X \longmapsto (X, 0)$. ■

Further examples of geometric morphisms will occur frequently in the succeeding paragraphs.

1.18 Warning. Some texts (notably [GV] adopt the "opposite" convention to 1.16 in defining the 2-arrows of $\mathfrak{Top}$; they define a natural transformation $f \longrightarrow g$ to be a natural transformation $f_* \longrightarrow g_*$, or equivalently $g^* \longrightarrow f^*$. Our reasons for adopting the convention of 1.16 will become apparent in §4.3.

1.2. EQUIVALENCE RELATIONS AND PARTIAL MAPS

1.21 Lemma. *In a topos, every monomorphism is an equalizer.*

Proof. The morphism $1 \xrightarrow{t} \Omega$ is split mono, and so is the equalizer of 1_Ω and $\Omega \longrightarrow 1 \xrightarrow{t} \Omega$. But every monomorphism is a pullback of t. ■

1.22 Corollary. *A topos is* balanced *(i.e. a morphism which is both mono and epi is an isomorphism).* ■

1.23 Proposition. *In a topos, equivalence relations are effective* (*cf.* 0.43).

Proof. Let $R \overset{a}{\underset{b}{\rightrightarrows}} X$ ve an equivalence relation. Let $X \times X \xrightarrow{\phi} \Omega$ be the classifying map of $R \xrightarrow{(a,b)} X \times X$, and $X \xrightarrow{\bar{\phi}} \Omega^X$ its exponential transpose. [Observe that in $\mathscr{S}$, Ω^X is the set of subsets of X, and $\bar{\phi}$ is the map which sends an element of X to its equivalence class modulo R]. We will show that $R \rightrightarrows X$ is a kernel-pair of $\bar{\phi}$.

Let $U \overset{f}{\underset{g}{\rightrightarrows}} X$ be a pair of morphisms coequalized by $\bar{\phi}$. Then applying the exponential adjunction, we have $\phi(f \times 1_X) = \phi(g \times 1_X): U \times X \longrightarrow \Omega$; and composing with $U \xrightarrow{(1_U, g)} U \times X$, we have $\phi(f, g) = \phi(g, g)$. But $U \xrightarrow{(g,g)} X \times X$ factors through R since R is reflexive, and so $\phi(g, g)$ classifies the maximal subobject $U \xrightarrow{1} U$. Hence $\phi(f, g)$ also classifies this subobject; so (f, g) factors through R.

Conversely, we must show that $\bar{\phi}a = \bar{\phi}b$, or equivalently that the subobjects of $R \times X$ classified by $\phi(a \times 1_X)$ and $\phi(b \times 1_X)$ are isomorphic. But if we compose these subobjects with the monomorphism $R \times X \xrightarrow{(a,b) \times 1} X \times X \times X$, then it follows easily from symmetry and transitivity of R that we obtain "the object of R-related triples" in each case, i.e. the intersection of the subobjects of $X \times X \times X$ obtained by pulling back $R \longrightarrow X \times X$ along (π_1, π_2) and (π_2, π_3). ■

1.24 Definition. In particular, we define the *singleton map* $\{\}: X \longrightarrow \Omega^X$ to be the exponential transpose of the classifying map $\delta: X \times X \longrightarrow \Omega$ of the diagonal map $X \overset{\Delta}{\rightarrowtail} X \times X$. It follows from 1.23 that $\{\}$ is mono, since its kernel-pair is $(1_X, 1_X)$. ■

1.25 Definition. A *partial map* $X \xrightarrow{f} Y$ in a category $\mathscr{E}$ is a diagram of the form

$$\begin{array}{ccc} X' & \xrightarrow{f} & Y. \\ \big\downarrow{\scriptstyle d} & & \\ X & & \end{array}$$

We say that partial maps with codomain Y are *representable* if there exists a mono $Y \overset{\eta}{\rightarrowtail} \tilde{Y}$ such that, for any $X \xrightarrow{f} Y$, there exists a unique $X \xrightarrow{\tilde{f}} \tilde{Y}$ making

$$\begin{array}{ccc} X' & \xrightarrow{f} & Y \\ \big\downarrow{\scriptstyle d} & & \big\downarrow{\scriptstyle \eta} \\ X & \xrightarrow{\tilde{f}} & \tilde{Y} \end{array}$$

a pullback diagram. (Note that the particular case $Y = 1$ reduces to the definition of Ω.) ■

1.26 Theorem (Lawvere–Tierney). *In a topos, all partial maps are representable.*

Proof. Let $\phi: \Omega^Y \times Y \longrightarrow \Omega$ classify the graph of the singleton map, i.e. the monomorphism $Y \overset{(\{\}, 1_Y)}{\rightarrowtail} \Omega^Y \times Y$; and let $\bar{\phi}: \Omega^Y \longrightarrow \Omega^Y$ be its exponential transpose. Define $\tilde{Y} \rightarrowtail \Omega^Y$ to be the equalizer of $\bar{\phi}$ and $1_{(\Omega^Y)}$. [We may think of $\tilde{Y}$ as "the set of subsets of Y having at most one element".]

Now since $\{\ \}$ is mono, it is easily seen that the square

$$\begin{array}{ccc} Y & \xrightarrow{1} & Y \\ \big\downarrow{\scriptstyle \Delta} & & \big\downarrow{\scriptstyle (\{\}, 1_Y)} \\ Y \times Y & \xrightarrow{\{\} \times 1_Y} & \Omega^Y \times Y \end{array}$$

is a pullback; but this says that $\phi(\{\ \} \times 1_Y)$ classifies the diagonal subobject of $Y \times Y$, and hence $\bar{\phi}.\{\} = \{\}$. So $\{\}$ factors through $\tilde{Y} \rightarrowtail \Omega^Y$, giving us a monomorphism $Y \overset{\eta}{\rightarrowtail} \tilde{Y}$.

Now let $X \xrightarrow{f} Y$ be a partial map. Then the graph of f (i.e. the mono $X' \overset{(d,f)}{\rightarrowtail} X \times Y$) is classified by a morphism $\psi: X \times Y \longrightarrow \Omega$. We wish to show that $\bar{\psi}: X \longrightarrow \Omega^Y$ factors through $\tilde{Y}$; but this amounts to showing that

$$\begin{array}{ccc} X' & \xrightarrow{f} & Y \\ \big\downarrow{\scriptstyle (d,f)} & & \big\downarrow{\scriptstyle (\{\},1_Y)} \\ X \times Y & \xrightarrow{\bar{\psi} \times 1} & \Omega^Y \times Y \end{array}$$

is a pullback, or equivalently that

$$\begin{array}{ccc} X' & \xrightarrow{f} & Y \\ \big\downarrow{\scriptstyle d} & & \big\downarrow{\scriptstyle \{\}} \\ X & \xrightarrow{\bar{\psi}} & \Omega^Y \end{array}$$

is a pullback. But suppose we are given $U \xrightarrow{a} X$, $U \xrightarrow{b} Y$ with $\bar{\psi}a = \{\}b$; then by transposition we obtain $\psi(a, b) = \delta(b, b) = (U \longrightarrow 1 \xrightarrow{t} \Omega)$ since (b, b) factors through $Y \overset{\Delta}{\rightarrowtail} Y \times Y$. Hence (a, b) factors (uniquely) through $X \overset{(d,f)}{\rightarrowtail} X \times Y$. So the square $(f, d, \{\}, \bar{\psi})$ is a pullback, and we have a factorization $\tilde{f}: X \longrightarrow \tilde{Y}$ of $\bar{\psi}$ through $\tilde{Y} \rightarrowtail \Omega^Y$.

It follows also from the above that

$$\begin{array}{ccc} X' & \xrightarrow{f} & Y \\ \big\downarrow{\scriptstyle d} & & \big\downarrow{\scriptstyle \eta} \\ X & \xrightarrow{\tilde{f}} & \tilde{Y} \end{array}$$

is a pullback; so it remains only to show uniqueness of $\tilde{f}$. Suppose that $\tilde{f}_1$ and $\tilde{f}_2$ both satisfy the conditions; then by working back through the above argument, we can show that the exponential transposes of the composites $X \xrightarrow{\tilde{f}_i} \tilde{Y} \rightarrowtail \Omega^Y$ both classify the same subobject of $X \times Y$, namely the graph of f. Hence they are equal; and since $\tilde{Y} \rightarrowtail \Omega^Y$ is mono we have $\tilde{f}_1 = \tilde{f}_2$. ■

Note that it follows from the uniqueness clause of 1.25 that $Y \longmapsto \tilde{Y}$ is in fact a functor $\mathscr{E} \longrightarrow \mathscr{E}$, and that η is a natural transformation.

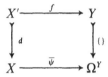

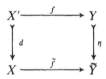

1.27. Corollary. *The objects $\tilde{Y}$ constructed in* 1.26 *are injective; hence in particular a topos has* enough injectives, *i.e. for any X we have a mono $X \rightarrowtail E$ with E injective.*

Proof. Suppose given a diagram

$$\begin{array}{ccc} X' & \rightarrowtail & X \\ \downarrow & \swarrow & \\ \tilde{Y} & & \end{array}$$

Form the pullback

$$\begin{array}{ccc} X'' & \rightarrowtail & X' \\ \downarrow & & \downarrow \\ Y & \xrightarrow{\eta} & \tilde{Y} \end{array} ;$$

then we have a unique $X \longrightarrow \tilde{Y}$ such that

$$\begin{array}{ccc} X'' & \rightarrowtail & X \\ \downarrow & & \downarrow \\ Y & \xrightarrow{\eta} & \tilde{Y} \end{array}$$

is also a pullback. And the triangle

$$\begin{array}{ccc} X' & \rightarrowtail & X \\ \downarrow & \swarrow & \\ \tilde{Y} & & \end{array}$$

commutes, since both ways round represent the same partial map $X' \longrightarrow Y$. ∎

1.28 Corollary. *Suppose we are given a pushout square*

$$\begin{array}{ccc} X & \longrightarrow & Y \\ {\scriptstyle f}\downarrow & & \downarrow{\scriptstyle g} \\ Z & \longrightarrow & T \end{array}$$

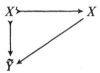

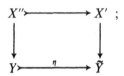

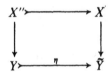

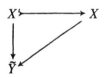

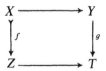

in a topos, with f mono. Then g is also mono, and the square is also a pullback.

Proof. Let $Z \xrightarrow{h} \tilde{Y}$ represent the partial map

$$\begin{array}{ccc} X & \longrightarrow & Y. \\ {\scriptstyle f}\big\downarrow & & \\ Z & & \end{array}$$

Then the cone $(Y \xrightarrow{\eta} \tilde{Y},\ Z \xrightarrow{h} \tilde{Y})$ factors through the pushout T, and in particular the monomorphism η factors through g. So g is mono; and since the square

$$\begin{array}{ccc} X & \longrightarrow & Y \\ {\scriptstyle f}\big\downarrow & & \big\downarrow{\scriptstyle \eta} \\ Z & \xrightarrow{\ h\ } & \tilde{Y} \end{array}$$

is a pullback, the original one is also. ■

1.3. THE CATEGORY $\mathscr{E}^{\mathrm{op}}$

The original definition of an elementary topos, as given by Lawvere and Tierney, included the condition that $\mathscr{E}$ should have finite colimits as well as finite limits. It was first shown by C. J. Mikkelsen [84] that this axiom can in fact be deduced from those of 1.11; the proof which we sketch in this paragraph is due to R. Paré [96].

Let $P: \mathscr{E}^{\mathrm{op}} \longrightarrow \mathscr{E}$ denote the "contravariant power-set functor"; i.e. $PX = \Omega^X$, and if $f: X \longrightarrow Y$, then $Pf: \Omega^Y \longrightarrow \Omega^X$ is the transpose of the composite

$$\Omega^Y \times X \xrightarrow{1 \times f} \Omega^Y \times Y \xrightarrow{\mathrm{ev}} \Omega.$$

(ev is the *evaluation map*, i.e. the counit of the exponential adjunction.) We write P_* for the same data considered as a functor $\mathscr{E} \longrightarrow \mathscr{E}^{\mathrm{op}}$.

1.31 LEMMA. *We have an adjunction* $(P_* \dashv P)$.

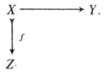

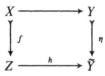

Proof. The statement asserts that we have a bijection (natural in X and Y) between morphisms $X \longrightarrow \Omega^Y$ and morphisms $Y \longrightarrow \Omega^X$. But both correspond naturally to morphisms $X \times Y \longrightarrow \Omega$, since $X \times Y \cong Y \times X$. ■

Now if $f: X \longrightarrow Y$ is mono, we can also define a morphism $\exists f: \Omega^X \longrightarrow \Omega^Y$, namely the transpose of the classifying map of the monomorphism

$$\in_X \rightarrowtail \Omega^X \times X \xrightarrow{1 \times f} \Omega^X \times Y.$$

($\in_X$ is the subobject classified by $\Omega^X \times X \xrightarrow{\text{ev}} \Omega$.)

1.32 Lemma ("Beck condition"). *Let*

$$\begin{array}{ccc} X & \xrightarrow{f} & Y \\ {\scriptstyle g}\downarrow & & \downarrow{\scriptstyle h} \\ Z & \xrightarrow{k} & T \end{array}$$

be a pullback diagram with g, h mono. Then the diagram

$$\begin{array}{ccc} \Omega^Y & \xrightarrow{Pf} & \Omega^X \\ {\scriptstyle \exists h}\downarrow & & \downarrow{\scriptstyle \exists g} \\ \Omega^T & \xrightarrow{Pk} & \Omega^Z \end{array}$$

commutes.

Proof. It is easily verified that both ways round, when transposed, are classifying maps of the same subobject of $\Omega^Y \times Z$, namely $E \rightarrowtail \Omega^Y \times X \xrightarrow{1 \times g} \Omega^Y \times Z$, where

$$\begin{array}{ccc} E & \longrightarrow & \in_Y \\ \downarrow & & \downarrow \\ \Omega^Y \times X & \xrightarrow{1 \times f} & \Omega^Y \times Y \end{array}$$

is a pullback. ■

1.33 Corollary. *If $X \xrightarrow{f} Y$ is mono, then $Pf . \exists f = 1_{(\Omega^X)}$.*

Proof. Apply 1.32 to the pullback

$$\begin{array}{ccc} X & \xrightarrow{\ 1\ } & X \\ {\scriptstyle 1}\big\downarrow & & \big\downarrow{\scriptstyle f} \\ X & \xrightarrow[\ f\ ]{} & Y \end{array} \quad . \blacksquare$$

1.34 THEOREM (Paré). *The functor P is monadic; i.e.* $\mathscr{E}^{\mathrm{op}}$ *is equivalent to the category of algebras for the monad on* $\mathscr{E}$ *induced by the adjunction of* 1.31.

Proof. By 0.13, it is sufficient to show that $\mathscr{E}^{\mathrm{op}}$ has coequalizers of reflexive pairs, that P preserves them, and that P reflects isomorphisms. But the first of these conditions is trivial, since coequalizers in $\mathscr{E}^{\mathrm{op}}$ are just equalizers in $\mathscr{E}$.

Let $X \xrightarrow{f} Y \underset{h}{\overset{g}{\rightrightarrows}} Z$ be a coreflexive equalizer diagram in $\mathscr{E}$, i.e. a reflexive coequalizer in $\mathscr{E}^{\mathrm{op}}$. Since we have a morphism $Z \xrightarrow{d} Y$ such that $dg = dh = 1_Y$, g and h are mono, and the diagram

$$\begin{array}{ccc} X & \xrightarrow{\ f\ } & Y \\ {\scriptstyle f}\big\downarrow & & \big\downarrow{\scriptstyle g} \\ Y & \xrightarrow[\ h\ ]{} & Z \end{array}$$

is a pullback. Applying 1.32 to it, we have $\exists f . Pf = Ph . \exists g$; and since f and g are mono, we have $Pf . \exists f = 1_{(\Omega^X)}$, $Pg . \exists g = 1_{(\Omega^Y)}$. So the diagram

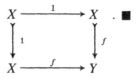

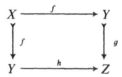

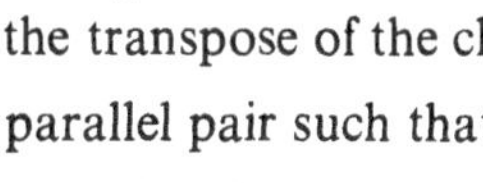
Ω^Z
Pg
Ph
Ω^Y
Pf
Ω^X
$\exists g$
$\exists f$

is a split coequalizer system in $\mathscr{E}$; in particular Pf is a coequalizer of Pg and Ph.

Now let $X \xrightarrow{f} Y$ be any morphism of $\mathscr{E}$. Then the composite $Y \xrightarrow{\{\}} \Omega^Y \xrightarrow{Pf} \Omega^X$ is easily seen to be the transpose of the classifying map of the graph of f. Hence if $X \underset{g}{\overset{f}{\rightrightarrows}} Y$ is a parallel pair such that $Pf = Pg$, then the graphs of f and g are isomorphic as subobjects of $X \times Y$, from which it follows easily that $f = g$. Thus P is faithful, and so in particular it reflects monos and epis; hence by 1.22 it reflects isomorphisms.

So the conditions of 0.13 are satisfied. $\blacksquare$

1.35 REMARK. In the particular case $\mathscr{E} = \mathscr{S}$, 1.34 can be regarded as a consequence of the Lindenbaum–Tarski theorem ([BA], Thm. 25.1) that $\mathscr{S}^{\mathrm{op}}$ is equivalent to the category of complete atomic Boolean algebras; for the latter is a varietal category in the sense of [177], and hence monadic over $\mathscr{S}$. A similar "algebraic" description of $\mathscr{E}^{\mathrm{op}}$, for a general topos $\mathscr{E}$, has been given by C. J. Mikkelsen [84] (see also [23]). ■

1.36 COROLLARY. *A topos has finite colimits. Moreover, if it has infinite limits of a particular type, then it has the corresponding colimits.*

Proof. It is well known that the forgetful functor for a category of algebras over $\mathscr{E}$ creates all types of limits which exist in $\mathscr{E}$. But limits in $\mathscr{E}^{\mathrm{op}}$ are just colimits in $\mathscr{E}$. ■

1.37 COROLLARY. *Let* $T: \mathscr{E} \longrightarrow \mathscr{F}$ *be a logical functor. Then*

(i) *T preserves finite colimits.*
(ii) *If T has a left adjoint, then it also has a right adjoint.*

Proof. By definition, T commutes up to isomorphism with the adjoint functors of 1.31, and preserves finite limits. So (i) follows immediately from 1.36; and (ii) is an application of 0.15. ■

In fact it can be shown (see [54]) that the converse of 1.37(ii) is also true; thus if the inverse image of a geometric morphism is logical, then the morphism is automatically essential.

1.38 EXAMPLE. To conclude this paragraph, we show how 1.31 and 1.34 may be used to give an explicit construction of coequalizers in $\mathscr{E}$.

Suppose given a parallel pair $X \underset{g}{\overset{f}{\rightrightarrows}} Y$; form the diagram

$$\begin{array}{ccccc} PX & \underset{Pg}{\overset{Pf}{\leftleftarrows}} & PY & \overset{m}{\longleftarrow} & Z \\ \uparrow{\scriptstyle P\alpha_X} & & \uparrow{\scriptstyle P\alpha_Y} & & \uparrow{\scriptstyle r} \\ PPPX & \underset{PPPg}{\overset{PPPf}{\leftleftarrows}} & PPPY & \overset{PPm}{\longleftarrow} & PPZ \end{array}$$

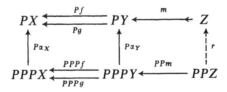

where m is the equalizer of Pf and Pg, α is the unit of $(P_* \dashv P)$, and r exists as a factorization through an equalizer. Now form

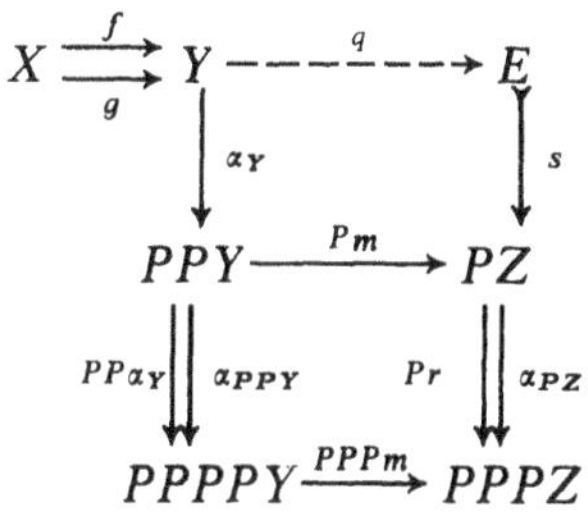

where s is the equalizer of Pr and α_{PZ}; then q exists as a factorization through an equalizer, and it is the coequalizer of f and g. ∎

1.4. PULLBACK FUNCTORS

Let $X \xrightarrow{f} Y$ be a morphism in a topos $\mathcal{E}$. Since $\mathcal{E}$ has (canonical) pullbacks, the operation of pulling back along f gives us a *pullback functor* $f^*: \mathcal{E}/Y \longrightarrow \mathcal{E}/X$.

1.41 LEMMA. *In any category $\mathcal{E}$ with pullbacks, f^* has a left adjoint $\Sigma_f: \mathcal{E}/X \longrightarrow \mathcal{E}/Y$.*

Proof. Define $\Sigma_f(Z \xrightarrow{h} X) = (Z \xrightarrow{h} X \xrightarrow{f} Y)$. The adjunction is trivial. ∎

1.42 THEOREM (Lawvere–Tierney). *Let $\mathcal{E}$ be a topos, X an object of $\mathcal{E}$. Then $\mathcal{E}/X$ is a topos, and the functor*

$$X^*: \mathcal{E} \longrightarrow \mathcal{E}/X;\ Y \longmapsto (Y \times X \xrightarrow{\pi_2} X)$$

(i.e. the pullback functor along $X \longrightarrow 1$) is logical.

Proof. (i) Products in $\mathcal{E}/X$ correspond to pullbacks over X in $\mathcal{E}$, and equalizers are the same as in $\mathcal{E}$ (i.e. Σ_X creates them). It is immediate from 1.41 that X^* preserves limits.

(ii) Since monos in $\mathcal{E}/X$ are the same as in $\mathcal{E}$, it is readily seen that $X^*(\Omega)$ is a subobject classifier for $\mathcal{E}/X$.

(iii) To form the exponential $(Z \xrightarrow{g} X)^{(Y \xrightarrow{f} X)}$ in $\mathcal{E}/X$, let $\theta: X \times Y \longrightarrow \tilde{X}$

represent the partial map

$$\begin{array}{ccc} Y & \xrightarrow{\quad f \quad} & X\ . \\ \big\downarrow{\scriptstyle (f,\,1_Y)} & & \\ X \times Y & & \end{array}$$

Now form the pullback

$$\begin{array}{ccc} E & \xrightarrow{\qquad\quad} & \tilde{Z}^Y\ ; \\ \big\downarrow & & \big\downarrow{\scriptstyle \tilde{g}^Y} \\ X & \xrightarrow{\quad \bar{\theta} \quad} & \tilde{X}^Y \end{array}$$

then for any $T \xrightarrow{h} X$ we have the following sequence of natural bijections:

$$\begin{array}{l} T \longrightarrow E \text{ over } X \\ \hline T \longrightarrow \tilde{Z}^Y \text{ over } T \xrightarrow{h} X \xrightarrow{\bar{\theta}} \tilde{X}^Y \\ \hline T \times Y \longrightarrow \tilde{Z} \text{ over } T \times Y \xrightarrow{h \times 1} X \times Y \xrightarrow{\theta} \tilde{X} \\ \hline T \times Y \longrightarrow Z \text{ over } \begin{array}{ccc} T \times_X Y & \xrightarrow{\qquad\quad} & X \\ \big\downarrow & & \\ T \times Y & & \end{array} \\ \hline T \times_X Y \longrightarrow Z \text{ over } X. \end{array}$$

So $E \longrightarrow X$ is the required exponential; and it is easily verified that if Y, Z have the form $A \times X$, $B \times X$, then $B^A \times X$ also has the properties of an exponential, and is thus isomorphic to E. So X^* preserves exponentials. ■

1.43 Corollary. *For any $X \xrightarrow{f} Y$ in $\mathscr{E}$ the pullback functor $f^*: \mathscr{E}/Y \longrightarrow \mathscr{E}/X$ is logical; and it has a right adjoint $\Pi_f: \mathscr{E}/X \longrightarrow \mathscr{E}/Y$.*

Proof. Regarding f as an object of $\mathscr{E}/Y$, we can identify $\mathscr{E}/X$ with the category $(\mathscr{E}/Y)/f$. So by working in the topos $\mathscr{E}/Y$ we can reduce to the case $Y = 1$. But in this case we already know f^* is logical, from 1.42; and existence of Π_f follows from 1.41 and 1.37(ii). ■

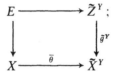

1.44 EXAMPLE. In the case $\mathscr{E} = \mathscr{S}$, it is convenient to regard objects of $\mathscr{S}/X$ as "X-indexed families of sets". In terms of this interpretation, the functor f^* may be described as "relabelling along f", i.e.

$$f^*(S_y \mid y \in Y) = (S_{f(x)} \mid x \in X).$$

And the functors Σ_f and Π_f correspond to forming coproducts and products over the fibres of f, i.e.

$$\Sigma_f(S_x \mid x \in X) = ((\coprod_{f(x)=y} S_x) \mid y \in Y)$$

and

$$\Pi_f(S_x \mid x \in X) = ((\prod_{f(x)=y} S_x) \mid y \in Y). \ \blacksquare$$

1.45 EXAMPLE. Consider the morphism $1 \xrightarrow{t} \Omega$. It is easily verified from 1.26 that $\Pi_t(X) = (\tilde{X} \xrightarrow{\phi} \Omega)$, where ϕ classifies $X \overset{\eta}{\rightarrowtail} \tilde{X}$. ■

1.46 COROLLARY. *Any morphism $X \xrightarrow{f} Y$ in $\mathscr{E}$ induces a geometric morphism $\mathscr{E}/X \xrightarrow{f} \mathscr{E}/Y$ with $f_* = \Pi_f$ and $f^* =$ pullback functor. We thus have a functor $\mathscr{E} \longrightarrow \mathfrak{Top}/\mathscr{E}$ for any topos $\mathscr{E}$, defined on objects by $X \longmapsto (\mathscr{E}/X \xrightarrow{X} \mathscr{E}/1 \simeq \mathscr{E})$.* ■

It should be noted that the geometric morphisms $\mathscr{E}/X \longrightarrow \mathscr{E}/Y$ which arise from morphisms of $\mathscr{E}$ are peculiar in two respects: they are essential and their inverse image functors are logical. In fact we have the following converse result:

1.47 THEOREM. *Let $\mathscr{F} \xrightarrow{f} \mathscr{E}$ be an essential geometric morphism such that f^* is logical, and such that the left adjoint $f_!$ of f^* preserves equalizers. Then there exists an object X of $\mathscr{E}$ (unique up to isomorphism) such that $\mathscr{F}$ is equivalent to $\mathscr{E}/X$ by an equivalence identifying $f_!$, f^* and f_* with Σ_X, X^* and Π_X respectively.*

Proof. Since f^* preserves Ω, we have a natural bijection between morphisms $Y \longrightarrow \Omega$ in $\mathscr{F}$ and morphisms $f_! Y \longrightarrow \Omega$ in $\mathscr{E}$, i.e. between subobjects of Y in $\mathscr{F}$ and subobjects of $f_! Y$ in $\mathscr{E}$. So if $Y' \xrightarrow{\sigma} Y$ is a morphism of $\mathscr{F}$ such that $f_!(\sigma)$ is iso, the pullback functor σ^* induces a bijection between subobjects of Y and subobjects of Y'; and in particular if σ is mono (so that $\sigma^*(\sigma) \cong 1_{Y'} \cong \sigma^*(1_Y)$ as subobjects of Y') we have $\sigma \cong 1_Y$ as subobjects of Y, i.e. σ is an isomorphism. Now let $Y \underset{\beta}{\overset{\alpha}{\rightrightarrows}} Z$ be a parallel pair in $\mathscr{F}$ such that $f_!(\alpha) = f_!(\beta)$; then by applying the above argument to the equalizer $Y' \overset{\sigma}{\rightarrowtail} Y$ of α and β, we deduce that $\alpha = \beta$. Thus $f_!$ is faithful, and in particular it reflects monos and epis; hence by 1.22 it reflects isomorphisms.

So by the opposite of 0.13 we deduce that $f_!$ is comonadic, i.e. that $\mathscr{F}$ is equivalent to the category $\mathscr{E}_{\mathbb{G}}$ of coalgebras for the comonad $\mathbb{G}$ induced by $(f_! \dashv f^*)$. Now define $X = f_!f^*(1_{\mathscr{E}})$: it will suffice to show that $\mathbb{G}$ is isomorphic to the comonad $\mathbb{C}$ whose functor part is $(-)\times X: \mathscr{E} \longrightarrow \mathscr{E}$, with counit and comultiplication given by $U \times X \xrightarrow{\pi_1} U$ and $U \times X \xrightarrow{1 \times \Delta} U \times X \times X$ respectively; for it is easily seen that $\mathscr{E}_{\mathbb{C}} \cong \mathscr{E}/X$.

But we have an obvious morphism of comonads $\theta: \mathbb{G} \longrightarrow \mathbb{C}$, defined by specifying that the components of $\theta_U: f_!f^*U \longrightarrow U \times X$ are the counit of $(f_! \dashv f^*)$ at U and $f_!f^*(U \longrightarrow 1)$ respectively. It therefore suffices to construct a two-sided inverse for θ_U. But this may be obtained by the following "deduction":

$$\frac{\dfrac{\dfrac{f^*U \xrightarrow{\alpha_{f^*U}} f^*f_!f^*U}{1 \longrightarrow (f^*f_!f^*U)^{f^*U} \cong f^*((f_!f^*U)^U)} \quad \text{since } f^* \text{ preserves exponentials}}{X \cong f_!1 \longrightarrow (f_!f^*U)^U}}{U \times X \xrightarrow{\psi} f_!f^*U}$$

where α is the unit of $(f_! \dashv f^*)$. The proof that ψ is a two-sided inverse for θ_U is a tedious but straightforward diagram-chase. ■

1.48 REMARK. The hypothesis that $f_!$ should preserve equalizers (which is of course always satisfied by geometric morphisms which arise as in 1.46) is necessary to the proof of 1.47, as the following example shows:

Let G be a nontrivial group, and consider the essential geometric morphism $\mathscr{S}^{\mathbf{G}^{\mathrm{op}}} \xrightarrow{\gamma} \mathscr{S}$ defined by $\gamma^*(S) = S$ with trivial G-action, $\gamma_*(M) =$ set of G-fixed elements of M, and $\gamma_!(M) =$ set of G-orbits of M. The adjunctions $(\gamma_! \dashv \gamma^* \dashv \gamma_*)$ are easily verified, and it is immediate from 1.14 that γ^* is logical. But $\mathscr{S}^{\mathbf{G}^{\mathrm{op}}}$ is clearly not equivalent to $\mathscr{S}/X$ for any X. (The reader may verify that the functor $\gamma_!$ does preserve coreflexive equalizers; so it is not sufficient to assume only that $f_!$ preserves coreflexive equalizers in 1.47.) ■

One particularly important consequence of 1.42 is the freedom which it gives us to use arguments involving "generalized elements". By analogy with the case $\mathscr{E} = \mathscr{S}$, we define an *element* (or sometimes *global element*) of an object X of $\mathscr{E}$ to be a morphism $1 \longrightarrow X$. But it is clear that in a general topos the object 1 is not a generator, and so an object cannot be determined by specifying its global elements; hence we are led by the principle of the Yoneda lemma to consider "elements of X defined over U", i.e. morphisms $U \longrightarrow X$, for a variable object U. (If $\mathscr{E}$ has a set of generators, then we can restrict U to lie in that set; but this possibility is not important for our present considerations.)

Now it follows at once from 1.41 that morphisms $U \longrightarrow X$ correspond bijectively to global elements of U^*X in the topos $\mathscr{E}/U$; and since U^* is logical, such elements can be "interpreted" in exactly the same way as global elements of X in $\mathscr{E}$. Thus for example, when we refer to Ω^X as "the object of subobjects of X", we mean not only that its global elements correspond bijectively to morphisms $X \cong 1 \times X \longrightarrow \Omega$, and hence to subobjects of X in $\mathscr{E}$, but also that its U-elements $U \longrightarrow \Omega^X$ correspond to subobjects of U^*X in $\mathscr{E}/U$, in a fashion which is natural in the variable U.

A particularly important generalized element of the object X is its *generic element* x, which can be described in $\mathscr{E}$ as the identity morphism $X \xrightarrow{1} X$, or in $\mathscr{E}/X$ as the global element

$$\begin{array}{ccc} X & \xrightarrow{\quad\Delta\quad} & X \times X \\ & {\scriptstyle 1_X}\searrow \qquad \swarrow{\scriptstyle \pi_2} & \\ & X & \end{array}$$

of X^*X. This has the universal property that $f^*(x) = f$ for any generalized element $U \xrightarrow{f} X$ of X; so if we have some assertion about generalized elements of X whose truth is preserved by pullback functors, and which is true for the generic element, then it is true for all generalized elements (and in particular for all global elements) of X.

1.49 Examples. We give a couple of typical examples to show how generic elements may be used in practice.

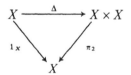

(i) We will construct a morphism $Y^X \times Y^X \xrightarrow{\text{eq}} \Omega^X$ which "internalizes" the operation of forming the equalizer of a parallel pair of morphisms $X \rightrightarrows Y$. Write Z for $Y^X \times Y^X$; then in the topos $\mathscr{E}/Z$, we have a generic element of $Z^*Z \cong Z^*Y^{Z^*X} \times Z^*Y^{Z^*X}$, which we can regard as a generic parallel pair $Z^*X \rightrightarrows Z^*Y$. Forming their equalizer in $\mathscr{E}/Z$ gives us a subobject of Z^*X, which we can regard as a global element of $\Omega^{Z^*X} \cong Z^*(\Omega^X)$. But such an element corresponds by 1.41 to a morphism $Z \longrightarrow \Omega^X$ in $\mathscr{E}$; and this is the required morphism. And since pullback functors preserve equalizers, it is easily seen that the map induced by eq on U-elements of $Y^X \times Y^X$ is the same as that induced by forming equalizers of parallel pairs in $\mathscr{E}/U$.

(ii) Similarly, we can internalize the operation of forming the intersection (= pullback) of a pair of subobjects of an object of $\mathscr{E}$, by a single morphism $\Omega \times \Omega \xrightarrow{\wedge} \Omega$. To do this, we take the generic pair of subobjects of 1 in $\mathscr{E}/\Omega \times \Omega$ (i.e. those classified by the two product projections $\Omega \times \Omega \longrightarrow \Omega$), form their intersection, and then take its classifying map. Or $\wedge$ may be described more explicitly as the classifying map of $1 \xrightarrow{(t,t)} \Omega \times \Omega$; it is easily verified that this map has the the right properties. ■

We shall return to the idea of generic elements, in the more formal context of the Mitchell–Bénabou language, in §5.4.

1.5. IMAGE FACTORIZATIONS

1.51 Lemma. *In a topos, colimits are universal (i.e. preserved by pullback functors, cf.* 0.42(ii)).

Proof. This is immediate from the existence of the functors Π_f (1.43). ■

From 1.51, we can deduce many important exactness properties, e.g.

1.52 Theorem (Kelly–Tierney). *Any morphism in a topos can be factored as an epi followed by a mono. (The factorization is unique up to isomorphism, by* 1.22.)

Proof. Given $X \xrightarrow{f} Y$, form the diagram

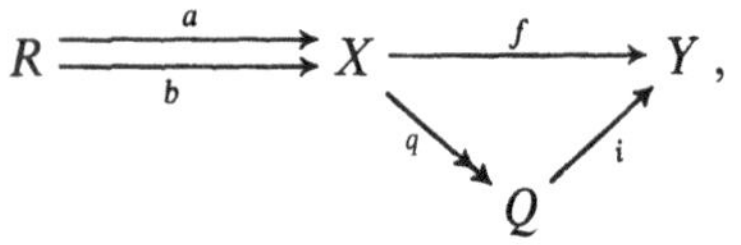

where (a, b) is the kernel-pair of f, and q is the coequalizer of (a, b). We need to show that i is mono; but suppose we have $T \underset{d}{\overset{c}{\rightrightarrows}} Q$ such that $ic = id$. Form the pullback

$$\begin{array}{ccc} S & \xrightarrow{e} & T; \\ \big\downarrow{\scriptstyle (g,h)} & & \big\downarrow{\scriptstyle (c,d)} \\ X \times X & \xrightarrow{q \times q} & Q \times Q \end{array}$$

then $fg = iqg = ice = ide = fh$, so (g, h) factors through $R \overset{(a,b)}{\rightarrowtail} X \times X$ (say by $S \xrightarrow{k} R$).

Now $ce = qg = qak = qbk = qh = de$; but $q \times q$ is epi, since we can factor it as $(q \times 1_Q)(1_X \times q)$ in which each factor is epi by 1.11(ii). So e is epi by 1.51, and $c = d$. ■

1.53 Corollary. *In a topos, every epimorphism is a coequalizer.*

Proof. Suppose f is epi in the diagram of 1.52; then so is i, and hence i is iso by 1.22. So f is a coequalizer of its own kernel-pair. ■

1.54 Remark. Combining 1.52 and 1.53 with 1.23, we have proved that a topos is an *exact category* in the sense of Barr [2]. ■

Since a topos has coproducts and image factorizations, it is easily seen that we can form finite unions of subobjects; i.e. given subobjects $X_1 \overset{a}{\rightarrowtail} Y$, $X_2 \overset{b}{\rightarrowtail} Y$, the image $X_1 \cup X_2 \rightarrowtail Y$ of $X_1 \amalg X_2 \xrightarrow{\binom{a}{b}} Y$ is the least upper bound of a and b in the poset of subobjects of Y. But the fact that coproducts are universal enables us to give an alternative description of $X_1 \cup X_2$, which is in many ways more convenient.

1.55 Proposition. *Let $X_1 \overset{a}{\rightarrowtail} Y$, $X_2 \overset{b}{\rightarrowtail} Y$ be two subobjects in a topos $\mathscr{E}$. Form their intersection (= pullback)*

$$\begin{array}{ccc} X_1 \cap X_2 & \overset{c}{\rightarrowtail} & X_1, \\ \big\downarrow{\scriptstyle d} & & \big\downarrow{\scriptstyle a} \\ X_2 & \overset{b}{\rightarrowtail} & Y \end{array}$$

and their union $X_1 \cup X_2$ *as above. Then the square*

$$\begin{array}{ccc} X_1 \cup X_2 & \xrightarrow{c} & X_1 \\ \downarrow{\scriptstyle d} & & \downarrow \\ X_2 & \longrightarrow & X_1 \cup X_2 \end{array}$$

is a pushout.

Proof. By definition, the pushout of c and d is the coequalizer of

$$X_1 \cap X_2 \underset{v_2 d}{\overset{v_1 c}{\rightrightarrows}} X_1 \amalg X_2;$$

whereas the epimorphism $X_1 \amalg X_2 \twoheadrightarrow X_1 \cup X_2$ is (by the construction of 1.52) the coequalizer of $R \underset{f}{\overset{e}{\rightrightarrows}} X_1 \amalg X_2$, where R is the kernel-pair of $X_1 \amalg X_2 \xrightarrow{\binom{a}{b}} Y$. But since coproducts in $\mathscr{E}$ are universal, the pullback

$$\begin{array}{ccc} R & \xrightarrow{e} & X_1 \amalg X_2 \\ \downarrow{\scriptstyle f} & & \downarrow{\scriptstyle \binom{a}{b}} \\ X_1 \amalg X_2 & \xrightarrow{\binom{a}{b}} & Y \end{array}$$

can be decomposed as a coproduct of four objects $R_{ij}(1 \leqslant i, j \leqslant 2)$, where R_{ij} is the pullback of X_i against X_j. Writing e_{ij}, f_{ij} for the composites of e, f with the inclusion $R_{ij} \longrightarrow R$, we thus have that $X_1 \amalg X_2 \twoheadrightarrow X_1 \cup X_2$ is the joint coequalizer of the four pairs (e_{ij}, f_{ij}). But a and b are mono, so $e_{11} = f_{11}$ and $e_{22} = f_{22}$; hence these two pairs do not affect the coequalizer. And we have $R_{12} = R_{21} = X_1 \cap X_2$, with $e_{12} = v_1 c = f_{21}, e_{21} = v_2 d = f_{12}$; so that the joint coequalizer of (e_{12}, f_{12}) and (e_{21}, f_{21}) is simply the coequalizer of $v_1 c$ and $v_2 d$ i.e. the pushout of c and d. ■

1.56 LEMMA. *The initial object* 0 *in a topos is* strict, *i.e. any morphism* $X \longrightarrow 0$ *is an isomorphism.*

Proof. Suppose given such a morphism; then trivially

$$\begin{array}{ccc} X & \longrightarrow & 0 \\ \downarrow{\scriptstyle 1_X} & & \downarrow{\scriptstyle 1_0} \\ X & \longrightarrow & 0 \end{array}$$

is a pullback. But 1_0 is initial in $\mathscr{E}/0$, so by 1.51 1_X is initial in $\mathscr{E}/X$, which implies that X is initial in $\mathscr{E}$. ■

1.57 Corollary. *Coproducts in a topos are disjoint* (*cf.* 0.42(*i*)).

Proof. By definition, the square

$$\begin{array}{ccc} 0 & \longrightarrow & X \\ \big\downarrow & & \big\downarrow{\scriptstyle \nu_1} \\ Y & \xrightarrow{\nu_2} & X \amalg Y \end{array}$$

is a pushout. But by 1.56 $0 \longrightarrow X$ and $0 \longrightarrow Y$ are trivially mono; hence by 1.28 the square is a pullback. ■

1.58 Remark. Combining 1.23, 1.36, 1.51, 1.53 and 1.57, we see that we have verified, from the axioms of 1.11, all those conditions of Giraud's theorem (0.45) which are not explicitly dependent on set theory. We are thus justified in regarding 1.11 as an appropriate "elementary" generalization of the notion of Grothendieck topos. ■

EXERCISES 1

†1. (A. Kock [66]) Show that axioms (ii) and (iii) of 1.11 may be replaced by the single axiom: "For every object X of $\mathscr{E}$, there exists a *power-object* PX and a subobject $\in_X \rightarrowtail PX \times X$ such that, for every object Y and every subobject $R \rightarrowtail Y \times X$, there exists a unique $Y \xrightarrow{r} PX$ such that

$$\begin{array}{ccc} R & \longrightarrow & \in_X \\ \big\downarrow & & \big\downarrow \\ Y \times X & \xrightarrow{r \times 1} & PX \times X \end{array}$$

is a pullback."

[Method: define the map $1 \xrightarrow{\ulcorner X \urcorner} PX$ to correspond to $X \overset{1}{\rightarrowtail} X \cong 1 \times X$, and define the singleton map $X \xrightarrow{\{\}} PX$ to correspond to $X \overset{\Delta}{\rightarrowtail} X \times X$.

Now let $P(X \times Y) \times X \xrightarrow{r} PY$ correspond to the subobject $\in_{(X \times Y)}$ of $P(X \times Y) \times X \times Y$, let

$$\begin{array}{ccc} Q & \longrightarrow & Y \\ \big\downarrow & & \big\downarrow {\scriptstyle \{\}} \\ P(X \times Y) \times X & \xrightarrow{\quad r \quad} & PY \end{array}$$

be a pullback, and let $P(X \times Y) \xrightarrow{q} PX$ correspond to the subobject Q. Then let

$$\begin{array}{ccc} Y^X & \longrightarrow & 1 \\ \big\downarrow & & \big\downarrow {\scriptstyle \ulcorner X \urcorner} \\ P(X \times Y) & \xrightarrow{\quad q \quad} & PX \end{array}$$

be a pullback, and prove that Y^X has the universal property of an exponential. Finally, prove that $P(1)$ is a subobject classifier.]

2. Show that axiom (i) of 1.11 may be reduced to the statement that $\mathscr{E}$ has finite products and a terminal object, and that pullbacks of diagrams of the form

$$\begin{array}{ccc} & & 1 \\ & & \big\downarrow {\scriptstyle t} \\ X & \longrightarrow & \Omega \end{array}$$

always exist. [Method: to construct the equalizer of $X \rightrightarrows Y$, form its classifying map $X \longrightarrow \Omega$.]

3. (D. Higgs) Let $\Omega \overset{\alpha}{\rightarrowtail} \Omega$ be a monomorphism in a topos, and let $U \overset{m}{\rightarrowtail} \Omega$ be the subobject classified by α. By considering the subobject of U classified by m, prove that the diagram

$$\begin{array}{ccc} U & \xrightarrow{\quad m \quad} & \Omega \\ \big\downarrow & & \big\uparrow {\scriptstyle \alpha} \\ 1 & \xrightarrow{\quad t \quad} & \Omega \end{array}$$

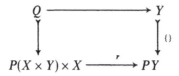

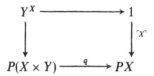

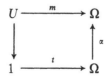

commutes, and deduce that

$$\begin{array}{ccc} U & \xrightarrow{1} & U \\ {\scriptstyle m}\downarrow & & \downarrow{\scriptstyle m} \\ \Omega & \xrightarrow{\alpha^2} & \Omega \end{array}$$

is a pullback. Hence show that $\alpha^2 = 1_\Omega$.

4. (H. Engenes [33]) Let **On** denote the class of ordinal numbers, regarded as a (large) category by means of its usual ordering. Prove that $\mathscr{S}^{\mathbf{On}^{\mathrm{op}}}$ is a topos, but $\mathscr{S}^{\mathbf{On}}$ is not. [Compare 1.14.]

†5. Prove that an object of a topos $\mathscr{E}$ is injective iff it is a retract of Ω^X for some X. Deduce that if E is injective then the functor $E^{(-)}: \mathscr{E}^{\mathrm{op}} \longrightarrow \mathscr{E}$ preserves reflexive coequalizers.

6. By analogy with 1.38, show how 1.34 may be used to construct coproducts in a topos.

7. Let X be a topological space, and $E \xrightarrow{p} X$ a local homeomorphism. Prove that the topos $\mathrm{Shv}(X)/\Gamma(E, p)$ is equivalent to $\mathrm{Shv}(E)$.

†8. Let $\mathscr{E}$ be a category with finite limits. Prove that $\mathscr{E}$ is cartesian closed iff the pullback functor X^* has a right adjoint Π_X, for every morphism X with codomain 1. [Hint: if $\mathscr{E}$ is cartesian closed, show that $\Pi_X(Y \xrightarrow{h} X)$ may be defined by the pullback

$$\begin{array}{ccc} \Pi_X(h) & \longrightarrow & 1 \\ \downarrow & & \downarrow{\scriptstyle \ulcorner 1_X \urcorner} \\ Y^X & \xrightarrow{h^X} & X^X \end{array}$$

.]

9. Let $\mathscr{F} \xrightarrow{f} \mathscr{E}$ be a geometric morphism, X an object of $\mathscr{E}$ and Y an object of $\mathscr{F}$. By considering generalized elements, show that the objects $f_*(Y^{f^*X})$ and $(f_*Y)^X$ are isomorphic in $\mathscr{E}$.

†10. Let $X \xrightarrow{f} Y$ be a morphism in a topos $\mathscr{E}$. Use image factorizations to

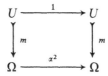

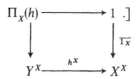

define the morphism $\exists f: \Omega^X \longrightarrow \Omega^Y$ without the assumption (made in §1.3) that f is mono. Show that $X \longmapsto \Omega^X$, $f \longmapsto \exists f$ is a covariant functor $\mathscr{E} \longrightarrow \mathscr{E}$; and that 1.32 remains true if we remove the restriction that g and h are mono. Show also that if f is epi then $\exists f \, . \, Pf = 1_{PY}$, and that the functor $f \longmapsto \exists f$ preserves pullback diagrams of the form

$$\begin{array}{ccc} X & \xrightarrow{\;f\;} & Y \\ {\scriptstyle g}\big\downarrow & & \big\downarrow{\scriptstyle h} \\ Z & \xrightarrow{\;k\;} & T \end{array}$$

with g, h mono.

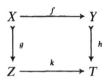

Chapter 2

Internal Category Theory

2.1. INTERNAL CATEGORIES AND DIAGRAMS

One of the major themes running through the whole of (elementary) topos theory is that a topos should be regarded as a "universe of discourse" within which we can carry out constructions in much the same way as we do within the category of sets, and with much the same results. We have already met some examples of this theme in §1.4; in this chapter, our aim is to develop category theory itself in a manner which is "internal" to an arbitrary topos. However, the basic definitions of internal category theory can be interpreted in a context considerably more general than that of a topos; so, for the time being, we shall assume simply that $\mathscr{E}$ is a category with finite limits. We shall re-adopt the other topos axioms as they become necessary to the argument.

2.11 DEFINITION. An *internal category* $\mathbf{C}$ in $\mathscr{E}$ consists of:

(i) a pair of objects C_0, C_1 (called the *object-of-objects* and *object-of-morphisms* of $\mathbf{C}$, respectively); and

(ii) four morphisms $C_1 \xrightarrow{d_0} C_0$, $C_1 \xrightarrow{d_1} C_0$, $C_0 \xrightarrow{i} C_1$ and $C_2 = C_1 \times_{C_0} C_1 \xrightarrow{m} C_1$;†

† *Convention.* Whenever C_1 appears as one factor of a pullback over C_0, we write it on the left of the symbol $\times_{C_0}$ if it is regarded as having structure map d_1, and on the right if it is regarded as having structure map d_0. The symbol $C_1 \times_{C_0} C_1$ therefore means the pullback

$$\begin{array}{ccc} C_2 & \xrightarrow{\pi_2} & C_1 \\ \downarrow{\scriptstyle \pi_1} & & \downarrow{\scriptstyle d_0} \\ C_1 & \xrightarrow{d_1} & C_0 \end{array}$$

(iii) such that $d_0 i = d_1 i = 1_{C_0}$, $d_0 m = d_0 \pi_1$, $d_1 m = d_1 \pi_2$,
$m(1 \times m) = m(m \times 1)\colon C_3 = C_1 \times_{C_0} C_1 \times_{C_0} C_1 \longrightarrow C_1$, and
$m(1 \times i) = m(i \times 1) = 1_{C_1}$.

An *internal functor* (or morphism of internal categories) $f\colon \mathbf{C} \longrightarrow \mathbf{D}$ consists of a pair of morphisms $C_0 \xrightarrow{f_0} D_0$, $C_1 \xrightarrow{f_1} D_1$ commuting with d_0, d_1, i and m. We write $\mathbf{cat}(\mathscr{E})$ for the category of internal categories and functors of $\mathscr{E}$. (In fact $\mathbf{cat}(\mathscr{E})$ is a 2-category, since we can also define a notion of *internal natural transformation*, but we shall not need this extra structure. For a detailed account of internal categories from the 2-categorical point of view, the reader is referred to the monumental work of J. W. Gray [168].) ■

2.12 DEFINITION. Let $\mathbf{C} \in \mathbf{cat}(\mathscr{E})$.

(i) We say $\mathbf{C}$ is a *poset* if $C_1 \xrightarrow{(d_0, d_1)} C_0 \times C_0$ is mono.
(ii) We say $\mathbf{C}$ is *discrete* if i is an isomorphism (which implies that d_0, d_1 and m are also isomorphisms).
(iii) We say $\mathbf{C}$ is a *monoid* if C_0 is the terminal object 1.
(iv) We define the *opposite category* $\mathbf{C}^{\text{op}}$ of $\mathbf{C}$ to be that obtained by interchanging d_0 and d_1, and "twisting" the definition of m. ■

2.13 REMARK. Recall that if Δ denotes the category of (nonzero) finite ordinals and order-preserving maps between them, then a *simplicial object* of $\mathscr{E}$ is a functor $\Delta^{\text{op}} \longrightarrow \mathscr{E}$. (Equivalently, a simplicial object may be specified by giving a sequence of objects C_n ($n \geqslant 0$) and morphisms $d_i^n\colon C_n \longrightarrow C_{n-1}$ ($0 \leqslant i \leqslant n$), $s_j^n\colon C_n \longrightarrow C_{n+1}$ ($0 \leqslant j \leqslant n$) satisfying certain relations; see [CW], §VII.5 or [CF], chapter II.) Now if we define C_n ($n \geqslant 2$) to be the n-fold pullback $C_1 \times_{C_0} C_1 \times_{C_0} \cdots \times_{C_0} C_1$, we see that the data of 2.11 are precisely what we need to make ($n \longmapsto C_{n-1}$) into a simplicial object of $\mathscr{E}$. [In the case $\mathscr{E} = \mathscr{S}$, C_n is the set of composable n-tuples of morphisms of $\mathbf{C}$, the morphisms d_i^n are obtained by dropping the first or last morphism of an n-tuple ($i = 0$ or n) or by composing two adjacent morphisms, and the s_j^n are obtained by inserting identity morphisms at one of the ($n + 1$) possible places.] Conversely, a simplicial object of $\mathscr{E}$ is derivable (up to isomorphism) from an internal category in this way iff it is left exact as a functor $\Delta^{\text{op}} \longrightarrow \mathscr{E}$. $\mathbf{cat}(\mathscr{E})$ may thus be identified (up to equivalence) with a full subcategory of the category $\mathbf{simpl}(\mathscr{E})$ of simplicial objects of $\mathscr{E}$. ■

In the case $\mathscr{E} = \mathscr{S}$, this "simplicial" description of small categories was first used by A. Grothendieck [43]. Unlike 2.11, it does not constitute an elementary definition of the concept of internal category (since the category $\mathbf{\Delta}$ is not finite, or even finitely-presented), but readers who are already familiar with the theory of simplicial objects may well find it easier to work with.

In the case $\mathscr{E} = \mathscr{S}$, we are accustomed to consider not only functors from one small category to another, but also functors from a small category into a large category, and in particular into $\mathscr{S}$ itself. To internalize this concept, we introduce the notion of internal diagram; an internal diagram on $\mathbf{C}$ should consist of a C_0-indexed family of objects of $\mathscr{E}$ (which, in accordance with the philosophy of § 1.4, we interpret as an object of $\mathscr{E}/C_0$), equipped with a morphism which describes how C_1 acts on the diagram. Formally, we have

2.14 Definition. Let $\mathbf{C} \in \mathbf{cat}(\mathscr{E})$. An *internal diagram* F on $\mathbf{C}$ consists of

(i) an object $F_0 \xrightarrow{\gamma_0} C_0$ of $\mathscr{E}/C_0$, and
(ii) a morphism $e\colon F_1 = F_0 \times_{C_0} C_1 \longrightarrow F_0$,
(iii) such that $\gamma_0 e = d_1 \pi_2$, $e(1 \times i) = 1_{F_0}$, and
$e(e \times 1) = e(1 \times m)\colon F_2 = F_0 \times_{C_0} C_1 \times_{C_0} C_1 \longrightarrow F_0$.

A *morphism of internal diagrams* $f\colon F \longrightarrow G$ is a morphism $F_0 \longrightarrow G_0$ over C_0, commuting with the structure morphisms e. We write $\mathscr{E}^{\mathbf{C}}$ for the category of internal diagrams on $\mathbf{C}$. ■

By interchanging the roles of d_0 and d_1 in 2.14, we obtain the notion of a contravariant internal diagram on $\mathbf{C}$ (commonly called an *internal presheaf* on $\mathbf{C}$). It should also be noted that some authors use the term "internal functor" for what we have called an internal diagram.

2.15 Remark. Let F be an internal diagram on $\mathbf{C}$. Then it is readily seen that if we define $F_n (n \geqslant 1)$ as $F_0 \times_{C_0} C_n$, and $\gamma_n = \pi_2\colon F_n \longrightarrow C_n$, the data of 2.14 are sufficient to make $\mathbf{F}$ into an internal category, in such a way that γ becomes an internal functor $\mathbf{F} \longrightarrow \mathbf{C}$. (For example, the maps $d_0, d_1\colon F_1 \rightrightarrows F_0$ are π_1 and e respectively.) Conversely, an object $\mathbf{F} \xrightarrow{\gamma} \mathbf{C}$ of $\mathbf{cat}(\mathscr{E})/\mathbf{C}$ is isomorphic to one constructed in this way iff the square

$$\begin{array}{ccc} F_1 & \xrightarrow{d_0} & F_0 \\ \downarrow{\scriptstyle \gamma_1} & & \downarrow{\scriptstyle \gamma_0} \\ C_1 & \xrightarrow{d_0} & C_0 \end{array}$$

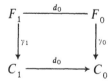

is a pullback. (An internal functor satisfying this condition is called a *discrete opfibration*; this is actually a special case of the notion of *split opfibration*, for which see Exercise 6 at the end of this chapter. The internal functors which correspond in the same way to internal presheaves are called *discrete fibrations*.) We may thus identify $\mathscr{E}^{\mathbf{C}}$ (up to equivalence) with the full subcategory of $\mathbf{cat}(\mathscr{E})/\mathbf{C}$ whose objects are discrete opfibrations. ∎

The next four results are all proved by straightforward diagram-chasing arguments.

2.16 LEMMA. (i) *The category* $\mathbf{cat}(\mathscr{E})$ *has finite limits; in fact the forgetful functor* $\mathbf{cat}(\mathscr{E}) \longrightarrow \mathscr{E} \times \mathscr{E}; C \longmapsto (C_0, C_1)$ *creates them.*

(ii) *The category* $\mathscr{E}^{\mathbf{C}}$ *has finite limits, and the inclusion functor* $\mathscr{E}^{\mathbf{C}} \longrightarrow \mathbf{cat}(\mathscr{E})/\mathbf{C}$ *creates them.* ∎

2.17 LEMMA. *Let* $\mathbf{C} \xrightarrow{\gamma} \mathbf{D} \xrightarrow{\delta} \mathbf{E}$ *be morphisms of* $\mathbf{cat}(\mathscr{E})$.

(i) *If* γ *and* δ *are discrete opfibrations, so is the composite* $\delta\gamma$.
(ii) *If* $\delta\gamma$ *and* δ *are discrete opfibrations, so is* γ. ∎

2.18 COROLLARY. *Let* F *be an internal diagram on* $\mathbf{C}$, *and* $\mathbf{F} \xrightarrow{\gamma} \mathbf{C}$ *the corresponding discrete opfibration. Then the category* $(\mathscr{E}^{\mathbf{C}})/F$ *is equivalent to* $\mathscr{E}^{\mathbf{F}}$. ∎

2.19 LEMMA. *Let* $\mathbf{C} \xrightarrow{f} \mathbf{D}$ *be a morphism of* $\mathbf{cat}(\mathscr{E})$. *Then the pullback functor* $f^*\colon \mathbf{cat}(\mathscr{E})/\mathbf{D} \longrightarrow \mathbf{cat}(\mathscr{E})/\mathbf{C}$ *preserves discrete opfibrations, and so induces a functor* $f^*\colon \mathscr{E}^{\mathbf{D}} \longrightarrow \mathscr{E}^{\mathbf{C}}$. [*In the case* $\mathscr{E} = \mathscr{S}$, *this functor corresponds to the operation of composing a functor* $\mathbf{D} \longrightarrow \mathscr{S}$ *with* $\mathbf{C} \xrightarrow{f} \mathbf{D}$.] ∎

2.2. INTERNAL LIMITS AND COLIMITS

2.21 PROPOSITION. *Let* $\mathbf{C} \in \mathbf{cat}(\mathscr{E})$. *Then the forgetful functor*

$$U\colon \mathscr{E}^{\mathbf{C}} \longrightarrow \mathscr{E}/C_0;\ F \longmapsto (F_0 \xrightarrow{\gamma_0} C_0)$$

is monadic.

Proof. Define $T_{\mathbf{C}}$ to be the composite

$$\mathscr{E}/C_0 \xrightarrow{d_0^*} \mathscr{E}/C_1 \xrightarrow{\Sigma d_1} \mathscr{E}/C_0.$$

Then for any $X \xrightarrow{\gamma} C_0$ in $\mathscr{E}/C_0$, the maps $1 \times i\colon X \longrightarrow X \times_{C_0} C_1$ and $1 \times m\colon X \times_{C_0} C_1 \times_{C_0} C_1 \longrightarrow X \times_{C_0} C_1$ define morphisms $\eta_\gamma\colon \gamma \longrightarrow T_{\mathbf{C}}(\gamma)$ and $\mu_\gamma\colon T_{\mathbf{C}} T_{\mathbf{C}}(\gamma) \longrightarrow T_{\mathbf{C}}(\gamma)$ in $\mathscr{E}/C_0$ which are natural in γ; and the equations of 2.11(iii) easily imply that $\mathbb{T}_{\mathbf{C}} = (T_{\mathbf{C}}, \eta, \mu)$ is a monad on $\mathscr{E}/C_0$. Now if $F = (F_0 \xrightarrow{\gamma_0} C_0, e)$ is an internal diagram on $\mathbf{C}$, the first equation of 2.14(iii) says that e is a morphism $T_{\mathbf{C}}(\gamma_0) \longrightarrow \gamma_0$ in $\mathscr{E}/C_0$, and the other two equations say precisely that it is a structure map for the monad $\mathbb{T}_{\mathbf{C}}$. So $\mathscr{E}^{\mathbf{C}}$ is isomorphic to the category of $\mathbb{T}_{\mathbf{C}}$-algebras. ■

2.22 Corollary. *In particular, the forgetful functor of* 2.21 *has a left adjoint* $R\colon \mathscr{E}/C_0 \longrightarrow \mathscr{E}^{\mathbf{C}}$. *If* $X \xrightarrow{\gamma} C_0$ *is an object of* $\mathscr{E}/C_0$, *then* $R(\gamma)$ *corresponds to the discrete opfibration*

$$\begin{array}{ccc} X \times_{C_0} C_1 \times_{C_0} C_1 & \overset{\pi_{12}}{\underset{1 \times m}{\rightrightarrows}} & X \times_{C_0} C_1 . \\ \downarrow{\scriptstyle \pi_3} & & \downarrow{\scriptstyle d_1 \pi_2} \\ C_1 & \overset{d_0}{\underset{d_1}{\rightrightarrows}} & C_0 \end{array}$$

Diagrams in the image of R *are said to be* representable. ■

2.23 Example. In the case $\mathscr{E} = \mathscr{S}$, let U be an object of the small category $\mathbf{C}$, regarded as a morphism $1 \xrightarrow{u} C_0$. Then $R(u)$ is easily seen to correspond to the representable functor $h^U = \hom_{\mathbf{C}}(U, —)$. More generally, $R(X \xrightarrow{\gamma} C_0)$ corresponds to an X-indexed coproduct of representable functors. It may be observed that the Yoneda Lemma (0.11) is simply the statement that R is left adjoint to the forgetful functor; and 0.12 similarly reduces to the statement that any object in a category of algebras has a free presentation ([CW], p. 149). ■

We now assume that $\mathscr{E}$ has reflexive coequalizers.

2.24 Definition. (i) We define a functor $\varinjlim\colon \mathbf{cat}(\mathscr{E}) \longrightarrow \mathscr{E}$ by $\varinjlim \mathbf{C} = \operatorname{coeq}(C_1 \overset{d_0}{\underset{d_1}{\rightrightarrows}} C_0)$. (Note that the pair (d_0, d_1) is reflexive, with common splitting i.)

(ii) We define a functor $\varinjlim_{\mathbf{C}}\colon \mathscr{E}^{\mathbf{C}} \longrightarrow \mathscr{E}$ by $\varinjlim_{\mathbf{C}} (\mathbf{F} \xrightarrow{\gamma} \mathbf{C}) = \varinjlim \mathbf{F}$. ■

2.25 Lemma. (i) $\varinjlim$ *is left adjoint to the functor* $\Delta\colon \mathscr{E} \longrightarrow \mathbf{cat}(\mathscr{E})$ *which sends* X *to the discrete category* $(X \overset{1}{\underset{1}{\rightrightarrows}} X)$.

(ii) $\varinjlim_{\mathbf{C}}$ *is left adjoint to the functor* $\mathbf{C}^*\colon \mathscr{E} \longrightarrow \mathscr{E}^{\mathbf{C}}$ *which sends* X *to the constant diagram* $(X \times \mathbf{C} \xrightarrow{\pi_2} \mathbf{C})$.

Proof. Straightforward verification. ∎

But in the case $\mathscr{E} = \mathscr{S}$, the adjunction of 2.25(ii) is simply the defining property of the colimit of a functor $\mathbf{C} \longrightarrow \mathscr{S}$ (see [CW], p. 67); so we are justified in calling $\varinjlim_{\mathbf{C}}$ the *colimit functor* for internal diagrams on $\mathbf{C}$. In fact we can paraphrase 2.25(ii) by saying that the category $\mathscr{E}$ is *internally cocomplete.*

The following technical lemma has a number of important applications:

2.26 LEMMA. *Let* $X \xrightarrow{\gamma} C_0$ *be an object of* $\mathscr{E}/C_0$. *Then* $\varinjlim_{\mathbf{C}}(R(\gamma)) \cong X$.

Proof. From 2.22, we have to show that

$$X \times_{C_0} C_1 \times_{C_0} C_1 \underset{1 \times m}{\overset{\pi_{12}}{\rightrightarrows}} X \times_{C_0} C_1 \xrightarrow{\pi_1} X$$

is a coequalizer diagram; but in fact it is a split coequilizer, with splittings given by

$$X \xrightarrow{1 \times i} X \times_{C_0} C_1 \xrightarrow{1 \times i \times 1} X \times_{C_0} C_1 \times_{C_0} C_1. \blacksquare$$

An alternative, more conceptual proof of 2.26 may be given by observing that the diagram

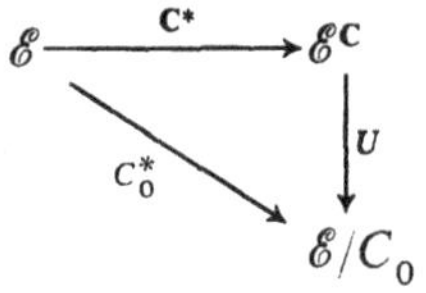

commutes. Hence the corresponding diagram

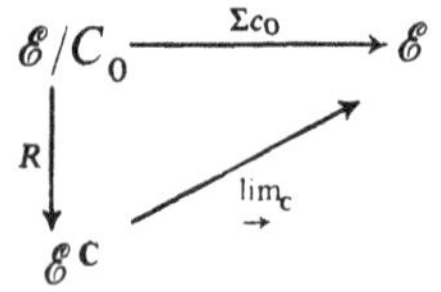

of left adjoints must also commute up to isomorphism.

In view of 2.25(ii), it is of interest to ask when the functor $\mathbf{C}^*$ has a right adjoint $\varprojlim_{\mathbf{C}}$. Now when $\mathbf{C}$ is a discrete category (so that $\mathscr{E}^{\mathbf{C}} \cong \mathscr{E}/C_0$), this is clearly equivalent to asking for the existence of the functor Π_{C_0} defined in

1.43. So from Ex. 1.8, we know that it is necessary for $\mathscr{E}$ to be cartesian closed. In fact this condition is also suffcient:

2.27 PROPOSITION. *A cartesian closed category is* internally complete, *i.e. for any* $\mathbf{C}$ *in* $\mathbf{cat}(\mathscr{E})$ *we have a functor* $\varprojlim_{\mathbf{C}}\colon \mathscr{E}^{\mathbf{C}} \longrightarrow \mathscr{E}$ *which is right adjoint to* $\mathbf{C}^*$.

Proof. Let $\mathbf{F} \xrightarrow{\gamma} \mathbf{C}$ be a discrete opfibration. We define a morphism $h\colon \Pi_{C_0}(\gamma_0) \longrightarrow \Pi_{C_1}(\gamma_1)$ as follows: let β be the counit of $(C_0^* \dashv \Pi_{C_0})$, and then define the transpose of h to be the composite

$$C_1^*(\Pi_{C_0}(\gamma_0)) \cong d_0^* C_0^*(\Pi_{C_0}(\gamma_0)) \xrightarrow{d_0^*\beta} d_0^*(\gamma_0) \cong \gamma_1.$$

Now define $\varprojlim_{\mathbf{C}}(F)$ to be the equalizer of

$$\begin{array}{ccccc}
\Pi_{C_0}(\gamma_0) & \xrightarrow{i_0} & F_0^{C_0} & \xrightarrow{\mathbf{F}_0^{d_2}} & F_0^{C_1}, \\
& \searrow^{h} & & & \nearrow_{d_1^{C_1}} \\
& & \Pi_{C_1}(\gamma_1) & \xrightarrow{i_1} & F_1^{C_1}
\end{array}$$

where the inclusions i_0 and i_1 arise from the construction of Ex. 1.8. [Note that in the case $\mathscr{E} = \mathscr{S}$, it follows from 1.44 that this is just the standard construction of limits via products and equalizers ([CW], p. 109).]

Now morphisms $X \xrightarrow{f} \Pi_{C_0}(\gamma_0)$ in $\mathscr{E}$ correspond to morphisms $X \times C_0 \xrightarrow{\bar{f}} F_0$ over C_0; and f factors through the above equalizer iff the square

$$\begin{array}{ccc}
X \times C_1 & \xrightarrow{d_0^*(\bar{f})} & F_1 \\
\downarrow{\scriptstyle 1 \times d_1} & & \downarrow{\scriptstyle d_1} \\
X \times C_0 & \xrightarrow{\bar{f}} & F_0
\end{array}$$

commutes, i.e. iff $\bar{f}$ induces a morphism of diagrams $\mathbf{C}^*(X) \longrightarrow F$. So we have the required adjunction. ■

2.3. DIAGRAMS IN A TOPOS

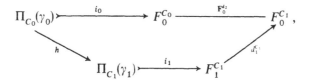

In this paragraph we investigate some of the special features of internal category theory in the case when the base category $\mathscr{E}$ is a topos. In fact, however, we shall not be using the full strength of the topos axioms; except in the proof of 2.32, the arguments we use are valid in any category $\mathscr{E}$ with the property that $\mathscr{E}/X$ is cartesian closed for every X. Such categories have been called *locally closed categories* by J. Penon [100] and *closed span categories* by B. Day [158]; apart from toposes, the principal examples are various cartesian-closed modifications of **esp**.

2.31 PROPOSITION. *Let $\mathscr{E}$ be a topos, $\mathbf{C}$ an internal category in $\mathscr{E}$. Then the forgetful functor $U: \mathscr{E}^{\mathbf{C}} \longrightarrow \mathscr{E}/C_0$ of 2.21 is comonadic.*

Proof. The functor $T_{\mathbf{C}}$ defined in 2.21 has a right adjoint $G_{\mathbf{C}}$, namely the composite

$$\mathscr{E}/C_0 \xrightarrow{d_1^*} \mathscr{E}/C_1 \xrightarrow{\Pi_{d_0}} \mathscr{E}/C_0.$$

It follows from 0.14 that $G_{\mathbf{C}}$ is the functor part of a comonad $\mathbb{G}_{\mathbf{C}}$, such that $\mathscr{E}^{\mathbf{C}}$ is isomorphic to the category of $\mathbb{G}_{\mathbf{C}}$-coalgebras.

2.32 THEOREM (Lawvere–Tierney). *Let $\mathbb{G} = (G, \varepsilon, \delta)$ be a comonad on a topos $\mathscr{E}$, such that the functor G is left exact. Then the category $\mathscr{E}_{\mathbb{G}}$ of $\mathbb{G}$-coalgebras is a topos, and there is a geometric morphism $\mathscr{E} \to \mathscr{E}_{\mathbb{G}}$, whose direct and inverse images are the cofree and forgetful functors respectively.*

Proof. (i) Since G is left exact, the forgetful functor $\mathscr{E}_{\mathbb{G}} \longrightarrow \mathscr{E}$ creates finite limits; so $\mathscr{E}_{\mathbb{G}}$ has them.

(ii) Given coalgebras $(X \xrightarrow{\theta} GX)$ and $(Y \xrightarrow{\phi} GY)$, we form the exponential $(Y, \phi)^{(X, \theta)}$ as the equalizer of

$$\begin{array}{ccccc} G(Y^X) & & \xrightarrow{\quad G(\phi^X) \quad} & & G(GY^X) \\ & \searrow^{\delta_{Y^X}} & & \nearrow_{G(GY^\theta)} & \\ & GG(Y^X) & \xrightarrow{\ G\rho\ } & G(GY^{GX}) & \end{array}$$

where $\rho: G(Y^X) \longrightarrow GY^{GX}$ is the exponential transpose of

$$G(Y^X) \times GX \cong G(Y^X \times X) \xrightarrow{G(\mathrm{ev})} GY.$$

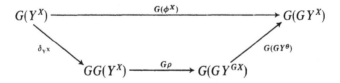

(Note that this equalizer has a coalgebra structure, since all the morphisms in the diagram are coalgebra homomorphisms, where the objects are regarded as cofree coalgebras.)

Now for any coalgebra (Z, ψ), we have a natural bijection between coalgebra homomorphisms $(Z, \psi) \longrightarrow (G(Y^X), \delta)$ and $\mathscr{E}$-morphisms $Z \longrightarrow Y^X$, i.e. $\mathscr{E}$-morphisms $Z \times X \longrightarrow Y$. But the statement that $Z \times X \xrightarrow{f} Y$ is a coalgebra homomorphism is equivalent to the statement that the diagram

$$\begin{array}{ccccc} Z & \xrightarrow{\bar{f}} & Y^X & \xrightarrow{\phi^X} & GY^X \\ \big\downarrow{\scriptstyle \psi} & & & & \big\uparrow{\scriptstyle GY^\theta} \\ GZ & \xrightarrow{G\bar{f}} & G(Y^X) & \xrightarrow{\rho} & GY^{GX} \end{array}$$

commutes, which in turn is equivalent to the statement that the corresponding coalgebra homomorphism $Z \longrightarrow G(Y^X)$ factors through the equalizer defined above. So the equalizer is indeed the required exponential.

(iii) Similarly, we can identify the subobject classifier $\Omega_{\mathbf{G}}$ of $\mathscr{E}_{\mathbf{G}}$ as a subcoalgebra of $(G\Omega, \delta)$, namely the equalizer of

$$\begin{array}{ccc} G\Omega & \xrightarrow{\quad 1_{G\Omega} \quad} & G\Omega, \\ & {\scriptstyle \delta_\Omega}\searrow \quad \nearrow{\scriptstyle G\tau} & \\ & GG\Omega & \end{array}$$

where τ is the classifying map of $G(t)\colon 1 \cong G(1) \rightarrowtail G\Omega$. Now coalgebra homomorphisms $(X, \theta) \longrightarrow (G\Omega, \delta)$ correspond to $\mathscr{E}$-morphisms $X \longrightarrow \Omega$, i.e. to arbitrary subobjects of X in $\mathscr{E}$. And as in (ii), we find that the homomorphisms which factor through the above equalizer correspond precisely to those subobjects of X which have a (necessarily unique) subcoalgebra structure.

(iv) The fact that the specified functors form a geometric morphism is immediate from (i). ■

2.33 Corollary. *If* $\mathscr{E}$ *is a topos and* $\mathbf{C} \in \mathbf{cat}(\mathscr{E})$, *then* $\mathscr{E}^{\mathbf{C}}$ *is a topos.*

Proof. Since the functor $G_{\mathbf{C}}$ has a left ajdoint, it is left exact; so this is immediate from 1.42, 2.31 and 2.32. ■

Moreover, the functor $\mathbf{C}^*\colon \mathscr{E} \longrightarrow \mathscr{E}^{\mathbf{C}}$ has both left and right adjoints by

2.25 and 2.27, so it is the inverse image of an essential geometric morphism $\mathscr{E}^{\mathbf{C}} \longrightarrow \mathscr{E}$. But in fact we have a more general result: first we recall from 2.19 that if $\mathscr{E}$ is any category with finite limits and $\mathbf{C} \xrightarrow{f} \mathbf{D}$ any morphism of $\mathbf{cat}(\mathscr{E})$, then pullback along f induces a functor $f^*: \mathscr{E}^{\mathbf{D}} \longrightarrow \mathscr{E}^{\mathbf{C}}$. Now we have

2.34 Theorem. (i) *If $\mathscr{E}$ has reflexive coequalizers and pullback functors in $\mathscr{E}$ preserve them, then f^* has a left adjoint $\varinjlim_f: \mathscr{E}^{\mathbf{C}} \longrightarrow \mathscr{E}^{\mathbf{D}}$.*

(ii) *If $\mathscr{E}$ is a topos, then f^* has a right adjoint $\varprojlim_f$.*

Proof. (i) It follows easily from 2.16 that the diagram

$$\begin{array}{ccc} \mathscr{E}^{\mathbf{D}} & \xrightarrow{f^*} & \mathscr{E}^{\mathbf{C}} \\ \downarrow{\scriptstyle U} & & \downarrow{\scriptstyle U} \\ \mathscr{E}/D_0 & \xrightarrow{f_0^*} & \mathscr{E}/C_0 \end{array}$$

commutes. Now since pullback functors preserve reflexive coequalizers, so does the functor $T_{\mathbf{D}}$; hence $U: \mathscr{E}^{\mathbf{D}} \longrightarrow \mathscr{E}/D_0$ creates them, and in particular $\mathscr{E}^{\mathbf{D}}$ has them. So by 1.41 and 2.21, we may apply the adjoint lifting theorem (0.15) to obtain a left adjoint for f^*.

The argument for (ii) is similar, using the opposite of 0.15, 1.43 and 2.31. ■

2.35 Corollary. *Let $\mathscr{E}$ be a topos. Then the assignment $\mathbf{C} \longmapsto \mathscr{E}^{\mathbf{C}}$, $f \longmapsto (\varprojlim_f, f^*)$ defines a functor (in the sense of 2-categories) $\mathbf{cat}(\mathscr{E}) \longrightarrow \mathfrak{Top}/\mathscr{E}$. Moreover, if we regard $\mathscr{E}$ as embedded in $\mathbf{cat}(\mathscr{E})$ via the functor Δ of 2.25(i), then the composite $\mathscr{E} \xrightarrow{\Delta} \mathbf{cat}(\mathscr{E}) \longrightarrow \mathfrak{Top}/\mathscr{E}$ is (up to isomorphism) the functor of 1.46.* ■

In the case $\mathscr{E} = \mathscr{S}$, the existence and properties of the adjoints of f^* were first considered by D. M. Kan [171]; for this reason, the functors $\varinjlim_f$ and $\varprojlim_f$ are known as the *left* and *right* (internal) *Kan extensions* along f. Note that if we take $\mathbf{D}$ to be the discrete category $\mathbf{1}$ (i.e. the terminal object of $\mathbf{cat}(\mathscr{E})$) in 2.34, we recover the functors $\varinjlim_{\mathbf{C}}$ and $\varprojlim_{\mathbf{C}}$ defined in 2.24 and 2.27 respectively. So we could have defined these functors by using 0.15; but the more explicit descriptions of them which we gave in the last paragraph are found to be useful in practice. In the next paragraph, we shall give a more explicit description of $\varinjlim_f$ for a general f, using the tensor product of internal profunctors; but there does not seem to be any particularly convenient description of $\varprojlim_f$ in general.

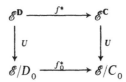

An alternative way of looking at left Kan extensions is provided by the following result:

2.36 PROPOSITION. *Let $\mathscr{E}$ be a category with finite limits, $\mathbf{D}$ an internal category in $\mathscr{E}$. Then the following statements are equivalent:*

(i) *For any $\mathbf{C} \xrightarrow{f} \mathbf{D}$, the left Kan extension $\varinjlim_f$ exists.*
(ii) *The inclusion functor $\mathscr{E}^{\mathbf{D}} \longrightarrow \mathbf{cat}(\mathscr{E})/\mathbf{D}$ has a left adjoint L.*

Proof. Given $\varinjlim_f$, we can define L by $L(\mathbf{C} \xrightarrow{f} \mathbf{D}) = \varinjlim_f(\mathbf{C} \xrightarrow{1} \mathbf{C})$. Then if $\mathbf{F} \xrightarrow{\gamma} \mathbf{D}$ is any discrete opfibration, morphisms $\mathbf{C} \longrightarrow \mathbf{F}$ over $\mathbf{D}$ correspond to global elements of the pullback $f^*(\gamma)$ in $\mathbf{cat}(\mathscr{E})/\mathbf{C}$ (or in $\mathscr{E}^{\mathbf{C}}$), and hence to morphisms $L(f) = \varinjlim_f(1_{\mathbf{C}}) \longrightarrow \gamma$ in $\mathscr{E}^{\mathbf{D}}$.

Conversely, given L we can define $\varinjlim_f$ as the composite

$$\mathscr{E}^{\mathbf{C}} \xrightarrow{\text{incl.}} \mathbf{cat}(\mathscr{E})/\mathbf{C} \xrightarrow{\Sigma_f} \mathbf{cat}(\mathscr{E})/\mathbf{D} \xrightarrow{L} \mathscr{E}^{\mathbf{D}},$$

and it is again easy to verify that we have the required adjunction. ■

It is clear, however, that we do not have an analogue of 2.36 for right adjoints. Part of the reason for this is that, even when $\mathscr{E}$ is a topos, the pullback functor $f^*\colon \mathbf{cat}(\mathscr{E})/\mathbf{D} \longrightarrow \mathbf{cat}(\mathscr{E})/\mathbf{C}$ need not have a right adjoint. It is of course a matter of interest to know when such a right adjoint exists; we now give the statement of a theorem, essentially due to F. Conduché [20], which provides an answer to this question .(We omit the details of the proof, since they are somewhat unsightly, and we shall not need to use the result anywhere in this book.)

2.37 THEOREM. *Let $\mathscr{E}$ be a topos, $\mathbf{C} \xrightarrow{f} \mathbf{D}$ a morphism of $\mathbf{cat}(\mathscr{E})$. Let P_f, Q_f be the pullbacks*

$$\begin{array}{ccc} P_f & \longrightarrow & D_2, \\ \downarrow & & \downarrow{\scriptstyle m} \\ C_1 & \xrightarrow{f_1} & D_1 \end{array} \qquad \begin{array}{ccc} Q_f & \longrightarrow & D_2 \\ \downarrow & & \downarrow{\scriptstyle 1\times i\times 1} \\ C_3 & \xrightarrow{f_3} & D_3 \end{array}$$

respectively, and let $C_2 \xrightarrow{q} P_f$ be the morphism induced by $C_2 \xrightarrow{(f_2, m)} D_2 \times C_1$. Then the pullback functor $f^*\colon$ $\mathbf{cat}(\mathscr{E})/\mathbf{D} \longrightarrow \mathbf{cat}(\mathscr{E})/\mathbf{C}$ has a right adjoint

Π_f *iff the diagram*

$$Q_f \underset{m \times 1}{\overset{1 \times m}{\rightrightarrows}} C_2 \xrightarrow{q} P_f$$

is a coequalizer. [*In the case* $\mathscr{E} = \mathscr{S}$, *this condition says that if* α *is a morphism of* $\mathbf{C}$ *and we are given a factorization of* $f(\alpha)$ *as a composite of two morphisms of* $\mathbf{D}$, *there is an "essentially unique" factorization of* α *in* $\mathbf{C}$ *mapping to the given one in* $\mathbf{D}$.] ■

The existence of additional structure on internal diagram categories, as exemplified by 2.33 and 2.35, is one aspect of the advantages which we gain by taking our base category $\mathscr{E}$ to be a topos. Another aspect is the ability which the topos axioms give us to construct particular internal categories having certain specified properties. One important example, which will be of use in both chapter 4 and chapter 6, is the following, due to J. Bénabou [7]:

2.38 EXAMPLE. Let $\mathscr{E}$ be a topos, $U \xrightarrow{f} X$ a morphism of $\mathscr{E}$. Regarding f as an "X-indexed family of objects of $\mathscr{E}$", we can construct an internal category $\mathrm{Full}_{\mathscr{E}}(f)$, which plays the rôle of "the full subcategory of $\mathscr{E}$ whose objects are (the fibres of) f", as follows: $\mathrm{Full}_{\mathscr{E}}(f)_0 = X$, and $\mathrm{Full}_{\mathscr{E}}(f)_1 \xrightarrow{(d_0, d_1)} X \times X$ is the exponential $\pi_2^*(f)^{\pi_1^*(f)}$, constructed in the topos $\mathscr{E}/X \times X$. Since pullback functors preserve exponentials, defining the multiplication in $\mathrm{Full}_{\mathscr{E}}(f)$ amounts to defining a morphism $\pi_2^*(f)^{\pi_1^*(f)} \times \pi_3^*(f)^{\pi_2^*(f)} \longrightarrow \pi_3^*(f)^{\pi_1^*(f)}$ in $\mathscr{E}/X \times X \times X$; and for this we take the internal composition map. Similarly, the inclusion-of-identities of $\mathrm{Full}_{\mathscr{E}}(f)$ is defined by the morphism $1_X \xrightarrow{\bar{1}_f} f^f$ in $\mathscr{E}/X$.

Note also that $\mathrm{Full}_{\mathscr{E}}(f)$ is equipped with a canonical internal diagram which plays the rôle of the "inclusion functor"; this is simply the object $U \xrightarrow{f} X$ of $\mathscr{E}/X$, equipped with a structure map for the appropriate monad which is essentially the evaluation map $\pi_2^*(f)^{\pi_1^*(f)} \times \pi_1^*(f) \longrightarrow \pi_2^*(f)$ in $\mathscr{E}/X \times X$. ■

J. Penon [100] has pointed out that 2.38 is actually a special case of a much more general construction involving locally internal categories; we shall discuss this generalization in the Appendix.

Another, more trivial (but equally useful), source of examples of internal categories is provided by

2.39 EXAMPLE. Let $\mathscr{E}$ be a topos. Then we can identify any finite (external) category $\mathbf{C}$ with an internal category $I(\mathbf{C})$, the "internalization" of $\mathbf{C}$, by identifying a p-element set with the p-fold coproduct of copies of 1 in $\mathscr{E}$, and a map between finite sets with the appropriate compound of coproduct inclusions. Moreover, if $\mathbf{F} \xrightarrow{\gamma} I(\mathbf{C})$ is a discrete opfibration, then by universality of coproducts in $\mathscr{E}$ we can regard $F_0 \xrightarrow{\gamma_0} I(C)_0 = \coprod_{u \in C_0} 1$ as a C_0-indexed family of objects $F(u) = v_u^*(\gamma_0)$, and the action map $F_0 \times_{I(C)_0} I(C)_1 \xrightarrow{e} F_0$ as a family of morphisms $F(u) \xrightarrow{F(f)} F(v)$ for each $u \xrightarrow{f} v$ in $\mathbf{C}$. Thus $\mathscr{E}^{I(\mathbf{C})}$ is equivalent to the category of (external) functors $\mathbf{C} \longrightarrow \mathscr{E}$. For this reason, we normally do not distinguish between a finite category and its internalization.

2.4. INTERNAL PROFUNCTORS

In this paragraph we introduce a notion of "generalized functor", due originally to J. Bénabou, which will play an important role in the proof of Diaconescu's Theorem (4.34). For the time being, we shall assume that $\mathscr{E}$ is a category with finite limits and reflexive coequalizers, and that the latter are universal (i.e. preserved by pullback functors).

2.41 DEFINITION. Let $\mathbf{C}$, $\mathbf{D}$ be internal categories in $\mathscr{E}$. By an *internal profunctor* from $\mathbf{C}$ to $\mathbf{D}$, we mean an internal diagram on $\mathbf{C}^{\text{op}} \times \mathbf{D}$. Equivalently, a profunctor G may be specified by giving an object $G_0 \xrightarrow{(\gamma_0, \delta_0)} C_0 \times D_0$ of $\mathscr{E}/C_0 \times D_0$, together with left and right action maps $C_1 \times_{C_0} G_0 \xrightarrow{\alpha} G_0$, $G_0 \times_{D_0} D_1 \xrightarrow{\beta} G_0$ over $C_0 \times D_0$, such that α and β are associative and unitary (in the sense of 2.14) and commute with each other. We write "$G\colon \mathbf{C} \dashrightarrow \mathbf{D}$" for "$G$ is a profunctor from $\mathbf{C}$ to $\mathbf{D}$"; and we also write $\text{Prof}_{\mathscr{E}}(\mathbf{C}, \mathbf{D})$ for the category $\mathscr{E}^{\mathbf{C}^{\text{op}} \times \mathbf{D}}$ of profunctors $\mathbf{C} \dashrightarrow \mathbf{D}$. ■

Note that a profunctor $\mathbf{C} \dashrightarrow \mathbf{D}$ may equivalently be regarded as an internal presheaf on $\mathbf{D}^*(\mathbf{C})$ in $\mathscr{E}^{\mathbf{D}}$, or as an internal diagram on $(\mathbf{C}^{\text{op}})^*(\mathbf{D})$ in $\mathscr{E}^{\mathbf{C}^{\text{op}}}$. In particular, we may identify $\mathscr{E}^{\mathbf{C}^{\text{op}}}$ and $\mathscr{E}^{\mathbf{C}}$ with $\text{Prof}_{\mathscr{E}}(\mathbf{C}, \mathbf{1})$ and $\text{Prof}_{\mathscr{E}}(\mathbf{1}, \mathbf{C})$ respectively.

2.42 DEFINITION. Let $\mathbf{B}, \mathbf{C}, \mathbf{D}$ be three internal categories, and $\mathbf{B} \overset{F}{\dashrightarrow} \mathbf{C} \overset{G}{\dashrightarrow} \mathbf{D}$ two profunctors. We define a profunctor

$$F \otimes_{\mathbf{C}} G\colon \mathbf{B} \dashrightarrow \mathbf{D}$$

as follows: $(F \otimes_{\mathbf{C}} G)_0$ is the coequalizer of

$$F_0 \times_{C_0} C_1 \times_{C_0} G_0 \underset{\beta_F \times 1}{\overset{1 \times \alpha_G}{\rightrightarrows}} F_0 \times_{C_0} G_0$$

in $\mathscr{E}/B_0 \times D_0$. Since this pair is reflexive, with splitting $F_0 \times_{C_0} G_0 \xrightarrow{1 \times i \times 1} F_0 \times_{C_0} C_1 \times_{C_0} G_0$, the hypotheses on $\mathscr{E}$ imply that

$$B_1 \times_{B_0} F_0 \times_{C_0} C_1 \times_{C_0} G_0 \underset{1 \times \beta_F \times 1}{\overset{1 \times 1 \times \alpha_G}{\rightrightarrows}} B_1 \times_{B_0} F_0 \times_{C_0} G_0 \longrightarrow B_1 \times_{B_0} (F \otimes_{\mathbf{C}} G)_0$$

is a coequalizer; and since $1 \times \alpha_G$ and $\beta_F \times 1$ are both equivariant for α_F, the latter induces a left action

$$B_1 \times_{B_0} (F \otimes_{\mathbf{C}} G)_0 \longrightarrow (F \otimes_{\mathbf{C}} G)_0.$$

Similarly, β_G induces a right action of $\mathbf{D}$ on $(F \otimes_{\mathbf{C}} G)_0$; and it is easy to check that these induced actions satisfy the necessary identities. Moreover, it is not hard to check that $\otimes_{\mathbf{C}}$ is in fact a bifunctor

$$\mathrm{Prof}_{\mathscr{E}}(\mathbf{B}, \mathbf{C}) \times \mathrm{Prof}_{\mathscr{E}}(\mathbf{C}, \mathbf{D}) \longrightarrow \mathrm{Prof}_{\mathscr{E}}(\mathbf{B}, \mathbf{D}). \blacksquare$$

2.43 PROPOSITION. *The bifunctors $\otimes_{\mathbf{C}}$ are associative up to coherent natural isomorphism.*

Proof. Suppose given $\mathbf{B} \overset{F}{\dashrightarrow} \mathbf{C} \overset{G}{\dashrightarrow} \mathbf{D} \overset{H}{\dashrightarrow} \mathbf{E}$. Form the diagram

$$\begin{array}{ccccc}
F_0 \times_{C_0} C_1 \times_{C_0} G_0 \times_{D_0} D_1 \times_{D_0} H_0 & \rightrightarrows & F_0 \times_{C_0} G_0 \times_{D_0} D_1 \times_{D_0} H_0 & \longrightarrow & (F \otimes_{\mathbf{C}} G)_0 \times_{D_0} D_1 \times_{D_0} H_0 \\
\downdownarrows & & \downdownarrows & & \downdownarrows \\
F_0 \times_{C_0} C_1 \times_{C_0} G_0 \times_{D_0} H_0 & \rightrightarrows & F_0 \times_{C_0} G_0 \times_{D_0} H_0 & \longrightarrow & (F \otimes_{\mathbf{C}} G)_0 \times_{D_0} H_0 \\
\downarrow & & \downarrow & & \\
F_0 \times_{C_0} C_1 \times_{C_0} (G \otimes_{\mathbf{D}} H)_0 & \rightrightarrows & F_0 \times_{C_0} (G \otimes_{\mathbf{D}} H)_0 & &
\end{array}$$

Here the first two rows and columns are coequalizers by the assumptions on $\mathscr{E}$; so the coequalizer of the third row is canonically isomorphic to that of the third column. But these are by definition equal to $(F \otimes_{\mathbf{C}} (G \otimes_{\mathbf{D}} H))_0$ and $((F \otimes_{\mathbf{C}} G) \otimes_{\mathbf{D}} H)_0$ respectively; and the isomorphism is clearly an isomorphism of profunctors $\mathbf{B} \dashrightarrow \mathbf{E}$. Moreover, the canonical nature of the isomorphism ensures that it is natural in F, G and H, and that it satisfies the usual pentagonal coherence condition ([CW], p. 158). $\blacksquare$

Now let $\mathbf{C} \xrightarrow{f} \mathbf{D}$ be an internal functor. We define a profunctor

$f_\#: \mathbf{C} \dashrightarrow \mathbf{D}$ as follows:

$$(f_\#)_0 = C_0 \times_{D_0} D_1 \xrightarrow{(\pi_1,\, d_1\pi_2)} C_0 \times D_0,$$

with left action of $\mathbf{C}$ given by

$$C_1 \times_{C_0} C_0 \times_{D_0} D_1 \xrightarrow{(d_0\pi_1,\, m(f_1\pi_1,\, \pi_3))} C_0 \times_{D_0} D_1,$$

and right action of $\mathbf{D}$ by

$$C_0 \times_{D_0} D_1 \times_{D_0} D_1 \xrightarrow{1 \times m} C_0 \times_{D_0} D_1.$$

Similarly, we make the object $D_1 \times_{D_0} C_0 \xrightarrow{(d_0\pi_1,\, \pi_2)} D_0 \times C_0$ of $\mathscr{E}/D_0 \times C_0$ into a profunctor $f^\#$: $\mathbf{D} \dashrightarrow \mathbf{C}$. Note in particular that $(1_{\mathbf{C}})_\#$ and $(1_{\mathbf{C}})^\#$ are both (isomorphic to) the object $C_1 \xrightarrow{(d_0,\, d_1)} C_0 \times C_0$ with left and right actions given by m; we denote this profunctor by $Y(\mathbf{C})$ and call it the *Yoneda profunctor* on $\mathbf{C}$.

2.44 LEMMA. *For any* $\mathbf{B}$, *the diagram*

$$\begin{array}{ccc} \mathrm{Prof}_{\mathscr{E}}(\mathbf{B}, \mathbf{D}) & \xrightarrow{(-)\otimes_{\mathbf{D}} f^\#} & \mathrm{Prof}_{\mathscr{E}}(\mathbf{B}, \mathbf{C}) \\ \| & & \| \\ \mathscr{E}^{\mathbf{B}^{\mathrm{op}} \times \mathbf{D}} & \xrightarrow{(1 \times f)^*} & \mathscr{E}^{\mathbf{B}^{\mathrm{op}} \times \mathbf{C}} \end{array}$$

commutes up to natural isomorphism. Similarly, $f_\# \otimes_{\mathbf{D}} (-)$ *may be identified with* $(f^{\mathrm{op}} \times 1)^*$.

Proof. Given $\mathbf{B} \overset{\mathbf{G}}{\dashrightarrow} \mathbf{D}$, we have a split coequalizer diagram

$$G_0 \times_{D_0} D_1 \times_{D_0} D_1 \times_{D_0} C_0 \underset{\beta_G \times 1 \times 1}{\overset{1 \times m \times 1}{\rightrightarrows}} G_0 \times_{D_0} D_1 \times_{D_0} C_0 \xrightarrow{\beta_G \times 1} G_0 \times_{D_0} C_0 = ((1 \times f)^* G)_0 .$$

(with splittings $1 \times 1 \times (if_0, 1)$ and $1 \times (if_0, 1)$)

But this is precisely the coequalizer which yields $(G \otimes_{\mathbf{D}} f^\#)_0$; and it is straightforward to check that the actions of $\mathbf{B}$ and $\mathbf{C}$ on $(G \otimes_{\mathbf{D}} f^\#)_0$ agree with those on $((1 \times f)^* G)_0$. ■

2.45 COROLLARY. *The Yoneda profunctor* $Y(\mathbf{C})$ *is (up to canonical isomor-*

phism) a two-sided unit for $\otimes_{\mathbf{C}}$. *Hence the internal profunctors in* $\mathscr{E}$ *are the 1-arrows of a bicategory* $\mathfrak{Prof}_{\mathscr{E}}$, *whose objects are the internal categories.*

Proof. After 2.44, it only remains to check the coherence conditions relating the unit isomorphism for $Y(\mathbf{C})$ to the associativity isomorphism of 2.43; the proof of these is tedious but straightforward. ∎

Now consider the profunctor $f_\# \otimes_{\mathbf{D}} f^\#: \mathbf{C} \dashrightarrow \mathbf{C}$. By the argument of 2.44, the object-of-objects of this profunctor is $C_0 \times_{D_0} D_1 \times_{D_0} C_0 \xrightarrow{(\pi_1, \pi_3)} C_0 \times C_0$; and the morphism $C_1 \xrightarrow{(d_0, f_1, d_1)} C_0 \times_{D_0} D_1 \times_{D_0} C_0$ induces a morphism of profunctors

$$Y(\mathbf{C}) \xrightarrow{\eta} f_\# \otimes_{\mathbf{D}} f^\#.$$

Similarly, $(f^\# \otimes_{\mathbf{C}} f_\#)_0$ is the coequalizer of

$$D_1 \times_{D_0} C_1 \times_{D_0} D_1 \overset{(\pi_1, d_0\pi_2, m(f_1\pi_2, \pi_3))}{\underset{(m(\pi_1, f_1\pi_2), d_1\pi_2, \pi_3)}{\rightrightarrows}} D_1 \times_{D_0} C_0 \times_{D_0} D_1;$$

and the morphism $D_1 \times_{D_0} C_0 \times_{D_0} D_1 \xrightarrow{m(\pi_1, \pi_3)} D_1$ factors through this coequalizer to induce

$$f^\# \otimes_{\mathbf{C}} f_\# \xrightarrow{\varepsilon} Y(\mathbf{D}).$$

It is now straightforward to verify that η and ε satisfy the triangular identities

$$\begin{array}{ccc} Y(\mathbf{C}) \otimes f_\# & \xrightarrow{\qquad\qquad\cong\qquad\qquad} & f_\# \\ \Big\downarrow{\scriptstyle \eta \otimes 1} & & \Big\downarrow{\scriptstyle \wr} \\ (f_\# \otimes f^\#) \otimes f_\# \xrightarrow{\cong} f_\# \otimes f_\#) & \xrightarrow{1 \otimes \varepsilon} & f_\# \otimes Y(\mathbf{D}) \end{array}$$

and

$$\begin{array}{ccc} f^\# \otimes Y(\mathbf{C}) & \xrightarrow{\qquad\qquad\cong\qquad\qquad} & f^\# \;; \\ \Big\downarrow{\scriptstyle 1 \otimes \eta} & & \Big\downarrow{\scriptstyle \wr} \\ f^\# \otimes (f_\# \otimes f^\#) \xrightarrow{\cong} (f^\# \otimes f_\#) \otimes f^\# & \xrightarrow{\varepsilon \otimes 1} & Y(\mathbf{D}) \otimes f^\# \end{array}$$

so $f^\#$ is left adjoint to $f_\#$ in the bicategory $\mathfrak{Prof}_{\mathscr{E}}$. Hence in particular we have

2.46 LEMMA. *For any* **B**, *the diagram*

$$\begin{array}{ccc} \mathrm{Prof}_{\mathscr{E}}(\mathbf{B},\mathbf{C}) & \xrightarrow{(-)\otimes_{\mathbf{C}} f_\#} & \mathrm{Prof}_{\mathscr{E}}(\mathbf{B},\mathbf{D}) \\ \| & & \| \\ \mathscr{E}^{\mathbf{B}^{\mathrm{op}}\times\mathbf{C}} & \xrightarrow{\varinjlim_{(1\times f)}} & \mathscr{E}^{\mathbf{B}^{\mathrm{op}}\times\mathbf{D}} \end{array}$$

commutes up to isomorphism. (Similarly for $f^\# \otimes_{\mathbf{C}} (-)$.)

Proof. Since $f_\#$ is right adjoint to $f^\#$ in $\mathfrak{Prof}_{\mathscr{E}}$, the functor $(-) \otimes f_\#$ is left adjoint to $(-) \otimes f^\#$. So this is immediate from 2.44. ∎

As a consequence of 2.46, we obtain the following technical result, which will be of importance in §4.4:

2.47 COROLLARY. *Let* $\mathbf{C} \xrightarrow{f} \mathbf{D}$ *be an internal functor, G an internal presheaf on* **C** *and F an internal diagram on* **D**. *Then we have a natural isomorphism*

$$(f^*F) \otimes_{\mathbf{C}} G \cong F \otimes_{\mathbf{D}} (\varinjlim_f G).$$

Proof. Combine 2.44 and 2.46 with the associativity isomorphism $(F \otimes_{\mathbf{D}} f^\#) \otimes_{\mathbf{C}} G \cong F \otimes_{\mathbf{D}} (f^\# \otimes_{\mathbf{C}} G)$. ∎

In 2.44 and 2.46 we saw that the functors f^* and $\varinjlim_f$ may be described in terms of the tensor product of profunctors. We shall now show that the converse is also true.

2.48 THEOREM (J. Bénabou [6]). *Let* $\mathbf{C} \overset{G}{\dashrightarrow} \mathbf{D}$ *be an internal profunctor. Then there exists a diagram*

$$\begin{array}{ccc} \mathbf{G} & \xrightarrow{\delta} & \mathbf{D} \\ \big\downarrow{\scriptstyle\gamma} & & \\ \mathbf{C} & & \end{array}$$

in **cat**($\mathscr{E}$) *such that* $G \cong \gamma^\# \otimes_{\mathbf{G}} \delta_\#$ *in* $\mathrm{Prof}_{\mathscr{E}}(\mathbf{C}, \mathbf{D})$. *Hence for any* **B**, *the functor* $(-) \otimes_{\mathbf{C}} G\colon \mathrm{Prof}_{\mathscr{E}}(\mathbf{B},\mathbf{C}) \longrightarrow \mathrm{Prof}_{\mathscr{E}}(\mathbf{B},\mathbf{D})$ *is isomorphic to the composite* $\mathscr{E}^{\mathbf{B}^{\mathrm{op}}\times\mathbf{C}} \xrightarrow{(1\times\gamma)^*} \mathscr{E}^{\mathbf{B}^{\mathrm{op}}\times\mathbf{G}} \xrightarrow{\varinjlim_{(1\times\delta)}} \mathscr{E}^{\mathbf{B}^{\mathrm{op}}\times\mathbf{D}}$.

Proof. Define G_1 by the pullback diagram

$$\begin{array}{ccc} G_1 & \xrightarrow{\mu} & C_1 \times_{C_0} G_0 \\ \Big\downarrow{\scriptstyle\lambda} & & \Big\downarrow{\scriptstyle\alpha} \\ G_0 \times_{D_0} D_1 & \xrightarrow{\beta} & G_0 \end{array} \quad ;$$

and let $d_0, d_1 : G_1 \rightrightarrows G_0$ be $\pi_1\lambda$ and $\pi_2\mu$ respectively. To define the multiplication $G_1 \times_{G_0} G_1 \xrightarrow{m} G_1$, consider the diagram

$$\begin{array}{ccccc} G_1 \times_{G_0} G_1 & \xrightarrow{\psi} & C_1 \times_{C_0} G_1 & \xrightarrow{\pi_2} & G_1 \\ \Big\downarrow{\scriptstyle\phi} & & \Big\downarrow{\scriptstyle 1\times\lambda} & & \Big\downarrow{\scriptstyle\lambda} \\ G_1 \times_{D_0} D_1 & \xrightarrow{\mu\times 1} & C_1 \times_{C_0} G_0 \times_{D_0} D_1 & \xrightarrow{\pi_{23}} & G_0 \times_{D_0} D_1 \\ \Big\downarrow{\scriptstyle\pi_1} & & \Big\downarrow{\scriptstyle\pi_{12}} & & \Big\downarrow{\scriptstyle\pi_1} \\ G_1 & \xrightarrow{\mu} & C_1 \times_{C_0} G_0 & \xrightarrow{\pi_2} & G_0 \end{array} \quad ,$$

where all the squares are pullbacks. Now $d_1\pi_2\lambda\pi_1\phi = \delta_0\beta\lambda\pi_1\phi = \delta_0\pi_2\mu\pi_1\phi = \delta_0\pi_1\lambda\pi_2\psi = d_0\pi_2\lambda\pi_2\psi$, so $(\pi_2\lambda\pi_1\phi, \pi_2\lambda\pi_2\psi)$ maps $G_1 \times_{G_0} G_1$ into $D_1 \times_{D_0} D_1$. Hence we can define a morphism

$$\rho = (\pi_1\lambda\pi_1\phi, m(\pi_2\lambda\pi_1\phi, \pi_2\lambda\pi_2\psi)) \colon G_1 \times_{G_0} G_1 \longrightarrow G_0 \times_{D_0} D_1 .$$

Similarly, we have

$$\sigma = (m(\pi_1\mu\pi_1\phi, \pi_1\mu\pi_2\psi), \pi_2\mu\pi_2\psi) \colon G_1 \times_{G_0} G_1 \longrightarrow C_1 \times_{C_0} G_0 ,$$

and these combine to give $m \colon G_1 \times_{G_0} G_1 \longrightarrow G_1$. The fact that m is unitary and associative follows from the corresponding laws for multiplication in $\mathbf{C}$ and $\mathbf{D}$.

Now if we define $\gamma_1 = \pi_1\mu \colon G_1 \longrightarrow C_1$ and $\delta_1 = \pi_2\lambda \colon G_1 \longrightarrow D_1$, we obtain internal functors $\mathbf{G} \xrightarrow{\gamma} \mathbf{C}$ and $\mathbf{G} \xrightarrow{\delta} \mathbf{D}$ as required. To establish the isomorphism $G \cong \gamma^{\#} \otimes \delta_{\#}$, we have to show that

$$C_1 \times_{C_0} G_1 \times_{D_0} D_1 \underset{(m(\pi_1, \pi_1\mu\pi_2), \pi_2\mu\pi_2, \pi_3)}{\overset{(\pi_1, \pi_1\lambda\pi_2, m(\pi_2\lambda\pi_2, \pi_3))}{\rightrightarrows}} C_1 \times_{C_0} G_0 \times_{D_0} D_1 \xrightarrow{\alpha(1\times\beta)} G_0$$

is a coequalizer. To do this, we use an argument reminiscent of 0.17; note first that we have a split coequalizer diagram

$$C_1 \times_{C_0} G_0 \times_{D_0} D_1 \times_{D_0} D_1 \underset{\underset{1 \times 1 \times (1, id_1)}{\longleftarrow}}{\overset{1 \times 1 \times m}{\underset{1 \times \beta \times 1}{\rightrightarrows}}} C_1 \times_{C_0} G_0 \times_{D_0} D_1 \underset{1 \times (1, i\delta_0)}{\overset{1 \times \beta}{\rightleftarrows}} C_1 \times_{C_0} G_0 .$$

Now the pair $(1 \times 1 \times m, 1 \times \beta \times 1)$ factors through the pair above by the morphism

$$C_1 \times_{C_0} G_0 \times_{D_0} D_1 \times_{D_0} D_1 \xrightarrow{1 \times \tau \times 1} C_1 \times_{C_0} G_1 \times_{D_0} D_1 ,$$

where τ is the splitting of λ induced by $G_0 \xrightarrow{(i\gamma_0, 1)} C_1 \times_{C_0} G_0$; so if $C_1 \times_{C_0} G_1 \times_{D_0} D_1 \xrightarrow{q} Q$ is the coequalizer of the given pair, then q factors through $1 \times \beta$. Similarly, q factors through $\alpha \times 1$; so it factors through the pushout

$$\begin{array}{ccc} C_1 \times_{C_0} G_0 \times_{D_0} D_1 & \xrightarrow{1 \times \beta} & C_1 \times_{C_0} G_0 \\ \Big\downarrow{\scriptstyle \alpha \times 1} & & \Big\downarrow{\scriptstyle \alpha} \\ G_0 \times_{D_0} D_1 & \xrightarrow{\beta} & G_0 \end{array} .$$

But we may verify directly that $\alpha(1 \times \beta)$ coequalizes the given pair; so it is isomorphic to q. The last sentence of the theorem follows at once from 2.44, 2.46 and associativity of $\otimes$. ■

Note that if we take $\mathbf{C} = \mathbf{1}$ in 2.48 (so that G is simply an internal diagram on $\mathbf{D}$), then $\mathbf{G} \overset{\delta}{\dashrightarrow} \mathbf{D}$ is the discrete obfibration corresponding to G as in 2.15. Similar remarks apply when $\mathbf{D} = \mathbf{1}$.

2.49 COROLLARY. *If $\mathscr{E}$ is a topos, then the bicategory $\mathfrak{Prof}_{\mathscr{E}}$ is* biclosed; *i.e. for a fixed* 1*-arrow* $\mathbf{C} \overset{G}{\dashrightarrow} \mathbf{D}$, *the functors* $(-) \otimes_{\mathbf{C}} G$ *and* $G \otimes_{\mathbf{D}} (-)$ *have right adjoints.*

Proof. From 2.34, we know that $(1 \times \gamma)^*$ and $\varinjlim_{(1 \times \delta)}$ have right adjoints $\varprojlim_{(1 \times \gamma)}$ and $(1 \times \delta)^*$ respectively; so this is immediate from 2.48. ■

2.5. FILTERED CATEGORIES

In 2.33 we saw that the functor $\mathbf{C}^*: \mathscr{E} \longrightarrow \mathscr{E}^{\mathbf{C}}$ is the inverse image of a geometric morphism. We might ask: when is $\mathbf{C}^*$ the direct image of a geometric morphism? In other words, when is its left adjoint $\varinjlim_{\mathbf{C}}$ left exact? It is a well-known result (see [CW], p. 211) that for $\mathscr{E} = \mathscr{S}$, the operation of taking colimits over a small category **C** commutes with finite limits iff **C** is filtered. In this paragraph we shall prove the corresponding result for internal colimits; we shall assume throughout that $\mathscr{E}$ is a topos, although in fact all the arguments are valid in the more general context of an exact category (see [40]).

2.51 Definition. (i) An internal category **C** in $\mathscr{E}$ is said to be *filtered* if

(a) $C_0 \longrightarrow 1$ is epi. [Interpretation in the case $\mathscr{E} = \mathscr{S}$: **C** is nonempty.]

(b) The map $P \xrightarrow{(d_0\pi_1, d_0\pi_2)} C_0 \times C_0$ is epi, where P is the pullback

$$\begin{array}{ccc} P & \xrightarrow{\pi_2} & C_1 \\ {\scriptstyle \pi_2}\downarrow & & \downarrow{\scriptstyle d_1} \\ C_1 & \xrightarrow{d_1} & C_0 \end{array} \quad .$$

[Interpretation: any pair of objects U, V can be embedded in a diagram

$$\begin{array}{ccc} U & \searrow & \\ & & W. \\ V & \nearrow & \end{array}]$$

(c) The map $T \xrightarrow{(\pi_1\pi_1, \pi_1\pi_2)} R$ is epi, where R and T are the pullbacks

$$\begin{array}{ccc} R & \xrightarrow{\pi_2} & C_1 \\ {\scriptstyle \pi_1}\downarrow & & \downarrow{\scriptstyle (d_0, d_1)} \\ C_1 & \xrightarrow{(d_0, d_1)} & C_0 \times C_0 \end{array} \quad \text{and} \quad \begin{array}{ccc} T & \xrightarrow{\pi_2} & C_2 \\ {\scriptstyle \pi_1}\downarrow & & \downarrow{\scriptstyle (\pi_2, m)} \\ C_2 & \xrightarrow{(\pi_2, m)} & C_1 \times C_1 \end{array} \quad .$$

[Interpretation: given a parallel pair of arrows $U \rightrightarrows V$, we can find $V \longrightarrow W$ coequalizing them.]

(ii) **C** is said to be *weakly filtered* if it satisfies condition (c) above, and also

(d) The map $S \xrightarrow{(\pi_1\pi_1, \pi_1\pi_2)} Q$ is epi, where Q and S are the pullbacks

$$\begin{array}{ccc} Q & \xrightarrow{\pi_2} & C_1 \\ {\scriptstyle \pi_1}\downarrow & & \downarrow{\scriptstyle d_0} \\ C_1 & \xrightarrow{d_0} & C_0 \end{array} \quad \text{and} \quad \begin{array}{ccc} S & \xrightarrow{\pi_2} & C_2 \\ {\scriptstyle \pi_1}\downarrow & & \downarrow{\scriptstyle m} \\ C_2 & \xrightarrow{m} & C_1 \end{array} \quad .$$

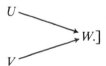

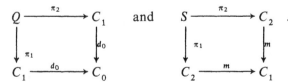

[Interpretation: any diagram $\begin{array}{ccc} & & V \\ & \nearrow & \\ U & & \\ & \searrow & \\ & & W \end{array}$ can be embedded in a commutative square.] ■

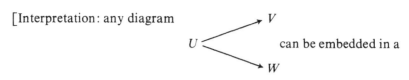

Throughout this paragraph, the letters P, Q, R, S, T will denote the pullbacks given those names in 2.51. If we wish to specify the category from which they are constructed, we shall write $P_{\mathbf{C}}$, etc. Note that if $\mathbf{C}$ is a poset, then condition (c) is vacuous, since we have $R_{\mathbf{C}} \cong C_1$ and $T_{\mathbf{C}} \cong C_2$; when we are dealing with posets, the term *directed* is commonly used as a synonym for filtered.

2.52 REMARK. Condition (b) of 2.51 is equivalent to the following statement: "For any object U of $\mathscr{E}$ and any pair of U-elements $U \underset{\gamma_2}{\overset{\gamma_1}{\rightrightarrows}} C_0$, there exists an epi $V \overset{\varepsilon}{\twoheadrightarrow} U$ and V-elements $V \underset{\lambda_2}{\overset{\lambda_1}{\rightrightarrows}} C_1$ such that $d_1\lambda_1 = d_1\lambda_2$ and $d_0\lambda_i = \gamma_i\varepsilon$ $(i = 1, 2)$."

And conditions (c) and (d) can be stated in similar "elementary" forms.

Proof. Suppose condition (b) is satisfied and we are given $U \underset{\gamma_2}{\overset{\gamma_1}{\rightrightarrows}} C_0$. Then we can construct V by forming the pullback

$$\begin{array}{ccc} V & \xrightarrow{(\lambda_1, \lambda_2)} & P \\ {\scriptstyle \varepsilon}\downarrow & & \downarrow \\ U & \xrightarrow{(\gamma_1, \gamma_2)} & C_0 \otimes C_0 \end{array} ;$$

and ε is epi by 1.51. Conversely, if we are given the above condition, then by applying it to the generic pair $C_0 \times C_0 \underset{\pi_2}{\overset{\pi_1}{\rightrightarrows}} C_0$ we obtain a commutative triangle

$$\begin{array}{ccc} V & \longrightarrow & P \\ & {\scriptstyle \varepsilon}\searrow & \downarrow \\ & & C_0 \times C_0 \end{array}$$

with ε epi; so $P \longrightarrow C_0 \times C_0$ is epi. ■

We shall in fact make use interchangeably of the "universal" forms of conditions (b), (c) and (d) as given in 2.51, and of the "elementary" forms.

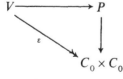

2.53 LEMMA. *A filtered category is weakly filtered.*

Proof. Suppose $\mathbf{C}$ is filtered. To prove the elementary form of condition (d), suppose we are given $U \overset{\gamma_1}{\underset{\gamma_2}{\rightrightarrows}} C_1$ coequalized by $C_1 \xrightarrow{d_0} C_0$. We must construct an epi $V \overset{\varepsilon}{\twoheadrightarrow} U$ and $V \overset{\delta_1}{\underset{\delta_2}{\rightrightarrows}} C_1$ such that $d_0\delta_i = d_1\gamma_i\varepsilon$ ($i = 1, 2$) and $m(\gamma_1\varepsilon, \delta_1) = m(\gamma_2\varepsilon, \delta_2)$. Applying (b) to the pair $(d_1\gamma_1, d_1\gamma_2)$, we have $W \overset{\varepsilon_1}{\twoheadrightarrow} U$ and $W \overset{\lambda_1}{\underset{\lambda_2}{\rightrightarrows}} C_1$ such that $d_0\lambda_i = d_1\gamma_i\varepsilon_1$ ($i = 1, 2$) and $d_1\lambda_1 = d_1\lambda_2$. Now the morphisms $m(\gamma_1\varepsilon_1, \lambda_1)$ and $m(\gamma_2\varepsilon_1, \lambda_2)$: $W \rightrightarrows C_1$ may not be equal, but they are coequalized by both d_0 and d_1, so we can apply (c) to obtain $V \overset{\varepsilon_2}{\twoheadrightarrow} W$ and $V \xrightarrow{\mu} C_1$ such that $m(m(\gamma_1\varepsilon_1, \lambda_1)\,\varepsilon_2, \mu) = m(m(\gamma_2\varepsilon_1, \lambda_2)\,\varepsilon_2, \mu)$. Then by associativity of m we see that $\varepsilon = \varepsilon_1\varepsilon_2$ and $\delta_i = m(\lambda_i\varepsilon_2, \mu)$ are the required morphisms. ■

2.54 LEMMA. *If* $\mathbf{C}$ *is filtered, then* $\varinjlim \mathbf{C} \cong 1$.

Proof. Let $C_0 \overset{\tau}{\twoheadrightarrow} L$ be the "colimiting cone" of $\mathbf{C}$, i.e. the coequalizer of d_0 and d_1. Now if $U \overset{\gamma_1}{\underset{\gamma_2}{\rightrightarrows}} C_0$ is any pair of morphisms, then by (b) we can find $V \overset{\varepsilon}{\twoheadrightarrow} U$ and $V \overset{\lambda_1}{\underset{\lambda_2}{\rightrightarrows}} C_1$ such that $\tau\gamma_1\varepsilon = \tau d_0\lambda_1 = \tau d_1\lambda_1 = \tau d_1\lambda_2 = \tau d_0\lambda_2 = \tau\lambda_2\varepsilon$. But ε is epi, so $\tau\gamma_1 = \tau\gamma_2$. Hence in particular τ coequalizes the kernel-pair of $C_0 \longrightarrow 1$, so by (a) and 1.53 we have a factorization

$$\begin{array}{ccc} C_0 & \longrightarrow & 1 \\ & \searrow^{\tau} & \downarrow{\scriptstyle\sigma} \\ & & L \end{array} .$$

And σ is epi since τ is, and mono since its domain is 1; hence $L \cong 1$. ■

Now for any $\mathbf{C} \in \mathbf{cat}(\mathscr{E})$ with $\varinjlim \mathbf{C} = L$, we observe that the morphisms defining $\mathbf{C}$ are all morphisms over L; so $\mathbf{C}$ can be regarded as an object of $\mathbf{cat}(\mathscr{E}/L)$. And the notion of "discrete opfibration over $\mathbf{C}$" is unaltered by this change in the status of $\mathbf{C}$.

2.55 PROPOSITION. *If* $\mathbf{C}$ *is weakly followed in* $\mathbf{cat}(\mathscr{E})$, *then it is filtered when regarded as an object of* $\mathbf{cat}(\mathscr{E}/L)$. *Hence* $\mathbf{C}$ *is filtered in* $\mathbf{cat}(\mathscr{E})$ *iff it is weakly filtered and* $\varinjlim \mathbf{C} \cong 1$.

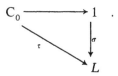

Proof. Suppose $\mathbf{C}$ is weakly filtered in $\mathscr{E}$. Now condition (c) is unaffected by the change from $\mathscr{E}$ to $\mathscr{E}/L$, since $\Sigma_L\colon \mathscr{E}/L \longrightarrow \mathscr{E}$ reflects pullbacks; and condition (a) in $\mathscr{E}/L$ is simply the statement that $C_0 \longrightarrow L$ is epi. It remains to prove (b) in $\mathscr{E}/L$, which is equivalent to the statement that the image (in $\mathscr{E}$) of $P \xrightarrow{(d_0\pi_1,\, d_0\pi_2)} C_0 \times C_0$ is precisely $C_0 \times_L C_0$. Write $I \rightarrowtail C_0 \times C_0$ for this image; then it is clear that $I \leqslant C_0 \times_L C_0$, since $P \xrightarrow{d_0\pi_1} C_0$ and $P \xrightarrow{d_0\pi_2} C_0$ are both morphisms over L. But I is an equivalence relation on C_0 (reflexivity and symmetry are obvious, and transitivity follows easily from (d)). So by 1.23 there is a morphism $C_0 \xrightarrow{f} Y$ whose kernel-pair is I. Now f coequalizes $C_1 \underset{d_1}{\overset{d_0}{\rightrightarrows}} C_0$, since $(1_{C_1}, id_1)\colon C_1 \longrightarrow P$ provides a factorization of this pair through I, and so we have a factorization

$$\begin{array}{ccc} C_0 & \longrightarrow\!\!\!\!\rightarrow & L \\ & \searrow^{f} & \downarrow \\ & & Y \end{array} .$$

Hence $C_0 \times_L C_0 \leqslant I$; so the first part of the proposition is established. And the second follows easily from it and the last two lemmas. ■

2.56 LEMMA. *Let* $\mathbf{B} \xrightarrow{\gamma} \mathbf{C}$ *be a discrete opfibration. If* $\mathbf{C}$ *is weakly filtered, then so is* $\mathbf{B}$.

Proof. A straightforward diagram-chase shows that the squares

$$\begin{array}{ccc} S_{\mathbf{B}} & \longrightarrow & Q_{\mathbf{B}} \\ \downarrow & & \downarrow \\ S_{\mathbf{C}} & \longrightarrow & Q_{\mathbf{C}} \end{array} \qquad \text{and} \qquad \begin{array}{ccc} T_{\mathbf{B}} & \longrightarrow & R_{\mathbf{B}} \\ \downarrow & & \downarrow \\ T_{\mathbf{C}} & \longrightarrow & R_{\mathbf{C}} \end{array}$$

are pullbacks, where the vertical arrows are induced by γ in the obvious way. So the result follows from 1.51. ■

2.57 PROPOSITION. *If* $\mathbf{C}$ *is filtered, then* $\varinjlim_{\mathbf{C}}\colon \mathscr{E}^{\mathbf{C}} \longrightarrow \mathscr{E}$ *preserves binary products.*

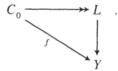

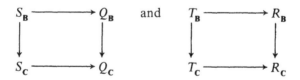

Proof. Let $\mathbf{F}^i \xrightarrow{\alpha^i} \mathbf{C}$ $(i = 1, 2)$ be discrete opfibrations with colimits $F^i_0 \xrightarrow{\sigma_i} L^i$, and let $\mathbf{G} \equiv \mathbf{F}^1 \times_{\mathbf{C}} \mathbf{F}^2$ have colimit $G_0 \xrightarrow{\tau} M$. Now the projections $\mathbf{G} \xrightarrow{\pi^i} \mathbf{F}^i$ induce morphisms $M \longrightarrow L^i$ and hence a morphism $M \xrightarrow{\phi} L^1 \times L^2$; we will show that ϕ is an isomorphism.

To show that ϕ is mono, it suffices by 1.52 to show that any pair of morphisms $U \overset{\gamma_1}{\underset{\gamma_2}{\rightrightarrows}} G_0$ coequalized by $\phi\tau$ is in fact coequalized by τ; for this will imply that $G_0 \xrightarrow{\tau} M \xrightarrow{\phi} L^1 \times L^2$ is the image factorization of $\phi\tau$. So suppose we have such a pair, and write $\gamma_j = (\gamma_j^1, \gamma_j^2)$ where $\gamma_j^i \colon U \longrightarrow F_0^i$.

Now γ_1^1 and γ_2^1 are coequalized by σ^1, and by 2.56 $\mathbf{F}^1$ is weakly filtered, so by 2.55 we can find an epi $V^1 \overset{\varepsilon^1}{\twoheadrightarrow} U$ and morphisms $V^1 \overset{\lambda_1^1}{\underset{\lambda_2^1}{\rightrightarrows}} F_1^1$ such that $d_1\lambda_1^1 = d_1\lambda_2^1$ and $d_0\lambda_j^1 = \gamma_j^1\varepsilon^1$. Similarly we can find $V^2 \xrightarrow{\varepsilon^2} U$ and $V^2 \overset{\lambda_1^2}{\underset{\lambda_2^2}{\rightrightarrows}} F_1^2$; and by forming a pullback over U, we can in fact assume that $V^1 = V^2 = V$ say and $\varepsilon^1 = \varepsilon^2 = \varepsilon$.

Now consider the four morphisms

$$\mu_j^i = \alpha_1^i \lambda_j^i \colon V \longrightarrow C_1 \qquad (i = 1, 2, j = 1, 2).$$

We have

$$d_0\mu_j^1 = d_0\alpha_1^1\lambda_j^1 = \alpha_0^1 d_0\lambda_j^1 = \alpha_0^1\gamma_j^1\varepsilon = \alpha_0^2\gamma_j^2\varepsilon = d_0\mu_j^2 \qquad (j = 1, 2),$$

since (γ_j^1, γ_j^2) maps into $F_0^1 \times_{C_0} F_0^2$. So by an easy extension of the argument of 2.53, we can find $W \overset{\zeta}{\twoheadrightarrow} V$ and $W \overset{\rho^1}{\underset{\rho^2}{\rightrightarrows}} C_1$ such that $m(\mu_j^1\zeta, \rho^1) = m(\mu_j^2\zeta, \rho^2)$ for *both* $j = 1$ and $j = 2$ simultaneously. Then using the fact that the α^j are discrete opfibrations, we can lift the ρ^i to morphisms $\theta^i \colon W \longrightarrow F_1^i$ such that $d_0\theta^i = d_1\lambda_1^i\zeta = d_1\lambda_2^i\zeta$. Define $\psi_j^i = m(\lambda_j^i\zeta, \theta^i) \colon W \longrightarrow F_1^i$; then we have $\alpha_1^1\psi_j^1 = m(\mu_j^1\zeta, \rho^1) = m(\mu_j^2\zeta, \rho^2) = \alpha_1^2\psi_j^2$, so that (ψ_j^1, ψ_j^2) is a morphism $W \longrightarrow F_1^1 \times_{C_1} F_1^2 = G_1$. And we have $d_1(\psi_1^1, \psi_1^2) = (d_1\theta^1, d_1\theta^2) = d_1(\psi_2^1, \psi_2^2)$ and $d_0(\psi_j^1, \psi_j^2) = (d_0\lambda_j^1, d_0\lambda_j^2)\,\zeta = (\gamma_j^1, \gamma_j^2)\,\varepsilon\zeta = \gamma_j\varepsilon\zeta$. So $\gamma_1\varepsilon\zeta$ and $\gamma_2\varepsilon\zeta$ are coequalized by $G_0 \overset{\tau}{\twoheadrightarrow} M$; but $\varepsilon\zeta$ is epi, so γ_1 and γ_2 are coequalized by τ.

To show that ϕ is epi, consider the morphisms $(\alpha_0^1\pi^1, \alpha_0^2\pi^2) \colon F_0^1 \times F_0^2 \rightrightarrows C_0$. By 2.51(b), we can find $V \overset{\varepsilon}{\twoheadrightarrow} F_0^1 \times F_0^2$ and $V \overset{\lambda^1}{\underset{\lambda^2}{\rightrightarrows}} C_1$ such that $d_1\lambda^1 = d_1\lambda^2$ and $d_0\lambda^i = \alpha_0^i\pi^i\varepsilon$. Now we can lift the λ^i to morphisms $\mu^i \colon V \longrightarrow F_1^i$ such that $d_0\mu^i = \pi^i\varepsilon$, and so define a map $\rho = (d_1\mu^1, d_1\mu^2) \colon V \longrightarrow F_0^1 \times_{C_0} F_0^2 = G_0$. Now $\phi\tau\rho = (\sigma^1 \times \sigma^2)\rho = (\sigma^1 d_1\mu^1, \sigma^2 d_1\mu^2) = (\sigma^1 d_0\mu^1, \sigma^2 d_0\mu^2) = (\sigma^1 \times \sigma^2)\,\varepsilon$, which is epi; so ϕ is epi. Hence by 1.22 ϕ is an isomorphism. ■

2.58 THEOREM. *If* $\mathbf{C}$ *is filtered, then* $\varinjlim_{\mathbf{C}} \colon \mathscr{E}^{\mathbf{C}} \longrightarrow \mathscr{E}$ *is left exact.*

Proof. In 2.54 we saw that $\varinjlim_{\mathbf{C}}$ preserves the terminal object; so it suffices to prove that it preserves pullbacks. But on combining 2.18, 2.55, 2.56 and 2.57, we see that $\varinjlim_{\mathbf{C}}$ transforms pullbacks over $(\mathbf{F} \xrightarrow{\gamma} \mathbf{C})$ in $\mathscr{E}^{\mathbf{C}}$ into products in $\mathscr{E}/\varinjlim \mathbf{F}$, i.e. pullbacks over $\varinjlim_{\mathbf{C}}(\gamma)$ in $\mathscr{E}$. ■

In fact we have a converse result, too:

2.59 THEOREM. *If* $\varinjlim_{\mathbf{C}}$ *is left exact, then* $\mathbf{C}$ *is filtered.*

Proof. (a) Since $\varinjlim_{\mathbf{C}}$ preserves the terminal object, we have $\varinjlim \mathbf{C} \cong 1$; so $C_0 \longrightarrow 1$ is epi.

(b) Let $U \overset{\gamma_1}{\underset{\gamma_2}{\rightrightarrows}} C_0$ be a pair of U-elements of C_0. Form the representable diagrams $R(\gamma_i)$; then $\varinjlim_{\mathbf{C}}(R(\gamma_i)) \cong U$ by 2.26, so $\varinjlim_{\mathbf{C}}(R(\gamma_1) \times R(\gamma_2)) \cong U \times U$. Form the pullback

$$\begin{array}{ccc} V & \xrightarrow{\quad \varepsilon \quad} & U \\ \big\downarrow & & \big\downarrow{\scriptstyle \Delta} \\ R(\gamma_1)_0 \times_{C_0} R(\gamma_2)_0 & \twoheadrightarrow & U \times U \end{array}$$

and let λ_i be the composite

$$V \longrightarrow R(\gamma_1)_0 \times_{C_0} R(\gamma_2)_0 \xrightarrow{\pi_1} R(\gamma_i)_0 = U \times_{C_0} C_1 \xrightarrow{\pi_2} C_1 \qquad (i = 1, 2).$$

Then it is readily checked that ε and the λ_i have the properties given in 2.52.

(c) Let $U \overset{\gamma_1}{\underset{\gamma_2}{\rightrightarrows}} C_1$ be two U-elements such that $d_i\gamma_1 = d_i\gamma_2 = \eta_i$ say $(i = 0, 1)$. Then the γ_i induce morphisms of representable diagrams $R(\eta_1) \overset{\bar{\gamma}_1}{\underset{2}{\rightrightarrows}} R(\eta_0)$ in an obvious way; define $E \rightarrowtail R(\eta_1)$ to be their equalizer in $\mathscr{E}^{\mathbf{C}}$. Now on applying the functor $\varinjlim_{\mathbf{C}}$ to the $\bar{\gamma}_i$, we obtain 1_U in each case; so we have $\varinjlim_{\mathbf{C}}(E) \cong \mathrm{eq}(1_U, 1_U) \cong U$, and in particular $E_0 \twoheadrightarrow U$ is epi. Now let δ be the composite

$$E_0 \rightarrowtail R(\eta_1)_0 = U \times_{C_0} C_1 \xrightarrow{\pi_2} C_1\,;$$

then it is easily checked that $E_0 \twoheadrightarrow U$ and δ have the properties required for the elementary form of condition (c). ■

EXERCISES 2

[In these exercises, $\mathscr{E}$ is any category with finite limits, unless otherwise stated.]

1. Give a precise definition of "internal natural transformation". To what concept in the theory of simplicial objects does it correspond? If $\mathbf{C} \underset{g}{\overset{f}{\rightrightarrows}} \mathbf{D}$ is a parallel pair of internal functors, show that each internal natural transformation $f \longrightarrow g$ induces an (external) natural transformation $f^* \longrightarrow g^*$, where $\mathscr{E}^{\mathbf{D}} \xrightarrow{f^*} \mathscr{E}^{\mathbf{C}}$ is the functor defined in 2.19. Deduce that if $\mathscr{E}$ is a topos, then the functor $\mathbf{cat}(\mathscr{E}) \longrightarrow \mathfrak{Top}/\mathscr{E}$ of 2.35 extends to a functor $\mathfrak{cat}(\mathscr{E}) \longrightarrow \mathfrak{Top}/\mathscr{E}$, where $\mathfrak{cat}(\mathscr{E})$ is the 2-category whose objects and 1-arrows are those of $\mathbf{cat}(\mathscr{E})$, and whose 2-arrows are internal natural transformations.

2. Give a definition of "internal groupoid in $\mathscr{E}$". Prove that if $\mathbf{G}$ is a groupoid, then so is the total category $\mathbf{F}$ of any discrete opfibration $\mathbf{F} \longrightarrow \mathbf{G}$.

3. Let $\mathbf{C}$ be an internal category in $\mathscr{E}$. Define the internal category $\mathbf{C}^2$ "whose objects are the morphisms of $\mathbf{C}$" [hint: the object-of-morphisms of $\mathbf{C}^2$ is the object $S_{\mathbf{C}}$ of § 2.5], and construct a pair of internal functors $\mathbf{C}^2 \underset{\partial_1}{\overset{\partial_0}{\rightrightarrows}} \mathbf{C}$. Given any pair of internal functors $\mathbf{B} \underset{\gamma_1}{\overset{\gamma_0}{\rightrightarrows}} \mathbf{C}$, show that internal natural transformations $\gamma_0 \longrightarrow \gamma_1$ correspond to internal functors $\mathbf{B} \xrightarrow{\delta} \mathbf{C}^2$ satisfying $\partial_0 \delta = \gamma_0$ and $\partial_1 \delta = \gamma_1$. [We say $\mathfrak{cat}(\mathscr{E})$ is a *representable 2-category*.] Show also that $\mathbf{C}^2 \underset{\partial_1}{\overset{\partial_0}{\rightrightarrows}} \mathbf{C}$ has the structure of an internal category in $\mathbf{cat}(\mathscr{E})$.

4. Suppose $\mathscr{E}$ is cartesian closed. For $\mathbf{C}$, $\mathbf{D} \in \mathbf{cat}(\mathscr{E})$, define the "object of internal functors $\mathbf{C} \longrightarrow \mathbf{D}$" as a subobject or $D_0^{C_0} \times D_1^{C_1} \times D_2^{C_2}$ obtained by intersecting certain equalizers. Use the idea of question 3 to show that this object is the object-of-objects of an internal category $\mathbf{D}^{\mathbf{C}}$ in $\mathscr{E}$; and deduce that $\mathbf{cat}(\mathscr{E})$ is cartesian closed.

5. Let $\mathbf{C}$ be an internal category in $\mathscr{E}$, and $\mathbf{D}$ an internal category in $\mathscr{E}^{\mathbf{C}}$,

represented by the diagram

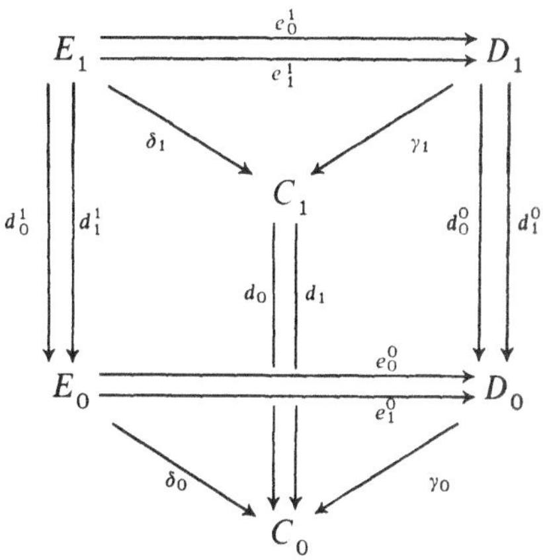

Show that $E_1 \xrightarrow{(d_0^1,\, e_1^1)} E_0 \times_{D_0} D_1$ is an isomorphism, and hence define an internal category structure on $G(\mathbf{D}) = (D_1 \times_{D_0} E_0 \overset{d_0^0 \pi_2}{\underset{e_1^0 \pi_2}{\rightrightarrows}} D_0)$. [The multiplication is given by

$$D_1 \times_{D_0} E_0 \times_{D_0} D_1 \times_{D_0} E_0 \xrightarrow{\cong} D_1 \times_0 E_1 \times_{D_0} E_0 \xrightarrow{1 \times (e_0^1,\, d_1^1) \times 1}$$

$$D_1 \times_{D_0} D_1 \times_{D_0} E_0 \times_{D_0} E_0 \xrightarrow{m \times m} D_1 \times_{D_0} E_0.]$$

Show also that $(\gamma_0, \gamma_1 \pi_1)$ defines an internal functor $G(\mathbf{D}) \longrightarrow \mathbf{C}$, and deduce that G is a functor $\mathbf{cat}(\mathscr{E}^{\mathbf{C}}) \longrightarrow \mathbf{cat}(\mathscr{E})/\mathbf{C}$. [$G(\mathbf{D})$ is called the *Grothendieck category* of $\mathbf{D}$; see [44], VI 9.]

6. Let $\mathbf{B} \xrightarrow{f} \mathbf{C}$ be an object of $\mathbf{cat}(\mathscr{E})/\mathbf{C}$. Show that f is isomorphic to an object in the image of the functor G iff there exist morphisms σ: $B_0 \times_{C_0} C_1 \longrightarrow B_1$ and $\tau: B_1 \longrightarrow B_1$ satisfying the following equations:

   (i) $d_0\sigma = \pi_1$ (ii) $f_1\sigma = \pi_2$ (iii) $\sigma(1 \times i) = i$
   (iv) $\sigma(1 \times m) = m(\sigma\pi_{12}, \sigma(d_1\sigma\pi_{12}, \pi_3)): B_0 \times_{C_0} C_1 \times_{C_0} C_1 \longrightarrow B_1$
   (v) $\pi = \tau$ (vi) $f_1\tau = if_0d_1$ (vii) $d_1\tau = d_1$
   (viii) $d_0\tau = d_1\sigma(d_0, f_1)$ (ix) $1_{B_1} = m(\sigma(d_0, f_1), \tau)$
   (x) $\tau\sigma = id_1\sigma$ (xi) $\tau m = m(\tau m(\pi_1, \sigma(d_0, f_1)\pi_2), \tau\pi_2): B_2 \longrightarrow B_1$.

   [Hint: given σ and τ, define $D_0 = B_0$, $D_1 = \mathrm{im}(\sigma)$ and $E_0 = \mathrm{im}(\tau)$. Equation (ix) then implies that $B_1 \cong D_1 \times_{D_0} E_0$. A morphism of $\mathbf{cat}(\mathscr{E})$ equipped with morphisms σ and τ as above is called a *split opfibration.*] Show also that f is a discrete opfibration iff it is a split opfibration and the corresponding internal category in $\mathscr{E}^{\mathbf{C}}$ is discrete.

†7. Let $\mathbf{C}$, $\mathbf{D}$ be as in question 5. Show that we have an equivalence of categories $(\mathscr{E}^{\mathbf{C}})^{\mathbf{D}} \simeq \mathscr{E}^{G(\mathbf{D})}$, obtained by applying the functor G to discrete opfibrations over $\mathbf{D}$.

8. Let $\mathbf{C}$ be a poset in $\mathscr{E}$, and $\mathbf{D}$ a poset in $\mathscr{E}^{\mathbf{C}}$. Show that $G(\mathbf{D})$ is a poset in $\mathscr{E}$.

9. Let $\mathbf{C}$ be a small category. Show that $\mathbf{C}$ is filtered (resp. weakly filtered) iff, given any finite diagram (resp. any finite connected diagram) in $\mathbf{C}$, there is a cone (not necessarily a colimiting cone) from the diagram to some object of $\mathbf{C}$.

†10. Let $\mathbf{C}$ be an internal category in $\mathscr{E}$. We say that $1 \xrightarrow{u} C_0$ is a *terminal object* of $\mathbf{C}$ if there exists a morphism $C_0 \xrightarrow{h} C_1$ such that $d_0 h = 1_{C_0}$ and

$$\begin{array}{ccc} C_0 & \longrightarrow & 1 \\ {\scriptstyle h}\downarrow & & \downarrow{\scriptstyle u} \\ C_1 & \xrightarrow{d_1} & C_0 \end{array}$$

is a pullback. If $\mathbf{C}$ has a terminal object, show that it is filtered. [Use u and h to construct splittings for the epimorphisms of 2.51.]

†11. Let $\mathscr{E}$ be a topos. An internal functor $\mathbf{C} \xrightarrow{f} \mathbf{D}$ between filtered categories in $\mathscr{E}$ is said to be *cofinal* if (a) $D_1 \times_{D_0} C_0 \xrightarrow{d_0\pi_1} D_0$ is epi, and (b) $T_{\mathbf{D}} \times_{D_1} C_1 \xrightarrow{(\pi_1\pi_1,\, \pi_1\pi_2)\times d_0} R_{\mathbf{D}} \times_{D_0} C_0$ is epi, where the pullbacks are defined using the structure morphisms $T_{\mathbf{D}} \xrightarrow{\pi_2\pi_1} D_1$ and $R_{\mathbf{D}} \xrightarrow{d_1\pi_1} D_0$ respectively. Write down elementary forms of these conditions, and hence show that if we have a composable pair $\mathbf{C} \xrightarrow{f} \mathbf{D} \xrightarrow{g} \mathbf{E}$ with f cofinal, then gf is cofinal iff g is. Show also that $\mathbf{C} \xrightarrow{f} \mathbf{D}$ is cofinal iff the diagram

$$\begin{array}{ccc} \mathscr{E}^{\mathbf{D}} & \xrightarrow{f^*} & \mathscr{E}^{\mathbf{C}} \\ & {\scriptstyle \varinjlim_{\mathbf{D}}}\searrow & \downarrow{\scriptstyle \varinjlim_{\mathbf{C}}} \\ & & \mathscr{E} \end{array}$$

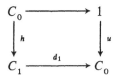

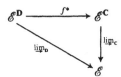

commutes up to isomorphism. [Hint: since all the functors involved preserve coequalizers, it is sufficient to check the commutativity on representables.]

Chapter 3

Topologies and Sheaves

3.1. TOPOLOGIES

We recall that the concept of Grothendieck topos was built up in two distinct steps: from $\mathscr{S}$, the "pre-existing" category of sets, we passed to the category $\mathscr{S}^{\mathbf{C}^{\mathrm{op}}}$ of presheaves on a small category $\mathbf{C}$, and then to the category $\mathrm{Shv}(\mathbf{C}, J)$ of sheaves for a Grothendieck topology J. In chapter 2, we saw how the first step may be done with $\mathscr{S}$ replaced by any elementary topos $\mathscr{E}$; our objective in the present chapter is to give a similar generalization of the second step.

3.11 Definition (Lawvere–Tierney [LH]). Let $\mathscr{E}$ be a topos. A *topology* in $\mathscr{E}$ is a morphism $j \colon \Omega \longrightarrow \Omega$ such that the diagrams

$$\begin{array}{ccc} 1 & \xrightarrow{\;t\;} & \Omega, \\ & \searrow^{t} & \downarrow{\scriptstyle j} \\ & & \Omega \end{array} \qquad \begin{array}{ccc} \Omega & \xrightarrow{\;j\;} & \Omega \\ & \searrow^{j} & \downarrow{\scriptstyle j} \\ & & \Omega \end{array} \qquad \text{and} \qquad \begin{array}{ccc} \Omega \times \Omega & \xrightarrow{\;\wedge\;} & \Omega \\ \downarrow{\scriptstyle j \times j} & & \downarrow{\scriptstyle j} \\ \Omega \times \Omega & \xrightarrow{\;\wedge\;} & \Omega \end{array}$$

commute, where $\wedge$ is the morphism defined in 1.49(ii). If j is a topology, we write $J \rightarrowtail \Omega$ for the subobject classified by j, and $\Omega_j \rightarrowtail \Omega$ for the equalizer of j and 1_Ω (equivalently, the image of j, since j is idempotent). ■

3.12 Example. Let $(\mathbf{C}, J)$ be a site (cf. 0.32). On comparing 0.32(ii) with 1.12(iii), we see that J is a subobject of Ω in $\mathscr{S}^{\mathbf{C}^{\mathrm{op}}}$; so it has a classifying map $\Omega \xrightarrow{\;j\;} \Omega$ in $\mathscr{S}^{\mathbf{C}^{\mathrm{op}}}$. Moreover, it is not hard to prove that a sub-presheaf J of Ω satisfies

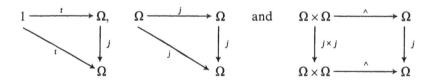

0.32(i) and (iii) iff its classifying map j satisfies the conditions of 3.11; so we have a bijection between topologies (in the sense of 3.11) in $\mathcal{S}^{\mathbf{C}^{op}}$ and Grothendieck topologies on **C**. ■

Although the overwhelming advantage of the elementary definition of a topology is its conciseness, there is another, more explicitly descriptive, way of defining a topology which has the advantage that it can be interpreted in categories more general than toposes.

3.13 Definition. Let $\mathcal{E}$ be any category with pullbacks. A *universal closure operation* on $\mathcal{E}$ is defined by specifying, for each $X \in \mathcal{E}$, a closure operation (i.e. an increasing, order-preserving, idempotent map) on the poset of subobjects of X—we denote the closure of $X' \rightarrowtail X$ by $\overline{X'} \rightarrowtail X$—in such a way that closure commutes with pullback along morphisms of $\mathcal{E}$; i.e. given $Y \xrightarrow{f} X$, we have $f^*(\overline{X'}) \cong \overline{(f^*X')}$ as subobjects of Y.

We shall use the words *dense* and *closed* with their usual meanings relative to a universal closure operation; i.e. $X' \rightarrowtail X$ is dense if $\overline{X'} \cong X$, and closed if $\overline{X'} \cong X'$. ■

3.14 Proposition. *Let $\mathcal{E}$ be a topos. Then there is a bijection between topologies in $\mathcal{E}$ and universal closure operations on $\mathcal{E}$.*

Proof. Let j be a topology in $\mathcal{E}$. We define the associated "j-closure" operation as follows; if $X' \rightarrowtail X$ has classifying map $X \xrightarrow{\phi} \Omega$, then $\overline{X'}$ is the subobject classified by $j\phi$. The first two diagrams of 3.11 trivially imply that this operation is increasing (i.e. $X' \leqslant \overline{X'}$) and idempotent; the fact that it preserves order is an easy consequence of the third diagram, since we have $X' \leqslant X''$ iff $X' \cap X'' \cong X'$; and universality is obvious from the form of the definition.

Conversely, suppose we are given a universal closure operation on $\mathcal{E}$. By applying it to the generic subobject $1 \overset{\top}{\rightarrowtail} \Omega$, we obtain a subobject $J \rightarrowtail \Omega$ with classifying map $\Omega \xrightarrow{j} \Omega$; and universality then says that the entire closure operation is induced by j in the above manner. To show that j is a topology, we need to show that any universal closure operation commutes with intersection of subobjects, which will imply the third condition of 3.11; the first two are immediate.

But it is clear that $\overline{X'}$ may be characterized as the unique subobject of X

such that $X' \rightarrowtail \overline{X'}$ is dense and $\overline{X'} \rightarrowtail X$ is closed. Now in the diagram

$$\begin{array}{ccccc}
X' & \longleftarrow & X' \cap X'' & & \\
\downarrow & & \downarrow & & \\
\overline{X'} & \longleftarrow & \overline{X'} \cap X'' & \longrightarrow & X'' \\
 & & \downarrow & & \downarrow \\
 & & \overline{X'} \cap \overline{X''} & \longrightarrow & \overline{X''}
\end{array}$$

both squares are pullbacks, from which it follows easily that

$$X' \cap X'' \rightarrowtail \overline{X'} \cap \overline{X''}$$

is dense; and similarly $\overline{X'} \cap \overline{X''} \rightarrowtail X$ is closed. So $\overline{X'} \cap \overline{X''} \cong \overline{X' \cap X''}$, as required. ∎

The reader is warned against the danger of confusing the universal closure operation induced by a topology in the sense of 3.11 with the "Kuratowskian" closure operation defined on the lattice of all subsets of a topological space [175]. It is perhaps unfortunate that the word "topology" has survived the process of repeated generalization outlined in §0.3 and the present paragraph, for there is in truth very little connection between a topology in the sense of 3.11 and a topology on a set. To emphasize the fact that the two types of closure operation are different, let us point out that Kuratowskian closure always commutes with finite unions but not, in general, with intersections; whereas we have just proved that a universal closure operation commutes with intersections, though it does not normally commute with unions.

3.15 LEMMA. *Let j be a topology in $\mathscr{E}$, and $X' \overset{\sigma}{\rightarrowtail} X$ a monomorphism with classifying map $X \overset{\phi}{\longrightarrow} \Omega$. Then σ is j-dense iff ϕ factors through $J \rightarrowtail \Omega$, and j-closed iff ϕ factors through $\Omega_j \rightarrowtail \Omega$.*

Proof. Trivial from the definitions of J and Ω_j. ∎

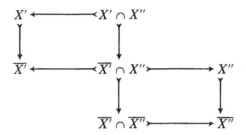

3.16 EXAMPLE. Let $(\mathbf{C}, J)$ be a site, and j the corresponding topology in $\mathscr{S}^{\mathbf{C}^{\mathrm{op}}}$. Then a sub-presheaf $R \rightarrowtail h_U$ is j-dense iff it is an element of $J(U)$, i.e. a J-covering sieve on U. In particular, consider the "classical" case of a topological space $(X, \mathbf{U})$: then a sieve on an open set $U \subseteq X$ is a family $\mathscr{V}$ of open subsets of U satisfying $V \in \mathscr{V}$, $V' \subseteq V \Rightarrow V' \in \mathscr{V}$. And $\overline{\mathscr{V}} = j_U(\mathscr{V})$ is the sieve

$$\{W \in U \mid W \subseteq \bigcup \mathscr{V}\};$$

so $\mathscr{V}$ is dense in h_U iff it covers U, and closed iff it has the form $\{V \in \mathbf{U} \mid V \subseteq V_0\}$ for some $V_0 \subseteq U$, i.e. iff it is representable. ∎

A third way in which we can describe a topology in a topos $\mathscr{E}$ is by describing the class of monomorphisms in $\mathscr{E}$ which are dense for it. If D is any subobject of Ω, we shall write Ξ_D for the class of monos whose classifying maps factor through D. (Note that it is immediate from the definition that the class Ξ_D is stable under pullback; but we cannot conversely assert that any pullback-stable class of monos is "classifiable" by a subobject of Ω.)

3.17 LEMMA. *Let D be a subobject of Ω, and suppose that $1 \xrightarrow{t} \Omega$ factors through D (equivalently, that Ξ_D contains all isomorphisms). Then the class Ξ_D is stable under the formation of pushouts.*

Proof. Let

$$\begin{array}{ccc} X & \xrightarrow{f} & Y \\ {\scriptstyle\sigma}\big\downarrow & & \big\downarrow{\scriptstyle\tau} \\ Z & \xrightarrow{g} & T \end{array}$$

be a pushout diagram with $\sigma \in \Xi_D$. By 1.28 the square is also a pullback, and so if $T \xrightarrow{\phi} \Omega$ is the classifying map of τ, the composite ϕg classifies $g^*(\tau) \cong \sigma$ and hence factors through D. But $\phi\tau$ classifies $\tau^*(\tau) \cong 1_Y$, so it too factors through D. Now using the fact that the square is a pushout, we obtain a factorization of ϕ through D. ∎

3.18 PROPOSITION. *Let J be a subobject of Ω. Then the classifying map $\Omega \xrightarrow{j} \Omega$ of $J \rightarrowtail \Omega$ is a topology iff the class Ξ_J is a* saturated multiplicative system *of monomorphisms of $\mathscr{E}$; i.e. Ξ_J contains all isomorphisms and satisfies $\sigma\tau \in \Xi_J \Leftrightarrow \sigma \in \Xi_J$ and $\tau \in \Xi_J$.*

Proof. Suppose j is a topology; then it is easily verified from 3.14 that the class of j-dense monomorphisms satisfies the given conditions.

Conversely, suppose Ξ_J satisfies the conditions. Then it is easily seen that the operation ρ on subobjects induced by j is describable as follows: if $X' \rightarrowtail X$ is a subobject of X, then $\rho(X')$ is the unique largest $X'' \rightarrowtail X$ for which $X' \cap X'' \rightarrowtail X''$ is in Ξ_J. Now since Ξ_J contains $1_{X'}$, we have at once that $X' \leqslant \rho(X')$. On the other hand, since the composite $X' \rightarrowtail \rho(X') \rightarrowtail \rho(\rho(X'))$ is in Ξ_J, we have $\rho(\rho(X')) \leqslant \rho(X')$, and so ρ is idempotent. It remains to show that ρ is order-preserving; consider a pair of subobjects $X' \rightarrowtail X$, $X'' \rightarrowtail X$ such that $X' \leqslant X''$, and form the diagram

$$\begin{array}{ccc} \rho(X') \cap \rho(X'') & \rightarrowtail & \rho(X'') \\ \downarrow{\scriptstyle \sigma} & & \downarrow{\scriptstyle \tau} \\ \rho(X') & \rightarrowtail & \rho(X') \cup \rho(X'') \end{array}$$

Now since $X' \leqslant X'' \leqslant \rho(X'')$, the inclusion $X' \rightarrowtail \rho(X')$ factors through σ, and so $\sigma \in \Xi_J$. But the square is a pushout, so by 3.17 $\tau \in \Xi_J$. But now the composite $X'' \rightarrowtail \rho(X'') \overset{\tau}{\rightarrowtail} \rho(X') \cup \rho(X'')$ is in Ξ_J, and so τ must be iso; i.e. we must have $\rho(X') \leqslant \rho(X'')$. So ρ is a universal closure operation, and by 3.14 j is a topology. ∎

We conclude this paragraph with a technical lemma about universal closure operations, which will be used repeatedly in the rest of this chapter.

3.19 LEMMA. *Suppose we have a commutative square*

$$\begin{array}{ccc} X & \longrightarrow & Y \\ \downarrow{\scriptstyle \sigma} & & \downarrow{\scriptstyle \tau} \\ Z & \xrightarrow{\phi} & T \end{array}$$

with σ dense, τ closed. Then ϕ factors (uniquely, since τ is mono) through τ.

Proof. The commutativity of the square implies that $\sigma \leqslant \phi^*(\tau)$ as subobjects of Z. So $1_Z \cong \bar{\sigma} \leqslant \overline{\phi^*(\tau)} \cong \phi^*(\bar{\tau}) \cong \phi^*(\tau)$, which is equivalent to saying that ϕ factors through τ. ∎

3.2. SHEAVES

3.21 DEFINITION. Let j be a topology in a topos $\mathscr{E}$, F an object of $\mathscr{E}$.

(i) We say F is (j-) *separated* if, given any j-dense $X' \overset{\sigma}{\rightarrowtail} X$ and any pair $X \overset{f}{\underset{g}{\rightrightarrows}} F$ such that $f\sigma = g\sigma$, we have $f = g$.

(ii) We say F is a (j-)*sheaf* if, given any j-dense $X' \overset{\sigma}{\rightarrowtail} X$ and any $X' \xrightarrow{f} F$, there exists a unique $X \xrightarrow{g} F$ such that $g\sigma = f$.

We write $\mathrm{sh}_j(\mathscr{E})$ for the full subcategory of $\mathscr{E}$ whose objects are sheaves. ■

3.22 EXAMPLE. Let $(\mathbf{C}, J)$ be a site, and j the corresponding topology in $\mathscr{S}^{\mathbf{C}^{\mathrm{op}}}$. Then the definitions of separated presheaf and sheaf given in 0.33 are easily seen to be equivalent to the above definitions, restricted to cases when the presheaf X is representable. But in $\mathscr{S}^{\mathbf{C}^{\mathrm{op}}}$, if the above conditions are satisfied for representable X then they are satisfied for all X, since we can use 0.12 to express X as a colimit of representable presheaves h_α, and then by universality of colimits X' is the colimit of the objects $X' \times_X h_\alpha$. So the definitions in 3.21 agree with those given earlier. ■

Our goal in the present paragraph is to prove that the category $\mathrm{sh}_j(\mathscr{E})$ is a topos. Note, however, that although we shall consider sheaves only in toposes, the definitions of 3.21 make sense for any universal closure operation on a category with finite limits; and in fact all the results of this paragraph which do not explicitly mention the topos structure, except 3.29, are true in this more general situation.

3.23 LEMMA. *$\mathrm{sh}_j(\mathscr{E})$ has finite limits, and the inclusion functor $\mathrm{sh}_j(\mathscr{E}) \longrightarrow \mathscr{E}$ preserves them.*

Proof. The conditions of 3.21 involve only morphisms with codomain F; so the limit in $\mathscr{E}$ of any finite diagram of sheaves is itself a sheaf. Hence it is also the limit in $\mathrm{sh}_j(\mathscr{E})$. ■

3.24 PROPOSITION. *If F is a sheaf and X is any object of $\mathscr{E}$, then F^X is a sheaf.*

Proof. Suppose given

$$\begin{array}{ccc} Y' & \xrightarrow{f} & F^X \\ \big\downarrow{\scriptstyle \sigma} & & \\ Y & & \end{array}$$

with σ dense. Then $Y' \times X \xrightarrow{[\sigma \times 1]} Y \times X$ is dense (since it is the pullback of σ along $Y \times X \xrightarrow{\pi_1} Y$), and so $\bar{f}: Y' \times X \longrightarrow F$ extends uniquely to $\bar{g}: Y \times X \longrightarrow F$. And then $g: Y \longrightarrow F^X$ is the unique extension of f. ■

3.25 COROLLARY. $\mathrm{sh}_j(\mathscr{E})$ *is cartesian closed, and the inclusion functor* $\mathrm{sh}_j(\mathscr{E}) \longrightarrow \mathscr{E}$ *preserves exponentials.*

Proof. Immediate from 3.23, 3.24 and the fact that the inclusion functor is full. ■

3.26 LEMMA. (i) *A subobject of a separated object is separated.*

(ii) *A closed subobject of a sheaf is a sheaf:*

(iii) *If F is a sheaf and G is separated, then any mono* $F \xrightarrow{\tau} G$ *is closed.*

Proof. (i) Suppose given

$$\begin{array}{ccc} X' & \xrightarrow{\sigma} & X \\ & & f \big\downarrow \big\downarrow g \\ & & F' \xrightarrow{\tau} F \end{array}$$

with σ dense, F separated, $f\sigma = g\sigma$. Then $\tau f\sigma = \tau g\sigma \Rightarrow \tau f = \tau g$; but τ is mono, so $f = g$.

(ii) Suppose given

$$\begin{array}{ccc} X' & \xrightarrow{f} & F' \\ \big\downarrow{\scriptstyle \sigma} & & \big\downarrow{\scriptstyle \tau} \\ X & & F \end{array}$$

with σ dense, τ closed, F a sheaf. Then there exists a unique $X \xrightarrow{g} F$ such that $g\sigma = \tau f$; but by 3.19 g factors uniquely through τ. So F' is a sheaf.

(iii) Consider the closure $\bar{F} \rightarrowtail^{\bar{\iota}} G$ of F in G. Now $F \rightarrowtail^{i} \bar{F}$ is dense, so there exists a unique $r : \bar{F} \longrightarrow F$ such that $ri = 1_F$. Now $iri = i$; but $\bar{F}$ is separated by (i), so $ir = 1_{\bar{F}}$. Hence r is a two-sided inverse for i, i.e. F is closed in G. ■

3.27 Lemma. *Ω_j is a sheaf.*

Proof. By 3.15, morphisms $X \longrightarrow \Omega_j$ correspond to closed subobjects of X; so it suffices to prove that if $X' \rightarrowtail X$ is dense and $Y' \rightarrowtail X'$ is closed, there is a unique closed subobject $Y \rightarrowtail X$ such that $Y \cap X' \cong Y'$. But if we define Y to be the closure of the composite $Y' \rightarrowtail X' \rightarrowtail X$, it is immediate that $Y \cap X' \cong Y'$; and conversely, if $Z \rightarrowtail X$ is any closed subobject with $Z \cap X' \cong Y'$, then $Y' \rightarrowtail Z$ is dense (being the pullback of $X' \rightarrowtail X$), and so Z is the closure of Y' in X. ■

3.28 Corollary. *$\mathrm{sh}_j(\mathscr{E})$ has a subobject classifier, namely Ω_j.*

Proof. By 3.23, monos in $\mathrm{sh}_j(\mathscr{E})$ agree with those in $\mathscr{E}$. So by 3.26(ii) and (iii) the subsheaves of a sheaf are precisely its closed subobjects; and by 3.15 these are classified by morphisms into Ω_j. ■

Combining 3.23, 3.25 and 3.28, we have now proved that $\mathrm{sh}_j(\mathscr{E})$ is a topos, as desired.

In conclusion, we give a couple of useful characterizations of separated objects:

3.29 Proposition. *Let F be an object of $\mathscr{E}$. The following are equivalent:*

(i) *F is separated.*
(ii) *The diagonal $F \rightarrowtail^{\Delta} F \times F$ is closed.*
(iii) *There exists a monomorphism $F \rightarrowtail G$, where G is a sheaf.*

Proof. (i) ⇒ (ii): Let $\bar{F} \rightarrowtail^{(a,b)} F \times F$ be the closure of Δ. Then a and b are equalized by the dense subobject $F \rightarrowtail \bar{F}$, so they are equal; hence $\bar{F} \cong F$.

(ii) ⇒ (iii): Since Δ is closed, its classifying map $F \times F \xrightarrow{\delta} \Omega$ factors through $\Omega_j \rightarrowtail \Omega$; so the singleton map $F \rightarrowtail^{\{\}} \Omega^F$ factors through Ω_j^F. And Ω_j^F is a sheaf by 3.24 and 3.27.

(iii) ⇒ (i) is immediate from 3.26(i). ■

It should be noted that the implication (ii) ⇒ (iii) of 3.29 does not hold for an arbitrary universal closure operation on a category with finite limits (see

Exercise 3 at the end of this chapter); but the equivalence of (i) and (ii) is still valid in this case. The proof that (ii) implies (i) is a straightforward application of 3.19.

3.3. THE ASSOCIATED SHEAF FUNCTOR

In this paragraph we will construct a left adjoint $L: \mathscr{E} \longrightarrow \mathrm{sh}_j(\mathscr{E})$ for the inclusion functor $\mathrm{sh}_j(\mathscr{E}) \longrightarrow \mathscr{E}$. The method we use (which was first described in [52]) is based on that of Grothendieck ([GV], II 3.4) for constructing the associated sheaf functor for sheaves on a site; it should be mentioned that there is an alternative method (due to F. W. Lawvere) of constructing L in the elementary case, which is indicated in Exercise 4 at the end of this chapter.

Let j be a topology in $\mathscr{E}$. It follows easily from the third diagram of 3.11 that $\wedge : \Omega \times \Omega \longrightarrow \Omega$ maps the subobject $J \times J$ into J. We define $J_1 \rightarrowtail J \times J$ to be the equalizer of $J \times J \underset{\pi_1}{\overset{\wedge}{\rightrightarrows}} J$; and we shall write $d: 1 \rightarrowtail J$ for the generic j-dense monomorphism, i.e. the factorization of $1 \overset{t}{\rightarrowtail} \Omega$ through $J \rightarrowtail \Omega$.

3.31 LEMMA. $\mathbf{J} = (J_1 \underset{\pi_2}{\overset{\pi_1}{\rightrightarrows}} J)$ *is an internal poset in* $\mathscr{E}$. *Moreover,* $\mathbf{J}^{\mathrm{op}}$ *is filtered in the sense of* 2.51.

Proof. It is readily seen that a morphism $X \xrightarrow{(\phi_1, \phi_2)} J \times J$ factors through J_1 iff the two subobjects Y_1, Y_2 of X classified by ϕ_1, ϕ_2 satisfy $Y_1 \leqslant Y_2$. Using this, it is easy to define morphisms $J \longrightarrow J_1$ and $J_1 \times_J J_1 \longrightarrow J_1$ making J into a poset. Now $J \longrightarrow 1$ is certainly epi, since it is split by d; and the morphisms $J \times J \xrightarrow{(\wedge, \pi_1)} J_1$ and $J \times J \xrightarrow{(\wedge, \pi_2)} J_1$ combine to provide a splitting for the morphism $P \longrightarrow J \times J$ of 2.51(b). And condition 2.51(c) is trivially satisfied, since $\mathbf{J}$ is a poset. ■

Now let X be any object of $\mathscr{E}$. Write $\hat{X} \rightarrowtail \tilde{X}$ for the closure of $X \overset{\eta}{\rightarrowtail} \tilde{X}$ (cf. 1.25), and $\phi: \hat{X} \longrightarrow J$ for the classifying map of the dense monomorphism $X \overset{i}{\rightarrowtail} \hat{X}$. It follows easily from 3.19 that a morphism $U \longrightarrow \tilde{X}$ factors through $\hat{X}$ iff the domain of the corresponding partial map $U \longrightarrow X$ is a dense subobject of U. From this we may deduce as in 1.45 that $(X \longmapsto (\hat{X}, \phi))$ is in fact the functor $\Pi_d: \mathscr{E} \longrightarrow \mathscr{E}/J$, and in particular that it has a left adjoint d^*.

3.32 PROPOSITION. *There is a unique structure map* $e: J_1 \times_J \hat{X} \longrightarrow \hat{X}$ *making* $\hat{X}$ *into an internal presheaf on* **J**.

Proof. Consider the diagram

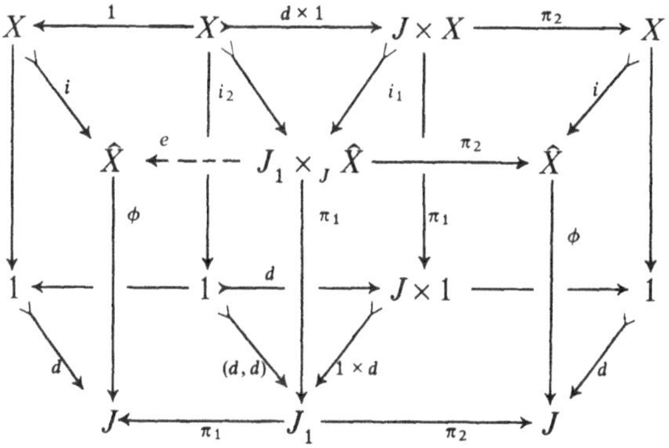

in which all faces of the right-hand cube are pullbacks, as are the square faces of the triangular prism. Hence in particular $X \overset{i_2}{\rightarrowtail} J_1 \times_J \hat{X}$ is dense, since it is obtained by pulling back and composing dense monomorphisms. So by the remarks above there is a unique $e: J_1 \times_J \hat{X} \longrightarrow \hat{X}$ making the top face of the left-hand cube a pullback; but the bottom face of the left-hand cube is also a pullback by the definition of J_1. So the front face of the left-hand cube commutes, since both ways round classify the dense subobject $X \overset{i_2}{\rightarrowtail} J_1 \times_J \hat{X}$. And we may similarly show that e satisfies the other two conditions of 2.14(iii), using the uniqueness of representations of partial maps. ■

We now define a functor $+ : \mathscr{E} \longrightarrow \mathscr{E}$ by

$$X^+ = \varinjlim_{\mathbf{J}}(\hat{X}, e).$$

(Note that we are here using $\varinjlim_{\mathbf{J}}$ to denote the internal colimit functor over $\mathbf{J}^{\mathrm{op}}$, for typographical reasons; we shall not be concerned at all with covariant diagrams on **J**.) The idea behind this definition is as follows: if $(\mathbf{C}, J)$ is a site, and X a presheaf on **C**, then for any $U \in C_0$, $\hat{X}(U)$ is the presheaf on $J(U)$ which sends a J-covering sieve $R \rightarrowtail h_U$ to the set hom(R, X). Thus the difference between $\hat{X}(U)$ and the constant presheaf $X(U)$ is a measure of "the amount by which X fails to satisfy the sheaf axiom at U". In general, we have the following result:

3.33 PROPOSITION. *There is a morphism of presheaves* $\alpha : \mathbf{J}^*(X) \longrightarrow (\hat{X}, e)$ *such that*

(i) *α is mono iff X is separated.*
(ii) *α is iso iff X is a sheaf.*

Proof. Define $\alpha_0 : X \times J \longrightarrow \hat{X}$ to be the representation of the partial map

$$\begin{array}{ccc} X & \xrightarrow{\ 1\ } & X\ . \\ \big\downarrow{\scriptstyle 1 \times d} & & \\ X \times J & & \end{array}$$

It is easily seen that this extends to a morphism of presheaves. Now a generalized element $U \longrightarrow X \times J$ corresponds to a diagram of the form $R \overset{\sigma}{\rightarrowtail} U \overset{f}{\longrightarrow} X$ with σ dense; and it is not hard to see that the operation of composing with α_0 sends the above diagram to the partial map

$$\begin{array}{ccc} R & \xrightarrow{\ f\sigma\ } & X\ . \\ \big\downarrow{\scriptstyle \sigma} & & \\ U & & \end{array}$$

So the statement that α_0 is mono (resp. iso) is precisely the definition of a separated object (resp. sheaf) as given in 3.21. But the forgetful functor $\mathscr{E}^{\mathbf{J}^{\mathrm{op}}} \longrightarrow \mathscr{E}/J$ reflects monos and isos; and hence α is mono (resp. iso) iff α_0 is. ■

3.34 COROLLARY. *If X is separated (resp. a sheaf), then the composite* $X \overset{i}{\rightarrowtail} \hat{X} \overset{u}{\twoheadrightarrow} X^+$ *is mono (resp. iso), where* $\hat{X} \overset{u}{\twoheadrightarrow} X^+$ *is the colimiting cone.*

Proof. Since $\mathbf{J}^{\mathrm{op}}$ is filtered, we have $\varinjlim \mathbf{J} \cong 1$ by 2.54, and so $\varinjlim_{\mathbf{J}} (\mathbf{J}^*(X)) \cong X$ since $X \times (-)$ preserves coequalizers. Also, $\varinjlim_{\mathbf{J}}$ preserves monos by 2.58; so we certainly have a mono (resp. iso) $X \longrightarrow X^+$ if X is separated (resp. a sheaf). The fact that this morphism is the given composite follows from the fact that the colimiting cone $X \times J \overset{\pi_1}{\longrightarrow} X$ of $\mathbf{J}^*(X)$ is split by $X \overset{1 \times d}{\longrightarrow} X \times J$. ■

3.35 COROLLARY. *Any morphism from X into a sheaf factors uniquely through $X \xrightarrow{ui} X^+$.*

Proof. Suppose we are given $X \xrightarrow{f} F$, F a sheaf. Then we can obtain the desired factorization by forming the diagram

$$\begin{array}{ccc} X & \xrightarrow{ui} & X^+ \\ {\scriptstyle f}\downarrow & & \downarrow{\scriptstyle f^+} \\ F & \xleftarrow{(ui)^{-1}} & F^+ \end{array} .$$

To show that it is unique, let g_1 and g_2 be two such factorizations; then $g_1ui = f = g_2ui \Rightarrow g_1u = g_2u$ since i is dense and F is a sheaf $\Rightarrow g_1 = g_2$ since u is epi. ∎

It follows from 3.35 that if $X^{(+)^n}$ is a sheaf for some $n \geqslant 0$, then it is the associated sheaf of X. In contrast to the situation considered by A. Heller and K. Rowe [46], in which it was necessary to iterate the functor $+$ a transfinite number of times before obtaining a sheaf, we shall shortly see that in our case it is sufficient to take $n = 2$.

3.36 THEOREM (Johnstone). *For any X, X^+ is separated.*

Proof. Suppose given $R \rightarrowtail^{\sigma} U \underset{\gamma_2}{\overset{\gamma_1}{\rightrightarrows}} X^+$ with σ dense, $\gamma_1\sigma = \gamma_2\sigma$. Form the pullbacks

$$\begin{array}{ccccc} Q & \overset{\rho}{\rightarrowtail} & T & \xrightarrow{(\beta_1, \beta_2)} & \hat{X} \times \hat{X} \\ {\scriptstyle w}\downarrow & & \downarrow{\scriptstyle v} & & \downarrow{\scriptstyle u \times u} \\ R & \overset{\sigma}{\rightarrowtail} & U & \xrightarrow{(\gamma_1, \gamma_2)} & X^+ \times X^+ \end{array} ;$$

then $u\beta_1\rho = \gamma_1\sigma w = \gamma_2\sigma w = u\beta_2\rho$, so $(\beta_1\rho, \beta_2\rho)$ factors through $\hat{X} \times_{X^+} \hat{X} \rightarrowtail \hat{X} \times \hat{X}$. But by 2.56 $(\hat{X}, e)^{\text{op}}$ is weakly filtered, so by 2.55 we can find an epi $V \overset{\varepsilon}{\twoheadrightarrow} Q$ and $V \xrightarrow{\delta} \hat{X}$ such that both the pairs $(\delta, \beta_1\rho\varepsilon)$ and $(\delta, \beta_2\rho\varepsilon)$ factor through $J_1 \times_J \hat{X} \xrightarrow{(e, \pi_2)} \hat{X} \times \hat{X}$. Let

$$\begin{array}{ccc} S_i & \xrightarrow{\alpha_i} & X \\ {\scriptstyle \tau_i}\downarrow & & \\ V & & \end{array}$$

($i = 1, 2, 3$) be the partial maps represented by $\beta_1\rho\varepsilon$, $\beta_2\rho\varepsilon$, δ respectively, and let

$$\begin{array}{ccc} P_i & \xrightarrow{\lambda_i} & X \\ \big\downarrow{\scriptstyle \mu_i} & & \\ T & & \end{array} \qquad (i = 1, 2)$$

be the maps represented by β_i. Then the given conditions on δ imply that $\tau_3 \leqslant \tau_1$, $\tau_3 \leqslant \tau_2$ and $\alpha_1|_{S_3} = \alpha_3 = \alpha_2|_{S_3}$; and we have pullback squares

$$\begin{array}{ccc} S_i & \xrightarrow{\zeta_i} & P_i \\ \big\downarrow{\scriptstyle \tau_i} & & \big\downarrow{\scriptstyle \mu_i} \\ V & \xrightarrow{\rho\varepsilon} & T \end{array} \qquad (i = 1, 2)$$

with $\lambda_i\zeta_i = \alpha_i$. Let $S_3 \overset{\theta}{\twoheadrightarrow} P_3 \overset{\kappa}{\rightarrowtail} Q$ be the image factorization of the composite $S_3 \overset{\tau_3}{\rightarrowtail} V \overset{\varepsilon}{\twoheadrightarrow} Q$; then by 3.19 ε factors through the closure of κ, and so κ is dense. Hence the composite $\rho\kappa$ is also dense; but we have $\rho\kappa \leqslant \mu_1$ and $\rho\kappa \leqslant \mu_2$ as subobjects of T, and the composites

$$\begin{array}{ccccc} & & P_1 & & \\ & \nearrow & & \searrow{\scriptstyle \lambda_1} & \\ P_3 & & & & X \\ & \searrow & & \nearrow{\scriptstyle \lambda_2} & \\ & & P_2 & & \end{array}$$

are equal, since they are equalized by the epimorphism $S_3 \overset{\theta}{\twoheadrightarrow} P_3$. So if $T \xrightarrow{\beta_3} \hat{X}$ represents the partial map

$$\begin{array}{ccccc} P_3 & \rightarrowtail & P_1 & \xrightarrow{\lambda_1} & X \; , \\ \big\downarrow{\scriptstyle \rho\kappa} & & & & \\ T & & & & \end{array}$$

then the pairs (β_3, β_1) and (β_3, β_2) both factor through $J_1 \times_J \hat{X} \xrightarrow{(e, \pi_2)} \hat{X} \times \hat{X}$, and hence $\gamma_1 v = u\beta_1 = u\beta_3 = u\beta_2 = \gamma_2 v$. But v is epi, so $\gamma_1 = \gamma_2$. ■

3.37 THEOREM. *If X is separated, then X^+ is a sheaf.*

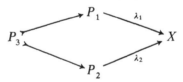

Proof. By 3.34 we know $X \xrightarrow{ui} X^+$ is mono; and it is dense by 3.19, since u factors through its closure. So we have a morphism $X^+ \xrightarrow{\theta} \hat{X}$ representing the partial map

$$\begin{array}{ccc} X & \xrightarrow{\quad 1 \quad} & X \\ \downarrow{\scriptstyle ui} & & \\ X^+ & & \end{array} .$$

Now $\hat{X} \xrightarrow{\theta u} \hat{X}$ represents the partial map

$$\begin{array}{ccc} \hat{X} \times_{X^+} X & \xrightarrow{\quad \pi_2 \quad} & X, \\ \downarrow{\scriptstyle \pi_1} & & \\ \hat{X} & & \end{array}$$

through which

$$\begin{array}{ccc} X & \xrightarrow{\quad 1 \quad} & X \\ \downarrow{\scriptstyle i} & & \\ \hat{X} & & \end{array}$$

factors by $X \xrightarrow{(i,1)} \hat{X} \times_{X^+} X$, so the pair $(1_{\hat{X}}, \theta u)$ factors through $J_1 \times_J \hat{X} \xrightarrow{(e, \pi_2)} \hat{X} \times \hat{X}$. Hence $u\theta u = u$; but y is epi, so $u\theta = 1_{X^+}$. Now suppose we are given

$$\begin{array}{ccc} R & \xrightarrow{\quad \gamma \quad} & X^+ \\ \downarrow{\scriptstyle \sigma} & & \\ U & & \end{array}$$

with σ dense; let

$$\begin{array}{ccc} S & \xrightarrow{\quad \beta \quad} & X \\ \downarrow{\scriptstyle \tau} & & \\ R & & \end{array}$$

be the partial map represented by $R \xrightarrow{\theta\gamma} \hat{X}$, and let $U \xrightarrow{\delta} \hat{X}$ represent

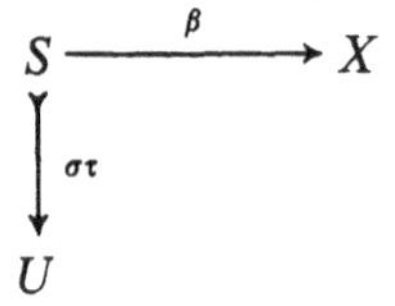

Then $\delta\sigma = \theta\gamma$, since both represent the same partial map; so $u\delta\sigma = u\theta\gamma = \gamma$, i.e. $u\delta$ is an extension of γ along σ. But we already know such an extension must be unique, by 3.36. ■

Combining 3.35, 3.36 and 3.37, we have proved that the functor $++:$ $\mathscr{E} \longrightarrow \mathscr{E}$ factors through the inclusion $\mathrm{sh}_j(\mathscr{E}) \longrightarrow \mathscr{E}$ to give us the *associated sheaf functor* $L: \mathscr{E} \longrightarrow \mathrm{sh}_j(\mathscr{E})$.

3.38 Proposition. *The functor* $+$ *is left exact.*

Proof. $+$ is defined as the composite $\mathscr{E} \longrightarrow \mathscr{E}^{\mathbf{J}^{\mathrm{op}}} \xrightarrow{\varinjlim_{\mathbf{J}}} \mathscr{E}$. But $\varinjlim_{\mathbf{J}}$ is left exact, by 2.58 and 3.31; and $\mathscr{E} \longrightarrow \mathscr{E}^{\mathbf{J}^{\mathrm{op}}}$ is left exact, since we have already noted that the composite $\mathscr{E} \longrightarrow \mathscr{E}^{\mathbf{J}^{\mathrm{op}}} \xrightarrow{U} \mathscr{E}/J$ has a left adjoint, namely d^*, and U creates limits. ■

3.39 Corollary. *We have a geometric morphism* $\mathrm{sh}_j(\mathscr{E}) \xrightarrow{i} \mathscr{E}$, *where* i_* *is the inclusion functor and* $i^* = L$.

Proof. The adjunction $(i^* \dashv i_*)$ follows as indicated from 3.35, 3.36 and 3.37; left exactness of i^* from 3.38 and the fact that i_* creates limits. ■

3.4. $\mathrm{sh}_j(\mathscr{E})$ AS A CATEGORY OF FRACTIONS

In this paragraph, we present an alternative description of the category $\mathrm{sh}_j(\mathscr{E})$, as a category of fractions (cf. 0.18) obtained by inverting a certain class of morphisms of $\mathscr{E}$. Once again, let j be a fixed topology in $\mathscr{E}$.

3.41 Definition. Let $X \xrightarrow{f} Y$ be a morphism of $\mathscr{E}$. Form the diagram

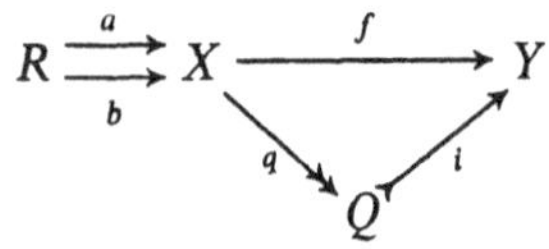

of 1.52, and let $X \overset{\tau}{\rightarrowtail} R$ denote the factorization of the diagonal $X \overset{\Delta}{\rightarrowtail} X \times X$ through $R \overset{(a,b)}{\rightarrowtail} X \times X$. We say f is (j-) *almost mono* (resp. *almost epi*) if τ (resp. i) is dense, and we say f is *bidense* if it is both almost mono and almost epi. (Note that f is mono (resp. epi) iff τ (resp. i) is an isomorphism.) ■

3.42 PROPOSITION. *Let $X' \overset{\sigma}{\rightarrowtail} X$ be a monomorphism in $\mathscr{E}$. Then $L(\sigma)$ is an isomorphism iff σ is dense.*

Proof. Suppose σ is dense. Then the universal morphism $X' \longrightarrow LX'$ (i.e. the unit of the adjunction of 3.39) factors uniquely through σ; so we have a morphism $X \longrightarrow LX'$ which in turn factors uniquely through the universal morphism $X \longrightarrow LX$. Thus we obtain a morphism $LX \longrightarrow LX'$, which is easily seen to be a two-sided inverse for $L(\sigma)$.

Conversely, suppose $L(\sigma)$ is an isomorphism. Then for any sheaf F, each morphism $X' \longrightarrow F$ factors uniquely through σ. In particular, this is true for the sheaf Ω_j; so each closed subobject of X' is the intersection of X' with a unique closed subobject of X. But the subobject $X' \overset{1}{\rightarrowtail} X'$ is the intersection of X' with either $X \overset{1}{\rightarrowtail} X$ or $\overline{X'} \overset{\bar{\sigma}}{\rightarrowtail} X$; so we must have $\bar{\sigma} \cong 1_X$, i.e. σ is dense. ■

3.43 COROLLARY. Let $X \overset{f}{\longrightarrow} Y$ *be any morphism of $\mathscr{E}$. Then $L(f)$ is an isomorphism (resp. mono, epi) iff f is bidense (resp. almost mono, almost epi).*

Proof. L preserves the exactness properties of the diagram of 3.41; so this is immediate from 3.42. ■

3.44 COROLLARY. *For any X, the universal morphism $X \longrightarrow LX$ is bidense.* ■

3.45 PROPOSITION. *Let Ξ_J denote the class of j-bidense morphisms of $\mathscr{E}$. Then Ξ_J admits both a left and a right calculus of fractions.*

Proof. Using the characterization of bidense morphisms provided by 3.43, and the fact that L is an exact functor, we can easily verify that

(i) A composite of bidense morphisms is bidense.

(ii) Pullbacks and pushouts of bidense morphisms are bidense.

(iii) Given a parallel pair $X \underset{g}{\overset{f}{\rightrightarrows}} Y$, f and g are equalized by a bidense morphism $\Leftrightarrow L(f) = L(g) \Leftrightarrow f$ and g are coequalized by a bidense morphism.

The conditions of 0.18, and their opposites, follow at once from these. ■

3.46 PROPOSITION. *The composite* $\mathrm{sh}_j(\mathscr{E}) \xrightarrow{i_*} \mathscr{E} \xrightarrow{P_\Xi} \mathscr{E}(\Xi_J)^{-1}$ *is an equivalence of categories.*

Proof. From 3.43 it follows that any morphism σ of $\mathrm{sh}_j(\mathscr{E})$ such that $i_*(\sigma)$ is bidense is already an isomorphism. Hence the composite $P_\Xi \cdot i_*$ is faithful; to prove that it is full, let

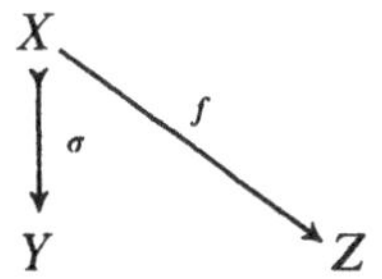

represent a morphism of $\mathscr{E}(\Xi_J)^{-1}$ with Y, Z sheaves. Then we can replace X by its associated sheaf without changing the equivalence class of the given morphism, whereupon σ becomes invertible in $\mathscr{E}$, and so (σ, f) is in the image of P_Ξ. Finally, we observe from 3.44 that the image of $P_\Xi . i_*$ meets every isomorphism class of objects of $\mathscr{E}(\Xi_J)^{-1}$. ■

The next result, though not a direct consequence of 3.46, is clearly closely related to it. It provides a useful criterion for an arbitrary geometric morphism to factor through one of the type described in 3.39.

3.47 THEOREM (Lawvere–Tierney). *Let* $\mathscr{F} \xrightarrow{f} \mathscr{E}$ *be a geometric morphism, j a topology in* $\mathscr{E}$. *The following conditions are equivalent:*

(i) *There exists a geometric morphism* $\mathscr{F} \xrightarrow{g} \mathrm{sh}_j(\mathscr{E})$ *such that* $ig \cong f$.

(ii) f^* *maps all j-bidense morphisms to isomorphisms.*

(iii) f^* *maps all j-dense monomorphisms to isomorphisms.*

(iv) $f^*(1 \overset{d}{\rightarrowtail} J)$ *is an isomorphism.*

(v) *For every object Y of* $\mathscr{F}$, $f_*(Y)$ *is a j-sheaf.*

Moreover, if g exists as in (i), *then it is unique up to canonical isomorphism.*

Proof. (i) ⇒ (ii) is immediate from 3.43, since we have $f^* \cong g^*i^*$. (ii) ⇒ (iii) ⇒ (iv) is trivial; and (iv) ⇒ (iii) since f^* preserves pullbacks.

(iii) ⇒ (v): Suppose we are given

$$\begin{array}{ccc} X' & \xrightarrow{\alpha} & f_*(Y) \\ \big\downarrow{\scriptstyle \sigma} & & \\ X & & \end{array}$$

with σ dense. Transposing, we obtain

$$\begin{array}{ccc} f^*(X') & \xrightarrow{\bar{\alpha}} & Y \\ \big\downarrow{\scriptstyle f^*(\sigma)} & & \\ f^*(X) & & \end{array}$$

and $\bar{\alpha}$ factors uniquely through $f^*(\sigma)$. Hence α factors uniquely through σ.

(v) ⇒ (i): Given (v), we clearly have a unique functor g_* such that $i_*g_* = f_*$; and we can define g^* to be the composite f^*i_*, since we then have natural bijections

$$\frac{\dfrac{F \longrightarrow g_*(Y)}{i_*(F) \longrightarrow f_*(Y)}}{g^*(F) \longrightarrow Y}$$

for $F \in \mathrm{sh}_j(\mathscr{E})$, $Y \in \mathscr{F}$. Moreover, the isomorphism $i_*g_* \cong f_*$ determines g_* (and hence g^*) up to canonical isomorphism, since i_* is full and faithful. ■

3.48 Remark. Yet another description of $\mathrm{sh}_j(\mathscr{E})$ has been given by *J.* Lambek and B. A. Rattray [70] using the ideas of § 1.3. If we replace Ω by Ω_j in the definition of the adjoint functors of 1.31, then the adjunction is still valid; let $\mathbb{G}_j$ denote the comonad it induces on $\mathscr{E}^{\mathrm{op}}$. Then the category $(\mathscr{E}^{\mathrm{op}})_{\mathbb{G}_j}$ of $\mathbb{G}_j$-coalgebras is equivalent to $\mathrm{sh}_j(\mathscr{E})$, by an equivalence which identifies the comparison functor $\mathscr{E} \longrightarrow (\mathscr{E}^{\mathrm{op}})_{\mathbb{G}_j}$ with the associated sheaf functor. ■

3.5. EXAMPLES OF TOPOLOGIES

In this paragraph we describe certain methods of constructing topologies in a topos, other than those which arise from Grothendieck topologies as in

3.12. Further examples will occur in later chapters (see, for example, 4.14 and 5.17). We begin by introducing two further binary operations on the subobject classifier Ω, similar to the morphism $\wedge$ introduced in 1.49(ii).

3.51 DEFINITION. (i) $\Omega \times \Omega \xrightarrow{\vee} \Omega$ is the classifying map of the union (1.55) of the two subobjects $\Omega \times 1 \rightarrowtail^{1 \times t} \Omega \times \Omega$ and $1 \times \Omega \rightarrowtail^{t \times 1} \Omega \times \Omega$.

(ii) $\Omega \times \Omega \xrightarrow{\Rightarrow} \Omega$ is the classifying map of the order relation $\Omega_1 \rightarrowtail \Omega \times \Omega$ (i.e. the equalizer of π_1 and $\wedge$, cf. 3.31). ■

Since pullback functors preserve unions, it is clear that $\vee$ "internalizes" the operation of forming unions in the same way that $\wedge$ internalizes that of forming intersections. The "external" description of $\Rightarrow$ is slightly less simple; if $Y_1 \rightarrowtail X$ and $Y_2 \rightarrowtail X$ are classified by ϕ_1 and ϕ_2 respectively, then $\Rightarrow(\phi_1, \phi_2)$ classifies the unique largest subobject $Z \rightarrowtail X$ such that $Z \cap Y_1 \leqslant Y_2$.

Now let $U \rightarrowtail 1$ be a subobject of 1 in $\mathscr{E}$, with classifying map $1 \xrightarrow{u} \Omega$. (U is sometimes called an *open object*, by analogy with the "classical" case of sheaves on a topological space.)

3.52 PROPOSITION. (i) *The composite* $\Omega \xrightarrow{u \times 1} \Omega \times \Omega \xrightarrow{\Rightarrow} \Omega$ *is a topology in* $\mathscr{E}$, *called the* open topology *defined by* U, *and denoted* j_U^o.

(ii) *The composite* $\Omega \xrightarrow{u \times 1} \Omega \times \Omega \xrightarrow{\vee} \Omega$ *is a topology in* $\mathscr{E}$, *called the* closed complement *of* j_U^o, *and denoted* j_U^c.

Proof. From the external descriptions of $\vee$ and $\Rightarrow$ given above, it is easily verified that the operations on subobjects induced by these two morphisms are increasing, order-preserving and idempotent; so by 3.14 the two morphisms are topologies. ■

The reason for the names "open and closed topologies" is to be found in the effect of these topologies in the case $\mathscr{E} = \mathrm{Shv}(X)$, X a topological space. If U is an open subset of X (i.e. an open object in Shv(X)), then the category of sheaves in Shv(X) for j_U^o (resp. j_U^c) is equivalent to Shv(U) (resp. Shv($X - U$)) by an equivalence which identifies the geometric morphism $\mathrm{sh}_j(\mathrm{Shv}(X)) \longrightarrow \mathrm{Shv}(X)$ of 3.39 with that induced (as in 0.26) by the inclusion map $U \longrightarrow X$ (resp. $X - U \longrightarrow X$).

In general, we have the following results:

3.53 PROPOSITION. (i) *A monomorphism* $X' \overset{\sigma}{\rightarrowtail} X$ *is* j_U^c*-dense iff the square*

$$\begin{array}{ccc} X' \times U & \xrightarrow{\pi_1} & X' \\ \downarrow{\scriptstyle \sigma \times 1} & & \downarrow{\scriptstyle \sigma} \\ X \times U & \xrightarrow{\pi_1} & X \end{array}$$

is a pushout.

(ii) *X is a j_U^c-sheaf iff $X \times U \cong U$.*

(iii) *The associated j_U^c-sheaf of an arbitrary object X is the pushout of*

$$\begin{array}{ccc} X \times U & \xrightarrow{\pi_1} & X \ . \\ \downarrow{\scriptstyle \pi_2} & & \\ U & & \end{array}$$

Proof. (i) By definition, σ is j_U^c-dense iff X is the union of the subobjects $X' \overset{\sigma}{\rightarrowtail} X$ and $X \times U \overset{\pi_1}{\rightarrowtail} X$; but this is the same as saying that the given square is a pushout.

(ii) Suppose X is a j_U^c-sheaf. Now if $Z \longrightarrow U$ is any object over U, we have $Z \times U \xrightarrow{\pi_1} Z$ iso, and so any subobject of Z (in particular $O \rightarrowtail Z$) is is j_U^c-dense. So by the sheaf axiom, we have a unique morphism $Z \longrightarrow X$ in $\mathscr{E}$, and hence a unique $Z \longrightarrow X \times U$ in $\mathscr{E}/U$. Hence $X \times U \cong U$, since both are terminal objects of $\mathscr{E}/U$. Conversely, if $X \times U \cong U$ and we are given

$$\begin{array}{ccc} Y' & \xrightarrow{f} & X \\ \downarrow{\scriptstyle \sigma} & & \\ Y & & \end{array}$$

with σ j_U^c-dense, then there is a unique $Y \times U \xrightarrow{g} X$ such that

$$\begin{array}{ccc} Y' \times U & \xrightarrow{\pi_1} & Y' \\ \downarrow{\scriptstyle \sigma \times 1} & & \downarrow{\scriptstyle f} \\ Y \times U & \xrightarrow{g} & X \end{array}$$

commutes, namely $Y \times U \xrightarrow{\pi_2} U \cong X \times U \xrightarrow{\pi_1} X$. So by (i) there is a unique $Y \xrightarrow{h} X$ extending f.

(iii) By an argument similar to that of (ii), any morphism from X into a sheaf factors uniquely through the given pushout. But since $(-) \times U$ preserves pushouts, it is easy to verify that the pushout is a sheaf. ■

3.54 PROPOSITION. (i) *A monomorphism* $X' \xrightarrow{\sigma} X$ *is* j^o_U*-dense iff* $X' \times U \xrightarrow{\sigma \times 1} X \times U$ *is an isomorphism.*

(ii) X *is a* j^o_U*-sheaf iff the transpose* $X \xrightarrow{\eta} X^U$ *of* $X \times U \xrightarrow{\pi_1} X$ *is an isomorphism; and in fact the associated sheaf functor for* j^o_U *is given by* $X \longmapsto X^U$.

(iii) *The category* $\mathrm{sh}_{j^o_U}(\mathscr{E})$ *is equivalent to* $\mathscr{E}/U$, *by an equivalence which identifies the inclusion functor with* Π_U *and the associated sheaf functor with* U^*.

Proof. (i) By definition, σ is j^o_U-dense iff $X \times U \xrightarrow{\pi_1} X$ factors through σ; but since

$$\begin{array}{ccc} X' \times U & \xrightarrow{\pi_1} & X' \\ {\scriptstyle \sigma \times 1}\downarrow & & \downarrow{\scriptstyle \sigma} \\ X \times U & \xrightarrow{\pi_1} & X \end{array}$$

is a pullback, this is clearly equivalent to $\sigma \times 1_U$ being iso.

(ii) It is clear from (i) that any object of the form X^U is a j^o_U-sheaf. But it is easily verified that $X \times U \xrightarrow{\eta \times 1} X^U \times U$ is an isomorphism, with inverse $X^U \times U \xrightarrow{(\mathrm{ev}, \pi_2)} X \times U$; so it follows from (i) that η is j^o_U-bidense. Hence X^U is the associated sheaf of X; and in particular η is iso if X is a sheaf.

(iii) Since $U \longrightarrow 1$ is mono, we have $U^U \cong 1$; so on applying the construction of Π_U given in Ex. 1.8, we have $\Pi_U(Z \longrightarrow U) \cong Z^U$. The result is thus immediate from (i) and (ii). ■

An alternative characterization of open topologies, due to M. Tierney, is provided by the following result:

3.55 PROPOSITION. *Let* j *be a topology in* $\mathscr{E}$. *The following conditions are equivalent:*

(i) j *is open.*
(ii) *The associated sheaf functor* $\mathscr{E} \xrightarrow{L} \mathrm{sh}_j(\mathscr{E})$ *is logical.*
(iii) $\mathscr{E} \xrightarrow{L} \mathrm{sh}_j(\mathscr{E})$ *preserves exponentials.*
(iv) *If* $X' \xrightarrow{\sigma} X$ *is* j*-dense, so is* $X'^Y \xrightarrow{\sigma^Y} X^Y$ *for all* Y.
(v) $\Pi_J(1 \xrightarrow{d} J)$ *is a* j*-dense subobject of* 1.
(vi) *The internal poset* $\mathbf{J}^{\mathrm{op}}$ *has a terminal object* (*Ex.* 2.10), *i.e. there exists* $1 \xrightarrow{u} J$ *such that* $J \xrightarrow{u \times 1} J \times J$ *factors through* $J_1 \rightarrowtail J \times J$.

Proof. (i) ⇒ (ii) is immediate from 3.54(iii) and 1.42; and (ii) ⇒ (iii) is trivial.

(iii) ⇒ (iv): If σ is j-dense, then $L(\sigma)$ is iso by 3.42, and hence so is

$LX'^{LY} \xrightarrow{L\sigma^{LY}} LX^{LY}$. But by (iii) this is isomorphic to $L(X'^{Y}) \xrightarrow{L(\sigma^{Y})} L(X^{Y})$; hence by 3.42 again σ^{Y} is j-dense.

(iv) ⇒ (v): From Ex. 1.8, we have a pullback square

$$\begin{array}{ccc} \Pi_J(d) & \longrightarrow & 1^J \; ; \\ \big\downarrow & & \big\downarrow{\scriptstyle d^J} \\ 1 & \xrightarrow{\ \bar{1}_J\ } & J^J \end{array}$$

but d^J is dense by (iv), and so $\Pi_J(d) \rightarrowtail 1$ is dense.

(v) ⇒ (vi): From the adjunction $(J^* \dashv \Pi_J)$, we have that $J^*\Pi_J(d) \leqslant d$ as subobjects of J; but the former is classified by $J \longrightarrow 1 \xrightarrow{u} J$, where u classifies $\Pi_J(d) \rightarrowtail 1$, and the latter by 1_J. So u is a terminal object of $\mathbf{J}^{\mathrm{op}}$.

(vi) ⇒ (i): Let $U \rightarrowtail 1$ be the subobject of 1 classified by u; then it is immediate from 3.54(i) that the j-dense monomorphisms are precisely the j_U^o-dense ones. ■

A problem which frequently arises is the following: given a subobject D of Ω in a topos $\mathscr{E}$, can we construct a topology j which is *generated* by D, in the sense that (i) all monos whose classifying maps factor through D are j-dense, and (ii) j is minimal with respect to (i)? We present here a particularly elegant solution to this problem, due to A. Joyal.

First we define, for each object X, a binary relation θ on the subobjects of X by saying that $\theta(\sigma, \tau)$ holds iff, for each $X' \overset{\rho}{\rightarrowtail} X$ such that $\sigma \cap \rho \leqslant \tau$, we actually have $\rho \leqslant \tau$. From the description of ⇒ given earlier, this is clearly equivalent to the statement that $(\sigma \Rightarrow \tau) \cong \tau$, and so θ is "represented" by a subobject $\Theta \rightarrowtail \Omega \times \Omega$, namely that classified by

$$\Omega \times \Omega \xrightarrow{1 \times \Delta} \Omega \times \Omega \times \Omega \xrightarrow{\Rightarrow \times 1} \Omega \times \Omega \xrightarrow{\delta} \Omega,$$

where δ classifies the diagonal $\Omega \overset{\Delta}{\rightarrowtail} \Omega \times \Omega$. Now given a subobject $D \rightarrowtail \Omega$, we define a new subobject $D^r \rightarrowtail \Omega$ to be

$$\Pi_{(\Omega \times \Omega \xrightarrow{\pi_2} \Omega)}\big((\pi_1^* D \Rightarrow \Theta) \rightarrowtail \Omega \times \Omega\big).$$

And we define $D^l \rightarrowtail \Omega$ by the corresponding formula with π_1 and π_2 interchanged.

3.56 LEMMA. *Let $X' \overset{\tau}{\rightarrowtail} X$ be a monomorphism in $\mathscr{E}$. Then τ is in Ξ_{D^r} (cf. 3.17)* *iff, for every* $Y \xrightarrow{\alpha} X$ *and every* $Y' \rightarrowtail^{\sigma} Y$ *in* Ξ_D, *the relation* $\theta(\sigma, \alpha^*\tau)$ *holds.* [*A similar criterion can be given for monomorphisms to be in* Ξ_{D^l}.]

Proof. Let $X \xrightarrow{\phi} \Omega$ classify τ, and let $D \overset{i}{\rightarrowtail} \Omega$ be the given monomorphism. Then from the adjunction $(\pi_2^* \dashv \Pi_{\pi_2})$, we have that ϕ factors through D^r iff $\Omega \times X \xrightarrow{1 \times \phi} \Omega \times \Omega$ factors through $(\pi_1^* D \Rightarrow \Theta)$. But from the definition of $\Rightarrow$, this is in turn equivalent to saying that $D \times X \xrightarrow{i \times \phi} \Omega \times \Omega$ factors through Θ. So if we take $\alpha = \pi_2 : D \times X \longrightarrow X$, and σ to be the subobject classified by $D \times X \xrightarrow{\pi_1} D \overset{i}{\rightarrowtail} \Omega$, we see at once that the given condition is sufficient. But it is also necessary, since any pair (α, σ) as given defines a Y-element $Y \xrightarrow{(\psi, \alpha)} D \times X$, where $i\psi$ is the classifying map of σ; and $\theta(\sigma, \alpha^*\tau)$ holds iff the composite $Y \xrightarrow{(\psi, \alpha)} D \times X \xrightarrow{i \times \phi} \Omega \times \Omega$ factors through Θ. ■

3.57 THEOREM (Joyal). (i) *The operations* $D \longmapsto D^r$ *and* $D \longmapsto D^l$ *form a* Galois connection *between the lattice of subobjects of* Ω *and itself; i.e. the operations are order-reversing, and we have* $D \leqslant D^{rl}$, $D \leqslant D^{lr}$.

(ii) *If* $\Omega \xrightarrow{j} \Omega$ *is a topology in* $\mathscr{E}$, *then* $J^r \cong \Omega_j$ *and* $(\Omega_j)^l \cong J$.

(iii) *A subobject* $D \rightarrowtail \Omega$ *is* left closed *for the connection (i.e.* $D^{rl} \cong D$*) iff the classifying map of* $D \rightarrowtail \Omega$ *is a topology in* $\mathscr{E}$.

Proof. (i) is immediate from 3.56 and its analogue, since we have $D \leqslant E$ iff $\Xi_D \subseteq \Xi_E$.

(ii) For any topology j, if σ is j-dense and τ is j-closed, then it follows easily from 3.19 that $\theta(\sigma, \tau)$ holds, since $\sigma \cap \rho$ is a dense subobject of ρ. Hence it is immediate that $J^r \geqslant \Omega_j$ and $(\Omega_j)^l \geqslant J$.

To prove equality, suppose that $X' \overset{\tau}{\rightarrowtail} X$ is in Ξ_{J^r}. Then if we apply 3.56 with $\alpha = \bar{\tau} : X' \rightarrowtail X$ and $\sigma = (X' \rightarrowtail \overline{X'})$, we obtain $\sigma \cong (\sigma \Rightarrow \sigma) \cong 1_{\overline{X'}}$. So τ is closed; hence $J^r \cong \Omega_j$. A similar argument using the analogue of 3.56 shows that if σ is in $\Xi_{(\Omega_j)^l}$, then $\bar{\sigma} \cong (\sigma \Rightarrow \bar{\sigma}) \cong 1_X$, and so $\sigma \in \Xi_J$.

(iii) If D is classified by a topology, the equality $D^{rl} \cong D$ follows at once from (ii). Suppose conversely that $D^{rl} \cong D$; then we need to verify the conditions of 3.18 for the class Ξ_D.

But if σ is an isomorphism, it is clear that $\theta(\sigma, \tau)$ holds for any τ; hence $\sigma \in \Xi_{E^l}$ for any E, and in particular $\sigma \in \Xi_{D^{rl}} = \Xi_D$. Now suppose we are given $X'' \overset{\sigma'}{\rightarrowtail} X' \overset{\sigma}{\rightarrowtail} X$ with $\sigma, \sigma' \in \Xi_D$. Then for any $Y \overset{\tau}{\rightarrowtail} X$ in Ξ_{D^r}, we have $\theta(\sigma, \tau)$ and $\theta(\sigma', \sigma^*\tau)$, from whence a simple diagram chase shows that $\theta(\sigma\sigma', \tau)$ holds. And this remains true upon pulling back along any $Z \xrightarrow{\alpha} X$, so $\sigma\sigma' \in \Xi_{D^{rl}} = \Xi_D$. A similar argument shows that if $\sigma\sigma' \in \Xi_D$, then σ and σ' are in Ξ_D. ■

3.58 COROLLARY. *Let $D \rightarrowtail \Omega$ be any subobject of Ω. Then there exists a topology j in $\mathscr{E}$ which is generated by D, in the sense that $D \rightarrowtail \Omega$ factors through $J \rightarrowtail \Omega$ and j is minimal with respect to this condition.*

Proof. Take $J = D^{rl}$ and apply 3.57(i) and (iii). ■

Using 3.58, we can construct a number of interesting examples of topologies, among them the following:

3.59 EXAMPLES. (i) Let j_1, j_2 be two topologies in $\mathscr{E}$. Then we can construct a topology $j_1 + j_2$ (the *join* of j_1 and j_2) which contains j_1 and j_2 and is minimal with respect to this condition, by applying 3.58 to the union of $J_1 \rightarrowtail \Omega$ and $J_2 \rightarrowtail \Omega$.

(ii) Let $X' \xrightarrow{\sigma} X$ be a monomorphism in $\mathscr{E}$. Then there is a unique smallest topology j such that σ is j-dense; to construct it, apply 3.58 to the image $I \rightarrowtail \Omega$ of the classifying map $X \rightarrowtail \Omega$ of σ.

(iii) Let $\mathscr{F} \xrightarrow{f} \mathscr{E}$ be a geometric morphism, j a topology in $\mathscr{E}$. Then there is a unique smallest topology j' in $\mathscr{F}$ such that the composite $\mathbf{sh}_{j'}(\mathscr{F}) \longrightarrow \mathscr{F} \xrightarrow{f} \mathscr{E}$ factors through $\mathbf{sh}_j(\mathscr{E}) \longrightarrow \mathscr{E}$. For by 3.42 and 3.47, such a factorization exists iff $f^*(1 \xrightarrow{d} J)$ is j'-dense; so we simply apply (ii) to this monomorphism in $\mathscr{F}$.

(iv) Let $X \xrightarrow{f} Y$ be a morphism of $\mathscr{E}$. Then we can construct the unique smallest topology in E for which $L(f)$ is mono (resp. epi), by applying (ii) to the monomorphism τ (resp. i) of 3.41. And by taking the join of these two topologies, we obtain the unique smallest topology for which $L(f)$ is iso.

(v) Let $X' \xrightarrow{\tau} X$ be a monomorphism in $\mathscr{E}$. By starting from the opposite side of the Galois connection of 3.57, we can form the unique largest topology for which τ is closed. For example, by taking τ to be the diagonal map of an object F, and applying 3.29(ii), we can construct the unique largest topology for which F is separated. ■

EXERCISES 3

1. If j_1 and j_2 are topologies in a topos $\mathscr{E}$, prove that $\Omega \xrightarrow{(j_1, j_2)} \Omega \times \Omega \xrightarrow{\wedge} \Omega$ is a topology.

2. Let $\mathscr{E}$ be a category with finite limits, and $\mathbb{L} = (L, \eta, \mu)$ a left exact idempotent monad on $\mathscr{E}$ (i.e. such that L is left exact and μ is an isomorphism).

Define an operation ρ_X on subobjects of X in $\mathscr{E}$ by $\rho_X(X' \rightarrowtail X) = \eta_X^*(LX' \rightarrowtail LX)$. Prove that ρ is a universal closure operation on $\mathscr{E}$. Prove also that X is separated (resp. a sheaf) for ρ iff η_X is mono (resp. iso) [hint: first prove that the diagonal $X \rightarrowtail R$ is ρ-dense, where $R \rightrightarrows X$ is the kernel-pair of η_X], and deduce that L is the associated sheaf functor for ρ. Deduce that if $\mathscr{E}$ is a topos, then there is a bijection between (isomorphism classes of) left exact idempotent monads on $\mathscr{E}$ and topologies in $\mathscr{E}$. [Use 3.42.]

3. Let G be an abelian group, H a subgroup of G. Define $\bar{H} \subseteq G$ to be the set of elements which are torsion modulo H, i.e. $\{g \in G \mid ng \in H$ for some $n \geqslant 1\}$. Prove that $\bar{H}$ is a subgroup of G, and that $H \mapsto \bar{H}$ defines a universal closure operation on the category **ab** of abelian groups. Show that G is separated for this operation iff it is torsion-free, and a sheaf iff it is torsion-free and divisible. What is the associated sheaf functor? By restricting this operation to the full subcategory **fgab** of finitely-generated abelian groups, show that 3.29(iii) and 3.42 are both false for a general universal closure operation.

4. (Lawvere's construction of L) Let j be a topology in a topos $\mathscr{E}$, X an object of $\mathscr{E}$. Let MX be the image of the composite $X \xrightarrow{\{\}} \Omega^X \xrightarrow{j^X} \Omega_j^X$, and let LX be the closure of MX in Ω_j^X. Prove that $X \twoheadrightarrow MX$ is universal among morphisms from X into separated objects [hint: what is its kernel-pair?], and that LX is the associated sheaf of X.

5. Let X be the discrete two-point topological space, and let U, V denote the two nontrivial open subsets of X. Show that for any presheaf P on X, $P^+(X)$ is the pullback of

$$\begin{array}{ccc} & & P(U) \\ & & \downarrow{\scriptstyle \rho^U_\varnothing} \\ P(V) & \xrightarrow{\rho^V_\varnothing} & P(\varnothing) \end{array}$$

Hence calculate P^+ and P^{++} when P is defined by $P(X) = \varnothing$, $P(U) = P(V) = P(\varnothing) =$ two-element set, with $\rho^U_\varnothing = \rho^V_\varnothing = 1$. Deduce that in general P, P^+, P^{++} and MP may all be distinct.

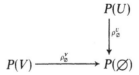

6. (Freyd's proof of left exactness of L; see [FK])

   (i) Prove directly that $L: \mathscr{E} \longrightarrow \mathrm{sh}_j(\mathscr{E})$ preserves finite products. [Hint: show that $X \times Y \longrightarrow LX \times LY$ is universal among maps from $X \times Y$ into sheaves, using 3.24.]

   (ii) If Y is a sheaf and X is any object of $\mathscr{E}$, prove that Y^X and Y^{LX} are naturally isomorphic.

   (iii) Prove that the inclusion $\mathrm{sh}_j(\mathscr{E}) \longrightarrow \mathscr{E}$ preserves injectives. [Use Ex. 1.5.]

   (iv) Use Ex. 1.5 to show that $\mathscr{E} \xrightarrow{L} \mathrm{sh}_j(\mathscr{E})$ preserves coreflexive equalizers.

7. Let j be a topology in $\mathscr{E}$, and suppose

$$\begin{array}{ccc} \Omega \times \Omega & \xrightarrow{\Rightarrow} & \Omega \\ {\scriptstyle 1 \times j}\downarrow & & \downarrow{\scriptstyle j} \\ \Omega \times \Omega & \xrightarrow{\Rightarrow} & \Omega \end{array}$$

commutes. Prove that the retraction $\Omega \twoheadrightarrow \Omega_j$ is bidense, and deduce that the associated sheaf functor for j preserves the subobject classifier. Verify that this condition is satisfied when j is an open topology.

8. Let F be an object of a topos $\mathscr{E}$. Let $\alpha: F \times \Omega \longrightarrow \tilde{F}$ represent the partial map

$$\begin{array}{ccc} F & \xrightarrow{1} & F \ , \\ {\scriptstyle 1 \times t}\downarrow & & \\ F \times \Omega & & \end{array}$$

and let $\tilde{F} \xrightarrow{\phi} \Omega$ classify $F \overset{\eta}{\rightarrowtail} \tilde{F}$. Define $W_F \rightarrowtail \tilde{F}$ to be "the largest subobject on which α is uniquely invertible", i.e. the pullback

$$\begin{array}{ccccc} W_F & \longrightarrow & & & F \times \Omega \ , \\ \downarrow & & & & \downarrow{\scriptstyle \{\}} \\ \tilde{F} & \xrightarrow{\{\}} & \Omega^{\tilde{F}} & \xrightarrow{\Omega^{\alpha}} & \Omega^{(F \times \Omega)} \end{array}$$

and define $J_F \rightarrowtail \Omega$ to be $\Pi_\phi(W_F \rightarrowtail \tilde{F})$. Show that a monomorphism

$X' \rightarrowtail X$ is in Ξ_{J_F} iff, for every $T \xrightarrow{f} X$, each morphism $f^*(X') \longrightarrow F$ extends uniquely to a morphism $T \longrightarrow F$. Deduce that the classifying map j_F of $J_F \rightarrowtail \Omega$ is a topology, and that it is the unique largest topology for which F is a sheaf.

9. Let j_1, j_2 be two topologies in $\mathscr{E}$. Using question 8, prove that $\text{sh}_{j_1+j_2}(\mathscr{E})$ is the intersection of $\text{sh}_{j_1}(\mathscr{E})$ and $\text{sh}_{j_2}(\mathscr{E})$.

10. Let j be a topology in $\mathscr{E}$. Define the *exterior* of j, ext(j), to be the j-closure of $0 \rightarrowtail 1$, and the *interior* of j, int(j), to be the equalizer of $1 \xrightarrow{\bar{j}} \Omega^\Omega$ and $1 \xrightarrow{\overline{1_\Omega}} \Omega^\Omega$. Show that $j \longmapsto \text{ext}(j)$ is an order-preserving map from the lattice of topologies in $\mathscr{E}$ to the lattice of subobjects of 1, and that it is right adjoint to the map $U \longmapsto j_U^c$. [Hint: if $U \leqslant \text{ext}(j)$ and $X' \overset{\sigma}{\rightarrowtail} X$ is j_U^c-dense, prove that

$$\begin{array}{ccc} X' \times \text{ext}(j) & \xrightarrow{\pi_1} & X' \\ \Big\downarrow{\scriptstyle \sigma \times 1} & & \Big\downarrow{\scriptstyle \sigma} \\ X \times \text{ext}(j) & \xrightarrow{\pi_1} & X \end{array}$$

is a pushout.] Establish a similar (contravariant) adjunction between the maps $j \longmapsto \text{int}(j)$ and $U \longmapsto j_U^o$.

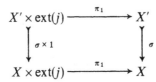

Chapter 4

Geometric Morphisms

4.1. THE FACTORIZATION THEOREM

This chapter is devoted to a more detailed study of the 2-category $\mathfrak{Top}$ introduced in 1.16.

4.11 Definition. Let $\mathscr{F} \xrightarrow{f} \mathscr{E}$ be a geometric morphism.

(i) We say f is an *inclusion* if the following equivalent conditions are satisfied:
   (a) f_* is full and faithful.
   (b) The counit of the adjunction $(f^* \dashv f_*)$ is an isomorphism
   (c) (if f is essential) The unit of $(f_! \dashv f^*)$ is an isomorphism, or $f_!$ is full and faithful.

(ii) We say f is a *surjection* if the following equivalent conditions are satisfied:
   (a) f^* reflects isomorphisms.
   (b) f^* is faithful.
   (c) The unit of $(f^* \dashv f_*)$ is mono. ■

4.12 Examples. (i) Let j be a topology in $\mathscr{E}$. Then the geometric morphism $\mathrm{sh}_j(\mathscr{E}) \xrightarrow{i} \mathscr{E}$ of 3.39 is an inclusion.

(ii) Let $\mathbb{G}$ be a left exact comonad on $\mathscr{F}$. Then the geometric morphism $\mathscr{F} \longrightarrow \mathscr{F}_{\mathbb{G}}$ of 2.32 is a surjection.

[We shall see shortly that examples (i) and (ii) are the most general possible.]

(iii) Let $X \xrightarrow{f} Y$ be a morphism of $\mathscr{E}$. Then the functor $\Sigma_f : \mathscr{E}/X \longrightarrow \mathscr{E}/Y$ is always faithful, and it is full iff f is mono. So we deduce from 4.11(i)(c) that the geometric morphism $\mathscr{E}/X \longrightarrow \mathscr{E}/Y$ of 1.46 is an inclusion iff f is mono.

Similarly, this morphism is a surjection iff f is epi. For if $I \rightarrowtail Y$ is the image of f, $f^*(I \rightarrowtail Y)$ is an isomorphism in $\mathscr{E}/X$; hence if f^* reflects isomorphisms f must be epi. Conversely, if f is epi and

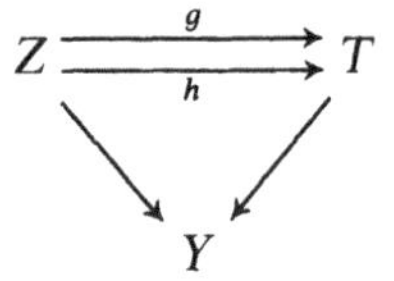

are two morphisms of $\mathscr{E}/Y$ with the same image under f^*, then the composites $X \times_Y Z \xrightarrow{\pi_2} Z \underset{h}{\overset{g}{\rightrightarrows}} T$ are equal; hence g and h are equal since π_2, being the pullback of an epi, is epi.

(iv) Let $X \xrightarrow{f} Y$ be a continuous map of topological spaces. Then arguments similar to those of (iii) show that the geometric morphism $\mathrm{Shv}(X) \longrightarrow \mathrm{Shv}(Y)$ of 1.17(i) is an inclusion (resp. a surjection) if f is the inclusion of a subspace (resp. surjective); and the converse implications hold if we impose the condition that X is a T_0-space (resp. Y is a T_1-space). ■

4.13 Lemma. *Let $\mathscr{F} \xrightarrow{f} \mathscr{E}$ be a geometric morphism. If f is both an inclusion and a surjection then (f_*, f^*) is an equivalence of categories (and hence f is an equivalence in $\mathfrak{Top}$).*

Proof. Let α, β be the unit and counit of $(f^* \dashv f_*)$. By 4.11(i)(b) we know that β is iso, so it suffices to prove that α is iso. But the "triangular identity"

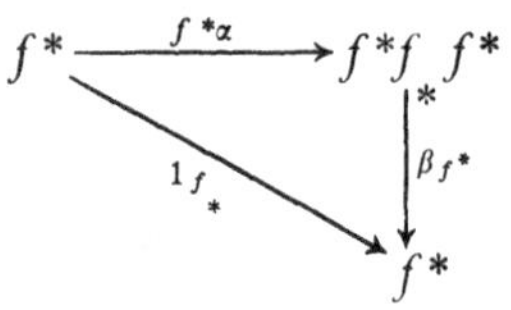

implies that $f^*\alpha$ is iso; and f^* reflects isos. ■

4.14 Theorem (Lawvere–Tierney). *Let $\mathscr{F} \xrightarrow{f} \mathscr{E}$ be a geometric morphism. Then there is a factorization*

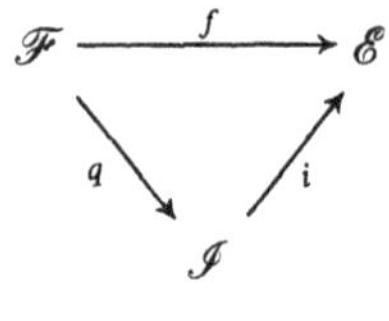

(up to natural isomorphism) of f, such that q is a surjection and i is an inclusion; and this factorization is unique up to equivalence.

Proof. Define $j : \Omega \longrightarrow \Omega$ in $\mathscr{E}$ to be the composite

$$\Omega \xrightarrow{\alpha} f_* f^* \Omega \xrightarrow{\phi} \Omega,$$

where ϕ classifies $f_* f^*(t)$: $1 \cong f_* f^*(1) \rightarrowtail f_* f^*(\Omega)$ and α is the unit of $(f^* \dashv f_*)$. Let $J \rightarrowtail \Omega$ be the subobject classified by j.

Now if $X' \overset{\sigma}{\rightarrowtail} X$ is a monomorphism in $\mathscr{E}$ with classifying map $X \xrightarrow{\theta} \Omega$, then

$$\begin{aligned} \theta \text{ factors through } J \rightarrowtail \Omega &\Leftrightarrow \alpha\theta \text{ factors through } f_* f^*(t) \\ &\Leftrightarrow f^*(\theta) \text{ factors through } f^*(t) \\ &\Leftrightarrow f^*(\sigma) \text{ is an isomorphism.} \end{aligned}$$

It follows easily from this that the class Ξ_J satisfies the conditions of 3.18, and so j is a topology in $\mathscr{E}$. Moreover, if we define $\mathscr{J} = \mathrm{sh}_j(\mathscr{E})$, then by 3.47 we have a factorization $\mathscr{F} \xrightarrow{q} \mathscr{J}$ of f through the inclusion $\mathscr{J} \xrightarrow{i} \mathscr{E}$.

Now suppose γ is a morphism of $\mathscr{J}$ such that $q^*(\gamma)$ is iso. Then $f^* i_*(\gamma)$ is iso, so $i_*(\gamma)$ is j-bidense; hence by 3.46 γ is iso. So q is a surjection.

Finally, suppose $\mathscr{F} \xrightarrow{r} \mathscr{L} \xrightarrow{l} \mathscr{E}$ is any other factorization of f into surjection and inclusion. Since r is surjective, l^* inverts all j-dense monomorphisms, and so we have a factorization

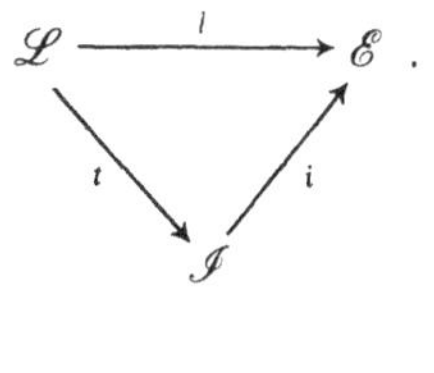

But now t must be both a surjection and an inclusion; hence by 4.13 it is an equivalence. ■

4.15 Proposition. (i) *Let $\mathscr{F} \xrightarrow{f} \mathscr{E}$ be an inclusion. Then there is a (unique) topology j in $\mathscr{E}$ such that $\mathscr{F} \simeq \mathrm{sh}_j(\mathscr{E})$.*

(ii) *Let $\mathscr{F} \xrightarrow{f} \mathscr{E}$ be a surjection. Then there is a comonad $\mathbb{G}$ on $\mathscr{F}$ (unique up to isomorphism) such that $\mathscr{E} \simeq \mathscr{F}_{\mathbb{G}}$.*

Proof. (i) is immediate from the construction of the image topos $\mathcal{I}$ in 4.14.

(ii) The functor f^* has a right adjoint, preserves equalizers and reflects isomorphisms; so by the opposite of 0.13 it is comonadic. ■

4.16 REMARK. It should be noted that, in the diagram of 4.14, the comonads on $\mathcal{F}$ induced by $(f^* \dashv f_*)$ and $(q^* \dashv q_*)$ are isomorphic, since that on $\mathcal{I}$ induced by $(i^* \dashv i_*)$ is trivial. In fact an alternative proof of 4.14 can be given using 2.32; we define the image topos $\mathcal{I}$ to be $\mathcal{F}_{\mathbb{G}}$, where $\mathbb{G}$ is the comonad induced by $(f^* \dashv f_*)$, and then use a more precise form of Beck's Tripleability Theorem (see [CW], p. 150, Exercise 2) to prove that the comparison functor $\mathcal{E} \longrightarrow \mathcal{I}$ has a full and faithful right adjoint. ■

4.17 REMARK. It is also worth noting that an alternative proof of 4.15(i) can be given using the ideas of Exercise 3.2, since the monad on $\mathcal{E}$ induced by $(f^* \dashv f_*)$ is idempotent by 4.11(i)(b). Indeed, in the more general situation of 4.14, the monad on $\mathcal{E}$ induced by $(i^* \dashv i_*)$ is the "associated idempotent monad" (in the sense of Fakir [163]) of the monad induced by $(f^* \dashv f_*)$; this idea can be used to give yet a third construction of the image topos. ■

4.18 EXAMPLES. (i) Let $X \xrightarrow{f} Y$ be a morphism in a topos $\mathcal{E}$. Then it follows easily from 4.12(iii) that the image topos $\mathcal{I}$ of $\mathcal{E}/X \longrightarrow \mathcal{E}/Y$ is (equivalent to) $\mathcal{E}/Q$, where Q is the image of f (1.52).

(ii) Similarly, if $X \xrightarrow{f} Y$ is a continuous map of topological spaces, then the image topos of $\mathrm{Shv}(X) \longrightarrow \mathrm{Shv}(Y)$ is $\mathrm{Shv}(I)$, where I is the image of f topologized as a subspace of Y.

(iii) Let $(X, \mathbb{U})$ be a topological space. Then it is not hard to check that we have a geometric morphism $\mathcal{S}/X \xrightarrow{f} \mathcal{S}^{\mathbb{U}^{\mathrm{op}}}$, given by $f_*(E \xrightarrow{p} X) =$ (pre)sheaf of all (not necessarily continuous) sections of p, and $f^*(P) =$ indexed family of stalks of P. (Compare 0.24).

Now the image of this morphism is just the topos $\mathrm{Shv}(X)$; and the two descriptions of the image topos given in 4.14 and 4.16 correspond to the two "classical" descriptions of sheaves on a space (i.e. as presheaves satisfying an exactness condition, or as sets over X equipped with a coalgebra structure (= a topology such that the projection is a local homeomorphism)). ■

As a corollary of 4.14, we may extend the universal properties of the "generated topologies" constructed in 3.59 from sheaf subtoposes of $\mathcal{E}$ to arbitrary toposes over $\mathscr{E}$. Specifically, we have

4.19 COROLLARY. *Let* $\mathscr{F} \xrightarrow{f} \mathscr{E}$ *be a geometric morphism,* $X' \overset{\sigma}{\rightarrowtail} X$ *a monomorphism in E, and j the topology in* $\mathscr{E}$ *generated by* σ *as in* 3.59(ii). *Then f factors through* $\mathrm{sh}_j(\mathscr{E}) \longrightarrow \mathscr{E}$ *iff* $f^*(\sigma)$ *is an isomorphism.*

Proof. Let k be the topology in $\mathscr{E}$ induced by f as in 4.14. Then each of the given conditions is equivalent to saying that $j \leqslant k$ in the lattice of topologies in $\mathscr{E}$. ■

4.2 THE GLUEING CONSTRUCTION

4.21 PROPOSITION. *The 2-category* $\mathfrak{Top}$ *has an initial object (namely the* degenerate topos **1** *with one object and one (identity) morphism), and it has finite coproducts (the underlying category of the coproduct of* $\mathscr{E}_1$ *and* $\mathscr{E}_2$ *is actually their product in* $\mathfrak{Cat}$; *cf.* 1.15).

Proof. Straightforward. ■

It is natural to ask whether more general finite colimits exist in $\mathfrak{Top}$. In general, the answer to this question seems to be unknown; but we do have the weaker notion of lax colimit, in which diagrams commuting up to a 2-isomorphism are replaced by diagrams commuting up to a 2-arrow which may not be invertible. In this paragraph we introduce a general "glueing construction", due to G. C. Wraith [129], which will enable us to construct lax colimits; but first we must digress to introduce the necessary 2-categorical terminology.

4.22 DEFINITION. Let $\mathfrak{C}$ be a 2-category, **D** a category.

(i) A *lax diagram* $\mathbf{D} \xrightarrow{(\Gamma, \alpha)} \mathfrak{C}$ consists of the following data:
 (a) for each object d of **D**, an object $\Gamma(d)$ of $\mathfrak{C}$,
 (b) for each morphism $d' \xrightarrow{f} d$ of **D**, a 1-arrow $\Gamma(f)\colon \Gamma(d') \longrightarrow \Gamma(d)$,
 (c) for each object d, a 2-arrow $\alpha_d\colon 1_{\Gamma(d)} \longrightarrow \Gamma(1_d)$,
 (d) for each composable pair $d'' \xrightarrow{f'} d' \xrightarrow{f} d$, a 2-arrow $\alpha_{f,f'}\colon \Gamma(f)\Gamma(f') \longrightarrow \Gamma(ff')$,

(e) satisfying the coherence conditions that

$$\begin{array}{ccccc}
\Gamma(f)\Gamma(1_d) & \xleftarrow{\Gamma(f) * \alpha_d} & \Gamma(f) & \xrightarrow{\alpha_{d'} * \Gamma(f)} & \Gamma(1_{d'})\Gamma(f) \\
& \searrow^{\alpha_{f,1_d}} & \downarrow{\scriptstyle 1_{\Gamma(f)}} & \swarrow_{\alpha_{1_{d'},f}} & \\
& & \Gamma(f) & &
\end{array}$$

and

$$\begin{array}{ccc}
\Gamma(f)\Gamma(f')\Gamma(f'') & \xrightarrow{\Gamma(f) * \alpha_{f',f''}} & \Gamma(f)\Gamma(f'f'') \\
\downarrow{\scriptstyle \alpha_{f,f'} * \Gamma(f'')} & & \downarrow{\scriptstyle \alpha_{f,f'f''}} \\
\Gamma(ff')\Gamma(f'') & \xrightarrow{\alpha_{ff',f''}} & \Gamma(ff'f'')
\end{array}$$

commute.

(ii) A *lax cone* (X, s, σ) under a lax diagram (Γ, α) consists of
   (a) an object X of $\mathfrak{C}$,
   (b) for each object d of $\mathbf{D}$, a 1-arrow $s_d\colon \Gamma(d) \longrightarrow X$,
   (c) for each morphism $d' \xrightarrow{f} d$ of $\mathbf{D}$, a 2-arrow $\sigma_f\colon s_d * \Gamma(f) \longrightarrow s_{d'}$,
   (d) satisfying the coherence conditions $\sigma_{1_d} . (s_d * \alpha_d) = 1_{s_d}$ and $\sigma_{f'} . (\sigma_f * \Gamma(f')) = \sigma_{ff'} . (s_d * \alpha_{f,f'})$.

   In the terminology of [184], a lax cone is a right lax natural transformation from (Γ, α) to a constant diagram. A *morphism of lax cones* $\gamma\colon (X, s, \sigma) \longrightarrow (X, t, \tau)$ is a modification of natural transformations, i.e. a family of 2-arrows $\gamma_d\colon s_d \longrightarrow t_d$ which are compatible with the σ's and τ's. We write $\mathrm{lcn}(\Gamma, \alpha; X)$ for the category of lax cones under (Γ, α) with vertex X.

(iii) A *lax colimit* of a lax diagram (Γ, α) is a representing object for the functor $\mathrm{lcn}(\Gamma, \alpha; -)\colon \mathfrak{C} \longrightarrow \mathfrak{Cat}$; i.e. it is a lax cone (L, t, τ) such that for any X, the functor $\mathfrak{C}(L, X) \longrightarrow \mathrm{lcn}(\Gamma, \alpha; X)$; $h \longmapsto (X, ht, h * \tau)$ is an equivalence of categories. ■

4.23 Example (Bénabou [155]). Let $\mathscr{E}$ be a category. Then specifying a monad $\mathbb{T}$ on $\mathscr{E}$ is precisely equivalent to specifying a lax diagram $\mathbf{1} \longrightarrow \mathfrak{Cat}$ which sends the unique object of $\mathbf{1}$ to $\mathscr{E}$. Moreover, the universal property of the Kleisli category $Kl(\mathbb{T})$ says that it is a lax colimit for this lax diagram. (The Eilenberg–Moore category $\mathscr{E}^{\mathbb{T}}$ is a lax limit for the same diagram.) ■

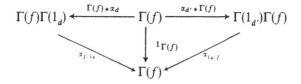

4.24 Construction. Let $\mathfrak{C}$ denote the 2-category whose objects and 1-arrows are toposes and left exact functors, and whose 2-arrows are the *opposites* of natural transformations. (If α is a 2-arrow of $\mathfrak{C}$, we shall write $\bar{\alpha}$ for the corresponding natural transformation in the opposite direction.) Let (Γ, α) be a lax diagram in $\mathfrak{C}$ over a finite category **D**. Define $\mathscr{E} = \prod_{d \in D_0} \Gamma(d)$; then on $\mathscr{E}$ we have a left exact comonad $\mathbb{G} = (G, \varepsilon, \delta)$ defined by

$$G(X_d \mid d \in D_0) = \Big(\prod_{d' \xrightarrow{f} d} \Gamma(f)(X_{d'}) \,\Big|\, d \in D_0 \Big),$$

$$(\varepsilon_X)_d : \prod_f \Gamma(f)(X_{d'}) \longrightarrow \Gamma(1_d)(X_d) \xrightarrow{(\bar{\alpha}_d)_{X_d}} X_d,$$

and

$$\begin{array}{ccc} (\delta_X)_d : \prod_f \Gamma(f)(X_{d'}) & \longrightarrow & \prod_{(f,f')} \Gamma(f)\Gamma(f')(X_{d''}) \\ \downarrow \pi_{ff'} & & \downarrow \pi_{(f,f')} \\ \Gamma(ff')(X_{d''}) & \xrightarrow{(\bar{\alpha}_{f,f'})_{X_{d''}}} & \Gamma(f)\Gamma(f')(X_{d''}) \end{array}$$

Since the $\Gamma(f)$ are left exact functors, so is G; and it is easily verified from 4.22(i)(e) that δ is coassociative and ε is a counit for it. So by 2.32 we have a topos $\mathscr{E}_{\mathbb{G}} = \mathrm{Gl}(\Gamma, \alpha)$; we call it the *glueing* of the lax diagram (Γ, α). An object of $\mathrm{Gl}(\Gamma, \alpha)$ may thus be described as a family of objects $X_d \in \Gamma(d)(d \in D_0)$, together with morphisms $\xi_f : X_d \longrightarrow \Gamma(f)(X_{d'})$ for each $d' \xrightarrow{f} d$, satisfying coherence conditions which relate the ξ's to the α's. ■

4.25 Theorem (Wraith). *The 2-category* $\mathfrak{Top}$ *has finite lax colimits.*

Proof. Let (Γ, α) be a finite lax diagram in $\mathfrak{Top}$. Compose with the forgetful functor $\mathfrak{Top} \longrightarrow \mathfrak{C}$ which sends a geometric morphism to its direct image, and apply the construction of 4.24. Since the functors $\Gamma(f)_*$ have left adjoints, so does the functor G constructed in 4.24; specifically, its left adjoint H is given by

$$H(X_d \mid d \in D_0) = \Big(\coprod_{d' \xrightarrow{f} d} \Gamma(f)^*(X_d) \,\Big|\, d' \in D_0 \Big).$$

So by the opposite of 0.14 H has a monad structure $\mathbb{H}$, and we can regard the objects of $\mathrm{Gl}(\Gamma, \alpha)$ as $\mathbb{H}$-algebras rather than $\mathbb{G}$-coalgebras—i.e. we may replace the structure morphisms $\xi_f: X_d \longrightarrow \Gamma(f)_*(X_{d'})$ with their transposes $\bar{\xi}_f: \Gamma(f)^*(X_d) \longrightarrow X_{d'}$.

Now let t_d be the geometric morphism $\Gamma(d) \xrightarrow{v_d} \mathscr{E} \longrightarrow \mathscr{E}_{\mathbb{G}} = \mathrm{Gl}(\Gamma, \alpha)$, where v_d is the coproduct inclusion in $\mathfrak{Top}$, i.e. the morphism of 1.17(iii); and let $\tau_f: \Gamma(f)^* t_d^* \longrightarrow t_{d'}^*$ be the natural transformation whose value at the object (X, ξ) is $\bar{\xi}_f$. Then $(\mathrm{Gl}(\Gamma, \alpha), t, \tau)$ clearly forms a lax cone under (Γ, α).

But given any lax cone $(\mathscr{F}, s, \sigma)$ under (Γ, α) in $\mathfrak{Top}$, we may define a geometric morphism $h: \mathrm{Gl}(\Gamma, \alpha) \longrightarrow \mathscr{F}$ by

$$h^*(Y) = (X, \xi) \quad \text{where} \quad X_d = s_d^*(Y) \text{ and } \bar{\xi}_f = (\sigma_f)_Y,$$

and

$$h_*(X, \xi) = \mathrm{eq}\Big(\prod_d s_{d*}(X_d) \overset{a}{\underset{b}{\rightrightarrows}} \prod_f s_{d*}\Gamma(f)_*(X_{d'})\Big),$$

where a and b are induced by the composites

$$\Pi s_{d*}(X_d) \xrightarrow{\pi_d} s_{d*}(X_d) \xrightarrow{s_{d*}(\xi_f)} s_{d*}\Gamma(f)_*(X_{d'})$$

and

$$\Pi s_{d*}(X_d) \xrightarrow{\pi_{d'}} s_{d'*}(X_{d'}) \xrightarrow{\overline{(\sigma_f)}_{X_{d'}}} s_{d*}\Gamma(f)_*(X_{d'})$$

respectively.

It is now straightforward to verify that h^* is left exact and left adjoint to h_*; and we clearly have a canonical isomorphism $(\mathscr{F}, s, \sigma) \cong (\mathscr{F}, ht, h * \tau)$ in $\mathrm{lcn}(\Gamma, \alpha; \mathscr{F})$ (in fact an equality on inverse image functors). So the assignment $(\mathscr{F}, s, \sigma) \longmapsto h$ defines an inverse (up to natural isomorphism) for the functor $\mathfrak{Top}(\mathrm{Gl}(\Gamma, \alpha), \mathscr{F}) \longrightarrow \mathrm{lcn}(\Gamma, \alpha; \mathscr{F})$ of 4.22(iii). ■

In certain cases, the glueing construction can be used to construct actual colimits (in the sense of 2-categories) in $\mathfrak{Top}$. An important example is the following:

4.26 PROPOSITION. *Pushouts of pairs of inclusions exist in* $\mathfrak{Top}$.

Proof. Let

$$\begin{array}{ccc} \mathscr{E}_1 & \xrightarrow{u} & \mathscr{E}_2 \\ \downarrow{\scriptstyle v} & & \\ \mathscr{E}_3 & & \end{array}$$

be a diagram of inclusions in $\mathfrak{Top}$. Apply the glueing construction to the lax diagram constructed as follows: **D** is the category $(d \underset{g}{\overset{f}{\rightleftarrows}} e)$, i.e. the indiscrete category (or trivial connected groupoid) with two objects d and e. $\Gamma(d) = \mathscr{E}_2$, $\Gamma(e) = \mathscr{E}_3$, $\Gamma(f) = v_*u^*$, $\Gamma(g) = u_*v^*$, $\Gamma(1_d) = 1_{\mathscr{E}_2}$, $\Gamma(1_e) = 1_{\mathscr{E}_3}$, $\alpha_d = \alpha_e = 1$, $\overline{\alpha_{f,g}}$ is the composite $1 \xrightarrow{\eta} v_*v^* \xrightarrow{\cong} v_*u^*u_*v^*$ (η = unit of $(v^* \dashv v_*)$), and $\alpha_{g,f}$ is similarly defined.

Then an object of $\mathrm{Gl}(\Gamma, \alpha)$ consists of a quadruple (X, Y, θ, ϕ) where $X \in \mathscr{E}_2$, $Y \in \mathscr{E}_3$ and $\theta: X \longrightarrow u_*v^*Y$, $\phi: Y \longrightarrow v_*u^*X$ satisfy appropriate coherence conditions. But these conditions simply say that $\bar{\theta}: u^*X \longrightarrow v^*Y$ and $\bar{\phi}: v^*Y \longrightarrow u^*X$ are mutually inverse isomorphisms in $\mathscr{E}_1$. It follows that $\mathrm{Gl}(\Gamma, \alpha)$ is the *limit* in $\mathfrak{Cat}$ of the diagram

$$\begin{array}{ccc} & & \mathscr{E}_2 \\ & & \downarrow{\scriptstyle u^*} \\ \mathscr{E}_3 & \xrightarrow{v^*} & \mathscr{E}_1 \end{array} ;$$

and from this we may verify by methods similar to those of 4.25 that it is a colimit in $\mathfrak{Top}$. ■

The fact that, in both 4.21 and 4.26, the underlying categories of colimits in $\mathfrak{Top}$ are actually limits in $\mathfrak{Cat}$ may be explained, at least for toposes defined over $\mathscr{S}$, by the fact that the forgetful functor $(\mathfrak{Top}/\mathscr{S})^{\mathrm{op}} \longrightarrow \mathfrak{Cat}$ which sends a geometric morphism to its inverse image is representable by the topos $\mathscr{S}[U]$ (see 4.37(iv) below). To construct the colimit of a more general diagram in $\mathfrak{Top}$, it should therefore be sufficient to show that the corresponding limit in $\mathfrak{Cat}$ has a topos structure; but this has not yet been done in general.

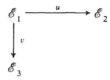

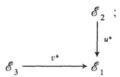

Another application of the construction of 4.24 concerns the problem of "Artin glueing" (see [GV], IV 9.5). Given a topos $\mathscr{E}$ and an open object U in $\mathscr{E}$, we have two sheaf subtoposes $\mathscr{E}_o$, $\mathscr{E}_c$ of E determined by the open and closed topologies j^o_U, j^c_U (3.52). What additional information do we need to reconstruct the topos $\mathcal{E}$ from $\mathcal{E}_o$ and $\mathcal{E}_c$? Recalling the topological analogy, when E corresponds to a space X and $\mathcal{E}_o$, $\mathcal{E}_c$ to complementary open and closed subspaces of X, it is clear that we need some information on how the "frontier" of $\mathcal{E}_o$ sits in $\mathcal{E}_c$. If we let $\mathcal{E}_o \xrightarrow{i_o} \mathcal{E}$, $\mathcal{E}_c \xrightarrow{i_c} \mathcal{E}$ be the inclusion morphisms, then it is easily verified that the composite $i_o^* i_{c*} : \mathcal{E}_c \longrightarrow \mathcal{E}_o$ is the trivial functor which sends every object of $\mathcal{E}_c$ to the terminal object. But the composite $i_c^* i_{o*}$ may be nontrivial; we call it the *fringe functor* associated with the open object U.

4.27 Theorem (Artin-Wraith). *Let $\mathcal{E}_1$, $\mathcal{E}_2$ be toposes and $L: \mathcal{E}_1 \longrightarrow \mathcal{E}_2$ any left exact functor. Then there exists a topos $\mathcal{E}$ (unique up to equivalence) containing an open object U, such that $\mathcal{E}_1$ and $\mathcal{E}_2$ are (equivalent to) the corresponding open and closed sheaf toposes, and L is the fringe functor.*

Proof. Apply the glueing construction to the diagram $(\mathcal{E}_1 \xrightarrow{L} \mathcal{E}_2)$ in $\mathfrak{C}$, regarded as a lax diagram in which the α's are identities. Then an object of $\mathcal{E} = \mathrm{Gl}(\Gamma, \alpha)$ consists of a triple (X_1, X_2, u) where $X_1 \in \mathcal{E}_1$, $X_2 \in \mathcal{E}_2$ and $u: X_2 \longrightarrow L(X_1)$ (the coherence conditions being vacuous); thus we may describe $\mathcal{E}$ as the comma category $(\mathcal{E}_2 \downarrow L)$. The open object U is $(1, 0, 0 \longrightarrow 1)$; it is easily verified from 3.53 and 3.54 that the j^c_U-sheaves are those objects (X_1, X_2, u) for which $X_1 \cong 1$ in $\mathcal{E}_1$, and the j^o_U-sheaves are those for which u is an isomorphism in $\mathcal{E}_2$. The remaining details are straightforward. ■

4.28 Example. Let $X \xrightarrow{f} Y$ be a continuous map of topological spaces, and let $S = \{a, b\}$ denote the Sierpinski space, where a is the open point. Define the *open mapping cylinder* of f to be the quotient space M^o_f obtained from $(X \times S) \amalg Y$ by identifying (x, b) with $f(x)$ for every $x \in X$. Then the topos $\mathrm{Shv}(M^o_f)$ may be obtained by applying the construction of 4.27 to the left exact functor $\mathrm{Shv}(X) \xrightarrow{f_*} \mathrm{Shv}(Y)$. Similarly, if we apply 4.27 to $\mathrm{Shv}(Y) \xrightarrow{f^*} \mathrm{Shv}(X)$, we obtain the topos of sheaves on the *closed mapping cylinder* M^c_f, which is defined similarly to M^o_f but with the orientation of S reversed.

4.3. DIACONESCU'S THEOREM

A most important tool in describing the structure of $\mathfrak{Top}$ is the theorem of R. Diaconescu [30] characterizing geometric morphisms whose codomain has the form $\mathcal{E}^{\mathbf{C}}$ for some $\mathbf{C} \in \mathbf{cat}(\mathcal{E})$. Throughout this and the next paragraph, we shall be working with toposes *defined over* a fixed "base topos" $\mathscr{E}$ (i.e. objects of $\mathfrak{Top}/\mathscr{E}$); we shall also use the term *$\mathscr{E}$-topos* for such an object.

4.31 DEFINITION. Let $\mathbf{C}$ be an internal category in $\mathscr{E}$, G an internal presheaf on $\mathbf{C}$, and $\mathbf{G} \xrightarrow{\gamma} \mathbf{C}$ the corresponding discrete fibration. We say that G is *flat* if $\mathbf{G}$ is a filtered category; and we write Flat($\mathbf{C}^{\mathrm{op}}, \mathscr{E}$) for the full subcategory of $\mathscr{E}^{\mathbf{C}^{\mathrm{op}}}$ whose objects are flat presheaves on $\mathbf{C}$. ■

4.32 LEMMA. *If G is a flat presheaf on $\mathbf{C}$, then the functor $(-)\otimes_{\mathbf{C}} G : \mathscr{E}^{\mathbf{C}} \longrightarrow \mathscr{E}$ is the inverse image of a geometric morphism.*

Proof. By 2.49 we know $(-)\otimes_{\mathbf{C}} G$ has a right adjoint; and by 2.48 we can write it as the composite $\mathscr{E}^{\mathbf{C}} \xrightarrow{\gamma^*} \mathscr{E}^{\mathbf{G}} \xrightarrow{\varinjlim_{\mathbf{G}}} \mathscr{E}$. Now γ^* is left exact since it has a left adjoint, by 2.34; and $\varinjlim_{\mathbf{G}}$ is left exact by 2.58. ■

4.33 EXAMPLE. Let $\mathbf{C}$ be a small category, and $G : \mathbf{C}^{\mathrm{op}} \longrightarrow \mathscr{S}$ a functor. If the presheaf corresponding to G is flat, then the functor G is left exact, since we may easily verify that it can be written as the composite $\mathbf{C}^{\mathrm{op}} \xrightarrow{h} \mathscr{S}^{\mathbf{C}} \xrightarrow{(-)\otimes G} \mathscr{S}$, where h is the Yoneda embedding. Conversely, if $\mathbf{C}$ has all finite colimits then any left exact functor $G : \mathbf{C}^{\mathrm{op}} \longrightarrow \mathscr{S}$ determines a flat presheaf. For example, let $x \in G(U)$, $y \in G(V)$ be two elements of $G_0 = \coprod_{U \in C_0} G(U)$; then both can be mapped in the category $\mathbf{G}$ into the element $\langle x, y\rangle$ of $G(U \amalg V) \cong G(U) \times G(V)$, so condition 2.51(b) is verified. (The arguments for (a) and (c) are similar.) ■

4.34 THEOREM (Diaconescu). *Let $\mathbf{C}$ be an internal category in $\mathscr{E}$, and let $\mathscr{F} \xrightarrow{f} \mathscr{E}$ be an $\mathscr{E}$-topos. Then there is an equivalence between the category $\mathfrak{Top}/\mathscr{E}(\mathscr{F}, \mathscr{E}^{\mathbf{C}})$ of geometric morphisms over $\mathscr{E}$ from $\mathscr{F}$ to $\mathscr{E}^{\mathbf{C}}$, and the category* Flat $(f^*\mathbf{C}^{\mathrm{op}}, \mathscr{F})$; *and this equivalence is natural in $\mathscr{F}$.*

Proof. (i) Consider first the case $\mathscr{F} = \mathscr{E}^{\mathbf{C}}$. Corresponding to the identity geometric morphism, we take the Yoneda profunctor $Y(\mathbf{C})$ (2.45), regarded as a presheaf on $\mathbf{C}^*(\mathbf{C})$ in $\mathscr{E}^{\mathbf{C}}$. To verify that $Y(\mathbf{C})$ is flat, observe that since the forgetful functor $U : \mathscr{E}^{\mathbf{C}} \longrightarrow \mathscr{E}/C_0$ preserves and reflects limits and epimorphisms, it is sufficient to verify that the total category of the discrete fibration

$$\begin{array}{ccc}
C_2 & \overset{m}{\underset{\pi_2}{\rightrightarrows}} & C_1 \\
\Big\downarrow{\scriptstyle (\pi_1, d_1\pi_2)} & & \Big\downarrow{\scriptstyle (d_0, d_1)} \\
C_1 \times C_0 & \overset{d_0 \times 1}{\underset{d_1 \times 1}{\rightrightarrows}} & C_0 \times C_0 \\
& {\scriptstyle \pi_2}\searrow \quad \swarrow{\scriptstyle \pi_2} & \\
& C_0 &
\end{array}$$

is filtered in $\mathscr{E}/C_0$. But the diagram

$$\begin{array}{ccc}
C_1 & \xrightarrow{d_1} & C_0 \\
\Big\downarrow{\scriptstyle (1, id_1)} & & \Big\downarrow{\scriptstyle i} \\
C_2 & \xrightarrow{\pi_2} & C_1
\end{array}$$

is easily seen to be a pullback, and $m(1_{C_1}, id_1) = 1_{C_1}$; so $1_{C_0} \xrightarrow{i} d_1$ is a terminal object of the above category in the sense of Ex. 2.10. Hence it is filtered, and $Y(\mathbf{C})$ is flat.

(ii) Let $g: \mathscr{F} \longrightarrow \mathscr{E}^{\mathbf{C}}$ be a morphism of $\mathscr{E}$-toposes. Since g^* preserves pullbacks and epimorphisms, it preserves discrete fibrations and filtered categories; hence $g^*(Y(\mathbf{C}))$ is a flat presheaf on $g^*\mathbf{C}^*(\mathbf{C}) \cong f^*(\mathbf{C})$ in $\mathscr{F}$. Moreover, a natural transformation $\alpha: g_1 \longrightarrow g_2$ over $\mathscr{E}$ clearly induces a morphism of presheaves $g_1^*(Y(\mathbf{C})) \longrightarrow g_2^*(Y(\mathbf{C}))$; so we have a functor

$$\Phi: \mathfrak{Top}/\mathscr{E}(\mathscr{F}, \mathscr{E}^{\mathbf{C}}) \longrightarrow \mathrm{Flat}(f^*\mathbf{C}^{\mathrm{op}}, \mathscr{F});\ g \longmapsto g^*(Y(\mathbf{C})).$$

(iii) Let G be a flat presheaf on $f^*\mathbf{C}$. Define $\Psi(G)^*$ to be the functor

$$\mathscr{E}^{\mathbf{C}} \xrightarrow{(f^{\mathbf{C}})^*} \mathscr{F}^{f^*\mathbf{C}} \xrightarrow{(-)\otimes G} \mathscr{F},$$

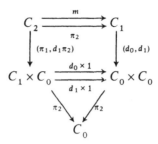

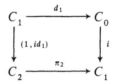

where $(f^{\mathbf{C}})^*$ denotes the functor f^* applied to discrete opfibrations over $\mathbf{C}$. Now it follows at once from 2.16 that $(f^{\mathbf{C}})^*$ is left exact; and it has a right adjoint $(f^{\mathbf{C}})_*$, which may be described as f_* followed by pullback along the unit $\mathbf{C} \xrightarrow{\alpha} f_*f^*\mathbf{C}$. So it follows from 4.32 that $\Psi(G)^*$ is the inverse image of a geometric morphism $\Psi(G)\colon \mathscr{F} \longrightarrow \mathscr{E}^{\mathbf{C}}$. Moreover, for any X in $\mathscr{E}$, we have

$$\begin{aligned}\Psi(G)^*\mathbf{C}^*(X) &\cong (f^*\mathbf{C})^*f^*(X) \otimes_{f^*\mathbf{C}} G \\ &\cong \varinjlim_{\mathbf{G}}(\mathbf{G}^*f^*(X)) \\ &\cong f^*(X) \times \varinjlim \mathbf{G} \\ &\cong f^*(X) \qquad \text{by 2.54.}\end{aligned}$$

So $\Psi(G)$ is a morphism over $\mathscr{E}$; and Ψ is clearly a functor

$$\mathrm{Flat}(f^*\mathbf{C}^{\mathrm{op}}, \mathscr{F}) \longrightarrow \mathfrak{Top}/\mathscr{E}(\mathscr{F}, \mathscr{E}^{\mathbf{C}}).$$

(iv) Finally, we must show that Φ and Ψ are inverse up to natural isomorphism. But it is easily verified that $(f^{\mathbf{C}})^*(Y(\mathbf{C})) \cong Y(f^*\mathbf{C})$; so $\Phi\Psi(G) \cong Y(f^*\mathbf{C}) \otimes_{f^*\mathbf{C}} G \cong G$ by 2.45.

Similarly, given a geometric morphism g and an object F of $\mathscr{E}^{\mathbf{C}}$, we have

$$\begin{aligned}\Psi\Phi(g)^*(F) &= (f^{\mathbf{C}})^*(F) \otimes g^*(Y(\mathbf{C})) \\ &\cong (g^{\mathbf{C}})^*(\mathbf{C}^{\mathbf{C}})^*(F) \otimes g^*(Y(\mathbf{C})) \\ &\cong g^*(\mathbf{C}^*(F) \otimes_{\mathbf{C}^*\mathbf{C}} Y(\mathbf{C})) \qquad \text{since } g^* \text{ preserves tensor products} \\ &\cong g^*(F) \qquad \text{by 2.45.}\end{aligned}$$

And this isomorphism is clearly natural in F and in g. ■

4.35 Corollary. *The diagram*

$$\begin{array}{ccc} \mathscr{F}^{f^*\mathbf{C}} & \xrightarrow{f^{\mathbf{C}}} & \mathscr{E}^{\mathbf{C}} \\ \downarrow & & \downarrow \\ \mathscr{F} & \xrightarrow{f} & \mathscr{E} \end{array}$$

is a pullback in $\mathfrak{Top}$, *where* $f^{\mathbf{C}}$ *is the geometric morphism constructed in* 4.34(iii).

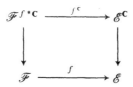

Proof. Suppose given a topos $\mathscr{G}$ and morphisms $\mathscr{G} \xrightarrow{g} \mathscr{F}$, $\mathscr{G} \xrightarrow{h} \mathscr{E}^{\mathbf{C}}$ forming a commutative square (up to isomorphism). Then by 4.34 h corresponds to a flat presheaf H on $g^*f^*\mathbf{C}$, which in turn corresponds to a geometric morphism $\mathscr{G} \xrightarrow{k} \mathscr{F}^{f^*\mathbf{C}}$ over $\mathscr{F}$. And since $f^{\mathbf{C}}$ corresponds to the flat presheaf $Y(f^*\mathbf{C})$, we have a canonical isomorphism $f^{\mathbf{C}}.k \cong h$ because both morphisms correspond to H. ■

4.36 COROLLARY. *The functor* $\mathbf{cat}(\mathscr{E}) \longrightarrow \mathfrak{Top}/\mathscr{E}$ *of* 2.35 *preserves finite products.*

Proof. We have to show that

$$\begin{array}{ccc} \mathscr{E}^{\mathbf{C}\times\mathbf{D}} & \longrightarrow & \mathscr{E}^{\mathbf{D}} \\ \downarrow & & \downarrow \\ \mathscr{E}^{\mathbf{C}} & \longrightarrow & \mathscr{E} \end{array}$$

is a product diagram in $\mathfrak{Top}/\mathscr{E}$, i.e. a pullback diagram in $\mathfrak{Top}$. But we have already remarked (in §2.4) that we have canonical isomorphisms $\mathscr{E}^{\mathbf{C}\times\mathbf{D}} \cong (\mathscr{E}^{\mathbf{C}})^{\mathbf{C}^*\mathbf{D}} \cong (\mathscr{E}^{\mathbf{D}})^{\mathbf{D}^*\mathbf{C}}$; so this follows from 4.34. ■

Unfortunately, however, the functor of 2.35 does not preserve all finite limits; we shall provide a counterexample in Ex. 6.7 below.

The importance of Diaconescu's Theorem is emphasized by its applications which occur throughout topos theory. Many of them will be of use in later chapters of this book; we give here some discussion of a number of interesting examples.

4.37 EXAMPLES. (i) Let $\mathbf{X} = (X \underset{1}{\overset{1}{\rightrightarrows}} X)$ be a discrete category. Then $\mathscr{E}^{\mathbf{X}} \cong \mathscr{E}^{\mathbf{X}^{\mathrm{op}}} \cong \mathscr{E}/X$, and since the total category of any discrete fibration $\mathbf{G} \xrightarrow{\gamma} \mathbf{X}$ is again a discrete category, it is easily seen that an object $G \xrightarrow{\gamma} X$ of $\mathscr{E}/X$ corresponds to a flat presheaf iff $G \longrightarrow 1$ is iso, i.e. iff γ is a global element of X. So $\mathfrak{Top}/\mathscr{E}(\mathscr{F}, \mathscr{E}^{\mathbf{X}})$ is equivalent to the discrete category of global elements of f^*X; and in particular note that $Y(\mathbf{X})$ is just the generic element of X as defined in §1.4. The generic property of this element can thus be extended from toposes of the form $\mathscr{E}/U$ to arbitrary $\mathscr{E}$-toposes.

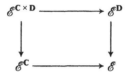

(ii) Let $\mathbf{G} = (G \rightrightarrows 1)$ be an internal group in $\mathscr{E}$ (i.e. a monoid equipped with an inverse map $G \longrightarrow G$). Now a presheaf on $\mathbf{G}$ is just a (left) G-object, i.e. an object M equipped with a unitary, associative action map $G \times M \xrightarrow{\alpha} M$. Using the fact that the morphisms of $\mathbf{G}$ are invertible, we may easily show that the category $(G \times M \underset{\pi_2}{\overset{\alpha}{\rightrightarrows}} M)$ satisfies 2.51(b) (resp. (c)) iff G acts transitively (resp, effectively) on M, i.e. iff $G \times M \xrightarrow{(\alpha, \pi_2)} M \times M$ is epi (resp. mono). So a presheaf on $\mathbf{G}$ is flat iff it is a (left) *G-torsor*, i.e. iff $G \times M \xrightarrow{(\alpha, \pi_2)} M \times M$ is iso and $M \longrightarrow 1$ is epi.

We shall see in §8.3 that (isomorphism classes of) f^*G-torsors in an $\mathscr{E}$-topos $\mathscr{F}$ correspond to elements of the first cohomology group $H^1(\mathscr{F}; G)$: for example if $\mathscr{E} = \mathscr{S}$, $G = \mathbb{Z}_2$ and $\mathscr{F}$ is the topos of sheaves on the unit circle S^1, then there are two isomorphism classes of G-torsors in $\mathscr{F}$, namely (the sheaves of sections of) the projection $S^1 \times G \xrightarrow{\pi_1} S^1$ and the double covering map $S^1 \longrightarrow S^1$; these correspond to the two elements of $\mathbb{Z}_2 \cong H^1(S^1; \mathbb{Z}_2)$. In this case, therefore, we can interpret Diaconescu's Theorem as the statement that the topos $\mathscr{E}^{\mathbf{G}}$ of right G-objects is a *classifying topos* for the functor $H^1(-; G)$. Note, however, that the group structure on this functor does not derive from the category structure on $\mathfrak{Top}/\mathscr{E}(-, \mathscr{E}^{\mathbf{G}})$, but from the fact that, in the case when G is abelian, $\mathscr{E}^{\mathbf{G}}$ is an abelian group object in $\mathfrak{Top}/\mathscr{E}$; this follows from 4.36 and the fact that $\mathbf{G}$ is an abelian group object in $\mathbf{cat}(\mathscr{E})$.

(iii) Let $\mathbf{2}$ denote the internalization (2.39) of the finite category represented diagrammatically by $(0 \longrightarrow 1)$. Then a presheaf on $\mathbf{2}$ corresponds to a diagram $D_1 \xrightarrow{f} D_0$ in $\mathscr{E}$; and it is easy to verify that such a presheaf is flat iff $D_0 \longrightarrow 1$ is iso and f is mono. So Flat($\mathbf{2}^{\mathrm{op}}$, $\mathscr{E}$) is equivalent to the poset of open objects of $\mathscr{E}$. Thus the topos $\mathscr{E}^{\mathbf{2}}$ is a classifying topos for open objects in $\mathscr{E}$-toposes; we call it the *Sierpinski topos* over $\mathscr{E}$, by analogy with the universal property of the Sierpinski space in **esp**. (Note also that $\mathscr{E}^{\mathbf{2}}$ may be obtained by applying Artin glueing (4.27) to the diagram $\mathscr{E} \xrightarrow{1} \mathscr{E}$; this is another aspect of the fact that it contains a generic open object.)

(iv) Let $\mathscr{E} = \mathscr{S}$, and take $\mathbf{C}$ to be the category $\mathscr{S}_f$ (which we can regard as a small category by choosing one set of each finite cardinality). Now $\mathscr{S}_f$ has finite colimits, so by 4.33 a flat presheaf on $\mathbf{C}$ is just a left exact functor $G: \mathscr{S}_f^{\mathrm{op}} \longrightarrow \mathscr{S}$. But any such functor is determined by the set $G(1)$, since every object of $\mathscr{S}_f$ is a finite coproduct of 1's; and conversely any set X determines the left exact functor $\hom_{\mathscr{S}}(-, X): \mathscr{S}_f^{\mathrm{op}} \longrightarrow \mathscr{S}$. So we have an equivalence Flat($\mathbf{C}^{\mathrm{op}}$, $\mathscr{S}$) $\simeq \mathscr{S}$; and more generally we can show for any $\mathscr{S}$-topos $\mathscr{F} \xrightarrow{\gamma} \mathscr{S}$ that Flat($\gamma^*\mathbf{C}^{\mathrm{op}}$, $\mathscr{F}$) $\simeq \mathscr{F}$. Thus the topos $\mathscr{S}^{\mathbf{C}}$ is an *object classifier* for $\mathscr{S}$-toposes, i.e. a representing object for the forgetful functor

$$(\mathfrak{Top}/\mathscr{S})^{\mathrm{op}} \longrightarrow \mathfrak{Cat}; (\mathscr{F} \xrightarrow{\gamma} \mathscr{S}) \longmapsto \mathscr{F},\ f \longmapsto f^*.$$

The *generic object* U of $\mathscr{S}^{\mathbf{C}}$ which is classified by the identity geometric morphism is readily checked to be the inclusion functor $\mathscr{S}_f \longrightarrow \mathscr{S}$. We normally denote the object classifier by $\mathcal{S}[U]$, and think of it as "$\mathcal{S}$ extended by an indeterminate object U"; we shall return to this example, in a more general form, in §6.3 below.

(v) By an *interval* in a topos $\mathcal{E}$, we mean a poset **I** which is totally ordered (i.e. $I_1 \amalg I_1^{\text{op}} \longrightarrow I_0 \times I_0$ is epi) and has distinct maximal and minimal elements $1 \xrightarrow{t} I_0$, $1 \xrightarrow{f} I_0$. A morphism of intervals is an order-preserving map (= internal functor) which preserves the distinguished elements; we write $\mathbf{int}(\mathcal{E})$ for the category of intervals in $\mathcal{E}$. Now if we define $I_{-1} = 1$, the assignment $n \longmapsto I_{n-2}$ has the structure of a *cosimplicial object* $S(\mathbf{I})$ in $\mathcal{E}$, i.e. a covariant functor $\Delta \longrightarrow \mathcal{E}$ (cf. 2.13). [The co-degeneracy maps of $S(\mathbf{I})$ are the face maps of **I**; the first and last co-face maps $S(I)_0 \rightrightarrows S(I)_{n+1}$ are obtained by adding f to the beginning (or t to the end) of an ordered n-tuple of objects of **I**, and the other co-face maps are the degeneracy maps of **I**.]

If $\mathcal{E}$ is defined over $\mathcal{S}$ by a geometric morphism γ, we can regard $S(\mathbf{I})$ as an internal diagram on $\gamma^*(\Delta)$ (cf. 2.39); it is not hard to show that this diagram is flat as a presheaf on $\gamma^*(\Delta)^{\text{op}}$, and conversely every flat presheaf on $\gamma^*(\Delta)^{\text{op}}$ must be of this form. So the topos $\mathcal{S}^{\Delta^{\text{op}}}$ of simplicial sets is a classifying topos for intervals in $\mathcal{S}$-toposes; i.e.

$$\mathfrak{Top}/\mathcal{S}\,(\mathcal{E}, \mathcal{S}^{\Delta^{\text{op}}}) \simeq \mathbf{int}(\mathcal{E})$$

for any $\mathcal{S}$-topos $\mathcal{E}$.

In particular, take $\mathcal{E}$ to be the topos of Ex. 0.10, and let $\mathcal{F}$ denote the category of sequential spaces, regarded as a full subcategory of $\mathcal{E}$. Now the last part of Ex. 0.10 implies that the unit interval $I = [0, 1] \subseteq \mathbb{R}$ is an interval in $\mathcal{E}$; let $\mathcal{E} \xrightarrow{p} \mathcal{S}^{\Delta^{\text{op}}}$ be its classifying map. Since I_n is easily seen to be isomorphic (as an object of $\mathcal{F}$) to the standard $(n+1)$-simplex, it is not hard to verify that $p^*(\mathbf{K})$, for a simplicial set **K**, is simply the geometric realization of **K**([CF], III 1.4). (Note that this is a CW-complex and hence an object of $\mathcal{F}$.) Similarly, if X is an object of $\mathcal{F}$, then $p_*(X)$ is the singular complex of X. (This example is due to A. Joyal; it is described in greater detail in [57].) ■

4.38 REMARK. As a special case of 4.37(i), note that $\mathfrak{Top}/\mathcal{E}(\mathcal{E}/X, \mathcal{E}/Y)$ is equivalent to the discrete category whose objects are the elements of $X^*(Y)$ in $\mathcal{E}/X$, i.e. the morphisms $X \longrightarrow Y$ in $\mathcal{E}$. So the functor $\mathcal{E} \longrightarrow \mathfrak{Top}/\mathcal{E}$ of 1.46 is a full embedding (in the sense appropriate to 2-categories). ■

Note, however, that 4.38 does not generalize to non-discrete categories and the functor $\mathbf{cat}(\mathscr{E}) \longrightarrow \mathfrak{Top}/\mathscr{E}$ of 2.35. For example, if we take $\mathscr{E} = \mathrm{Shv}(S^1)$, $\mathbf{G}$ to be the constant sheaf $\Delta(\mathbb{Z}_2)$ (regarded as an internal group in $\mathscr{E}$), and $\mathbf{1}$ to be the terminal object of $\mathbf{cat}(\mathscr{E})$, then there is only one internal functor (= group homomorphism) $\mathbf{1} \longrightarrow \mathbf{G}$ in $\mathscr{E}$, but there are two non-isomorphic morphisms $\mathscr{E}^{\mathbf{1}} \longrightarrow \mathscr{E}^{\mathbf{G}}$, corresponding to the two torsors of 4.37(ii). Again, if we take the example of 4.37(iv), the only geometric morphisms $\mathscr{S} \longrightarrow \mathscr{S}[U]$ induced by internal functors $\mathbf{1} \longrightarrow \mathbf{C}$ are those which classify *finite* sets.

We do, however, have the following description of $\mathfrak{Top}/\mathscr{E}(\mathscr{E}^{\mathbf{C}}, \mathscr{E}^{\mathbf{D}})$:

4.39 PROPOSITION. *Let* $\mathbf{C}$, $\mathbf{D}$ *be two internal categories in* $\mathscr{E}$. *Define a profunctor* $\mathbf{D} \overset{G}{\dashrightarrow} \mathbf{C}$ *to be* left flat *if the functor* $(-)\otimes_{\mathbf{D}} G\colon \mathscr{E}^{\mathbf{D}} \longrightarrow \mathscr{E}^{\mathbf{C}}$ *is left exact. (This is easily proved equivalent to the condition that* G *is flat as a presheaf on* $\mathbf{C}^*(\mathbf{D})$ *in* $\mathscr{E}^{\mathbf{C}}$.) *Then* $\mathfrak{Top}/\mathscr{E}(\mathscr{E}^{\mathbf{C}}, \mathscr{E}^{\mathbf{D}})$ *is equivalent to the full subcategory of* $\mathrm{Prof}_{\mathscr{E}}(\mathbf{D}, \mathbf{C})$ *whose objects are left flat profunctors. Moreover, this equivalence identifies the operation of composing geometric morphisms* $\mathscr{E}^{\mathbf{B}} \longrightarrow \mathscr{E}^{\mathbf{C}} \longrightarrow \mathscr{E}^{\mathbf{D}}$ *with the tensor product of profunctors.*

Proof. Straightforward application of 4.34. ■

Note in particular that if $\mathbf{C} \xrightarrow{f} \mathbf{D}$ is an internal functor, then the geometric morphism induced by f corresponds to the profunctor $f^{\#}$.

4.4 BOUNDED MORPHISMS

In contrast to the situation of 4.21, when we come to discuss the construction of limits in $\mathfrak{Top}$, we are immediately faced with the difficulty that it does not have a terminal object. The "best approximation" to a terminal object is given by the following result:

4.41 PROPOSITION. *Let* $\mathscr{E}$ *be a topos. If there exists a geometric morphism* $\mathscr{E} \longrightarrow \mathscr{S}$ *(or* $\mathscr{E} \longrightarrow \mathscr{S}_f$*), then it is unique up to canonical isomorphism. Moreover, there exists a morphism* $\mathscr{E} \longrightarrow \mathscr{S}$ *iff* $\mathscr{E}$ *has arbitrary* $\mathscr{S}$*-indexed copowers and "small hom-sets", and there exists* $\mathscr{E} \longrightarrow \mathscr{S}_f$ *iff* $\mathscr{E}$ *has finite hom-sets.*

Proof. Suppose given $\mathscr{E} \xrightarrow{f} \mathscr{S}$. Then it is clear that, for any object X of $\mathscr{E}$, the elements of the set $f_*(X)$ are in natural 1–1 correspondence with morphisms $1 \cong f^*(1) \longrightarrow X$ in $\mathscr{E}$; so f_* (and hence also f^*) is determined up to natural isomorphism. Moreover, the morphisms $X \longrightarrow Y$ in $\mathscr{E}$ are indexed by the set $f_*(Y^X)$; and for any set S, $f^*(S)$ is an S-indexed copower of 1 in $\mathscr{E}$, so that $f^*(S) \times X$ is an S-indexed copower of X.

Conversely, if $\mathscr{E}$ satisfies the conditions, we can simply define $f_*(X)$ = set of morphisms $1 \longrightarrow X$, and $f^*(S)$ = S-indexed copower of 1. It is then easy to verify that these functors form a geometric morphism.

The argument for $\mathscr{S}_f$ is similar. ■

On comparing 4.41 with Giraud's Theorem (0.45), we observe that one talks about $\mathscr{S}$-indexed coproducts of objects of $\mathscr{E}$, and the other about $\mathscr{S}$-indexed copowers. Now we cannot deduce from 4.41 that an $\mathscr{S}$-topos has $\mathscr{S}$-indexed coproducts (see Example 4.49(ii) below); but if we have a family of objects $(G_\alpha | \alpha \in A)$ indexed by a set A, and a single object G containing each G_α as a subobject, then we can easily identify $\coprod_{\alpha \in A} G_\alpha$ as a subobject of the copower $\coprod_{\alpha \in A} G$, simply by constructing its classifying map from the classifying maps of the $G_\alpha \rightarrowtail G$. Moreover, if $\mathscr{E}$ has a set of generators, it is clearly sufficient to be able to construct $\mathscr{S}$-indexed coproducts of the generators; and conversely if $\{G_\alpha | \alpha \in A\}$ is a set of generators for a Grothendieck topos, then the object $G = \coprod_{\alpha \in A} G_\alpha$ contains each G_α as a subobject. So on combining 1.58 with 4.41, we can reduce Giraud's Theorem to the statement:

> "*A topos $\mathscr{E}$ is a Grothendieck topos iff it is defined over $\mathscr{S}$ and contains an object G whose subobjects generate $\mathscr{E}$.*"

We now give a proof of the "relative" version of this theorem (again due to R. Diaconescu [30]), in which the particular topos $\mathscr{S}$ is replaced by an arbitrary base topos.

4.42 LEMMA. *Let $\mathscr{F} \xrightarrow{f} \mathscr{E}$ be a geometric morphism, G an object of $\mathscr{F}$. The following three statements are equivalent:*

(i) *For any $X \in \mathscr{F}$, the composite $f^*f_*(\tilde{X}^G) \times G \xrightarrow{\beta \times 1} \tilde{X}^G \times G \xrightarrow{\text{ev}} \tilde{X}$ is epi, where β is the counit of $(f^* \dashv f_*)$.*
(ii) *For any $X \in \mathscr{F}$, there exists $Y \in \mathscr{E}$ and an epi $f^*(Y) \times G \twoheadrightarrow \tilde{X}$.*
(iii) *For any $X \in \mathscr{F}$, there exists an object $Y \in \mathscr{E}$, a subobject $S \rightarrowtail f^*(Y) \times G$ and an epi $S \twoheadrightarrow X$.*

Proof. (i) ⇒ (ii) is immediate if we take $Y = f_*(\tilde{X}^G)$.

(ii) $\Rightarrow$ (iii) is immediate if we form the pullback

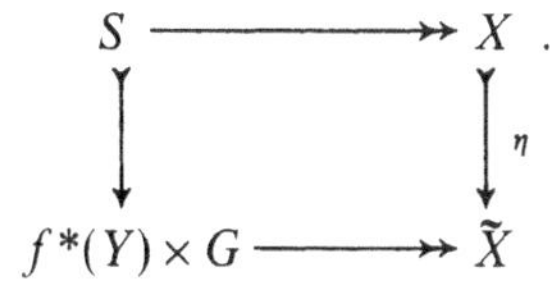

(iii) $\Rightarrow$ (ii): Apply (iii) to the object $\tilde{X}$. Then we have a diagram

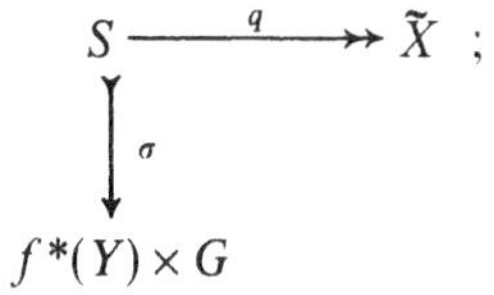

but $\tilde{X}$ is injective by 1.27, so we obtain a factorization $f^*(Y) \times G \xrightarrow{p} \tilde{X}$ of q through σ. Then $p\sigma = q$ is epi, so p is epi.

(ii) $\Rightarrow$ (i): Suppose given $f^*(Y) \times G \xrightarrow{p} \tilde{X}$. By transposing across the adjunctions $((-) \times G \dashv (-)^G)$ and $(f^* \dashv f_*)$, we obtain $Y \xrightarrow{\bar{p}} f_*(\tilde{X}^G)$; and the diagram

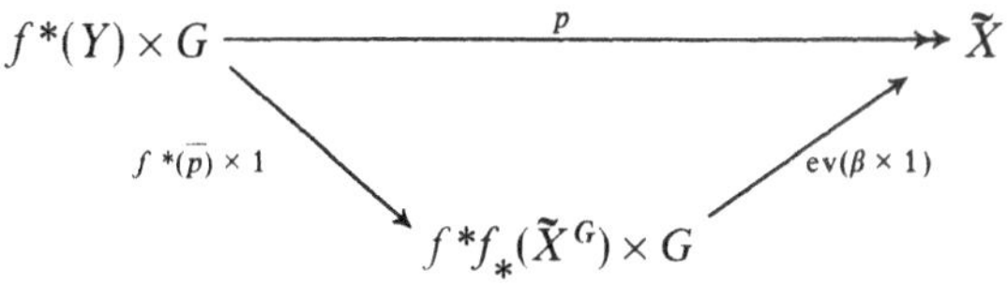

commutes, so $\text{ev}(\beta \times 1)$ is epi. ■

4.43 Definition (W. Mitchell). If G satisfies the equivalent conditions of 4.42, it is said to be an *object of generators* for $\mathscr{F}$ over $\mathscr{E}$. If $\mathscr{F}$ has an object of generators over $\mathscr{E}$, we say the geometric morphism f is *bounded.* ■

Now in the case $\mathscr{E} = \mathscr{S}$, $f^*(Y) \times G$ is a Y-indexed copower of G in $\mathscr{F}$, and so we can think of the diagram

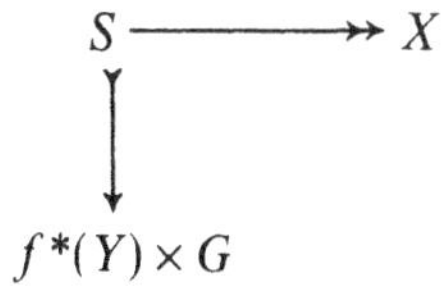

of 4.42(iii) as a "Y-indexed cover of X by subobjects of G". This provides the link between 4.43 and the usual definition of a set of generators.

4.44 LEMMA. *Let* $\mathscr{G} \xrightarrow{g} \mathscr{F} \xrightarrow{f} \mathscr{E}$ *be two geometric morphisms. Then*

(i) *if f and g are bounded, the composite fg is bounded, and*

(ii) *if fg is bounded, g is bounded.*

Proof. (i) Let F (resp. G) be an object of generators for $\mathscr{F}$ over $\mathscr{E}$ (resp. $\mathscr{G}$ over $\mathscr{F}$). Now for any $X \in \mathscr{G}$, we can find $Y \in \mathscr{F}$ and a diagram

$$\begin{array}{ccc} S & \twoheadrightarrow & X \\ \downarrow & & \\ g^*(Y) \times G & & \end{array} .$$

And then we can find $Z \in \mathscr{E}$ and a diagram

$$\begin{array}{ccc} T & \twoheadrightarrow & Y \\ \downarrow & & \\ f^*(Z) \times F & & \end{array} .$$

Combining the two diagrams and forming a pullback, we have

$$\begin{array}{ccccc} U & \twoheadrightarrow & S & \twoheadrightarrow & X \\ \downarrow & & \downarrow & & \\ g^*(T) \times G & \twoheadrightarrow & g^*(Y) \times G & & \\ \downarrow & & & & \\ g^*f^*(Z) \times g^*(F) \times G & & & & \end{array} ;$$

so $g^*(F) \times G$ is an object of generators for $\mathscr{G}$ over $\mathscr{E}$.

(ii) is immediate from 4.42(i), since $g^*f^*f_*g_*(\tilde{X}^G) \times G \longrightarrow \tilde{X}$ factors through $g^*g_*(\tilde{X}^G) \times G \longrightarrow \tilde{X}$. ■

4.45 LEMMA. (i) *Let* $\mathbf{C}$ *be an internal category in* $\mathscr{E}$. *Then the geometric morphism* $\mathscr{E}^{\mathbf{C}} \longrightarrow \mathscr{E}$ *of* 2.33 *is bounded.*

(ii) *Any inclusion is bounded.*

Proof. (i) Let G be the internal diagram on $\mathbf{C}$ defined by the object $C_1 \xrightarrow{d_1} C_0$ of $\mathscr{E}/C_0$, with right action of $\mathbf{C}$ given by m. Now if

$$X = (X_0 \xrightarrow{\gamma_0} C_0,\ X_0 \times_{C_0} C_1 \xrightarrow{\alpha} X_0)$$

is an internal diagram on $\mathbf{C}$, we have an epimorphism $R(\gamma_0) \twoheadrightarrow X$ in $\mathscr{E}^{\mathbf{C}}$ (this is just the counit of $(R \dashv U)$, cf. 2.22). But the canonical monomorphism

$$X_0 \times_{C_0} C_1 \rightarrowtail X_0 \times C_1 \cong (X_0 \times C_0) \times_{C_0} C_1$$

is easily seen to be a morphism $R(\gamma_0) \rightarrowtail \mathbf{C}^*(X_0) \times G$ in $\mathscr{E}^{\mathbf{C}}$, so we have a diagram of the form required for 4.42(iii).

(ii) Let $\mathscr{F} \xrightarrow{f} \mathscr{E}$ be an inclusion. Then the counit of $(f^* \dashv f_*)$ is iso: so if we take $G = 1$ we have $f^*f_*(\tilde{X}^G) \times G \xrightarrow{\cong} \tilde{X}$ for any X. ■

4.46 Theorem (Giraud–Mitchell–Diaconescu). *Let $\mathscr{F} \xrightarrow{f} \mathscr{E}$ be a geometric morphism. Then the following are equivalent:*

(i) *f is bounded.*

(ii) *There exists an internal category $\mathbf{C}$ in $\mathscr{E}$ and an inclusion $\mathscr{F} \xrightarrow{i} \mathscr{E}^{\mathbf{C}}$ such that*

$$\begin{array}{ccc} \mathscr{F} & \xrightarrow{\quad f \quad} & \mathscr{E} \\ & {\scriptstyle i}\searrow \quad \nearrow & \\ & \mathscr{E}^{\mathbf{C}} & \end{array}$$

commutes up to natural isomorphism.

Proof. (ii) $\Rightarrow$ (i) is immediate from 4.45 and 4.44(i).

(i) $\Rightarrow$ (ii) Let G be an object of generators for $\mathscr{F}$ over $\mathscr{E}$, and let

$$\in_G \xrightarrow{(n,\, e)} \Omega^G \times G$$

be the membership-relation on G, i.e. the subobject classified by $\Omega^G \times G \xrightarrow{\text{ev}} \Omega$. Define $\mathbf{D} \in \mathbf{cat}(\mathscr{F})$ to be $\text{Full}_{\mathscr{F}}(\in_G \xrightarrow{n} \Omega^G)^{\text{op}}$ in the notation of 2.38, and let $\mathbf{E} \xrightarrow{\gamma} \mathbf{D}$ be the discrete fibration corresponding to the "inclusion functor" defined there. Let $\mathbf{C} = f_*(\mathbf{D}) \in \mathbf{cat}(\mathscr{E})$, and let $f^*\mathbf{C} \xrightarrow{\beta} \mathbf{D}$

be the counit of $(f^* \dashv f_*)$. Form the pullback

$$\begin{array}{ccc} \mathbf{F} & \xrightarrow{\delta} & \mathbf{E} \\ {\scriptstyle\zeta}\downarrow & & \downarrow{\scriptstyle\gamma} \\ f^*\mathbf{C} & \xrightarrow{\beta} & \mathbf{D} \end{array}$$

in $\mathbf{cat}(\mathscr{F})$; then ζ is a discrete fibration by 2.19. We shall show first that $\mathbf{F}$ is filtered, and so by 4.34 ζ determines a geometric morphism $\mathscr{F} \xrightarrow{i} \mathscr{E}^{\mathbf{C}}$ over $\mathscr{E}$.

First observe that we have a pullback diagram

$$\begin{array}{ccccc} F_0 & \xrightarrow{\delta_0} & E_0 = \in_G & \longrightarrow & 1 \;; \\ \downarrow & & \downarrow{\scriptstyle(n,e)} & & \downarrow{\scriptstyle t} \\ f^*C_0 \times G & \xrightarrow{\beta \times 1} & D_0 \times G = \Omega^G \times G & \xrightarrow{\text{ev}} & \Omega \end{array}$$

but $\text{ev}(\beta \times 1)$ is epi by 4.42(i) with $X = 1$. So $F_0 \longrightarrow 1$ is epi, i.e. 2.51(a) is verified.

Now let

$$\begin{array}{ccc} B & \xrightarrow{(q_1, q_2)} & F_0 \times F_0 \\ {\scriptstyle(a,b)}\downarrow & & \\ f^*A \times G & & \end{array}$$

be a cover of $F_0 \times F_0$ by subobjects of G. Then the morphism

$$B \xrightarrow{(\zeta \circ q_1, \zeta \circ q_2, a, b)} f^*C_0 \times f^*C_0 \times f^*A \times G \cong f^*(C_0 \times C_0 \times A) \times G$$

is clearly mono, since (a, b) is; let $f^*(C_0 \times C_0 \times A) \xrightarrow{p} \Omega^G$ be the transpose of its classifying map. And the morphism

$$f^*(C_0 \times C_0 \times A) \xrightarrow{f^*(\pi_1)} f^*C_0 \xrightarrow{\beta_0} D_0 = \Omega^G$$

clearly corresponds to the subobject

$$F_0 \times f^*(C_0 \times A) \rightarrowtail^{(\zeta\circ\pi_1, \pi_2, e\delta_0\pi_1)} f^*C_0 \times f^*(C_0 \times A) \times G,$$

since β_0 corresponds to $F_0 \xrightarrow{(\zeta_0, e\delta_0)} f^*C_0 \times G$.

Now the morphism

$$B \xrightarrow{(q_1, \zeta_0 q_2, a)} F_0 \times f^*(C_0 \times A)$$

is a morphism of objects over $f^*(C_0 \times C_0 \times A)$; so by the construction of $\mathbf{D}^{\mathrm{op}}$ as an internal full subcategory of $\mathscr{F}$, it induces an $f^*(C_0 \times C_0 \times A)$-element r_1 of D_1 such that $d_0 r_1 = \beta_0 . f^*(\pi_1)$ and $d_1 r_1 = p$.

Similarly, we may construct $f^*(C_0 \times C_0 \times A) \xrightarrow{r_2} D_1$ with $d_0 r_2 = \beta_0 . f^*(\pi_2)$ and $d_1 r_2 = p$. But now the diagram

$$\begin{array}{ccc} f^*(C_0 \times C_0 \times A) & \xrightarrow{\quad r_i \quad} & D_1 \\ & {\scriptstyle f^*(\hat{r}_i)} \searrow \quad \nearrow {\scriptstyle \beta_1} & \\ & f^*C_1 & \end{array}$$

commutes, where ^ denotes transposition across the adjunction $(f^* \dashv f_*)$; and we have $d_0 \hat{r}_i = \pi_i$, $d_1 \hat{r}_i = \hat{p}$.

Now we have a pullback diagram

$$\begin{array}{ccc} B & \xrightarrow{\quad q_3 \quad} & F_0 \\ \downarrow {\scriptstyle (\zeta_0 q_1, \zeta_0 q_2, a, b)} & & \downarrow {\scriptstyle (\zeta_0, e\delta_0)} \\ f^*(C_0 \times C_0 \times A) \times G & \xrightarrow{f^*\hat{p} \times 1} & f^*C_0 \times G \end{array},$$

since the composite $\mathrm{ev}(\beta_0 \times 1)(f^*\hat{p} \times 1) = \mathrm{ev}(p \times 1)$ classifies the subobject $(\zeta_0 q_1, \zeta_0 q_2, a, b)$. Using the fact that ζ is a discrete fibration, we can lift the composites

$$B \xrightarrow{(\zeta_0 q_1, \zeta_0 q_2, a)} f^*(C_0 \times C_0 \times A) \xrightarrow{f^*(\hat{r}_i)} f^*C_1$$

to morphisms $B \xrightarrow{s_i} F_1$ with $d_1 s_i = q$. To determine $d_0 s_i$, note first that

$$\zeta_0 d_0 s_i = d_0 \zeta_1 s_i = d_0 f^*(\hat{r}_i)(\zeta_0 q_1, \zeta_0 q_2, a) = f^*(\pi_i)(\zeta_0 q_1, \zeta_0 q_2, a) = \zeta_0 q_i.$$

And from the definition of the action of $\mathbf{D}$ on E_0 (cf. 2.38), we may easily verify that $e\delta_0 d_0 s_1 = ed_0 \delta_1 s_1$ is the B-element

$$B \xrightarrow{(q_1, \zeta_0 q_2, a)} F_0 \times f^*(C_0 \times A) \xrightarrow{e\delta_0 \pi_1} G$$

of G, i.e. we have $e\delta_0 d_0 s_1 = e\delta_0 q_1$. But $F_0 \rightarrowtail^{(\zeta_0,\, e\delta_0)} f^*C_0 \times G$ is mono, so we have $d_0 s_1 = q_1$; and similarly $d_0 s_2 = q_2$. Thus the morphism $B \xrightarrow{(s_1, s_2)} P_{\mathbf{F}}$ is a factorization of $B \xrightarrow{(q_1, q_2)} F_0 \times F_0$ through the morphism $P_{\mathbf{F}} \longrightarrow F_0 \times F_0$ of 2.51(b); so the latter is epi, and condition 2.51(b) is verified.

The verification of 2.51(c) is simpler, since it does not make use of the fact that G is an object of generators. Since $\mathbf{D}^{\mathrm{op}}$ is an internal full subcategory of $\mathscr{F}$, it is easily seen that U-elements of $T_{\mathbf{D}}$ (resp. $R_{\mathbf{D}}$) correspond to diagrams

$$\begin{array}{ccc} X & \overset{\theta}{\underset{\phi}{\rightrightarrows}} & Y \\ \downarrow{\scriptstyle \mu} & & \downarrow{\scriptstyle \nu} \\ U^*G & & U^*G \end{array} \qquad \left(\text{resp.} \quad \begin{array}{ccccc} Z & \xrightarrow{\psi} & X & \overset{\theta}{\underset{\phi}{\rightrightarrows}} & Y \\ \downarrow{\scriptstyle \lambda} & & \downarrow{\scriptstyle \mu} & & \downarrow{\scriptstyle \nu} \\ U^*G & & U^*G & & U^*G \end{array} \text{ with } \theta\psi = \phi\psi\right)$$

in $\mathscr{F}/U$, where the horizontal arrows are not required to be morphisms over U^*G. But since $\mathscr{F}/U$ has equalizers, we may construct a diagram of the latter type from one of the former type, by defining $Z \xrightarrow{\psi} X$ to be the equalizer of θ and ϕ, and λ to be the composite $\mu\psi$. By applying this construction to the generic element of $R_{\mathbf{D}}$ in $\mathscr{F}/R_{\mathbf{D}}$, we obtain a splitting $\sigma_{\mathbf{D}}$ for the morphism $T_{\mathbf{D}} \longrightarrow R_{\mathbf{D}}$ of 2.51(c).

Similarly, U-elements of $R_{\mathbf{E}}$ correspond to diagrams

$$\begin{array}{ccc} & 1_U & \\ {\scriptstyle x}\swarrow & & \searrow{\scriptstyle y} \\ X & \overset{\theta}{\underset{\phi}{\rightrightarrows}} & Y \\ \downarrow{\scriptstyle \mu} & & \downarrow{\scriptstyle \nu} \\ U^*G & & U^*G \end{array}$$

in $\mathscr{F}/U$ with $\theta x = y = \phi x$; but then we can factor x through the equalizer of θ and ϕ to obtain a diagram

$$\begin{array}{ccccc} & & 1_U & & \\ & {\scriptstyle z}\swarrow & \downarrow{\scriptstyle x} & \searrow{\scriptstyle y} & \\ Z & \xrightarrow{\psi} & X & \overset{\theta}{\underset{\phi}{\rightrightarrows}} & Y \\ \downarrow{\scriptstyle \lambda} & & \downarrow{\scriptstyle \mu} & & \downarrow{\scriptstyle \nu} \\ U^*G & & U^*G & & U^*G \end{array}$$

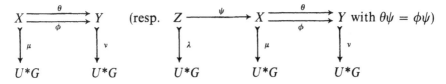

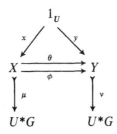

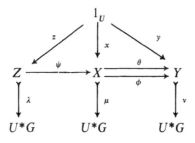

i.e. a U-element of $T_{\mathbf{E}}$. In "generic" terms, this says that $\sigma_{\mathbf{D}}$ can be lifted to a splitting $\sigma_{\mathbf{E}}$ of $T_{\mathbf{E}} \longrightarrow R_{\mathbf{E}}$.

And since f_* and f^* are both left exact, they preserve the constructions of R and T, so $f^*f_*\sigma_{\mathbf{D}}$ is a splitting for $T_{f^*\mathbf{C}} \longrightarrow R_{f^*\mathbf{C}}$. So we have a diagram.

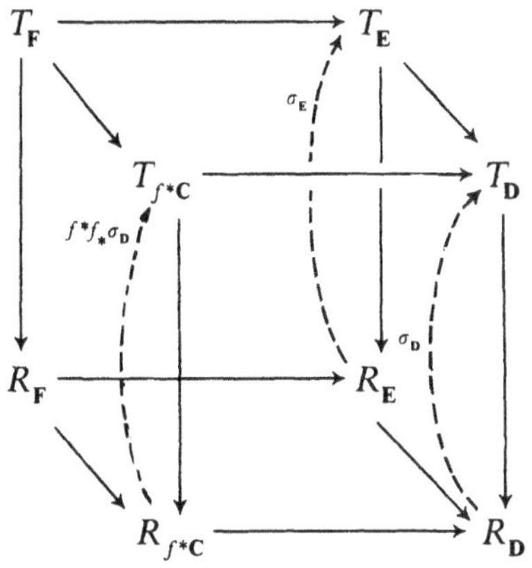

in which the top and bottom faces are pullbacks since the constructions of R and T commute with finite limits. Thus from the splittings $\sigma_{\mathbf{D}}$, $\sigma_{\mathbf{E}}$ and $f^*f_*\sigma_{\mathbf{D}}$ we may construct a splitting $\sigma_{\mathbf{F}}$ of $T_{\mathbf{F}} \longrightarrow R_{\mathbf{F}}$; so the latter is epi, and $\mathbf{F}$ satisfies 2.51(c).

We are now halfway through the proof of 4.46!

To show that the geometric morphism $\mathscr{F} \xrightarrow{i} \mathscr{E}^{\mathbf{C}}$ which we have constructed is an inclusion, we will show that the counit of $(i^* \dashv i_*)$ is iso. First observe that in the diagram

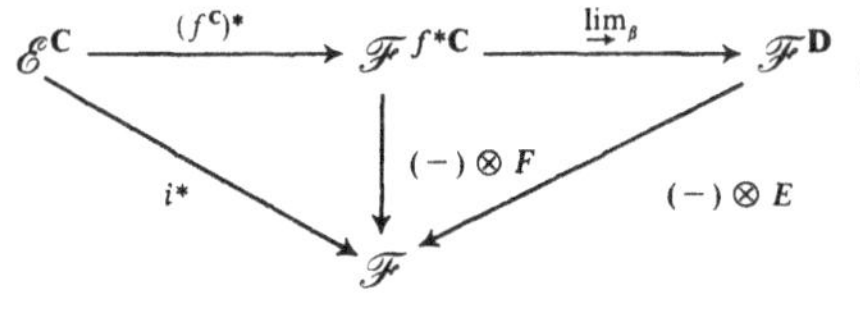

the left-hand triangle commutes by the definition of i^* (4.34), and the right-hand triangle commutes up to isomorphism by 2.47, since $F = \beta^*E$. It follows that the diagram

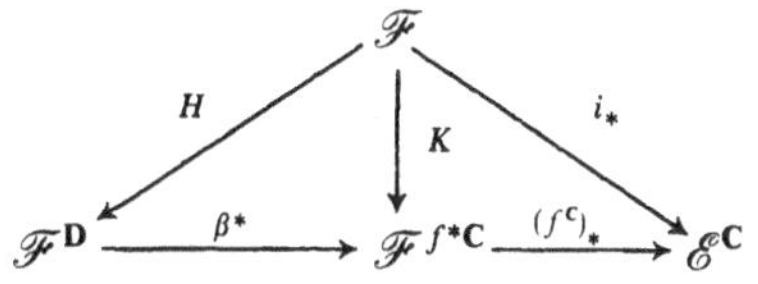

commutes up to isomorphism, where H, K are the right adjoints of $(-) \otimes E$ and $(-) \otimes F$ respectively. But the diagram

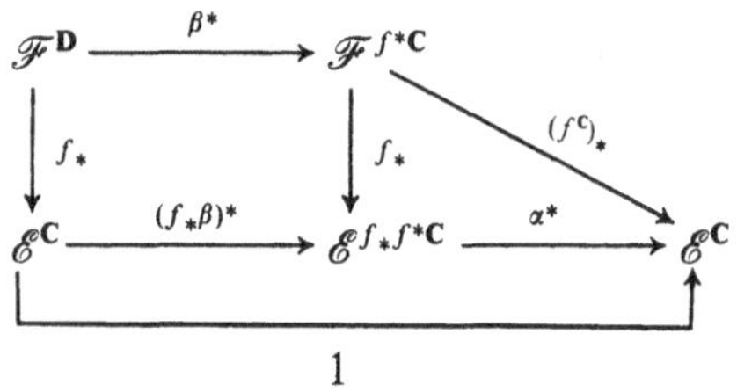

commutes, where $\mathbf{C} \xrightarrow{\alpha} f_* f^* \mathbf{C}$ is the unit of $(f^* \dashv f_*)$ (this uses the definition of $(f^{\mathbf{C}})_*$ in 4.34, the fact that f_* preserves pullbacks, and one of the triangular identities between α and β). Combining the three diagrams, we deduce that $i^* i_*$ is isomorphic to the composite

$$\mathscr{F} \xrightarrow{H} \mathscr{F}^{\mathbf{D}} \xrightarrow{f_*} \mathscr{E}^{\mathbf{C}} \xrightarrow{f^*} \mathscr{F}^{f^*\mathbf{C}} \xrightarrow{(-)\otimes F} \mathscr{F}.$$

Now let X be an object of $\mathscr{F}$. Then we may verify directly that HX is the internal diagram defined by $(HX_0 \longrightarrow D_0) = (\tilde{X}^G \xrightarrow{\phi^G} \Omega^G)$ (where ϕ classifies $X \xrightarrow{\eta} \tilde{X}$), with structure map $HX_0 \times_{D_0} D_1 \xrightarrow{\xi} HX_0$ which internalizes the operation of composing a partial map

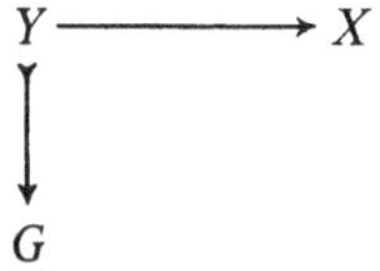

with a "partial endomorphism of G"

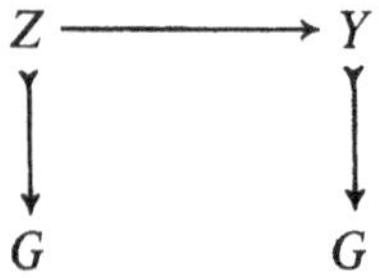

For if W is any internal diagram on $\mathbf{D}$, then morphisms $W_0 \xrightarrow{g} HX_0$ over $D_0 = \Omega^G$ correspond to morphisms $W_0 \times G \xrightarrow{\bar{g}} \tilde{X}$ over Ω, and hence to morphisms $W_0 \times_{\Omega^G} \in_G = W_0 \times_{D_0} E_0 \xrightarrow{g'} X$ in $\mathscr{F}$; and an easy diagram-chase shows that g is a morphism of internal diagrams iff g' factors through the canonical epimorphism $W_0 \times_{D_0} E_0 \longrightarrow W \otimes_{\mathbf{D}} E$.

So i^*i_*X is defined by the coequalizer diagram

$$f^*f_*\tilde{X}^G \times_{f^*C_0} f^*C_1 \times_{f^*C_0} F_0 \underset{1\times\omega}{\overset{f^*f_*\xi\times 1}{\rightrightarrows}} f^*f_*\tilde{X}^G \times_{f^*C_0} F_0 \overset{v}{\twoheadrightarrow} i^*i_*X,$$

where ω denotes the action of $f^*\mathbf{C}$ on F_0. Moreover, it is not hard to verify that the counit $i^*i_*X \xrightarrow{c} X$ is obtained by factoring $f^*f_*\tilde{X}^G \times_{f^*C_0} F_0 \xrightarrow{u} X$ through v, where u is obtained from the pullback diagram

$$\begin{array}{ccccc} f^*f_*\tilde{X}^G \times_{f^*C_0} F_0 & \longrightarrow & \tilde{X}^G \times_{D_0} E_0 & \longrightarrow & X \\ \big\downarrow{\scriptstyle 1\times e\zeta_0} & & \big\downarrow{\scriptstyle 1\times e} & & \big\downarrow{\scriptstyle \eta} \\ f^*f_*\tilde{X}^G \times G & \xrightarrow{\beta\times 1} & \tilde{X}^G \times G & \xrightarrow{\text{ev}} & \tilde{X} \end{array} \quad (\text{composite of top row } = u).$$

But u is epi, since $\text{ev}(\beta \times 1)$ is epi by 4.42(i); so c is epi. It thus remains to show that c is mono.

Let

$$K \underset{(x_2,\, w'_2)}{\overset{(x_1,\, w'_1)}{\rightrightarrows}} f^*f_*\tilde{X}^G \times_{f^*C_0} F_0$$

be the kernel-pair of u, and let

$$\begin{array}{ccc} L & \overset{\kappa}{\twoheadrightarrow} & K \\ \big\downarrow{\scriptstyle (h,g)} & & \\ f^*M \times G & & \end{array}$$

be a cover of K by subobjects of G. Write N for the object $f_*\tilde{X}^G \times f_*\tilde{X}^G \times M$ of $\mathscr{E}$, and j for the morphism

$$L \xrightarrow{(x_1\kappa,\, x_2\kappa,\, h)} f^*f_*\tilde{X}^G \times f^*f_*\tilde{X}^G \times f^*M \cong f^*N.$$

Then $L \xrightarrow{(j,g)} f^*N \times G$ is mono since (h, g) is; let $f^*N \xrightarrow{l} \Omega^G = D_0$ be the transpose of its classifying map. Proceeding as we did in verifying 2.51(b), we may now construct morphisms $f^*N \xrightarrow{\lambda_i} D_1$ $(i = 1, 2)$ with $d_1\lambda_i = l$ and

$$d_0\lambda_i = (f^*N \xrightarrow{f^*\pi_i} f^*f_*\tilde{X}^G \xrightarrow{f^*f_*\phi G} f^*C_0 \xrightarrow{\beta_0} D_0),$$

corresponding to the diagram

$$\begin{array}{ccc} L & \xrightarrow{(j,\, w_i\kappa)} & f^*N \times_{f^*C_0} F_0 \\ \Big\downarrow {\scriptstyle (j,g)} & & \Big\downarrow {\scriptstyle 1 \times e\delta_0} \\ f^*N \times G & & f^*N \times G \end{array}$$

in $\mathscr{F}/f^*N$.

But from the defiiition of ξ, it is easily verified that the composite

$$f^*N \xrightarrow{(\beta\,.\,f^*\pi_i,\, \lambda_i)} \tilde{X}^G \times_{D_0} D_1 \xrightarrow{\xi} \tilde{X}^G$$

corresponds to the partial map

$$\begin{array}{l} L \xrightarrow{\kappa} K \xrightarrow{(x_i,\, w_i)} f^*f_*\tilde{X}^G \times_{f^*C_0} F_0 \xrightarrow{u} X \ : \\ \Big\downarrow {\scriptstyle (j,g)} \\ f^*N \times G \end{array}$$

so by the definition of K we have $\xi(\beta\,.\,f^*\pi_1, \lambda_1) = \xi(\beta\,.\,f^*\pi_2, \lambda_2)$, and hence by transposition $f_*\xi(\pi_1, \hat{\lambda}_1) = f_*\xi(\pi_2, \hat{\lambda}_2)$. But now we have a pullback diagram

$$\begin{array}{ccc} L & \xrightarrow{w_3} & F_0\ , \\ \Big\downarrow {\scriptstyle (j,g)} & & \Big\downarrow {\scriptstyle (\zeta_0,\, e\delta_0)} \\ f^*N \times G & \xrightarrow{f^*(\hat{l}) \times 1} & f^*C_0 \times G \end{array}$$

and it is again easy to verify that

$$\omega(f^*\hat{\lambda}_i\,.\,j, w_3) = w_1\kappa \colon L \longrightarrow F_0$$

for $i = 1, 2$. So we can construct morphisms

$$k_i \colon L \xrightarrow{(x_i\kappa,\, f^*\lambda_i\,.\,j,\, w_3)} f^*f_*\tilde{X}^G \times_{f^*C_0} f^*C_1 \times_{f^*C_0} F_0$$

with the property that $(1 \times \omega)k_i = (x_i, w_i)\kappa$ and $(f^*f_*\xi \times 1)k_1 = (f^*f_*\xi \times 1)k_2$. It follows at once that $f^*f_*X^G \times_{f^*C_0} F_0 \xrightarrow{v} i^*i_*X$ coequalizes $(x_1, w_1)\kappa$ and $(x_2, w_2)\kappa$; but κ is epi, so $v(x_1, w_1) = v(x_2, w_2)$.

Thus we obtain a factorization of v through u; so c is an isomorphism. ■

4.47 PROPOSITION. *Let*

$$\begin{array}{ccc} & & \mathscr{G} \\ & & \big\downarrow g \\ \mathscr{F} & \xrightarrow{\ f\ } & \mathscr{E} \end{array}$$

be a diagram in $\mathfrak{Top}$ *with* g *bounded. Then the pullback* $\mathscr{F} \times_{\mathscr{E}} \mathscr{G}$ *exists in* $\mathfrak{Top}$ *and is bounded over* $\mathscr{F}$.

Proof. Since pullbacks can be constructed in stages, it is sufficient to consider the two cases (a) when $\mathscr{G}$ has the form $\mathscr{E}^{\mathbf{C}}$ for some $\mathbf{C} \in \mathbf{cat}(\mathscr{E})$ and (b) when g is an inclusion.

But (a) is just 4.35; for (b), let j be the topology in $\mathscr{E}$ induced by g, and let j' be the topology in $\mathscr{F}$ induced by j as in 3.59(iii). Then it follows easily from 4.19 that the square

$$\begin{array}{ccc} \mathrm{sh}_{j'}(\mathscr{F}) & \longrightarrow & \mathrm{sh}_j(\mathscr{E}) \simeq \mathscr{G} \\ \big\downarrow & & \big\downarrow g \\ \mathscr{F} & \xrightarrow{\ f\ } & \mathscr{E} \end{array}$$

is a pullback in $\mathfrak{Top}$. ■

4.48 COROLLARY. *Let* $\mathfrak{BTop}/\mathscr{E}$ *denote the full subcategory of* $\mathfrak{Top}/\mathscr{E}$ *whose objects are bounded* $\mathscr{E}$*-toposes. Then* $\mathfrak{BTop}/\mathscr{E}$ *has finite limits.*

Proof. $\mathfrak{BTop}/\mathscr{E}$ clearly has a terminal object, namely $\mathscr{E} \xrightarrow{1} \mathscr{E}$. But by 4.44(ii) any morphism in $\mathfrak{BTop}/\mathscr{E}$ is bounded, so by 4.47 we can construct the pullback in $\mathfrak{Top}/\mathscr{E}$ of any pair of morphisms in $\mathfrak{BTop}/\mathscr{E}$, and by 4.44(i) the pullback is bounded over $\mathscr{E}$. ■

We conclude this chapter with two different, but closely related, examples of geometric morphisms which are not bounded.

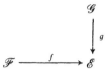

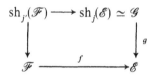

4.49 EXAMPLES. (i) Let G be an infinite group; then it is easily seen from 4.41 that the topos $(\mathscr{S}_f)^{\mathbf{G}}$ of Example 1.14 is defined over $\mathscr{S}_f$. (Explicitly, the geometric morphism γ is given by $\gamma^*(S) = S$ with trivial G-action, and $\gamma_*(M) =$ set of G-invariant elements of M.) But it is clear from the remarks after 4.43 that an $\mathscr{S}_f$-topos is bounded iff it has a finite set of generators. Now take G to be the group of integers, and let M_n denote the integers modulo n with G acting by translation; then a finite G-set X can be mapped into M_n iff all its G-orbits have length divisible by n. So each X can be mapped into only finitely many of the M_n; hence $(\mathscr{S}_f)^{\mathbf{G}}$ cannot have a finite set of generators. By choosing a suitable "large" group (i.e. one with a proper class of elements), we may construct a similar example with $\mathscr{S}_f$ replaced by $\mathscr{S}$ (see [GV], IV2.8).

(ii) Once again, let G be the group of integers, and let $\mathscr{E}$ be the full subcategory of $\mathscr{S}^{\mathbf{G}}$ whose objects are G-sets M with the property that multiplication by n is the identity on M for some positive integer n, i.e. all G-orbits of M have length dividing n. Now if M_1, M_2 are G-sets which satisfy this condition for n_1, n_2 respectively, it is easily seen that the product $M_1 \times M_2$ satisfies the condition for $\operatorname{lcm}(n_1, n_2)$; and the same is true for exponentials. Hence $\mathscr{E}$ is a topos, and the inclusion functor $\mathscr{E} \longrightarrow \mathscr{S}^{\mathbf{G}}$ is logical. Moreover, it is easily seen from 4.41 that $\mathscr{E}$ is defined over $\mathscr{S}$, but $\mathscr{E}$ does *not* have $\mathscr{S}$-indexed coproducts (e.g. we cannot construct $\coprod_n M_n$, where M_n is defined as in (i)), and so cannot be a Grothendieck topos. This explains why we had to introduce the distinction between coproducts and copowers in the discussion after 4.41; note also that $\mathscr{E}$ does have a set of generators (namely the M_n), but not an object of generators in the sense of 4.43. ■

EXERCISES 4

1. Let $\mathscr{F} \xrightarrow{q} \mathscr{I} \xrightarrow{i} \mathscr{E}$ be the image factorization of a geometric morphism $\mathscr{F} \xrightarrow{f} \mathscr{E}$.

   (i) If f is essential, prove that both q and i are. [Hint: it is convenient to use the description of $\mathscr{I}$ given in 4.16, and the opposite of 0.14.]

   (ii) If f^* preserves exponentials, prove that both q^* and i^* do. [Use the fact that i_* preserves exponentials, by 3.24.]

   (iii) If f^* is logical, prove that both q^* and i^* are. [Use (ii) and 3.55.]

†2. Let $\mathscr{E}$ be a topos, $\mathbf{C} \xrightarrow{f} \mathbf{D}$ a morphism of $\mathbf{cat}(\mathscr{E})$. Prove that the geometric morphism $\mathscr{E}^{\mathbf{C}} \longrightarrow \mathscr{E}^{\mathbf{D}}$ of 2.35 is a surjection if f_0 is epi, and an

inclusion if f is full and faithful (i.e. if

$$\begin{array}{ccc} C_1 & \xrightarrow{(d_0,d_1)} & C_0 \times C_0 \\ \downarrow{\scriptstyle f_1} & & \downarrow{\scriptstyle f_0 \times f_0} \\ D_1 & \xrightarrow{(d_0,d_1)} & D_0 \times D_0 \end{array}$$

is a pullback). [Hint: in the latter case, use internal profunctors to prove that the unit of $(\varinjlim_f \dashv f^*)$ is iso.] Deduce that, for any f, the image topos of $\mathscr{E}^{\mathbf{C}} \longrightarrow \mathscr{E}^{\mathbf{D}}$ can be written in the form $\mathscr{E}^{\mathbf{E}}$ for a suitable internal category $\mathbf{E}$.

3. Prove that an inclusion is a regular monomorphism in $\mathfrak{Top}$ (i.e. is the equalizer of its own cokernel-pair). Show, however, that not every equalizer in $\mathfrak{Top}$ is an inclusion. [Consider the morphism $\mathscr{S} \longrightarrow \mathscr{S}^{\mathbf{M}}$ induced by $\mathbf{1} \xrightarrow{e} \mathbf{M}$, where M is a nontrivial monoid.]

4. Let X, Y be the two non-isomorphic $\mathbb{Z}_2$-torsors in $\mathrm{Shv}(S^1)$. Show that there exists a surjection $\mathscr{E} \xrightarrow{u} \mathrm{Shv}(S^1)$ such that u^*X and u^*Y are isomorphic. By considering the morphisms $\mathrm{Shv}(S^1) \rightrightarrows \mathscr{S}[U]$ classifying X and Y (cf. 4.37(iv)), deduce that a surjection need not be an epimorphism in $\mathfrak{Top}$.

†5. Let X be a topological space. Show that a sheaf on X is locally constant (i.e. locally isomorphic to a constant sheaf) iff its representation as a local homeomorphism is a *covering projection* in the sense of [AT], §2.1. Deduce that if X is path-connected, then a locally constant sheaf may be completely specified by giving its stalk at some point of x of X, together with an action of the fundamental group $G = \Pi_1(X, x)$ on this stalk; and that if X has a universal covering space $\tilde{X} \longrightarrow X$, then every G-set arises from a locally constant sheaf in this way. Show also that (the sheaf of sections of) $\tilde{X} \longrightarrow X$ is a G-torsor in $\mathrm{Shv}(X)$, and that its classifying map $\mathrm{Shv}(X) \xrightarrow{u} \mathscr{S}^{\mathbf{G}}$ may be described as follows: $u^*(M, \alpha)$ is the locally constant sheaf on X with stalk M and $\Pi_1(X, x)$-action α, and $u_*(E \xrightarrow{p} X)$ is the set of morphisms $\tilde{X} \longrightarrow E$ in $\mathrm{Shv}(X)$, with G acting via automorphisms (= covering translations) of $\tilde{X}$.

6. Let $\mathbf{P}$ be an internal poset in $\mathscr{E}$, and $\mathscr{F}$ an $\mathscr{E}$-topos. Prove that $\mathfrak{Top}/\mathscr{E}(\mathscr{F}, \mathscr{E}^{\mathbf{P}})$ is a poset. [Hint: show that every flat presheaf on $\mathbf{P}$ is a subobject of 1 in $\mathscr{E}^{\mathbf{P}^{\mathrm{op}}}$.]

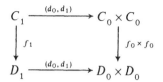

†7. Prove that the pullback square

$$\begin{array}{ccc} \mathscr{F}^{f^*\mathbf{C}} & \xrightarrow{f^{\mathbf{C}}} & \mathscr{E}^{\mathbf{C}} \\ \downarrow & & \downarrow \\ \mathscr{F} & \xrightarrow{f} & \mathscr{E} \end{array}$$

of 4.35 satisfies the "Beck condition" $\mathbf{C}^*.f_* \cong (f^{\mathbf{C}})_*.(f^*\mathbf{C})^*$. Is the same result true if we replace $\mathscr{E}^{\mathbf{C}}$ by an arbitrary bounded $\mathscr{E}$-topos? [Hint: consider the square

$$\begin{array}{ccc} \mathbf{1} & \longrightarrow & \mathrm{sh}_{j_U^c}(\mathscr{E}) \ , \\ \downarrow & & \downarrow \\ \mathrm{sh}_{j_U^o}(\mathscr{E}) & \longrightarrow & \mathscr{E} \end{array}$$

where U is an open object in $\mathscr{E}$.]

8. Let $\mathscr{E} \xrightarrow{\gamma} \mathscr{S}$ be a non-degenerate $\mathscr{S}$-topos. Prove that the following conditions are equivalent:

(i) γ^* is full.
(ii) γ_* preserves coproducts.
(iii) $\mathscr{E}$ is *connected*, i.e. if $U \amalg V \cong 1$ in $\mathscr{E}$, then either $U \cong 1$ and $V \cong 0$, or vice versa. (Cf. definition of connectedness for topological spaces.)

†9. Let $\mathscr{E} \xrightarrow{\gamma} \mathscr{S}$ be an S-topos such that γ is essential. (We say that $\mathscr{E}$ is *locally connected*, by analogy with Ex. 0.7). Show that γ^* preserves exponentials. Show also that if $X = \gamma_!(1)$, then there is a bijection between direct summands of 1 in $\mathscr{E}$ and subsets of X [use the fact that $\gamma_!$ preserves coproducts], and deduce that if X is finite then $\mathscr{E}$ can be written as an X-indexed coproduct in $\mathfrak{Top}$ of connected $\mathscr{S}$-toposes.

10. Give an alternative proof of 4.44(i) using the characterization of bounded morphisms provided by 4.46 (i.e. taking 4.46(ii) as the definition of boundedness). [Hint: you will need to use the result of Ex. 2.7.]

11. Let α be a regular cardinal. We say that a functor $\mathscr{C} \xrightarrow{T} \mathscr{D}$ between cocomplete categories has *rank* $< \alpha$ if it preserves α-filtered colimits

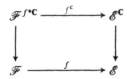

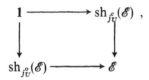

(cf. [165], § 5). If $\mathscr{E}$ is a Grothendieck topos, show that the direct image of any geometric morphism $\mathscr{S} \xrightarrow{p} \mathscr{E}$ has a rank [take $\alpha >$ card p^*G, where G is an object of generators for $\mathscr{E}$ over $\mathscr{S}$]. Hence show that if $\mathscr{S} \xrightarrow{L} \mathscr{S}$ is a left exact functor, the topos $\mathscr{E}$ obtained by Artin glueing along L is a Grothendieck topos iff L has a rank. [If L has rank $< \alpha$, show that $\mathscr{E}$ is generated by $(1, 0, 0 \longrightarrow 1)$ and all objects of the form $(Y, 1, 1 \longrightarrow L(Y))$ where Y is a set of cardinality $< \alpha$. The existence of left exact functors $\mathscr{S} \longrightarrow \mathscr{S}$ without rank appears to be closely related to the existence of measurable cardinals; see [10].]

Chapter 5

Logical Aspects of Topos Theory

5.1. BOOLEAN TOPOSES

So far, we have mainly considered toposes from the geometrical viewpoint, initiated by Grothendieck and his followers [GV], that a topos is a "generalized space". Thus, for example, the definition of a geometric morphism was governed by considerations of "what happens to sheaves when we are given a continuous map of topological spaces?"—although we have since seen that geometric morphisms arise naturally in a host of other situations, not all of them noticeably geometrical. In this chapter, our aim is to present some of the features of topos theory which derive from its alternative "logical" aspect, i.e. the idea that a topos is a "generalized set theory".

We begin with some detailed consideration of the albegraic structure of Ω. Recall that in chapter 3 we defined a partial order Ω_1 on Ω, and three binary operations $\wedge$, $\vee$ and $\Rightarrow$: $\Omega \times \Omega \longrightarrow \Omega$. We also have two distinguished elements of Ω, namely $1 \xrightarrow{t} \Omega$ (which is of course the classifying map of $1 \longrightarrow 1$), and $1 \xrightarrow{f} \Omega$ ("false"), which is defined as the classifying map of $0 \rightarrowtail 1$; and a unary operation $\Omega \xrightarrow{\neg} \Omega$, defined as the classifying map of $1 \overset{f}{\rightarrowtail} \Omega$.

5.11 Definition. We recall that a *lattice* may be defined as a set L equipped with two binary operations $\wedge$ and $\vee$, and two distinguished elements t and f, satisfying the equations $x \wedge x = x$, $x \wedge t = x$, $x \wedge y = y \wedge x$, $x \wedge (y \wedge z) = (x \wedge y) \wedge z$, $x \wedge (x \vee y) = x$, and their duals, for all $x, y, z \in L$. Clearly, we can reinterpret this definition in any category with finite products, replacing the above equations by commutative diagrams in the usual way. We say that a lattice is *distributive* if it satisfies the additional equation $x \wedge (y \vee z) = (x \wedge y) \vee (x \wedge z)$; it is well known that this equation is equivalent to its dual in a lattice. (We refer the reader to [LT] for an account of the fundamentals of lattice theory.) ■

It is well known that we may define a partial order L_1 on any lattice L, as the equalizer of $L \times L \underset{\pi_1}{\overset{\wedge}{\rightrightarrows}} L$ (or equivalently of $\vee$ and π_2); and conversely a poset $(L_1 \rightrightarrows L)$ is a lattice iff it is *antisymmetric* (i.e.

$$\begin{array}{ccc} L & \xrightarrow{\;i\;} & L_1 \\ {\scriptstyle i}\downarrow & & \downarrow{\scriptstyle (d_0,\, d_1)} \\ L_1 & \xrightarrow{(d_1,\, d_0)} & L \times L \end{array}$$

is a pullback) and has all finite limits and colimits ($\wedge$ and $\vee$ being the binary product and coproduct operators, and t, f the terminal and initial objects). This equivalence is again easy to prove in any category with finite limits.

5.12 DEFINITION. (i) *A Heyting algebra* (also called a Brouwerian lattice [LT] or pseudo-Boolean algebra [MM]) is a lattice which is cartesian closed as a poset; i.e. it has an additional binary operation $\Rightarrow$ with the property that $(x \wedge y) \leqslant z$ iff $x \leqslant (y \Rightarrow z)$. ($\Rightarrow$ may also be characterized equationally, by the conditions $x \Rightarrow x = t$, $x \wedge (x \Rightarrow y) = x \wedge y$, $y \wedge (x \Rightarrow y) = y$, and $x \Rightarrow (y \wedge z) = (x \Rightarrow y) \wedge (x \Rightarrow z)$; see [FK], Proposition 1.22.)

(ii) A *Boolean algebra* may be defined as a distributive lattice L equipped with a unary operation $\neg$, such that $x \wedge (\neg x) = f$ and $x \vee (\neg x) = t$ for all $x \in L$. ■

5.13 PROPOSITION. *Let $\mathscr{E}$ be a topos. Then with the operations t, f, $\wedge$, $\vee$, $\Rightarrow$ defined above, the poset $\mathbf{\Omega} = (\Omega_1 \rightrightarrows \Omega)$ is an internal Heyting algebra in $\mathscr{E}$.*

Proof. It is clearly sufficient to verify that, for each object X of $\mathscr{E}$, the given operations make the class of morphisms $X \longrightarrow \Omega$ into an (external) Heyting algebra. But this is immediate from the external descriptions of the operations which we gave in chapter 3. ■

Now any Boolean algebra becomes a Heyting algebra if we define the operation $\Rightarrow$ by $x \Rightarrow y = (\neg x) \vee y$; but the converse is not true. A Heyting algebra is clearly distributive (since $x \wedge (-)$, having a right adjoint, preserves coproducts), and we can always define a unary operation $\neg$ by $\neg x = x \Rightarrow f$; then $\neg x$ is the unique largest element of L satisfying $x \wedge (\neg x) = f$, but we do not necessarily have $x \vee (\neg x) = t$. In fact it is not hard to see that we have this equation iff $\neg\neg x = x$ for all $x \in L$; for the latter equation implies that

$$x \vee (\neg x) = \neg\neg(x \vee (\neg x)) = \neg(\neg x \wedge \neg\neg x) = \neg f = t,$$

and conversely we have

$$\begin{aligned}\neg\neg x = \neg\neg x \wedge t = \neg\neg x \wedge (x \vee \neg x) &= (\neg\neg x \wedge x) \vee (\neg\neg x \wedge \neg x)\\ &= (\neg\neg x \wedge x) \vee f = \neg\neg x \wedge x,\end{aligned}$$

so that $\neg\neg x \leqslant x$; but $x \leqslant \neg\neg x$ is easily seen to be true in any Heyting algebra.

5.14 PROPOSITION. *Let $\mathscr{E}$ be a topos. The following conditions are equivalent:*

(i) $\boldsymbol{\Omega}$ *is an internal Boolean algebra.*
(ii) $\neg\neg = 1_{\Omega}$.
(iii) *All subobjects in $\mathscr{E}$ have* complements, *i.e. given $X' \rightarrowtail X$ there exists $X'' \rightarrowtail X$ such that $X' \amalg X'' \longrightarrow X$ is an isomorphism.*
(iv) $1 \amalg 1 \xrightarrow{\binom{t}{f}} \Omega$ *is an isomorphism.*

If these conditions are satisfied, we say $\mathscr{E}$ is a Boolean topos.

Proof. (i) $\Leftrightarrow$ (ii) is immediate from the corresponding "external" result.

(i) $\Rightarrow$ (iii): Let $X \xrightarrow{\phi} \Omega$ be the classifying map of $X' \rightarrowtail X$, and let X'' be the subobject classified by $\neg\phi$. From the definition of a Boolean algebra, we then have $X' \cap X'' \cong 0$ and $X' \cup X'' \cong X$ as subobjects of X; but this is equivalent to saying that

$$\begin{array}{ccc} 0 & \longrightarrow & X' \\ \downarrow & & \downarrow \\ X'' & \rightarrowtail & X \end{array}$$

is a pushout diagram, i.e. $X' \amalg X'' \cong X$.

(iii) $\Rightarrow$ (iv): As a particular case, the generic subobject $1 \overset{t}{\rightarrowtail} \Omega$ has a complement, which must be $1 \overset{f}{\rightarrowtail} \Omega$ since the latter is the largest subobject whose intersection with t is 0.

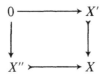

(iv) $\Rightarrow$ (ii) since, in any topos, the restriction of $\neg$ to the subobject $1 \amalg 1 \xrightarrow{\binom{t}{f}} \Omega$ is easily seen to be the involution which interchanges the factors of $1 \amalg 1$. ∎

5.15 EXAMPLES. (i) Let X be a topological space, and consider the topos $\mathrm{Shv}(X)$. For any open $U \subseteq X$, $\Omega(U)$ is the lattice of open subsets of U. In particular for $U = X$ and $V \in \Omega(X)$, we have $\neg V = \mathrm{int}(X - V)$, and so $\neg\neg V$ is the interior of the closure of V, which is not equal to V in general. In fact if X is a T_0-space, then $\mathrm{Shv}(X)$ is Boolean iff X is discrete.

(ii) Let G be a monoid, regarded as a small category with one object. Then the topos $\mathscr{S}^{\mathbf{G}}$ of G-sets is Boolean iff G is a group. For if G is a group, then any sub-G-set of a G-set M is simply a union of G-orbits of M, and its complement is the union of the remaining obits; but if not, then the set of non-invertible elements of G is a sub-G-set of G itself, which clearly does not have a complement. Slightly more generally, we can show for any small category $\mathbf{C}$ that $\mathscr{S}^{\mathbf{C}}$ is Boolean iff $\mathbf{C}$ is a groupoid. ∎

The question now arises: given a topos which is not Boolean, can we find a "best approximation" to it which is Boolean? One way of answering this question is suggested by the following:

5.16 REMARK. Let H be a Heyting algebra. We say x is a *regular element* of H if $\neg\neg x = x$. Let $B \subseteq H$ be the set of regular elements; then the partial order restricted to B makes it a Boolean algebra. (The operations $\wedge$ and $\neg$ in B agree with those in H, but $\vee_B = \neg\neg(\vee_H)$.) For example, if H is the lattice of open subsets of a topological space, then B is the algebra of regular open subsets. ∎

5.17 THEOREM (Lawvere–Tierney). *Let $\mathscr{E}$ be a topos. Then $\neg\neg$ is a toplogy in $\mathscr{E}$, and* $\mathrm{sh}_{\neg\neg}(\mathscr{E})$ *is Boolean.*

Proof. Let X be any object of $\mathscr{E}$. Then it is easily seen that the unary operation on subobjects of X induced by $\neg$ is order-reversing, and that $X' \leqslant \neg X''$ iff $X'' \leqslant \neg X'$. From this it follows at once that $X' \leqslant \neg\neg X'$ (take $X'' = \neg X'$) and that $\neg X' \cong \neg\neg\neg X'$; hence $\neg\neg$ induces a closure operation on subobjects of X. And it is clearly universal; so by 3.14 $\neg\neg$ is a topology. Now since $\Omega_{\neg\neg}$ is defined as the equalizer of 1_Ω and $\neg\neg$, it follows from 5.16 that it is an internal Boolean algebra in $\mathscr{E}$; hence it is also Boolean in $\mathrm{sh}_{\neg\neg}(\mathscr{E})$. ∎

It must be said that the topos $\mathrm{sh}_{\neg\neg}(\mathscr{E})$ is a "best Boolean approximation" to $\mathscr{E}$ only in a very weak sense. It is not, for example, true that it is the maximal Boolean sheaf subtopos of $\mathscr{E}$; in general no such maximal subtopos exists. We can, however, describe $\neg\neg$ by the following universal property:

5.18 PROPOSITION. *$\neg\neg$ is the unique largest topology in $\mathscr{E}$ for which $0 \rightarrowtail 1$ is closed.*

Proof. Since $0 \rightarrowtail 1$ has a complement (namely $1 \longrightarrow 1$) it is certainly $\neg\neg$-closed. Conversely, let j be a topology for which $0 \rightarrowtail 1$ is closed, and let $X' \overset{\sigma}{\rightarrowtail} X$ be a j-dense monomorphism. Let $X'' \rightarrowtail X$ be the negation of X' (i.e. the subobject classified by $\neg\phi_\sigma$); then from the pullback diagram

$$\begin{array}{ccc} 0 & \rightarrowtail & X' \\ \downarrow & & \downarrow \\ X'' & \rightarrowtail & X \end{array}$$

we see that $0 \rightarrowtail X''$ is j-dense. But it is also j-closed, since

$$\begin{array}{ccc} 0 & \longrightarrow & 0 \\ \downarrow & & \downarrow \\ X'' & \longrightarrow & 1 \end{array}$$

is a pullback. Hence $0 \rightarrowtail X''$ is iso, and so σ is $\neg\neg$-dense; i.e. $j \leqslant \neg\neg$. ■

Note that we could alternatively have proved 5.18 by using the ideas of 3.57 and 3.59(v). It follows in particular from 5.18 that 0 is always a $\neg\neg$-sheaf; hence if $\mathscr{E}$ is non-degenerate (i.e. satisfies $0 \not\cong 1$), so is $\mathrm{sh}_{\neg\neg}(\mathscr{E})$.

5.2. THE AXIOM OF CHOICE

Let $\mathscr{E}$ be a topos, X an object of $\mathscr{E}$, and write $\mathscr{P}X$ for the (external) poset of subobjects of X (i.e. monomorphisms with codomain X). Then we have an inclusion functor i_X: $\mathscr{P}_X \longrightarrow \mathscr{E}/X$; and this has a left adjoint, the *support functor* σ_X: $\mathscr{E}/X \longrightarrow \mathscr{P}X$, which sends a morphism $Y \overset{f}{\longrightarrow} X$ to its image (1.52) considered as a subobject of X.

5.21 DEFINITION. (i) We say that *supports split* in $\mathscr{E}$ (or $\mathscr{E}$ satisfies (SS)) if, for every $X \in \mathscr{E}$, the canonical epimorphism $X \twoheadrightarrow \sigma_1(X)$ (i.e. the unit of $(\sigma_1 \dashv i_1)$) is split. (This is equivalent to the statement that every subobject of 1 is projective in $\mathscr{E}$.)

(ii) We say that $\mathscr{E}$ satisfies the *axiom of choice* (AC) if supports split in $\mathscr{E}/X$ for every X. (Equivalently, every object of $\mathscr{E}$ is projective, or every epi in $\mathscr{E}$ splits.) ■

5.22 EXAMPLES. (i) Let $\mathbf{P}$ be a well-ordered set. Then supports split in the topos $\mathscr{S}^{\mathbf{P}}$; for if X is any nonzero diagram on $\mathbf{P}$, its support

$$\sigma_1(X) = \{p \in P \mid X(p) \neq \emptyset\}$$

has a minimal element p_0, and then it is clear that

$$\sigma_1(X) = \{p \in P \mid p_0 \leqslant p\} = h^{p_0}.$$

Then by the Yoneda Lemma (0.11) any element of $X(p_0)$ determines a splitting for $X \twoheadrightarrow \sigma_1(X)$.

But $\mathscr{S}^{\mathbf{P}}$ does not in general satisfy (AC); for let $\mathbf{P}$ be the ordinal $\mathbf{2} = \{0, 1\}$, and take Y = constant diagram 2, $X(0) = 2$, $X(1) = 1$. Then the obvious epimorphism $Y \twoheadrightarrow X$ cannot be split.

(ii) Let X be a topological space. Then $\mathrm{Shv}(X)$ does not in general satisfy (SS); for example if we take $X = S^1$ and Y = sheaf of sections of the double covering map (cf. 4.37(ii)), then $\sigma_1(Y) = 1$ but Y has no global elements. ■

An important consequence of (AC) is the following, due to R. Diaconescu [31]:

5.23 THEOREM. *If a topos satisfies* (AC), *then it is Boolean.*

Proof. Let $Y \overset{\sigma}{\rightarrowtail} X$ be a monomorphism in $\mathscr{E}$. Form the pushout

$$\begin{array}{ccc} Y & \overset{\sigma}{\rightarrowtail} & X \\ {\scriptstyle\sigma}\downarrow & & \downarrow{\scriptstyle q_1} \\ X & \xrightarrow{q_2} & Q \end{array} ;$$

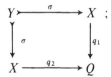

then by 1.28 q_1 and q_2 are mono, and the square is also a pullback. Now $X \amalg X \xrightarrow{\binom{q_1}{q_2}} Q$ is clearly epi; let $Q \xrightarrow{\tau} X \amalg X$ be a splitting for it.

For $i = 1, 2$ and $j = 1, 2$, form the pullbacks

$$\begin{array}{ccccc} T_{ij} & \xrightarrow{t_{ij}} & Z_i & \xrightarrow{w_i} & X \\ \big\downarrow{\scriptstyle s_{ij}} & & \big\downarrow{\scriptstyle z_i} & & \big\downarrow{\scriptstyle v_i} \\ X & \xrightarrow{q_j} & Q & \xrightarrow{\tau} & X \amalg X \end{array} \quad ;$$

then by the universality of coproducts in $\mathscr{E}$ we have $Q \cong Z_1 \amalg Z_2$ and $X \cong T_{1j} \amalg T_{2j}$. Now we can write $T_{11} \cong U_{11} \amalg U_{12}$, $T_{21} \cong U_{21} \amalg U_{22}$, where

$$\begin{array}{ccc} U_{ij} & \longrightarrow & T_{i1} \\ \big\downarrow & & \big\downarrow{\scriptstyle s_{i1}} \\ T_{j2} & \xrightarrow{s_{j2}} & X \end{array}$$

is a pullback; and so

$$X \cong U_{11} \amalg U_{12} \amalg U_{21} \amalg U_{22}.$$

But now the pair $X \underset{\tau q_j}{\overset{v_i}{\rightrightarrows}} X \amalg X$ is coreflexive, with common splitting given by the codiagonal map $X \amalg X \xrightarrow{\nabla} X$; hence the equality $v_i w_i t_{ij} = \tau q_j s_{ij}$ implies that $w_i t_{ij} = s_{ij}$. Hence the square

$$\begin{array}{ccc} U_{ii} & \longrightarrow & T_{i1} \\ \big\downarrow & & \big\downarrow{\scriptstyle t_{i1}} \\ T_{i2} & \xrightarrow{t_{i2}} & Z_i \end{array}$$

commutes, since the two ways round are coequalized by $Z_i \xrightarrow{w_i} X$; and it is not hard to see that this square must also be a pullback.

So on rewriting our original pushout square as

$$\begin{array}{ccc} Y & \longrightarrow & T_{11} \amalg T_{21} \\ \big\downarrow & & \big\downarrow{\scriptstyle t_{11} \amalg t_{21}} \\ T_{12} \amalg T_{22} & \xrightarrow{t_{12} \amalg t_{22}} & Z_1 \amalg Z_2 \end{array}$$

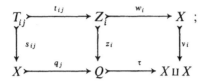

we obtain $Y \cong U_{11} \amalg U_{22}$ as a subobject of $T_{11} \amalg T_{21} \cong X$. But this subobject clearly has a complement, namely $U_{12} \amalg U_{21}$. ■

The definition of (AC) given in 5.21 is easily seen to be equivalent in $\mathcal{S}$ to any of the set-theoretic forms of the axiom of choice. However, it is clear that the topos $\mathcal{S}$ satisfies (SS) even if we do not assume choice, since every non-empty set has an element. It turns out that there is a weaker form of (AC), applicable to toposes which do not satisfy (SS), which is often useful in practice.

5.24 Definition. (i) We say an object X of a topos $\mathcal{E}$ is *internally projective* if the functor $(-)^X : \mathcal{E} \longrightarrow \mathcal{E}$ preserves epimorphisms.

(ii) We say an epimorphism $X \xrightarrow{f} Y$ in $\mathcal{E}$ is *locally split* if there exists an object V of $\mathcal{E}$ with global support (i.e. such that $V \longrightarrow 1$ is epi), such that $V^*(f)$ is split epi in $\mathcal{E}/V$. ■

5.25 Proposition. *The following conditions are equivalent:*

(*i*) *Every object of* $\mathcal{E}$ *is internally projective.*
(*ii*) *Every epi in* $\mathcal{E}$ *is locally split.*
(iii) *If* $X \xrightarrow{f} Y$ *is an epi in* $\mathcal{E}$, *then* $\Pi_Y(f)$ *has global support.*

If these conditions hold, we say $\mathcal{E}$ *satisfies the* implicit axiom of choice (IC).

Proof. (i) ⇒ (iii): From Ex. 1.8, we have a pullback diagram

$$\begin{array}{ccc} \Pi_Y(f) & \longrightarrow & X^Y \\ \downarrow & & \downarrow{\scriptstyle f^Y} \\ 1 & \xrightarrow{1_Y} & Y^Y \end{array}$$

so Y internally projective $\Rightarrow f^Y$ epi $\Rightarrow \Pi_Y(f) \longrightarrow 1$ epi.

(iii) ⇒(ii): By definition, global elements of $\Pi_Y(f)$ correspond to morhisms $1_Y \longrightarrow f$ in $\mathcal{E}/Y$, i.e. splittings of f in $\mathcal{E}$; so in the topos $\mathcal{E}/\Pi_Y(f)$ we have a generic such splitting.

(ii) ⇒(i): Let $X \xrightarrow{f} Y$ be an epi in $\mathcal{E}$, and V an object with global support such that $V^*(f)$ is split. Now V^* preserves exponentials, so for any Z we have $V^*(f^Z) \cong V^*(f)^{V^*Z}$ which is split epi; hence f^Z is epi, since V^* reflects epimorphisms by 4.12(iii). ■

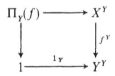

5.26 Example. Let G be a group. Then the topos $\mathscr{S}^G$ satisfies (IC), since the forgetful functor $\mathscr{S}^G \longrightarrow \mathscr{S}$ preserves exponentials and preserves and reflects epimorphisms. But $\mathscr{S}^G$ does not satisfy (AC) unless G is trivial. ■

5.27 Lemma. *If* $\mathscr{E}$ *satisfies* (IC), *so does* $\mathscr{E}/X$ *for any* X.

Proof. Let

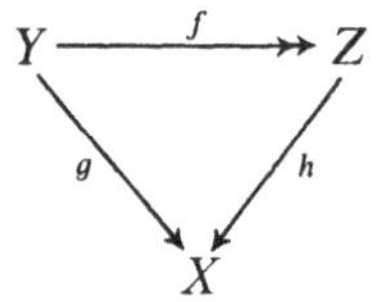

be an epimorphism in $\mathscr{E}/X$. Now the canonical isomorphism $\Pi_Z(f) \cong \Pi_X \Pi_h(f)$ transposes to give a morphism $X^*\Pi_Z(f) \longrightarrow \Pi_h(f)$ in $\mathscr{E}/X$; hence if $\Pi_Z(f)$ has global support in $\mathscr{E}$, $\Pi_h(f)$ has global support in $\mathscr{E}/X$. ■

5.28 Lemma. (AC) *is equivalent to the conjunction of* (IC) *and* (SS).

Proof. Suppose $\mathscr{E}$ satisfies (IC) and (SS); let $X \xrightarrow{f} Y$ be an epimorphism in $\mathscr{E}$. Then by (IC) $\Pi_Y(f) \longrightarrow 1$ is epi; but by (SS) this epi is split. So we have a global element of $\Pi_Y(f)$, i.e. a splitting of f in $\mathscr{E}$. The converse is trivial. ■

Note that we did not use the full strength of (SS) in 5.28, but merely the fact that 1 is projective. Hence if $\mathscr{E}$ satisfies (IC) and X is (externally) projective in $\mathscr{E}$, it follows from 5.27 that $\mathscr{E}/X$ satisfies (AC). Note also that 5.23 can be strengthened to give a proof that (IC) implies Booleanness; for if the negation of a subobject $Y \overset{\sigma}{\rightarrowtail} X$ becomes an "honest" complement after pullback along an epi $V \twoheadrightarrow 1$, it must have been so already in $\mathscr{E}$.

Now let $X \xrightarrow{f} Y$ be any morphism in $\mathscr{E}$. The functors $f^*\colon \mathscr{E}/Y \longrightarrow \mathscr{E}/X$ and $\Pi_f\colon \mathscr{E}/X \longrightarrow \mathscr{E}/Y$ preserve the terminal objects and monos, so they restrict to functors, $\mathscr{P}Y \longrightarrow \mathscr{P}X$ and $\mathscr{P}X \longrightarrow \mathscr{P}Y$ which we denote by f^{-1} and $\forall_f$ respectively; and clearly $f^{-1} \dashv \forall_f$, since $\mathscr{P}X$ is a full subcategory of $\mathscr{E}/X$. The functor Σ_f does not preserve the terminal object, but we can nevertheless obtain a left adjoint $\exists_f$ for f^{-1} by forming the composite $\sigma_Y.\Sigma_f.i_X$ (i.e. $\exists_f(T \rightarrowtail X) = \mathrm{im}\,(T \longrightarrow X \xrightarrow{f} Y)$). Note that $\exists_f$ is the "externalization" of the morphism $\exists f\colon \Omega^X \longrightarrow \Omega^Y$ defined in 1.32 and Ex. 1.10.

Summarizing our information about this situation, we have

5.29 PROPOSITION. *In the diagram*

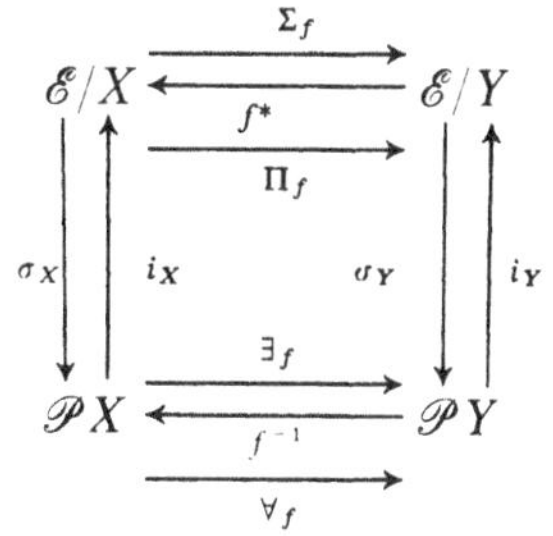

we have the adjunctions $\Sigma_f \dashv f^* \dashv \Pi_f$, $\exists_f \dashv f^{-1} \dashv \forall_f$, $\sigma \dashv i$, *and the commutative squares (up to natural isomorphism):*

$f^*i_Y = i_X f^{-1}$, $\Pi_f i_X = i_Y \forall_f$ *(by definition),*

and

$\sigma_Y \Sigma_f \cong \exists_f \sigma_Y$, $\sigma_X f^* \cong f^{-1}\sigma_Y$ *(by uniqueness of adjoints).*

We have $\Sigma_f i_X \cong i_Y \exists_f$ iff f *is mono; and we have* $\sigma_Y \Pi_f \cong \forall_f \sigma_X$ *for all* f *iff* $\mathscr{E}$ *satisfies* (IC).

Proof. Only the last line requires further comment. But it follows easily from 5.25 and 5.27 that $\mathscr{E}$ satisfies (IC) iff the functors Π_f preserve epis; and since we know they preserve monos, this is equivalent to saying that they commute with support functors. ■

5.3. THE AXIOM (SG)

5.31 DEFINITION. We say that a topos $\mathscr{E}$ satisfies the axiom (SG) if the subobjects of 1 in $\mathscr{E}$ form a class of generators (0.44); i.e. if, given any pair of morphisms $X \underset{g}{\overset{f}{\rightrightarrows}} Y$ such that $f \neq g$, we can find $U \rightarrowtail 1$ and $U \xrightarrow{h} X$ such that $fh \neq gh$. ■

Note that if $\mathscr{E}$ is defined over $\mathscr{S}$, then axiom (SG) is equivalent to the assertion that 1 is an object of generators for $\mathscr{E}$ over $\mathscr{S}$ (cf. 4.43). More generally, if

$\mathscr{E}$ satisfies (SG) then any geometric morphism $\mathscr{E} \longrightarrow \mathscr{F}$ is bounded, with 1 as object of generators; but the converse need not be true unless $\mathscr{F}$ satisfies (SG).

From a logical point of view, the significance of axiom (SG) is that morphisms $U \longrightarrow X$, where U is a subobject of 1, correspond to partial maps $1 \longrightarrow X$, i.e. to *global* elements of $\tilde{X}$. So by applying the functor $\tilde{}$ we can reduce "internal" questions about objects and morphisms of $\mathscr{E}$ to "external" questions about global elements. Another useful consequence of (SG), noted by F. Borceux [12], is the following result:

5.32 LEMMA. *If $\mathscr{E}$ satisfies* (SG), *then Ω is a cogenerator for $\mathscr{E}$.*

Proof. Suppose given $X \overset{f}{\underset{g}{\rightrightarrows}} Y$ with $f \neq g$. Then we can find $U \rightarrowtail 1$ and $U \xrightarrow{h} X$ with $fh \neq gh$. Now since $U \longrightarrow 1$ is mono, so is $U \xrightarrow{fh} Y$; let $Y \xrightarrow{\phi} \Omega$ be its classifying map. Also, in the pullback diagram

$$\begin{array}{ccc} V & \overset{p}{\rightarrowtail} & U \\ {\scriptstyle q}\downarrow & & \downarrow{\scriptstyle fh} \\ U & \underset{gh}{\rightarrowtail} & Y \end{array}$$

we have $p = q$, and so $V \overset{q}{\rightarrowtail} U$ is the equalizer of fh and gh; hence q is not iso. But now we have $\phi fh \neq \phi gh$, since the former classifies 1_U and the latter classifies q; hence $\phi f \neq \phi g$. ■

The next two results provide us with a supply of examples of toposes satisfying (SG).

5.33 LEMMA. *If $\mathscr{E}$ is Boolean and satisfies* (SS), *then $\mathscr{E}$ satisfies* (SG).

Proof. Suppose given $X \overset{f}{\underset{g}{\rightrightarrows}} Y$ with $f \neq g$. Then the equalizer $E \rightarrowtail X$ of f and g is not an isomorphism, and so its complement $E' \rightarrowtail X$ is nonzero. Let $U = \sigma_1(E')$; then U is nonzero by 1.55, and by (SS) we have a morphism $U \longrightarrow E'$. Let h be the composite $U \longrightarrow E' \longrightarrow X$; then the equalizer of fh and gh is $h^*(E \rightarrowtail X) \cong 0 \rightarrowtail U$, so $fh \neq gh$. ■

5.34 PROPOSITION. *Let $\mathscr{E}$ be a topos satisfying* (SG). *Then*

(i) *If $\mathbf{P}$ is an internal poset in $\mathscr{E}$, $\mathscr{E}^{\mathbf{P}}$ satisfies* (SG).
(ii) *If j is a topology in $\mathscr{E}$, $\mathrm{sh}_j(\mathscr{E})$ satisfies* (SG).

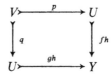

Proof. (i) Suppose given $X \overset{f}{\underset{g}{\rightrightarrows}} Y$ in $\mathscr{E}^{\mathbf{P}}$ with $f \neq g$, where

$$X = (X_0 \xrightarrow{\gamma_0} P_0, X_0 \times_{P_0} P_1 \xrightarrow{a} X_0),$$

etc. Then we can find $U \rightarrowtail 1$ in $\mathscr{E}$ and $U \xrightarrow{h} X_0$ with $f_0 h \neq g_0 h$; let δ denote the composite $U \xrightarrow{h} X_0 \xrightarrow{\gamma_0} P_0$, so that h is a morphism $\delta \longrightarrow \gamma_0$ in $\mathscr{E}/P_0$. Transposing across the adjunction of 2.22, we obtain $R(\delta) \xrightarrow{\bar{h}} X$ with $f\bar{h} \neq g\bar{h}$; but from the fact that $U \longrightarrow 1$ and $P_1 \xrightarrow{(d_0, d_1)} P_0 \times P_0$ are mono in $\mathscr{E}$, we readily deduce that $U \times_{P_0} P_1 \xrightarrow{d_1 \pi_2} P_0$ is mono; i.e. $R(\delta) \longrightarrow 1$ is mono in $\mathscr{E}^{\mathbf{P}}$.

(ii) is immediate from the fact that the associated sheaf functor preserves 1 and monomorphisms. ■

Our next objective is to prove a theorem characterizing those toposes defined over $\mathscr{S}$ which satisfy (SG). Before we can do this, however, we must devote some attention to the "infinitary" structure of the internal poset $\mathbf{\Omega}$.

Let $\mathbf{P} = (P_1 \rightrightarrows P)$ be an internal poset in a topos. Then we have morphisms $\uparrow$seg: $P \longrightarrow \Omega^P$, and $\downarrow$seg: $P \longrightarrow \Omega^P$, whose exponential transposes are respectively the classifying maps of $P_1 \rightarrowtail P \times P$ and $P_1{}^{\mathrm{op}} \rightarrowtail P \times P$. [The symbol $\downarrow$seg should be read as "down-segment"; in $\mathscr{S}$ it is the map which sends $p \in P$ to the subset $\{p' \in P \mid p' \leqslant p\}$.] It is readily seen that $\downarrow$seg is order-preserving (i.e. is an internal functor) and $\uparrow$seg is order-reversing, where Ω^P is partially ordered by $\Omega_1{}^P \rightarrowtail (\Omega \times \Omega)^P \cong \Omega^P \times \Omega^P$. If $\mathbf{P} \xrightarrow{f} \mathbf{Q}$ and $\mathbf{Q} \xrightarrow{g} \mathbf{P}$ are order-preserving maps between internal posets, we say that f is *internally left adjoint* to g if $P \xrightarrow{(1, gf)} P \times P$ factors through $P_1 \rightarrowtail P \times P$ and $Q \xrightarrow{(fg, 1)} Q \times Q$ through Q_1. (This is clearly equivalent to saying that the maps on generalized elements induced by f and g are adjoint in the external sense.)

We say that an internal poset $\mathbf{P}$ is *internally complete* if there exists an order-preserving map $\Omega^P \xrightarrow{\bigcup} P$ which is internally left adjoint to $\downarrow$seg. It is clear that in the case $\mathscr{E} = \mathscr{S}$, this is equivalent to the usual definition of completeness for lattices; and as in $\mathscr{S}$, it is not hard to prove that the condition is equivalent to its dual, i.e. the statement that $\uparrow$seg: $\mathbf{P}^{\mathrm{op}} \longrightarrow \mathbf{\Omega}^P$ has an internal left adjoint $\bigcap$. To see this, form the pullback

$$\begin{array}{ccc} B & \longrightarrow & \Omega_1^P \\ \downarrow & & \downarrow \\ \Omega^P \times P & \xrightarrow{1 \times \downarrow \mathrm{seg}} & \Omega^P \times \Omega^P \end{array},$$

and let $\Omega^P \xrightarrow{ubd} \Omega^P$ be the transpose of the classifying map of $B \rightarrowtail \Omega^P \times P$. Then by computing its effect on generalized elements, we may easily verify that the composite $\Omega^P \xrightarrow{ubd} \Omega^P \xrightarrow{\bigcap} P$ is an internal left adjoint for $\downarrow$seg; and we may similarly reconstruct $\bigcap$ from $\bigcup$.

5.35 LEMMA. *$\mathbf{\Omega}$ is an internally complete poset in any topos.*

Proof. Define $\bigcup$ to be the map $\Omega^t \colon \Omega^\Omega \longrightarrow \Omega^1 \cong \Omega$. Now let $X \xrightarrow{\phi} \Omega$, $X \xrightarrow{\psi} \Omega^\Omega$ be any two generalized elements, and $X' \overset{\sigma}{\rightarrowtail} X$, $Y \overset{\tau}{\rightarrowtail} X \times \Omega$ the corresponding subobjects. Form the pullback

$$\begin{array}{ccc} T & \rightarrowtail & Y \\ \downarrow & & \downarrow \tau \\ X & \xrightarrow{1 \times t} & X \times \Omega \end{array} ;$$

now since $\Omega_1 \rightarrowtail \Omega \times \Omega$ is classified by $\Rightarrow$, it is clear that the subobject $Z \rightarrowtail X \times \Omega$ corresponding to $\downarrow$seg . ϕ is

$$(X \overset{1 \times t}{\rightarrowtail} X \times \Omega) \Rightarrow (X' \times \Omega \overset{\sigma \times 1}{\rightarrowtail} X \times \Omega).$$

So we have

$$\begin{aligned} \psi \leqslant \downarrow\text{seg}\,.\,\phi \ &\text{iff}\ Y \leqslant Z \text{ as subobjects of } X \times \Omega \\ &\text{iff}\ Y \cap X = T \leqslant (X' \times \Omega) \text{ as subobjects of } X \times \Omega \\ &\text{iff}\ T = (1 \times t)^* Y \leqslant X' \text{ as subobjects of } X \\ &\text{iff}\ \Omega^t\,.\,\psi \leqslant \phi. \end{aligned}$$

Hence we have the desired internal adjunction. ■

If $\mathscr{E} \xrightarrow{T} \mathscr{F}$ is a logical functor and $\mathbf{P}$ is an internally complete poset in $\mathscr{E}$, then it is clear from the definition that $T(\mathbf{P})$ is internally complete in $\mathscr{F}$. More importantly, we have the following result:

5.36 PROPOSITION (C. J. Mikkelsen). *Let $\mathscr{E}$, $\mathscr{F}$ be toposes, and $\mathscr{E} \xrightarrow{T} \mathscr{F}$ a functor having a left adjoint L which preserves pullbacks. If $\mathbf{P}$ is an internally complete poset in $\mathscr{E}$, then $T\mathbf{P}$ is internally complete in $\mathscr{F}$.*

Proof. Consider a generalized element $U \xrightarrow{x} (\Omega_{\mathscr{F}})^{TP}$. This corresponds to a subobject $X \overset{(a,b)}{\rightarrowtail} U \times TP$ in $\mathscr{F}$; let $I \rightarrowtail LU \times P$ denote the image in $\mathscr{E}$ of $LX \xrightarrow{(La,\hat{b})} LU \times P$, where $\hat{}$ denotes transposition across $(L \dashv T)$. Let $LU \times P \xrightarrow{\phi} \Omega_{\mathscr{E}}$ classify $I \rightarrowtail LU \times P$, and let $U \xrightarrow{\bigcup x} TP$ be the transpose of the composite $LU \xrightarrow{\bar{\phi}} (\Omega_{\mathscr{E}})^{P} \xrightarrow{\bigcup} P$.

Applying this construction to the generic element of $(\Omega_{\mathscr{F}})^{TP}$, we obtain a morphism $(\Omega_{\mathscr{F}})^{TP} \xrightarrow{\bigcup} TP$; and since all the constructions involved are stable under pullback, we deduce that $\bigcup x$ is indeed the composite of x and $\bigcup$.

To prove that $\bigcup$ is internally left adjoint to $\downarrow$seg, let $U \xrightarrow{p} TP$ be a U-element of TP, and let $Y \rightarrowtail U \times TP$ be the subobject corresponding to $\downarrow \mathrm{seg}\,.\,p$. Now we also have a subobject $Z \rightarrowtail LU\,.\,P$ corresponding to $LU \xrightarrow{p} P \xrightarrow{\downarrow \mathrm{seg}} \Omega^{P}$; and since T preserves pullbacks, we have a diagram

$$\begin{array}{ccccc}
Y & \longrightarrow & TZ & \longrightarrow & TP_1^{\mathrm{op}} \\
\downarrow & & \downarrow & & \downarrow \\
U \times TP & \xrightarrow{\alpha \times 1} & TLU \times TP & \xrightarrow{T\hat{p} \times 1} & TP \times TP
\end{array}$$

where α is the unit of $(L \dashv T)$), in which both squares are pullbacks. So we have

$$\begin{aligned}
x \leqslant \downarrow \mathrm{seg}\,.\,p \ &\text{iff } X \leqslant Y \text{ as subobjects of } U \times TP \\
&\text{iff } LX \xrightarrow{(La,\hat{b})} LU \times P \text{ factors through } Z \rightarrowtail LU \times P \\
&\text{iff } I \leqslant Z \text{ as subobjects of } LU \times P \\
&\text{iff } \bar{\phi} \leqslant \downarrow \mathrm{seg}\,.\,\hat{p} \\
&\text{iff } \textstyle\bigcup \bar{\phi} \leqslant \hat{p} \text{ by internal completeness of } \mathbf{P} \\
&\text{iff } \textstyle\bigcup x \leqslant p. \ \blacksquare
\end{aligned}$$

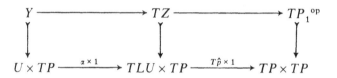

The most important application of 5.36 is, of course, the case when T is the direct image of a geometric morphism, with inverse image L. However, the advantage of the slightly more general formulation which we have given is that we can apply to it the case $\mathscr{E} = \mathscr{F}$, $T = (-)^{X}$, $L = (-) \times X$, to deduce that if $\mathbf{P}$ is internally complete then so is $\mathbf{P}^{X}$ for any X. (This could, however, also be deduced by factoring $(-)^{X}$ into the logical functor X^{*} followed by the direct-image functor Π_{X}.) In particular, the poset Ω^{X} is internally complete for any X. It is worth noting that the internal union map $\Omega^{\Omega^{X}} \longrightarrow \Omega^{X}$ is in fact the multiplication part of a monad structure on the covariant power-set functor ($X \longmapsto \Omega^{X}$, $f \longmapsto \exists f$) of Ex. 1.10; the unit of this monad is the singleton map (1.24). Moreover, the category of algebras for this monad is isomorphic to the category whose objects are (antisymmetric) internally complete posets in $\mathscr{E}$, and whose morphisms are sup–preserving maps (i.e. morphisms $P \xrightarrow{f} Q$ such that

$$\begin{array}{ccc} \Omega^P & \xrightarrow{\exists f} & \Omega^Q \\ \downarrow{\scriptstyle \cup} & & \downarrow{\scriptstyle \cup} \\ P & \xrightarrow{f} & Q \end{array}$$

commutes). We refer the reader to [1] or [84] for details of the proof.

5.37 THEOREM. *Let $\mathscr{E} \xrightarrow{\gamma} \mathscr{S}$ be an $\mathscr{S}$-topos. Then the following conditions are equivalent:*

(i) *$\mathscr{E}$ satisfies* (SG).
(ii) *There exists a complete Heyting algebra H in $\mathscr{S}$ such that $\mathscr{E} \simeq \mathrm{Shv}(\mathbf{H}, C)$, where C is the canonical Grothendieck topology on* $\mathbf{H}$ (*cf.* 0.35).

Moreover, the Heyting algebra H in (ii) *is determined up to isomorphism by $\mathscr{E}$.*

Proof. (ii) $\Rightarrow$ (i) is immediate from 5.34, since $\mathscr{S}$ satisfies (SG) and $\mathbf{H}$ is a poset.

(i) $\Rightarrow$ (ii): Suppose $\mathscr{E}$ satisfies (SG). Then γ is certainly bounded, since 1 is an object of generators for $\mathscr{E}$ over $\mathscr{S}$; and since $\gamma_*(\mathbf{\Omega}_{\mathscr{E}})$ is (equivalent to) the full subcategory of open objects in $\mathscr{E}$, it follows from 4.46 that we can write $\mathscr{E} \simeq \mathrm{Shv}(\gamma_*(\mathbf{\Omega}), J)$, where J is the Grothendieck topology on $\gamma_*(\mathbf{\Omega})$ defined as in 0.45.

But by 5.13, 5.35 and 5.36. $\gamma_*(\mathbf{\Omega})$ is a complete Heyting algebra; so it remains to prove that the topology J is canonical. (In fact, by 0.46, we know J is subcanonical; so it suffices to prove $C \leqslant J$.) But in any Heyting algebra, finite meets are distributive over arbitrary joins (indeed, for a complete lattice this condition is *equivalent* to cartesian closedness, by the Adjoint Functor Theorem), and so we may easily deduce from 0.35 that the C-covering sieves on $\gamma_*(\mathbf{\Omega})$ are precisely those generated by families $(U_\alpha \rightarrowtail U \mid \alpha \in A)$ for which $U = \bigcup_{\alpha \in A} U_\alpha$. But this implies that $\coprod_{\alpha \in A} U_\alpha \longrightarrow U$ is epi in $\mathscr{E}$, and so every object of $\mathscr{E}$ satisfies the sheaf axiom relative to this family. Hence every C-covering sieve is J-covering.

Finally, let F be a subobject of 1 in $\mathrm{Shv}(\mathbf{H}, C)$ where H is a complete Heyting algebra. Then for every $x \in H$ we have $F(x) = 0$ or 1, and from the above description of C-covering sieves on $\mathbf{H}$ it is easily seen that F must be representable, by the object $\bigcup\{x \in H \mid F(x) = 1\}$ of $\mathbf{H}$. So $\gamma_*(\boldsymbol{\Omega}_{\mathrm{Shv}(\mathbf{H}, C)}) \cong \mathbf{H}$; i.e. H is determined up to isomorphism by $\mathscr{E}$. ■

5.38 Remark. There is also a "relative" version of 5.37, due to W. Mitchell; if $\mathscr{F} \xrightarrow{f} \mathscr{E}$ is a bounded geometric morphism having 1 as object of generators, then there exists an internally complete Heyting algebra H in $\mathscr{E}$ (namely $H = f_*(\Omega_{\mathscr{F}})$) such that $\mathscr{F} \simeq \mathrm{sh}_j(\mathscr{E}^{\mathbf{H}^{\mathrm{op}}})$, where j is the internal equivalent of the canonical topology on $\mathbf{H}$. In particular, taking $f = 1_{\mathscr{E}}$, we have for any topos $\mathscr{E}$ that $\mathscr{E} \simeq \mathrm{sh}_j(\mathscr{E}^{\mathbf{\Omega}^{\mathrm{op}}})$; and in fact the inclusion $\mathscr{E} \longrightarrow \mathscr{E}^{\mathbf{\Omega}^{\mathrm{op}}}$ is that induced by the full-and-faithful internal functor $1 \xrightarrow{t} \mathbf{\Omega}$ (cf. Ex. 4.2). ■

5.39 Theorem. *Let $\mathscr{E} \xrightarrow{\gamma} \mathscr{S}$ be an $\mathscr{S}$-topos (and assume the axiom of choice for $\mathscr{S}$). Then the following conditions are equivalent:*

(i) *$\mathscr{E}$ satisfies* (AC).
(ii) *$\mathscr{E}$ is Boolean and satisfies* (SG).
(iii) *There exists a complete Boolean algebra B in $\mathscr{S}$ such that $\mathscr{E} \simeq$* Shv($\mathbf{B}$, C), *where C is the canonical Grothendieck topology on* $\mathbf{B}$.

Proof. (i) ⇒ (ii) is immediate from 5.23 and 5.33, and (ii) ⇔ (iii) is a special case of 5.37. It thus remains to show (ii) ⇒ (i).

Let $Y \xrightarrow{f} X$ be an epimorphism in $\mathscr{E}$. Define a *partial section* of f to be a commutative triangle

$$\begin{array}{ccc} X' & \xrightarrow{g} & Y \\ \downarrow & \swarrow_{f} & \\ X & & \end{array} ;$$

then it is clear that we can construct an "object of partial sections of f" as a subobject E of $\tilde{Y}^X$, and hence that the partial sections are parametrized by the set $S = \gamma_*(E)$. Moreover, S is nonempty, since we always have the partial section

$$\begin{array}{ccc} 0 & \longrightarrow & Y \\ \downarrow & \swarrow & \\ X & & \end{array} ;$$

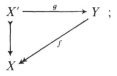

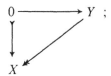

and if we partially order the partial sections by

$$(X', g) \leqslant (X'', g') \quad \text{if} \quad X' \leqslant X'' \quad \text{and} \quad g'|_{X'} = g,$$

then S satisfies the hypotheses of Zorn's Lemma, since $\mathscr{E}$ is a Grothendieck topos and we can therefore form arbitrary $\mathscr{S}$-indexed unions of subobjects of X. So S has a maximal element

$$\begin{array}{ccc} X_0 & \xrightarrow{\;g\;} & Y \, . \\ {\scriptstyle i}\big\downarrow & \swarrow{\scriptstyle f} & \\ X & & \end{array}$$

Suppose i is not an isomorphism; then the complement $X_1 \rightarrowtail X$ of X_0 is nonzero, and hence so is the pullback

$$\begin{array}{ccc} Y_1 & \rightarrowtail & Y \, . \\ \big\downarrow & & \big\downarrow{\scriptstyle f} \\ X_1 & \rightarrowtail & X \end{array}$$

So the two coproduct inclusions $Y_1 \rightrightarrows Y_1 \amalg Y_1$ are distinct; hence by (SG) there exists a nonzero $U \rightarrowtail 1$ and a morphism $U \longrightarrow Y_1$. But the composite $U \longrightarrow Y_1 \longrightarrow Y \xrightarrow{f} X$ is mono, and $U \cap X_0 \leqslant X_1 \cap X_0 = 0$. So we can extend our partial section (X_0, g) to one with domain $X_0 \amalg U$, contradicting the assumption that (X_0, g) is maximal. Hence i must be an isomorphism, and g is a splitting for f. ■

5.4. THE MITCHELL–BÉNABOU LANGUAGE

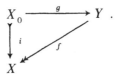

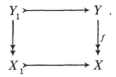

One of the most important ways in which we can formalize the ideas of a topos as a "generalized set theory" is by associating with a given topos $\mathscr{E}$ a language $L_{\mathscr{E}}$, which we can use as a convenient way of making statements about the objects and morphisms of $\mathscr{E}$, or even for proving theorems about them. This idea was first made explicit by W. Mitchell [85]; subsequently, various "improved" descriptions of $L_{\mathscr{E}}$ have been given by J. Bénabou [24], G. Osius [94] and M. Fourman [35], among others. The version which we describe here does not correspond precisely to any of these; it has been designed to emphasize the convenience of $L_{\mathscr{E}}$ as a simple mathematical shorthand for describing $\mathscr{E}$, rather than as a formal system for proving theorems about $\mathscr{E}$.

5.41 DEFINITION. The language $L_{\mathscr{E}}$ consists of the following symbols:

(i) *Types* $X, Y, \ldots$ are all objects of $\mathscr{E}$.
(ii) *Terms* of each type are constructed as follows:
  (a) For each type X we have a stock of *variables* $\mathbf{x}, \mathbf{x}', \mathbf{x}'', \ldots$.
  (b) For each morphism $X \xrightarrow{f} Y$ of $\mathscr{E}$ and each term τ of type X, we have a term $f(\tau)$ of type Y.
  (c) Given terms σ of type X, τ of type Y, we have a term $\langle\sigma, \tau\rangle$ of type $X \times Y$.
(iii) *Formulae* are terms of type Ω. More specifically, we have
  (a) For each type X, a binary predicate "$=_X$" of signature (X, X). [If σ, τ are terms of type X, then $(\sigma =_X \tau)$ is $\delta_X(\langle\sigma, \tau\rangle)$, where $\delta_X\colon X \times X \longrightarrow \Omega$ classifies the diagonal subobject.]
  (b) For each type X, a binary predicate "$\in_X$" of signature (X, Ω^X) [similarly defined using the evaluation map $X \times \Omega^X \longrightarrow \Omega$].
  (c) Logical connectives $\wedge$, $\vee$, $\Rightarrow$, $\neg$ [induced by the Heyting algebra operations on Ω].
(iv) *Quantifiers and descriptions.* Let ϕ be a formula and $\mathbf{x}$ a variable. Then
  (a) $\forall \mathbf{x}\phi$ and $\exists \mathbf{x}\phi$ are formulae.
  (b) $\{\mathbf{x} \mid \phi\}$ is a term of type Ω^X.
  (c) $\iota \mathbf{x}\phi$ is a term of type $\tilde{X}$. ■

The notions of *free* and *bound* occurrences of variables in a term of $L_{\mathscr{E}}$ are defined in the usual way; i.e. a variable is free unless it is bound by a quantifier or description-operator.

5.42 DEFINITION. We define an *interpretation* of $L_{\mathscr{E}}$ which assigns to a term τ of type X with free variables $(\mathbf{u}_1, \mathbf{u}_2, \ldots, \mathbf{u}_n)$ a morphism

$$|\tau|\colon \prod_{i=1}^{n} U_i \longrightarrow X,$$

as follows:

(a) Each variable of type X is interpreted as the identity morphism $X \xrightarrow{1} X$.
(b) $f(\tau)$ is interpreted as the composite $\prod U_i \xrightarrow{|\tau|} X \xrightarrow{f} Y$.

(c) Suppose σ has free variables $(\mathbf{u}_1, \ldots, \mathbf{u}_n, \mathbf{v}_1, \ldots, \mathbf{v}_m)$ and τ has free variables $(\mathbf{u}_1, \ldots, \mathbf{u}_n, \mathbf{w}_1, \ldots, \mathbf{w}_p)$, where the $\mathbf{v}_i$ and $\mathbf{w}_j$ are all distinct. Then $\langle \sigma, \tau \rangle$ is interpreted as the morphism whose components are

$$U \times V \times W \xrightarrow{\pi_{12}} U \times V \xrightarrow{|\sigma|} X$$

and

$$U \times V \times W \xrightarrow{\pi_{13}} U \times W \xrightarrow{|\tau|} Y,$$

where $U = \prod_{i=1}^{n} U_i$, etc. [Note that we regard the variables $\mathbf{x}$ and $\mathbf{x}'$ as distinct in this context, although they have the same type. Thus for example the interpretation of $\langle \mathbf{x}, \mathbf{x}' \rangle$ is $X \times X \xrightarrow{1} X \times X$, whereas $\langle \mathbf{x}, \mathbf{x} \rangle$ is interpreted as $X \xrightarrow{\Delta} X \times X$.]

(d) Suppose ϕ has free variables $(\mathbf{x}, \mathbf{u}_1, \mathbf{u}_2, \ldots, \mathbf{u}_n)$. Then

$$|\{\mathbf{x} \mid \phi\}| : \prod U_i \longrightarrow \Omega^X$$

is the exponential transpose of

$$|\phi| : X \times \prod U_i \longrightarrow \Omega.$$

And $|\iota\mathbf{x}\phi|$ is the representation of the partial map $\prod U_i \longrightarrow X$ given by the pullback

$$\begin{array}{ccc} P & \longrightarrow & X \\ \big\downarrow & & \big\downarrow {\scriptstyle \{\}} \\ \prod U_i & \xrightarrow{|\{\mathbf{x} \mid \phi\}|} & \Omega^X \end{array}$$

(e) Again, suppose ϕ has free variables $(\mathbf{x}, \mathbf{u}_1, \ldots, \mathbf{u}_n)$. Write f for the product projection $X \times \prod U_i \longrightarrow \prod U_i$, and let

$$\|\phi\| \rightarrowtail X \times \prod U_i$$

be the subobject classified by $|\phi|$ (the *extension* of ϕ). Then $|\forall \mathbf{x}\phi|$ (resp. $|\exists \mathbf{x}\phi|$) is the classifying map of $\forall_f(\|\phi\| \rightarrowtail X \times \Pi\, U_i)$ (resp. $\exists_f(\|\phi\| \rightarrowtail X \times \Pi\, U_i)$) (*cf.* 5.29).

(f) If $\mathbf{x}$ does not occur free in ϕ, then $\forall\mathbf{x}\phi$, $\exists\mathbf{x}\phi$, $\{\mathbf{x}|\phi\}$ and $\iota\mathbf{x}\phi$ may be interpreted by replacing ϕ by $(\mathbf{x} =_X \mathbf{x} \wedge \phi)$ (equivalently, by replacing $|\phi|$ by $X \times \prod U_i \xrightarrow{f} \prod U_i \xrightarrow{|\phi|} \Omega$). ■

5.43 Definition. Let ϕ be a formula of $L_{\mathscr{E}}$ with free variables $(\mathbf{u}_1, \ldots, \mathbf{u}_n)$. We say ϕ is *universally valid* in $\mathscr{E}$ (and write $\mathscr{E} \models \phi$) if $|\phi|$ is the composite $\prod U_i \longrightarrow 1 \xrightarrow{t} \Omega$ (equivalently, if $\|\phi\|$ is the maximal subobject of $\prod U_i$). ■

Note that any logical functor $T: \mathscr{E} \longrightarrow \mathscr{F}$ induces an obvious interpretation-preserving map from $L_{\mathscr{E}}$ to $L_{\mathscr{F}}$; so $\mathscr{E} \models \phi$ implies $\mathscr{F} \models T(\phi)$. Moreover, if T is faithful, then the converse implication holds.

5.44 Examples. The following assertions are valid for any topos $\mathscr{E}$. (The proofs are all easy exercises.)

(i) $\mathscr{E} \models \mathbf{x} =_X \mathbf{x}$ [but $\mathscr{E} \models \mathbf{x} =_X \mathbf{x}'$ iff $X \longrightarrow 1$ is mono].
(ii) $\mathscr{E} \models (\mathbf{x} \in_X \{\mathbf{x}|\phi\}) \Leftrightarrow \phi$.
(iii) If τ has type Ω^X, then $\mathscr{E} \models \tau =_{\Omega^X} \{\mathbf{x}|\mathbf{x} \in_X \tau\}$.
(iv) $\mathscr{E} \models \phi$ iff $\mathscr{E} \models \forall\mathbf{x}\phi$.
(v) $\mathscr{E} \models \exists\mathbf{x}\phi \Rightarrow \exists\mathbf{x}(\mathbf{x} =_X \mathbf{x})$. [Note that $\|\exists\mathbf{x}(\mathbf{x} =_X \mathbf{x})\|$ is precisely $\sigma_1(X)$.]
(vi) If σ and τ have the same free variables, then $\mathscr{E} \models \sigma =_X \tau$ iff $|\sigma| = |\tau|$.
(vii) $\mathscr{E} \models \phi \Rightarrow \neg\neg\phi$.
(viii) If ω is a variable of type Ω, then $\mathscr{E} \models (\omega \vee \neg\omega)$ iff $\mathscr{E}$ is Boolean. ■

In a similar vein to (vii) above, we may prove that all the tautologies of intuitionistic propositional logic (cf. [MM], chapter IX) are universally valid in $\mathscr{E}$. More generally, we may verify all the usual axioms and rules-of-inference of intuitionistic predicate logic, with the single exception of *modus ponens*: if $\models\phi$ and $\models\phi \Rightarrow \psi$ then $\models\psi$. The reason why this rule fails to hold is essentially that the product projection $X \times Y \xrightarrow{\pi_2} Y$ need not be an epimorphism unless X has global support; so a formula $\psi(\mathbf{y})$ which is *not* universally valid may become valid when we add the "dummy variable" $\mathbf{x}$. To take an extreme case, if ϕ contains a free variable of type 0, then $\mathscr{E} \models \phi$ and $\mathscr{E} \models \phi \Rightarrow \psi$ are always true, where ψ may be anything at all!

One solution to this difficulty, which has been proposed by M. Fourman [35], is to work with "partially defined terms" which are regarded as denoting generalized elements of $\tilde{X}$ rather than of X; the interpretation of a term in Fourman's language thus becomes a morphism $\prod \tilde{U}_i \longrightarrow \tilde{X}$. Since $\tilde{X}$ is injective, it always has global support; and so in this interpretation modus ponens becomes valid. However, the extra complications involved in this approach are considerable; so we have adopted the more straightforward interpretation outlined above, which is essentially the same as those of Bénabou and Osius. Note that this interpretation *does* satisfy the rule of inference which Osius [94] has called *restricted modus ponens*: if all free variables of ϕ are also free in ψ, $\mathscr{E} \models \phi$ and $\mathscr{E} \models \phi \Rightarrow \psi$, then $\mathscr{E} \models \psi$.

The principal respect in which our description of $L_{\mathscr{E}}$ differs from those of Bénabou and Osius is the *identification* of formulae with terms of type Ω. This seems a reasonable simplification from our point of view, since we should in any case have an interpretation-preserving bijection between (equivalence classes of) formulae and terms of type Ω, given by $\phi \longmapsto \{\ |\phi\}$ and $\tau \longmapsto (\ \in_1 \tau)$. (Here the blank space denotes the variable of type 1; it is universally valid that this variable is unique, and since it has no effect on the interpretation of any term in which it appears, it seems a work of supererogation to dignify it by giving it a symbol.)

5.45 Remark. We list here some of the commoner abuses of notation which we shall perpetrate when using $L_{\mathscr{E}}$.

(i) We shall drop the subscripts from the predicates $=$ and $\in$ whenever possible.

(ii) If $1 \xrightarrow{\gamma} X$ is a global element of X, we simply write γ for the term $\gamma(\ )$ of type X. In particular, if $X' \overset{\sigma}{\rightarrowtail} X$ is a subobject of X, we write $\ulcorner X' \urcorner$ or $\ulcorner \sigma \urcorner$ for the corresponding global element of Ω^X, or for the term which it defines.

(iii) If σ is a term of type Y^X, τ is a term of type X, we may write $\sigma(\tau)$ for $\mathrm{ev}(\langle \sigma, \tau \rangle)$, where $\mathrm{ev}\colon Y^X \times X \longrightarrow Y$ is the evaluation map.

(iv) If ϕ is a formula, and τ is a term none of whose free variables occurs bound in ϕ, we write $\phi[\tau/\mathbf{x}]$ for the formula obtained by substituting τ for all free occurrences of $\mathbf{x}$ in ϕ.

(v) We use the unique-existentiation quantifier $\exists!\mathbf{x}\phi$ to mean $\exists \mathbf{x}(\phi \wedge \forall \mathbf{x}'(\phi[\mathbf{x}'/\mathbf{x}] \Rightarrow \mathbf{x} = \mathbf{x}'))$.

(vi) In general, we shall feel free to use familiar set-theoretic abbreviations without further comment. For example, if $\mathbf{P} = (P_1 \rightrightarrows P)$ is an internal poset and $\mathbf{p}$, $\mathbf{p}'$ are two variables of type P, we shall write $\mathbf{p} \leqslant \mathbf{p}'$ for the formula $\langle \mathbf{p}, \mathbf{p}' \rangle \in_{P \times P} \ulcorner P_1 \urcorner$. ■

5.46 Examples. We now give some examples which illustrate the convenience of $L_{\mathscr{E}}$ as a means of "internalizing" familiar set-theoretic definitions.

(i) The *image map* im: $Y^X \longrightarrow \Omega^Y$ is $|\{y|\exists\mathbf{x}(\mathbf{f}(\mathbf{x}) = y)\}|$, where $\mathbf{f}$ is a variable of type Y^X.
(ii) The *object of epimorphisms* $\mathrm{Epi}(X, Y) \rightarrowtail Y^X$ is $\|\mathrm{im}(\mathbf{f}) = \ulcorner 1_Y \urcorner\|$, or equivalently $\|\forall\mathbf{y}\,\exists\mathbf{x}(\mathbf{f}(\mathbf{x}) = \mathbf{y})\|$, where $\mathbf{f}$ is again of type Y^X.
(iii) The *equalizer map* eq: $Y^X \times Y^X \longrightarrow \Omega^X$ is $|\{\mathbf{x}|\mathbf{f}(\mathbf{x}) = \mathbf{f}'(\mathbf{x})\}|$ (compare with the definition of this map in 1.49(ii)); and the *kernel-pair map* ker: $Y^X \longrightarrow \Omega^{X \times X}$ is $|\{\mathbf{t}|\mathbf{f}(\pi_1(\mathbf{t})) = \mathbf{f}(\pi_2(\mathbf{t}))\}|$, where $\mathbf{t}$ is of type $X \times X$.
(iv) The *object of monomorphisms* $\mathrm{Mono}(X, Y) \rightarrowtail Y^X$ is $\|\mathrm{ker}(\mathbf{f}) = \ulcorner\Delta\urcorner\|$. And the *object of isomorphisms* $\mathrm{Iso}(X, Y) \rightarrowtail Y^X$ may (by 1.22) be defined either as $\mathrm{Epi}(X, Y) \cap \mathrm{Mono}(X, Y)$, or as $\|\exists!\mathbf{g}(c_1(\langle\mathbf{f}, \mathbf{g}\rangle) = \bar{1}_X \wedge c_2(\langle\mathbf{g}, \mathbf{f}\rangle) = \bar{1}_Y)\|$, where $\mathbf{g}$ is of type X^Y and $c_1\colon Y^X \times X^Y \longrightarrow X^X$, $c_2\colon X^Y \times Y^X \longrightarrow Y^Y$ are the internal composition maps. ■

There is an alternative interpretation of the formulae of $L_{\mathscr{E}}$ (called the "external interpretation" by Osius) in which a formula ϕ with free variables $(\mathbf{x}_1, \mathbf{x}_2, \ldots, \mathbf{x}_n)$ is interpreted as a statement $\phi(x_1, x_2, \ldots, x_n)$ about generalized elements

$$U \xrightarrow{(x_1, \ldots, x_n)} \prod_{i=1}^{n} X_i.$$

(Specifically, $\phi(x_1, \ldots, x_n)$ means "$(x_1, \ldots, x_n)$ factors through $\|\phi\|$", or equivalently "$\mathscr{E} \models \phi[x_i(\mathbf{u})/\mathbf{x}_i]_{i=1}^{n}$".) In this context, the logical connective $\wedge$ is interpreted as conjunction in the usual way, but the other connectives and quantifiers involve an additional quantification over "change of the domain of definition of $(x_1, \ldots, x_n)$". The rules for this interpretation are commonly known as *Kripke–Joyal semantics*; they are a development of S. Kripke's original semantics [173] for modelling intuitionistic logic in what was essentially a topos of presheaves. We give the rules below; for convenience we work with formulae having only one or two free variables.

5.47 LEMMA. *Let ϕ, ψ be formulae of $L_{\mathscr{E}}$ with one free variable $\mathbf{x}$, χ a formula with free variables $(\mathbf{x}, \mathbf{y})$ and $U \xrightarrow{x} X$ a generalized element of X. Then*

(i) $(\phi \wedge \psi)(x)$ *iff* $\phi(x)$ *and* $\psi(x)$.
(ii) $(\phi \vee \psi)(x)$ *iff there exist* $V_1 \xrightarrow{\alpha_1} U$ *and* $V_2 \xrightarrow{\alpha_2} U$ *such that* $V_1 \amalg V_2 \longrightarrow U$ *is epi and we have* $\phi(x\alpha_1)$ *and* $\psi(x\alpha_2)$.
(iii) $(\phi \Rightarrow \psi)(x)$ *iff, for all* $V \xrightarrow{\alpha} U$ *such that* $\phi(x\alpha)$, *we have* $\psi(x\alpha)$.
(iv) $(\neg\phi)(x)$ *iff, for all* $V \xrightarrow{\alpha} U$ *such that* $\phi(x\alpha)$, *we have* $V \cong 0$.
(v) $(\exists\mathbf{y}\chi)(x)$ *iff there exists an epi* $V \xrightarrow{\alpha} U$ *and a V-element* $V \xrightarrow{y} Y$ *such that* $\chi(x\alpha, y)$.

(vi) $(\forall \mathbf{y}\chi)(x)$ *iff, for all* $V \xrightarrow{\alpha} U$ *and all* V*-elements* $V \xrightarrow{y} Y$, *we have* $\chi(x\alpha, y)$.

Proof. We give the argument for (ii) and (iii); the others are similar.

(ii) Suppose $(\phi \vee \psi)(x)$. Form the pullbacks

$$\begin{array}{ccc} V_1 & \longrightarrow & \|\phi\| \\ {\scriptstyle \alpha_1}\downarrow & & \downarrow \\ U & \xrightarrow{x} & \|\phi \vee \psi\| \end{array} \quad \text{and} \quad \begin{array}{ccc} V_2 & \longrightarrow & \|\psi\| \\ {\scriptstyle \alpha_2}\downarrow & & \downarrow \\ U & \xrightarrow{x} & \|\phi \vee \psi\| \end{array} ;$$

then since $\|\phi\| \amalg \|\psi\| \longrightarrow \|\phi \vee \psi\|$ is epi and coproducts in $\mathscr{E}$ are universal, we deduce that α_1 and α_2 have the required properties. Conversely, if the given conditions are satisfied, we deduce that x must factor through the image of $\|\phi\| \amalg \|\psi\| \longrightarrow X$, which is precisely $\|\phi \vee \psi\|$.

(iii) If $(\phi \Rightarrow \psi)(x)$, then clearly $(\phi \Rightarrow \psi)(x\alpha)$ for any α; so if we also have $\phi(x\alpha)$, then $x\alpha$ factors through $\|\phi \Rightarrow \psi\| \cap \|\phi\| \leqslant \|\psi\|$. Conversely, if the given condition is satisfied, let $I \rightarrowtail X$ be the image of x. Then if we form the pullback

$$\begin{array}{ccccc} V & \twoheadrightarrow & I \cap \|\phi\| & \rightarrowtail & \|\phi\| \\ {\scriptstyle \alpha}\downarrow & & \downarrow & & \downarrow \\ U & \twoheadrightarrow & I & \rightarrowtail & X \end{array} ,$$

the composite $x\alpha$ factors through $\|\psi\|$. which implies that $I \cap \|\phi\| \leqslant \|\psi\|$ and so $I \leqslant \|\phi \Rightarrow \psi\|$. ■

The observant reader will note that we have been using Kripke–Joyal semantics in an informal way in earlier chapters of this book. In particular, what we called in §2.5 the "universal" and "elementary" forms of the definition of filteredness were simply the internal and external interpretations of certain formulae of $L_{\mathscr{E}}$ involving existential quantification. And Lemma 3.56 is an example of the semantics for universal quantification; the $\prod$-functor occurring in the definition of D^r is of course a universal quantifier of $L_{\mathscr{E}}$. The proof of 5.47(v) and (vi) is thus implicit in the particular cases considered in 2.52 and 3.56.

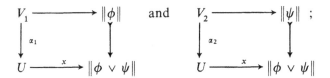

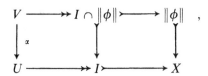

It should be pointed out that in Kripke's original semantics, disjunction and existential quantification were given "on-the-spot" interpretations similar to that for conjunction. The reason is that Kripke was concerned with a topos of the form $\mathscr{S}^{\mathbf{C}}$, and therefore restricted the objects U and V to be *representable* functors; the Yoneda Lemma easily implies that these are projective in $\mathscr{S}^{\mathbf{C}}$, so that the epis appearing in 5.47(ii) and (v) are automatically split. For a general topos, the need for a "local" interpretation of existentiation, as given in 5.47, was first emphasized by A. Joyal; however, it has been pointed out by A. Kock [61] that the unique-existential quantifier *does* have a global interpretation. Rather than prove this formally, we given an illustrative example.

5.48 DEFINITION. Let $\mathbf{P} = (P_1 \rightrightarrows P)$ be an internal poset in $\mathscr{E}$. We say that $\mathbf{P}$ is *well-ordered* if it is antisymmetric (5.11) and we have

$$\mathscr{E} \models (\exists \mathbf{p}(\mathbf{p} \in \mathbf{q})) \Rightarrow (\exists \mathbf{p}(\mathbf{p} \in \mathbf{q} \wedge \forall \mathbf{p}'(\mathbf{p}' \in \mathbf{q} \Rightarrow \mathbf{p} \leqslant \mathbf{p}'))),$$

where $\mathbf{q}$ is a free variable of type Ω^P. ■

Using 5.47, we can translate this condition as follows: "For any $U \xrightarrow{q} \Omega^P$, if there exists $V \overset{\alpha}{\twoheadrightarrow} U$ and $V \xrightarrow{p} P$ such that $(q\alpha, p)$ factors through $\in_P \rightarrowtail \Omega^P \times P$, then there exists $V_0 \overset{\alpha_0}{\twoheadrightarrow} U$ and $V_0 \xrightarrow{p_0} P$ such that $(q\alpha_0, p_0)$ factors through $\in_P$, and such that for all $W \xrightarrow{\beta} V_0$ and all $W \xrightarrow{p_1} P$, if $(q\alpha_0\beta, p_1)$ factors through $\in_P$, then $(p_0\beta, p_1)$ factors through $P_1 \rightarrowtail P \times P$". We hope that this example is sufficient to convince the reader of the need for a shorthand system such as $L_{\mathscr{E}}$!

5.49 PROPOSITION. *If* $\mathbf{P}$ *is well-ordered, then every epimorphism* $P \overset{f}{\twoheadrightarrow} X$ *is split.*

Proof. Let $X \xrightarrow{q} \Omega^P$ be the interpretation of the term $\{\mathbf{p} | f(\mathbf{p}) = \mathbf{x}\}$, i.e. the transpose of the classifying map of the graph of f. Then it is easily seen that $(qf, 1p)$ factors through $\in_P \rightarrowtail \Omega^P \times P$; so by the definition we can find $V \overset{\alpha}{\twoheadrightarrow} X$ and $V \xrightarrow{p} P$ such that "p is a minimal element of $q\alpha$". Now let $W \underset{\gamma_2}{\overset{\gamma_1}{\rightrightarrows}} V$ be the kernel-pair of α; then $q\alpha\gamma_1 = q\alpha\gamma_2$, so both morphisms define the same subobject of $W \times P$. But $p\gamma_1$ and $p\gamma_2$ are both minimal elements of this subobject, so by antisymmetry they are equal; hence p factors through the coequalizer of γ_1 and γ_2. But by 1.53 this coequalizer is (isomorphic to) α: so we can write $p = g\alpha$, and then $(q, g)\alpha$ factors through

$\in_P \rightarrowtail \Omega^P \times P$. But α is epi, so (q, g) factors through $\in_P$; which is clearly equivalent to saying that g is a splitting of f. ∎

The basic idea behind 5.49 is really nothing more than (an internal statement of) the sheaf axiom (0.23). In other words, the process of factoring the generalized element $V \xrightarrow{p} P$ through the epimorphism α is analogous to the process of constructing a global section of a sheaf by "glueing together" a family of local sections which agree on the overlapping parts of their domains. Alternatively, we may think of the slogan "Unique existence implies global existence" as a restatement of 1.22; whilst $\mathscr{E} \models \exists \mathbf{x}\phi$ asserts that a certain morphism of $\mathscr{E}$ is epi (but not necessarily split epi), $\mathscr{E} \models \exists !\mathbf{x}\phi$ asserts that it is both epi and mono, and therefore an isomorphism.

As a tailpiece to this paragraph, we give proofs of two "set-theoretic" lemmas which we shall require in the next chapter.

5.50 LEMMA (Tarski fixed-point theorem [185]). *Let* $\mathbf{P}$ *be an antisymmetric, internally complete poset in* $\mathscr{E}$, *and let* $P \xrightarrow{r} P$ *be an order-preserving map. Then there exists a global element* p_0 *of* P *such that* $rp_0 = p_0$.

Proof. Form the pullback

$$\begin{array}{ccc} Y & \longrightarrow & P_1 \\ {\scriptstyle i}\big\downarrow & & \big\downarrow \\ P & \xrightarrow{(1,r)} & P \times P \end{array}$$

and let p_0 be the composite $1 \xrightarrow{\ulcorner i \urcorner} \Omega^P \xrightarrow{\bigcup} P$. Now we have the following universally valid formulae:

$$\mathscr{E} \models \mathbf{p} \in \ulcorner i \urcorner \Leftrightarrow \mathbf{p} \leqslant r(\mathbf{p})$$

$$\mathscr{E} \models \mathbf{p} \in \ulcorner i \urcorner \Rightarrow \mathbf{p} \leqslant p_0$$

$$\mathscr{E} \models \mathbf{p} \leqslant p_0 \Rightarrow r(\mathbf{p}) \leqslant rp_0$$

So by restricted modus ponens and transitivity of $\leqslant$ we obtain

$$\mathscr{E} \models \mathbf{p} \in \ulcorner i \urcorner \Rightarrow \mathbf{p} \leqslant rp_0,$$

i.e. $\ulcorner i \urcorner \leqslant \downarrow\text{seg}.rp_0$, or equivalently $rp_0 \geqslant \bigcup \ulcorner i \urcorner = p_0$. But now we have $rrp_0 \geqslant rp_0$, so $rp_0 \in \ulcorner i \urcorner$ and hence $rp_0 \leqslant p_0$. ∎

Let X be an object of a topos $\mathscr{E}$, $S \overset{(a,b)}{\rightarrowtail} X \times X$ a binary relation on X. We say a subobject $X' \overset{i}{\rightarrowtail} X$ is *S-closed* if $X' \times_X S \xrightarrow{b\pi_2} X$ factors through i, i.e. if

$$\mathscr{E} \models (\mathbf{x} \in \ulcorner i \urcorner \wedge \langle \mathbf{x}, \mathbf{x}' \rangle \in \ulcorner S \urcorner) \Rightarrow \mathbf{x}' \in \ulcorner i \urcorner.$$

5.51 LEMMA. *Let $S \overset{(a,b)}{\rightarrowtail} X \times X$ be a symmetric relation on an object X of $\mathscr{E}$, and let $R \overset{(c,d)}{\rightarrowtail} X \times X$ be the kernel-pair of the coequilizer of (a, b). Then any S-closed subobject of X is R-closed.*

Proof. We define a binary relation $T \overset{(e,f)}{\rightarrowtail} X \times X$ to be the extension of the formula

$$\forall \mathbf{y}\ (\mathbf{y} \text{ is } S\text{-closed}) \Rightarrow (\mathbf{x} \in \mathbf{y} \Leftrightarrow \mathbf{x}' \in \mathbf{y}),$$

where $\mathbf{y}$ is of type Ω^X. It is clear from the nature of this formula that T is an equivalence relation on X, and so by 1.23 we have a morphism $X \xrightarrow{h} Y$ whose kernel-pair is (e, f). But since S is symmetric, we clearly have

$$\mathscr{E} \models (\langle \mathbf{x}, \mathbf{x}' \rangle \in \ulcorner S \urcorner \wedge \mathbf{y} \text{ is } S\text{-closed}) \Rightarrow (\mathbf{x} \in \mathbf{y} \Leftrightarrow \mathbf{x}' \in \mathbf{y}),$$

from which we deduce

$$\mathscr{E} \models \langle \mathbf{x}, \mathbf{x}' \rangle \in \ulcorner S \urcorner \Rightarrow \langle \mathbf{x}, \mathbf{x}' \rangle \in \ulcorner T \urcorner,$$

i.e. $S \leqslant T$ as subobjects of $X \times X$. But now we have $ha = hb$, so h factors through the coequalizer of (a, b) and hence $R \overset{(c,d)}{\rightarrowtail} X \times X$ factors through (e, f). And since

$$\mathscr{E} \models (\mathbf{y} \text{ is } S\text{-closed} \wedge \mathbf{x} \in \mathbf{y} \wedge \langle \mathbf{x}, \mathbf{x}' \rangle \in \ulcorner T \urcorner) \Rightarrow \mathbf{x}' \in \mathbf{y},$$

we clearly have that any S-closed subobject of X is also T-closed, and hence R-closed. ■

EXERCISES 5

1. Let $\mathscr{E}$ be a Boolean topos. Show that every topology in $\mathscr{E}$ is both open and closed, and deduce that every sheaf subtopos of $\mathscr{E}$ is Boolean. [Hint: if

j is a topology in $\mathscr{E}$, observe that j is determined by the composite $1 \xrightarrow{f} \Omega \xrightarrow{j} \Omega$.] Conversely, if every topology in $\mathscr{E}$ is open, show that $\mathscr{E}$ is Boolean. [Hint: if $\neg\neg = j^o_U$, then $U \rightarrowtail 1$ must be $\neg\neg$-dense; now consider j^c_U.]

2. Let **C** be a small category, and suppose the set of all morphisms of **C** admits a calculus of right fractions. (Verify that this condition is satisfied if $\mathbf{C}^{\mathrm{op}}$ is weakly filtered, or if **C** is a commutative monoid.) Show that any two nonempty sieves on an object of **C** have nonempty intersection, and deduce that every nonempty sieve is $\neg\neg$-covering. Hence show that $\mathrm{sh}_{\neg\neg}(\mathscr{S}^{\mathbf{C}^{\mathrm{op}}})$ is equivalent to $\mathscr{S}^{\mathbf{G}^{\mathrm{op}}}$, where **G** is the free groupoid generated by **C**, i.e. the category of fractions obtained by inverting every morphism of **C**.

3. Recall that a Heyting algebra H is said to be a *Stone lattice* ([LT], p. 130) if the equation $\neg x \vee \neg\neg x = t$ holds for all $x \in H$. If $\mathscr{E}$ is a topos, show that the following conditions are equivalent:

(a) Ω is an internal Stone lattice in $\mathscr{E}$.
(b) Every $\neg\neg$-closed subobject in $\mathscr{E}$ has a complement.
(c) $1 \overset{f}{\rightarrowtail} \Omega$ has a complement.
(d) An object X of $\mathscr{E}$ is $\neg\neg$-separated iff the diagonal map $X \overset{\Delta}{\rightarrowtail} X \times X$ has a complement. (An object satisfying this condition is said to be *decidable*.)
(e) Every $\neg\neg$-sheaf in $\mathscr{E}$ is decidable.
(f) $\Omega_{\neg\neg}$ is decidable.
(g) $1 \amalg 1 \xrightarrow{\binom{t}{f}} \Omega_{\neg\neg}$ is an isomorphism in $\mathscr{E}$.
(h) $\Omega_{\neg\neg}$ is a sublattice of Ω (in particular, the inclusion preserves $\vee$).
(j) (if $\mathscr{E} = \mathscr{S}^{\mathbf{C}^{\mathrm{op}}}$) $\mathbf{C}^{\mathrm{op}}$ satisfies condition (d) of 2.51.
(k) (if $\mathscr{E} = \mathrm{Shv}(X)$) X is extremally disconnected ([GT], p. 106).

[Hints: for (b) $\Rightarrow$ (d), use 3.29(ii). For (f) $\Rightarrow$ (g), show that any global element of a decidable object has a complement, by forming a suitable pullback diagram. For (c) $\Leftrightarrow$ (j), observe that (j) says that a pullback of a nonempty sieve is nonempty, i.e. the nonempty sieves form a sub-presheaf of Ω.]

†4. Recall that a topological space is said to be *separable* if it has a countable dense subset, and *zero-dimensional* if the subsets which are both open

and closed ("clopen sets") form a base for the topology ([GT], pp. 109 and 210). Show that both these properties are inherited by open subspaces, and hence prove that if X is separable and zero-dimensional, then Shv(X) satisfies (SS). [Method: given a sheaf F on X, let $\{x_0, x_1, x_2, \ldots\}$ be a countable dense subset of $\sigma_1(F)$. Now define an increasing sequence of clopen sets $U_n \subseteq \sigma_1(F)$ such that (i) $x_n \in U_n$; (ii) F has a section s_n over U_n; (iii) $s_n|_{U_{n-1}} = s_{n-1}$.]

†5. Let G denote the additive group of integers, and let $\mathscr{E}$ denote the full subcategory of $\mathscr{S}^{\mathbf{G}}$ consisting of those G-sets in which every G-orbit is finite. [Warning: this is *not* the same as the topos of Example 4.49(ii).] Show that the inclusion functor $\mathscr{E} \longrightarrow \mathscr{S}^{\mathbf{G}}$ does not reflect exponentials, but that it is left exact and has a right adjoint [namely, the functor which sends a G-set to the union of its finite orbits]. Deduce, using 2.32, that $\mathscr{E}$ is a topos; and show also that it is Boolean and defined over $\mathscr{S}$ (in fact $\mathscr{E}$ is a Grothendieck topos, by 4.46). Give an explicit description of the exponential functor in $\mathscr{E}$, and hence show that $\mathscr{E}$ does not satisfy (IC), even if we assume (AC) for $\mathscr{S}$.

6. Let M be the monoid with two elements 1 and e, such that $e^2 = e$. Show that the topos of M-sets satisfies (SS) but not (SG).

†7. Let $\mathscr{E}$ be a topos, X an object of $\mathscr{E}$, and suppose we have a monomorphism $\Omega^X \rightarrowtail X$. Prove that $\mathscr{E}$ is degenerate. [Hint: since Ω^X is injective, the mono must have a splitting $X \overset{f}{\twoheadrightarrow} \Omega^X$; now consider $\{\mathbf{x} \mid \neg(\mathbf{x} \in f(\mathbf{x}))\}$.] (Compare [73].)

†8. Let N denote the set of natural numbers, $s\colon N \longrightarrow N$ the successor map, and let $\mathscr{E} = \mathscr{S}^2$ be the Sierpinski topos (4.37(ii)) over $\mathscr{S}$. By considering the objects $N \xrightarrow{1} N$ and $N \xrightarrow{s} N$ of $\mathscr{E}$, show that the Schröder–Bernstein Theorem is false for $\mathscr{E}$. Prove, however, that the theorem holds in any Boolean topos. [Use the Tarski fixed-point theorem.]

9. (J. R. Isbell [51]). Let j and k be topologies in a topos $\mathscr{E}$. Let $l\colon \Omega \longrightarrow \Omega$ be the interpretation of the formula $\forall \omega'((\omega \leqslant \omega' \wedge \omega' \in \ulcorner \mathrm{J} \urcorner) \Rightarrow \omega' \in \ulcorner \mathrm{K} \urcorner)$, where ω, ω' are variables of type Ω. Prove that l is a topology. [Hint: use Kripke–Joyal semantics to describe the class Ξ_L, and then use 3.18.] Deduce that the lattice of topologies in $\mathscr{E}$ is a Heyting algebra, and in particular that it is distributive (cf. [GV], IV 9.1.14(a)).

†10. Let X be an object of a topos $\mathscr{E}$. Define the *unordered-pair map* $\mathrm{pr}_X \colon X \times X \longrightarrow \Omega^X$ to be

$$|\{\mathbf{x} \mid \mathbf{x} = \mathbf{x}' \vee \mathbf{x} = \mathbf{x}''\}|,$$

and the *Kuratowskian ordered-pair map* kpr_X to be the composite

$$X \times X \xrightarrow{(\{\}\,.\,\pi_1,\ \mathrm{pr}_X)} \Omega^X \times \Omega^X \xrightarrow{\mathrm{pr}_{\Omega^X}} \Omega^{\Omega^X}.$$

Show that kpr_X is mono. [Method: first observe that $\bigcup\,.\,\mathrm{kpr}_X = \mathrm{pr}_X$ and $\bigcap.\,\mathrm{kpr}_X = \{\}.\pi_1$. Now show that

$$\mathscr{E} \models \mathrm{pr}_X(\langle \mathbf{x}, \mathbf{x}' \rangle) = \mathrm{pr}\,(\langle \mathbf{x}, \mathbf{x}'' \rangle) \Rightarrow (\mathbf{x}'' = \mathbf{x} \vee \mathbf{x}'' = \mathbf{x}')$$

and

$$\mathscr{E} \models (\mathrm{pr}_X(\langle \mathbf{x}, \mathbf{x}' \rangle) = \mathrm{pr}_X(\langle \mathbf{x}, \mathbf{x}'' \rangle) \wedge \mathbf{x}'' = \mathbf{x}) \Rightarrow \mathbf{x} = \mathbf{x}'.]$$

Chapter 6

Natural Number Objects

6.1 DEFINITION AND BASIC PROPERTIES

6.11 Definition (Lawvere [71]). By a *natural number object* in a topos $\mathscr{E}$, we mean an object N together with morphisms $1 \xrightarrow{o} N \xrightarrow{s} N$ such that, for any diagram $1 \xrightarrow{x} X \xrightarrow{u} X$ in $\mathscr{E}$, there exists a unique $N \xrightarrow{f} X$ such that

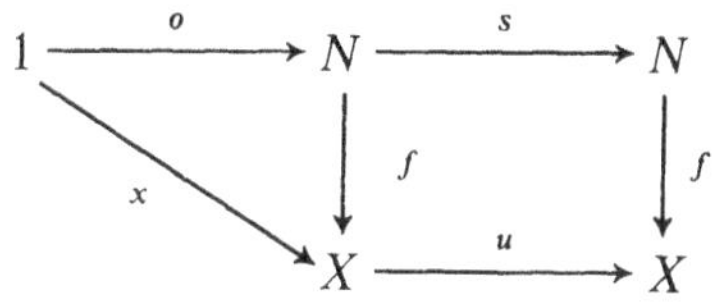

commutes. ■

The definition tells us that morphisms whose domain is N (or, using the exponential adjunction, an object of the form $N \times X$) may be defined "by recursion". For example, we have

6.12 Definition. The *arithmetic operations* $+$, . and $\exp: N \times N \longrightarrow N$ are defined by requiring their exponential transposes $N \longrightarrow N^N$ to satisfy the commutative diagrams

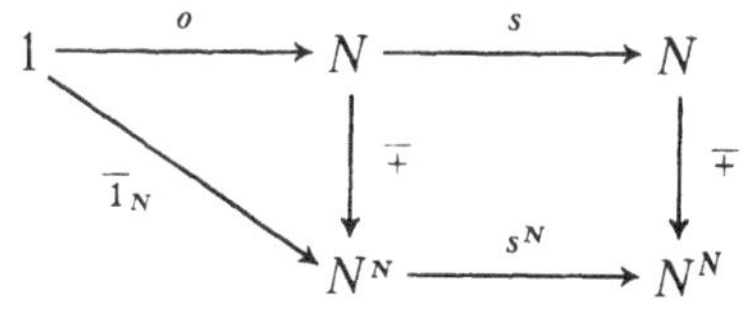

[i.e. $o + q = q$, $(sp) + q = s(p + q)$]

$$\begin{array}{ccccccc}
1 & \xrightarrow{o} & N & & \xrightarrow{s} & & N \\
\Big\downarrow{\scriptstyle o} & & \Big\downarrow{\scriptstyle \overline{\cdot}} & & & & \Big\downarrow{\scriptstyle \overline{\cdot}} \\
N & \xrightarrow{N^{(N \to 1)}} & N^N & \xrightarrow{1_{N^N} \times \overline{1}_N} & N^N \times N^N \cong (N \times N)^N & \xrightarrow{+^N} & N^N
\end{array}$$

[i.e. $o \cdot q = o$, $(sp) \cdot q = (p \cdot q) + q$], and

$$\begin{array}{ccccccc}
1 & \xrightarrow{o} & N & & \xrightarrow{s} & & N \\
\Big\downarrow{\scriptstyle so} & & \Big\downarrow{\scriptstyle \overline{\exp}} & & & & \Big\downarrow{\scriptstyle \overline{\exp}} \\
N & \xrightarrow{N^{(N \to 1)}} & N^N & \xrightarrow{1_{N^N} \times \overline{1}_N} & N^N \times N^N \cong (N \times N)^N & \xrightarrow{\cdot^N} & N^N
\end{array}$$

[i.e. $q^o = 1$, $q^{(sp)} = (q^p) \cdot q$]. ■

"The standard inductive arguments" (i.e. the uniqueness clause of 6.11) show that these operations satisfy the usual laws of arithmetic.

An important characterization of natural number objects in a topos was first given by P. Freyd [FK]:

6.13 PROPOSITION. *Let N be a natural number object in a topos. Then*

(i) $1 \xrightarrow{o} N \xleftarrow{s} N$ *is a coproduct diagram.* (*Hence in particular s is mono.*)

(ii) $N \underset{1_N}{\overset{s}{\rightrightarrows}} N \longrightarrow 1$ *is a coequalizer diagram.*

Proof. (i) Consider the morphisms

$$1 \xrightarrow{\nu_1} 1 \amalg N \xrightarrow{\nu_2 \binom{o}{s}} 1 \amalg N$$

It is not hard to see that these make $1 \amalg N$ into a natural number object in $\mathscr{E}$; for if we are given $1 \xrightarrow{x} X \xrightarrow{u} X$, then the diagram

$$\begin{array}{ccccc}
1 & \xrightarrow{\nu_1} & 1 \amalg N & \xrightarrow{\nu_2 \binom{o}{s}} & 1 \amalg N \\
& {\scriptstyle x} \searrow & \Big\downarrow{\scriptstyle \binom{x}{f}} & & \Big\downarrow{\scriptstyle \binom{x}{f}} \\
& & X & \xrightarrow{u} & X
\end{array}$$

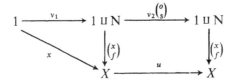

commutes iff f is the unique morphism such that

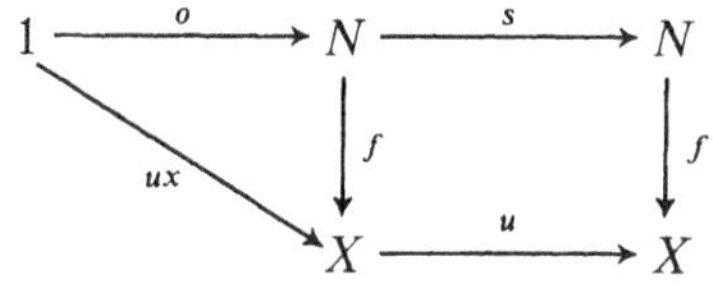

commutes. But it is clear from the definition that a natural number object, if it exists, is unique up to canonical isomorphism; so we have an isomorphism $1 \amalg N \xrightarrow{\cong} N$, which must be $\binom{o}{s}$ since the latter makes the appropriate diagram commute.

(ii) Let $N \xrightarrow{f} X$ be any morphism such that $fs = f$. Then we must have $f = (N \longrightarrow 1 \xrightarrow{fo} X)$, since both morphisms make the diagram

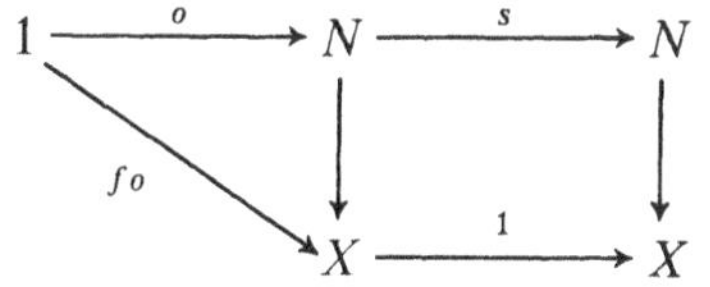

commute. Hence f factors (uniquely, since $N \longrightarrow 1$ is split epi) through $N \longrightarrow 1$. ■

Before proving the converse of 6.13, we introduce some special terminology. Suppose we are given an object X with morphisms $1 \xrightarrow{x} X \xrightarrow{u} X$; we will say that a subobject $X' \rightarrowtail X$ is (x, u)-*closed* if there exist (necessarily unique) morphisms $1 \xrightarrow{x'} X' \xrightarrow{u'} X'$ such that

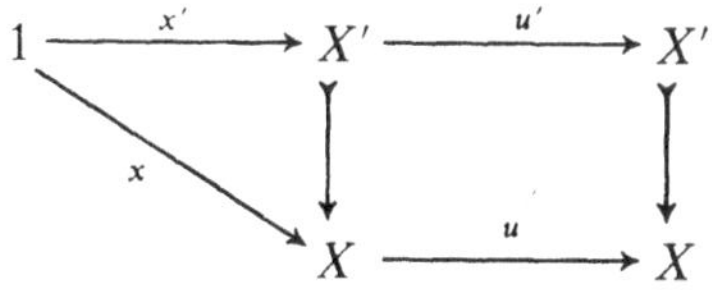

commutes, and that X' is (x, u)-*recursive* if it is (x, u)-closed and in addition $1 \amalg X' \xrightarrow{\binom{x'}{u'}} X'$ is epi. Now if we define $r: \Omega^X \longrightarrow \Omega^X$ to be the composite $\Omega^X \xrightarrow{\ulcorner x \urcorner \times \exists u} \Omega^X \times \Omega^X \xrightarrow{\cup} \Omega^X$, it is easy to see that X' is (x, u)-closed (resp. recursive) iff $\ulcorner X' \urcorner \geqslant r . \ulcorner X' \urcorner$ (resp. $\ulcorner X' \urcorner = r . \ulcorner X' \urcorner$). But r is clearly order-preserving; so by the Tarski fixed-point theorem (5.50) we deduce that any object X as above has a recursive subobject.

6.14 THEOREM (Freyd). *Let N be an object of a topos $\mathscr{E}$ equipped with morphisms $1 \xrightarrow{o} N \xrightarrow{s} N$ such that $1 \xrightarrow{o} N \xleftarrow{s} N$ is a coproduct and $N \underset{1}{\overset{s}{\rightrightarrows}} N \longrightarrow 1$ is a coequalizer. Then*

(i) *Any (o, s)-closed subobject of N is the whole of N (i.e. N satisfies Peano's fifth postulate).*
(ii) *N is a natural number object in $\mathscr{E}$.*

Proof. (i) Let $N' \overset{i}{\rightarrowtail} N$ be a (o, s)-closed subobject. By applying the argument above to $1 \xrightarrow{o'} N' \xrightarrow{s'} N'$, we may assume N' itself is (o, s)-recursive.

Now let $S \overset{(a,b)}{\rightarrowtail} N \times N$ be the union of the subobjects $N \overset{(1,s)}{\rightarrowtail} N \times N$ and $N \overset{(s,1)}{\rightarrowtail} N \times N$. We will show first that N' is S-closed (in the sense of 5.51). But we clearly have

$$\mathscr{E} = (\mathbf{n} \in \ulcorner N' \urcorner \Rightarrow s(\mathbf{n}) \in \ulcorner N' \urcorner);$$

and

$$\mathscr{E} \models s(\mathbf{n}) \in \ulcorner N' \urcorner \Rightarrow (s(\mathbf{n}) = o \vee \exists \mathbf{n}'(\mathbf{n}' \in \ulcorner N' \urcorner \wedge s(\mathbf{n}) = s(\mathbf{n}'))) .$$

However, from the given coproduct diagram we obtain

$$\mathscr{E} \models \neg(s(\mathbf{n}) = o)$$

and

$$\mathscr{E} \models (s(\mathbf{n}) = s(\mathbf{n}') \Rightarrow \mathbf{n} = \mathbf{n}') ;$$

so we deduce

$$\mathscr{E} \models s(\mathbf{n}) \in \ulcorner N' \urcorner \Rightarrow \mathbf{n} \in \ulcorner N' \urcorner.$$

Thus N' is S-closed.

Now the coequalizer of $S \underset{b}{\overset{a}{\rightrightarrows}} N$ is $N \longrightarrow 1$, since S contains the graph of s; so by 5.51 N' is closed under the kernel-pair of $N \longrightarrow 1$, i.e. the maximal relation $N \times N \overset{1}{\rightarrowtail} N \times N$. But since $N' \longrightarrow 1$ is epi (because N' has a global element), it follows at once that $N' \longrightarrow N$ must be epi.

(ii) Suppose given a diagram $1 \xrightarrow{x} X \xrightarrow{u} X$. Applying the argument above the statement of the theorem to $1 \xrightarrow{(o,x)} N \times X \xrightarrow{s \times u} N \times X$, we obtain a diagram

$$\begin{array}{ccccc} 1 & \xrightarrow{\;y\;} & Y & \xrightarrow{\;v\;} & Y \\ & {\scriptstyle (o,x)}\searrow & \big\downarrow{\scriptstyle (a,b)} & & \big\downarrow{\scriptstyle (a,b)} \\ & & N\times X & \xrightarrow{\;s\times u\;} & N\times X \end{array}$$

with $\begin{pmatrix} y \\ v \end{pmatrix}$ epi. We shall show that Y is the graph of a morphism $N \longrightarrow X$, i.e. that a is an isomorphism; this will clearly be sufficient for the existence clause of 6.11.

But the fact that a is epi follows at once from (i), since its image is a (o, s)-closed subobject of N; to show that it is mono, let $N' \rightarrowtail N$ be the extension of the formula $\exists !\mathbf{y}(a(\mathbf{y}) = \mathbf{n})$. Now since $\mathrm{im}(v) \leqslant a^{-1}(\mathrm{im}(s))$, the diagram

$$\begin{array}{ccc} 1 & \xrightarrow{\;y\;} & Y \\ \big\downarrow & & \big\downarrow{\scriptstyle a} \\ 1 & \xrightarrow{\;o\;} & N \end{array}$$

is a pullback; i.e. $\mathscr{E} \vDash a(\mathbf{y}) = o \Leftrightarrow \mathbf{y} = y$, and so $o \in \ulcorner N' \urcorner$. Similarly, we can show $\mathscr{E} \vDash \mathbf{n} \in \ulcorner N' \urcorner \Rightarrow s(\mathbf{n}) \in \ulcorner N' \urcorner$; thus N' is (o, s)-closed, and so $N' = N$, i.e. a is an isomorphism.

Finally, to establish the uniqueness clause of 6.11, suppose $N \overset{f}{\underset{f'}{\rightrightarrows}} X$ are both recursively defined by the diagram $1 \xrightarrow{x} X \xrightarrow{u} X$. Then the equalizer of f and f' is easily seen to be (o, s)-closed, and so $f = f'$. ■

6.15 Corollary. *Let $\mathscr{E}$ be a topos, and X an object of $\mathscr{E}$ such that $1 \amalg X \cong X$. Then there exists a subobject of X which is a natural number object in $\mathscr{E}$.*

Proof. Let $1 \xrightarrow{x} X$ and $X \xrightarrow{u} X$ be the morphisms defining the given isomorphism. Let $X \xrightarrow{h} Y$ be the coequalizer of u and 1_X, and let y denote the global element hx of Y. Now form the pullback

$$\begin{array}{ccc} N & \rightarrowtail & X\;. \\ \big\downarrow & & \big\downarrow{\scriptstyle h} \\ 1 & \overset{y}{\rightarrowtail} & Y \end{array}$$

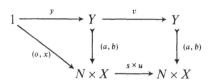

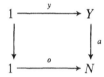

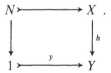

We observe that since u is a morphism over Y, the given isomorphism can be interpreted as an isomorphism $h \cong y \amalg h$ in $\mathscr{E}/Y$. But y is mono, so $y^*(y) = 1$; hence on applying y^* to this isomorphism we obtain $N \cong 1 \amalg N$ in $\mathscr{E}$. But we also have

$$\mathrm{coeq}(y^*(u), 1_N) \cong y^*(\mathrm{coeq}(u, 1_h)) \cong y^*(1_Y) \cong 1 \ ;$$

so by 6.14(ii) N is a natural number object in $\mathscr{E}$. ■

6.16 Proposition. *Let $\mathscr{E}$ and $\mathscr{F}$ be toposes and $T: \mathscr{E} \longrightarrow \mathscr{F}$ a functor preserving* 1 *and finite colimits. If $1 \xrightarrow{o} N \xrightarrow{s} N$ is a natural number object in $\mathscr{E}$, then $1 \cong T1 \xrightarrow{To} TN \xrightarrow{Ts} TN$ is a natural number object in $\mathscr{F}$.*

Proof. Immediate from 6.13 and 6.14(ii). ■

It should be noted that if T has a right adjoint R (e.g. if T is the inverse image of a geometric morphism), then 6.16 can be proved directly from the definition of N, without invoking 6.14. For if we are given $1 \xrightarrow{x} X \xrightarrow{u} X$ in $\mathscr{F}$, we can apply the definition to $1 \cong R1 \xrightarrow{Rx} RX \xrightarrow{Ru} RX$ to obtain a morphism $N \longrightarrow RX$, and then transpose to obtain $TN \longrightarrow X$ with the required properties.

An important application of 6.13 is in defining the order structure of N. A curious feature of this ordering is that, even when the topos $\mathscr{E}$ is not Boolean, the natural number object retains a certain amount of Booleanness in that all its "interesting" subobjects have complements. This will become even more apparent in § 6.2 below.

6.17 Proposition. (i) *The morphism $N \times N \xrightarrow{(\pi_1, +)} N \times N$ is mono (i.e. $(N, +, o)$ is a monoid with cancellation), and defines an antisymmetric partial order on N, which we denote by $W \rightarrowtail N \times N$.*

(ii) *Similarly, $N \times N \xrightarrow{(\pi_1, s+)} N \times N$ is mono and defines a strict ordering on N, which we denote by $S \rightarrowtail N \times N$.*

(iii) *We have coproduct decompositions $W \cong S \amalg \Delta$ and $N \times N \cong S \amalg \Delta \amalg S^{\mathrm{op}}$, where Δ as usual denotes the diagonal subobject.*

Proof. (i) Let $N' \rightarrowtail N$ be the extension of the formula

$$\forall \mathbf{n}' \forall \mathbf{n}''((\mathbf{n} + \mathbf{n}' = \mathbf{n} + \mathbf{n}'') \Rightarrow (\mathbf{n}' = \mathbf{n}'')).$$

Then from the definition of $+$ and the fact that s is mono, it is easy to see that $o \in \ulcorner N' \urcorner$ and

$$\mathscr{E} \models \mathbf{n} \in \ulcorner N' \urcorner \Rightarrow s(\mathbf{n}) \in \ulcorner N' \urcorner \ ,$$

so $N' = N$. But this is equivalent to saying that $(\pi_1, +)$ is mono.

To show that W is reflexive, we observe that $N \cong N \times 1 \xrightarrow{1 \times o} N \times N$ gives a factorization of the diagonal through it; transitivity is similarly implied by the existence of the map

$$W \times_N W \xrightarrow{(\pi_1\pi_1,\ +(\pi_2\pi_1,\ \pi_2\pi_2))} N \times N$$

and the associativity of $+$. And antisymmetry follows from the fact that

$$\begin{array}{ccc} 1 & \longrightarrow & 1 \\ {\scriptstyle (o,o)}\downarrow & & \downarrow{\scriptstyle o} \\ N \times N & \xrightarrow{+} & N \end{array}$$

is a pullback, which in turn follows easily from the coproduct decomposition $N \times N \cong 1 \amalg (N \times 1) \amalg (1 \times N) \amalg (N \times N)$.

(ii) is immediate from (i) and the fact that s is mono.

(iii) The coproduct $W \cong S \amalg \Delta$ follows at once from 6.13(i). From it and from the antisymmetry of W, we deduce that the three subobjects S, Δ and S^{op} are disjoint; so it remains to prove that their union is the whole of $N \times N$. Write $\phi(\mathbf{n}, \mathbf{n}')$ for the formula

$$(\mathbf{n} < \mathbf{n}') \vee (\mathbf{n} = \mathbf{n}') \vee (\mathbf{n} > \mathbf{n}') ,$$

and let $N' \rightarrowtail N$ be the extension of the formula $\forall \mathbf{n}' \phi(\mathbf{n}, \mathbf{n}')$. Now it is easy to see that we have

$$\mathscr{E} \models o \leqslant \mathbf{n}'$$

$$\mathscr{E} \models \mathbf{n} > \mathbf{n}' \Rightarrow s(\mathbf{n}) > \mathbf{n}',$$

$$\mathscr{E} \models \mathbf{n} = \mathbf{n}' \Rightarrow s(\mathbf{n}) > \mathbf{n}',$$

and

$$\mathscr{E} \models \mathbf{n} < \mathbf{n}' \Rightarrow (s(\mathbf{n}) < \mathbf{n}' \vee s(\mathbf{n}) = \mathbf{n}') ;$$

from which it follows easily that N' is (o, s)-closed. ■

So far we have considered natural number objects in terms of recursive definition of *morphisms* in a topos; but in mathematics there is another way in which we frequently wish to use the idea of recursion, namely the construction of *objects* in the topos $\mathscr{E}/N$. More precisely, suppose we are given an object X of $\mathscr{E}$ and a "process" T for constructing new objects out of old ones. (We shall loosely refer to this information as "recursion data in $\mathscr{E}$".) Then we wish to find an object $F \longrightarrow N$ of $\mathscr{E}/N$ which internalizes the notion of the sequence $(X, TX, TTX, \ldots)$; i.e. such that $o^*(F) \cong X$ and $s^*(F) \cong T_N(F)$, where T_N denotes the process T "applied fibrewise" to objects of $\mathscr{E}/N$.

Now in the case of the topos $\mathscr{S}$, which has no internal structure, it is clear that we can solve this problem when T is any function; but for a more general topos $\mathscr{E}$, it is clearly essential that T should "respect the internal structure of $\mathscr{E}$". In §6.3 below, we shall give an existence theorem which tells us that any recursion problem in a suitably-defined class has a solution; for the present, we give a uniqueness theorem (due to G. C. Wraith) which has many useful applications.

Recall that if $\mathscr{E}$ is any cartesian closed category, a functor $T: \mathscr{E} \longrightarrow \mathscr{E}$ is said to be *strong* if it is an $\mathscr{E}$-functor in the sense of closed categories [160]; i.e. if we are given morphisms $T_{X,Y}: Y^X \longrightarrow TY^{TX}$ for every pair of objects (X, Y) of $\mathscr{E}$, which internalize the effect of T on morphisms of $\mathscr{E}$.

6.18 Lemma. *Let $F \xrightarrow{\gamma} N$ be an object of $\mathscr{E}/N$, and suppose we are given*

(i) *a global element $1 \xrightarrow{x} o^*(\gamma)$ in $\mathscr{E}$, and*
(ii) *a morphism $\gamma \xrightarrow{t} s^*(\gamma)$ in $\mathscr{E}/N$.*

Then there exists a unique global element f of γ in $\mathscr{E}/N$ (i.e. section of γ in $\mathscr{E}$) such that $o^*(f) = x$ and $s^*(f) = tf$.

Proof. Define f by the diagram

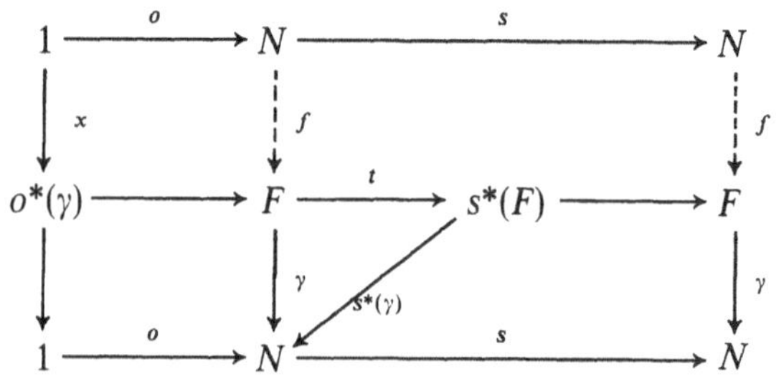

Then we have $\gamma f = 1_N$ by the uniqueness clause of 6.11; and f is clearly the unique element of γ satisfying the given conditions. ■

6.19 THEOREM (Wraith). *Let X be an object of $\mathscr{E}$, and $T: \mathscr{E}/N \longrightarrow \mathscr{E}/N$ a strong functor. Then if there exists an object $F \xrightarrow{\gamma} N$ of E/N satisfying $o^*(\gamma) \cong X$ and $s^*(\gamma) \cong T(\gamma)$, it is unique up to canonical isomorphism.*

Proof. Suppose γ, γ' are two such objects of $\mathscr{E}/N$. Form the object of isomorphisms $\mathrm{Iso}(\gamma, \gamma')$ in $\mathscr{E}/N$ (cf. 5.46(iv)); then it is clear that the strength of T induces a morphism $T_{\gamma,\gamma'}: \mathrm{Iso}(\gamma, \gamma') \longrightarrow \mathrm{Iso}(T\gamma, T\gamma')$. And the pullback functors o^* and s^*, being logical, preserve objects of isomorphisms; so we have an element

$$1 \xrightarrow{\overline{1_X}} \mathrm{Iso}(X, X) \cong o^*(\mathrm{Iso}(\gamma, \gamma'))$$

in $\mathscr{E}$, and a morphism

$$\mathrm{Iso}(\gamma, \gamma') \xrightarrow{T_{\gamma,\gamma'}} \mathrm{Iso}(T\gamma, T\gamma') \cong s^*(\mathrm{Iso}(\gamma, \gamma'))$$

in $\mathscr{E}/N$.

On applying the construction of 6.18, we obtain a canonical element of $\mathrm{Iso}(\gamma, \gamma')$, i.e. a canonical isomorphism $\gamma \xrightarrow{\cong} \gamma'$ in $\mathscr{E}/N$. ■

6.2. FINITE CARDINALS

One important way in which we can use natural number objects is to give a definition of "finite objects" in a topos. In this paragraph we investigate the basic properties of these objects.

6.21 DEFINITION (J. Bénabou [7]). Let $p: 1 \longrightarrow N$ be a natural number in a topos $\mathscr{E}$ (i.e. a global element of the natural number object). By the *cardinal* of p we mean the object $[p]$ in the pullback diagram

$$\begin{array}{ccc} [p] & \longrightarrow & 1 \\ \downarrow & & \downarrow p \\ N \times N & \xrightarrow{+} N \xrightarrow{s} & N \end{array}$$ ■

We think of $[p]$ as a finite object with p elements; clearly in $\mathcal{S}$ it is just the set $\{\langle a, b\rangle \in N \times N \mid a + b + 1 = p\}$. In fact it is not hard to prove in any topos that the composite $[p] \longrightarrow N \times N \xrightarrow{\pi_1} N$ is mono, and identifies $[p]$ with the down-segment of p for the strict order-relation of 6.17.

6.22 LEMMA. *Let $T: \mathscr{E} \longrightarrow \mathscr{F}$ be a functor preserving finite limits and colimits (e.g. a logical functor, or the inverse image of a geometric morphism), and p a natural number in $\mathscr{E}$. Then $T([p]) \cong [T(p)]$.*

Proof. From 6.16, we know that T preserves N; hence it also preserves recursively defined morphisms, including $+$, and so it preserves the pullback diagram of 6.21. ■

Finite cardinals behave as we would expect with respect to addition, multiplication and exponentiation:

6.23 THEOREM. *Let p, q be natural numbers in a topos $\mathscr{E}$. Then we have isomorphisms*

(i) $[o] \cong 0$
(ii) $[sp] \cong [p] \amalg 1$
(iii) $[p + q] \cong [p] \amalg [q]$
(iv) $[p \cdot q] \cong [p] \times [q]$
(v) $[q^p] \cong [q]^{[p]}$.

Proof. (i) is immediate from 6.13(i) and disjointness of coproducts in $\mathscr{E}$. To prove (ii), observe that since s is mono, $[sp]$ is the pullback of

$$\begin{array}{ccc} & & 1 \\ & & \downarrow{\scriptstyle p} \\ N \times N & \xrightarrow{+} & N \end{array} .$$

But by 6.13(i) again, we can replace $(N \times N \xrightarrow{+} N)$ by

$$(N \amalg 1) \times N \cong (N \times N) \amalg N \xrightarrow{\binom{s+}{1_N}} N \ ;$$

so the result follows from the universality of coproducts in $\mathscr{E}$.

It now follows from (i) and (ii) that the cardinal of the generic natural number n (which is of course the object $N \times N \xrightarrow{s+} N$ of $\mathscr{E}/N$) satisfies the recursion data

$$o^*[n] \cong 0 \quad \text{and} \quad s^*[n] \cong [n] \amalg 1.$$

To prove (iii), (iv) and (v), it suffices to consider a pair of natural numbers of the form (n, N^*q) in $\mathscr{E}/N$ (where q is a natural number in $\mathscr{E}$), since the pullback functor $p^*: \mathscr{E}/N \longrightarrow \mathscr{E}$ preserves coproducts, products and exponentials. But for (iii) we observe that the objects on each side of the isomorphism satisfy the recursion data

$$o^*[n + N^*q] \cong [o{+}q] \cong [q],$$

$$s^*[n + N^*q] \cong [sn + N^*q] \cong [s(n + N^*q)] \cong [n + N^*q] \amalg 1,$$

and

$$o^*([n] \amalg [N^*q]) \cong [o] \amalg [q] \cong 0 \amalg [q] \cong [q],$$

$$s^*([n] \amalg [N^*q]) \cong [sn] \amalg [N^*q] \cong [n] \amalg 1 \amalg [N^*q] \cong [n] \amalg [N^*q] \amalg 1.$$

But the functor $(-) \amalg 1$ is easily seen to be strong in any topos (its strength $Y^X \longrightarrow (Y \amalg 1)^{(X \amalg 1)}$ may be defined as the transpose of

$$Y^X \times (X \amalg 1) \cong (Y^X \times X) \amalg (Y^X \times 1) \xrightarrow{\text{ev} \amalg \pi_2} Y \amalg 1\,),$$

and so by 6.19 we obtain an isomorphism $[n + N^*q] \cong [n] \amalg [N^*q]$ in $\mathscr{E}/N$, as required.

The proofs of (iv) and (v) follow similarly from the recursive definitions of multiplication and exponentiation given in 6.12, and the fact that the functors $(-) \amalg [q]$ and $(-) \times [q]$ are strong. ■

One aspect of the finiteness of cardinals is that exponentiation to the power of a cardinal behaves in many respects like a *finite* limit. The next two propositions are both examples of this behaviour.

6.24 Proposition. *Let $\mathscr{E} \xrightarrow{T} \mathscr{F}$ be a functor between toposes preserving finite limits and colimits, X an object of $\mathscr{E}$ and p a natural number in $\mathscr{E}$. Then $T(X^{[p]}) \cong TX^{T[p]}$.*

Proof. Consider the object $(N^*X)^{[n]}$ of $\mathscr{E}/N$. By 6.23, this satisfies the recursion data

$$o^*((N^*X)^{[n]}) \cong X^{[o]} \cong X^0 \cong 1$$

and

$$s^*((N^*X)^{[n]}) \cong (N^*X)^{([n]\sqcup 1)} \cong (N^*X)^{[n]} \times N^*X \ .$$

But T (regarded as a functor $\mathscr{E}/N \longrightarrow \mathscr{F}/TN$) preserves this data, and so we have $T((N^*X)^{[n]}) \cong T(N^*X)^{T[n]}$ by 6.19 and 6.22. And since T preserves pullbacks, the square

$$\begin{array}{ccc} \mathscr{E}/N & \xrightarrow{\ T\ } & \mathscr{F}/TN \\ \Big\downarrow{\scriptstyle p^*} & & \Big\downarrow{\scriptstyle (Tp)^*} \\ \mathscr{E} & \xrightarrow{\ T\ } & \mathscr{F} \end{array}$$

commutes up to isomorphism; so on applying the functor $(Tp)^*$ to the above isomorphism, we obtain the required result. ■

6.25 PROPOSITION. *Let p be a natural number in a topos $\mathscr{E}$. Then the functor $(-)^{[p]}: \mathscr{E} \longrightarrow \mathscr{E}$ preserves reflexive coequalizers.*

Proof. Let $X \rightrightarrows Y \longrightarrow Z$ be a reflexive coequalizer diagram in $\mathscr{E}$. As usual, it suffices to consider the case $p = n$ and then apply the pullback functor p^*. So let $(N^*Y)^{[n]} \longrightarrow W$ be the coequalizer of the reflexive pair $(N^*X)^{[n]} \rightrightarrows (N^*Y)^{[n]}$ in $\mathscr{E}/N$. Now W satisfies the recursion data

$$o^*W \cong \operatorname{coeq}(1 \rightrightarrows 1) \cong 1$$

and

$$s^*W \cong \operatorname{coeq}(N^*X^{[n]} \times N^*X \rightrightarrows N^*Y^{[n]} \times N^*Y) \cong W \times N^*Z$$

by Ex. 0.1. So by 6.19 we have $W \cong (N^*Z)^{[n]}$, as required. ■

Note that it follows from 6.25 that finite cardinals are internally projective (5.24(i)), since every epimorphism in $\mathscr{E}$ can be expressed as a coequalizer of a reflexive pair (specifically, its own kernel-pair).

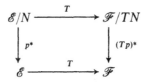

The "Booleanness" of the order-relation on N, as described in 6.17, has its effect on the structure of monomorphisms between finite cardinals.

6.26 LEMMA. *Let p be a natural number in a topos. Then*

(i) *$[p]$ is decidable (cf. Ex. 5.3); i.e. the diagonal $[p] \overset{\Delta}{\rightarrowtail} [p] \times [p]$ has a complement.*
(ii) *If $[p] \longrightarrow 1$ is mono, it has a complement.*
(iii) *Any complemented subobject of $[p]$ is (isomorphic to) a finite cardinal.*

Proof. (i) In 6.17(iii), we saw that N was decidable. But the property of decidability is easily seen to be inherited by finite products and subobjects (the latter since $X \overset{f}{\rightarrowtail} Y$ mono implies that

$$\begin{array}{ccc} X & \xrightarrow{f} & Y \\ \downarrow{\scriptstyle \Delta} & & \downarrow{\scriptstyle \Delta} \\ X \times X & \xrightarrow{f \times f} & Y \times Y \end{array}$$

is a pullback); and by definition $[p]$ is a subobject of $N \times N$.

(ii) Let p be any natural number, and define the subobjects U_0, U_1, U_2 of 1 to be the extensions of the formulae "$p < so$", "$p = so$", and "$p > so$" respectively. Then 6.17(iii) implies that $U_0 \amalg U_1 \amalg U_2 \cong 1$. But on pulling back to $\mathscr{E}/U_2$ the statement $p \geqslant sso$ becomes true, and so we have a monomorphism $U_2^*[sso] \cong U_2 \amalg U_2 \rightarrowtail [p]$. Hence if $[p] \rightarrowtail 1$ is mono, we must have $U_2 \cong 0$, and so $U_0 \amalg U_1 \cong 1$. But it is now easy to see that $U_1^*[p] \cong 1$ and $U_0^*[p] \cong 0$, so we must have $[p] \cong U_1$.

(iii) Consider the object $(M \xrightarrow{\gamma} N) = (1 \amalg 1)^{[n]}$ of $\mathscr{E}/N$. In the topos $\mathscr{E}/M$, we have a generic morphism $\gamma^*[n] \longrightarrow 1 \amalg 1$, and hence a generic coproduct decomposition $\gamma^*[n] \cong X \amalg Y$. It therefore suffices to find a natural number p in $\mathscr{E}/M$ whose cardinal is isomorphic to X. But such a natural number corresponds to a morphism $M \xrightarrow{p} N$ in $\mathscr{E}$, or equivalently to an element of the object $N^{(2^{[n]})}$ in $\mathscr{E}/N$ (where we write 2 for $1 \amalg 1$, and N denotes the natural number object of $\mathscr{E}/N$, i.e. the object N^*N). And we may construct an element of this object by the method of 6.18, using the data

$$1 \xrightarrow{o} N \cong N^1 \cong o^*(N^{(2^{[n]})})$$

and

$$N^{(2^{[n]})} \xrightarrow{\theta} N^{(2^{[n]} \times 2)} \cong s^*(N^{(2^{[n]})}),$$

where the transpose of θ is the composite

$$N^{(2^{[n]})} \times 2^{[n]} \times 2 \xrightarrow{(\text{ev.}\,\pi_{12},\binom{so}{o}.\,\pi_2)} N \times N \xrightarrow{+} N.$$

It is then straightforward to verify that $[p]$ satisfies the same recursion data as X. ■

We shall see in §6.4 that the object M constructed in 6.26(iii) is in fact the free monoid generated by $1 \amalg 1$; and in fact p is the monoid homomorphism $M \longrightarrow (N, +, 0)$ induced by $1 \amalg 1 \xrightarrow{\binom{so}{o}} N$. Note also that, for any monoid (G, m, e) in $\mathscr{E}$, the counit of the free-monoid adjunction at G can be regarded as a morphism $N^*G^{[n]} \xrightarrow{\mu_n} N^*G$ in $\mathscr{E}/N$; and pullback along $1 \xrightarrow{p} N$ gives us a morphism $G^{[p]} \xrightarrow{\mu_p} G$ in $\mathscr{E}$ which internalizes the operation of "multiplying a p-tuple of elements of G", in the sense that the diagrams

$$\begin{array}{ccc} G^{[o]} & \xrightarrow{\mu_o} & G \\ \wr\downarrow & \nearrow_{e} & \\ 1 & & \end{array} \quad \text{and} \quad \begin{array}{ccc} G^{[sp]} & \xrightarrow{\mu_{sp}} & G \\ \wr\downarrow & & \uparrow m \\ G^{[p]} \times G & \xrightarrow{\mu_p \times 1} & G \times G \end{array}$$

commute.

6.27 Theorem. *Let $X \xrightarrow{f} [p]$ be a morphism in $\mathscr{E}$ whose codomain is a finite cardinal. Then X is isomorphic to a finite cardinal in $\mathscr{E}$ iff f is isomorphic to a finite cardinal in $\mathscr{E}/[p]$.*

Proof. (i) Suppose $X \cong [q]$ for some natural number q. Now the pullback diagram

$$\begin{array}{ccc} [q] & \xrightarrow{f} & [p] \\ \downarrow{\scriptstyle (1,f)} & & \downarrow{\scriptstyle \Delta} \\ [q] \times [p] & \xrightarrow{f \times 1} & [p] \times [p] \end{array}$$

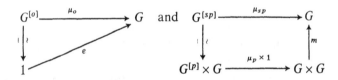

in $\mathscr{E}$ may be reinterpreted as a pullback diagram

$$\begin{array}{ccc} f & \longrightarrow & 1_{[p]} \\ \Big\downarrow x & & \Big\downarrow y \\ [p]^*[q] & \xrightarrow{[p]^*f} & [p]^*[p] \end{array}$$

in $\mathscr{E}/[p]$, where y is the generic element of $[p]$. But y has a complement by 6.26(i); hence x has a complement, and so f is isomorphic to a finite cardinal in $\mathscr{E}/[p]$ by 6.26(iii).

(ii) Conversely, let $[p] \xrightarrow{r} N$ be a natural number in $\mathscr{E}/[p]$ whose cardinal is isomorphic to f. Then a straightforward induction on p shows that X is isomorphic to the cardinal of $1 \xrightarrow{\bar{r}} N^{[p]} \xrightarrow{\mu_p} N$, where μ is defined using the monoid structure $(N, +, o)$. ■

6.28 Corollary. *A pullback of finite cardinals is a finite cardinal.*

Proof. Suppose we have a pullback diagram

$$\begin{array}{ccc} X & \xrightarrow{f} & [p] \\ \Big\downarrow g & & \Big\downarrow h \\ [q] & \xrightarrow{k} & [r] \end{array}$$

in $\mathscr{E}$. Then by 6.27(i) h is (isomorphic to) a finite cardinal in $\mathscr{E}/[r]$, so by 6.22 $g \cong k^*(h)$ is a finite cardinal in $\mathscr{E}/[q]$. Hence by 6.27(ii) is a finite cardinal in $\mathscr{E}$. ■

6.29 Theorem. *Let $\mathscr{E}_{fc}$ denote the full subcategory of $\mathscr{E}$ whose objects are finite cardinals. Then $\mathscr{E}_{fc}$ is a topos, and satisfies* (AC); *and the inclusion functor $\mathscr{E}_{fc} \longrightarrow \mathscr{E}$ is logical iff $\mathscr{E}$ is Boolean.*

Proof. From 6.28 and the fact that $[so] \cong 1$, $\mathscr{E}_{fc}$ has finite limits and the inclusion functor preserves them. And from 6.23(v), $\mathscr{E}_{fc}$ has exponentials and the inclusion functor preserves them. Combining 6.26(ii) and (iii) with 6.27, we deduce that the subobjects of a cardinal $[p]$ which are cardinals are precisely its complemented subobjects; and so $[sso] \cong 1 \amalg 1$ is a subobject classifier for $\mathscr{E}_{fc}$. But $1 \amalg 1$ is a subobject classifier for $\mathscr{E}$ iff $\mathscr{E}$ is Boolean.

Finally, it follows from the remark after 6.25 that $\mathscr{E}_{fc}$ satisfies (IC); so by

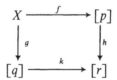

5.28 it is sufficient to show that 1 is projective in $\mathscr{E}_{fc}$, i.e. that any cardinal with global support has a global element. But if $[p]$ has global support, it is easy to see by the argument of 6.26(ii) that the formula "$p \geqslant so$" is valid, and hence there is a (unique) natural number q with $so + q = p$. Then $[p]$ has the global element $1 \cong [so] \xrightarrow{\nu_1} [so] \amalg [q] \cong [p]$. ■

It should be noted, however, that even in the case when $\mathscr{E}$ is Boolean the category $\mathscr{E}_{fc}$ does not necessarily reflect the internal structure of $\mathscr{E}$. For example, if G is a group, then the natural number object in $\mathscr{S}^{\mathbf{G}}$ is the set of natural numbers with *trivial* G-action, and so every finite cardinal in $\mathscr{S}^{\mathbf{G}}$ has trivial G-action. Thus $(\mathscr{S}^{\mathbf{G}})_{fc}$ is equivalent to $\mathscr{S}_f$ and not (as one might have hoped) to $(\mathscr{S}_f)^{\mathbf{G}}$; to recover the latter subtopos of $\mathscr{S}^{\mathbf{G}}$ one must use either the notion of local finiteness developed in §8.4 below, or the notion of Kuratowski-finiteness described in §9.1.

6.3. THE OBJECT CLASSIFIER

In the previous paragraph, we considered the category of finite cardinals in a topos $\mathscr{E}$ as an external category; the present paragraph will be devoted to the study of "the same" category regarded as an *internal* category in $\mathscr{E}$. It follows at once from 6.22 that the object $[n] = (N \times N \xrightarrow{s+} N)$ of $\mathscr{E}/N$ may be regarded as the indexed union of all finite cardinals in $\mathscr{E}$; and so the category $\text{Full}_{\mathscr{E}}([n])$ defined as in 2.38 "internalizes" the category $\mathscr{E}_{fc}$ of 6.29, in the sense that, for any object X of $\mathscr{E}$, the (external) category of X-elements of $\text{Full}_{\mathscr{E}}([n])$ is isomorphic to $(\mathscr{E}/X)_{fc}$. We shall write $\mathbf{E}_{\text{fin}}$ for $\text{Full}_{\mathscr{E}}([n])$; it follows from 6.29 that $\mathbf{E}_{\text{fin}}$ is in fact an internal topos in $\mathscr{E}$ (we leave the precise definition of this concept to the reader!).

At this point we shall find it convenient to introduce a notational convention which will greatly simplify the expressions with which we have to deal in this and the following paragraphs. The convention is simply that, when we are working in $\mathfrak{Top}/\mathscr{E}$ for some fixed $\mathscr{E}$, we shall no longer distinguish between an object X of $\mathscr{E}$ and the corresponding "constant object" f^*X in an $\mathscr{E}$-topos $(\mathscr{F} \xrightarrow{f} \mathscr{E})$, unless it is necessary to do so to avoid ambiguity. (This may be thought of as analogous to the convention whereby one identifies an element a of a ring A with the element $a.1_B$ in an arbitrary A-algebra B.) Thus, for example, the object of $\mathscr{E}/N$ which we have hitherto denoted by $(N^*X)^{[n]}$ will henceforth be simply $X^{[n]}$.

6.31 Lemma. *Let $\mathscr{E}$ and $\mathscr{F}$ be toposes with natural number objects, and $T: \mathscr{E} \longrightarrow \mathscr{F}$ an exact functor. Then $T(\mathbf{E}_{\text{fin}}) \cong \mathbf{F}_{\text{fin}}$.*

Proof. From 6.22 and 6.24, we know that T preserves $[n]$ and all constructions involved in the definition of $\mathbf{E}_{\text{fin}}$ (in particular, the exponential $[\pi_2^* n]^{[\pi_1^* n]}$ used to construct the object-of-morphisms; cf. 2.38). ■

6.32 Proposition. *We have an equivalence of categories*

$$\mathscr{E} \simeq \operatorname{Flat}(\mathbf{E}_{\text{fin}}{}^{\text{op}}, \mathscr{E}).$$

Proof. Given an object X of $\mathscr{E}$, we define a presheaf HX on $\mathbf{E}_{\text{fin}}$ by

$$(HX_0 \longrightarrow N) = X^{[n]},$$

with action of $\mathbf{E}_{\text{fin}}$ given by the internal composition map

$$[\pi_2^* n]^{[\pi_1^* n]} \times X^{[\pi_2^* n]} \longrightarrow X^{[\pi_1^* n]}$$

in $\mathscr{E}/N \times N$. Then it is easily seen that U-elements of HX_0 (resp. HX_1) correspond to morphisms $[p] \longrightarrow X$ (resp. commutative triangles

$$\begin{array}{ccc} [p] & \longrightarrow & X \\ \downarrow & \nearrow & \\ [q] & & \end{array} \quad)$$

in $\mathscr{E}/U$, where p and q are arbitrary natural numbers in $\mathscr{E}/U$. So to verify that **HX** is filtered, it is sufficient to show that $\mathscr{E}_{fc}$ has finite colimits, and that the inclusion functor $\mathscr{E}_{fc} \longrightarrow \mathscr{E}$ preserves them. Now 6.29 clearly implies the first half of this statement; but unfortunately we cannot directly deduce second half unless $\mathscr{E}$ happens to be Boolean. However, the fact that the initial object and coproducts are preserved follows at once from 6.23(i) and (iii); the proof that coequalizers are preserved is a tedious but straightforward inductive argument similar to 6.26(iii). So HX is a flat presheaf on $\mathbf{E}_{\text{fin}}$.

Conversely, let S be any flat presheaf on $\mathbf{E}_{\text{fin}}$, and define $V(S)$ to be the pullback of $S_0 \longrightarrow N$ along $1 \xrightarrow{S_0} N$. Then if we apply the forgetful functor

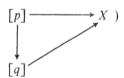

$\mathscr{E}^{\mathbf{E}_{\text{fin}}{}^{\text{op}}} \longrightarrow \mathscr{E}/N$ to the isomorphism $S \cong Y(\mathbf{E}_{\text{fin}}) \otimes_{\mathbf{E}_{\text{fin}}} S$, we obtain

$$(S_0 \longrightarrow N) \cong G \otimes_{\mathbf{E}_{\text{fin}}} S,$$

where $N \overset{G}{\dashrightarrow} \mathbf{E}_{\text{fin}}$ is the profunctor from the discrete category N to $\mathbf{E}_{\text{fin}}$ obtained by "forgetting" the left action of $\mathbf{E}_{\text{fin}}$ on the Yoneda profunctor $Y(\mathbf{E}_{\text{fin}})$. But now it is not hard to see that $G \cong R(so)^{[n]}$, where $R: \mathscr{E}/N \longrightarrow \mathscr{E}^{\mathbf{E}_{\text{fin}}}$ is the left adjoint of the forgetful functor; and since S is flat the functor $(-)\otimes S$ preserves exponentiation to the power of a finite cardinal, by 4.32 and 6.24. So we have

$$(S_0 \longrightarrow N) \cong (R(so) \otimes S)^{[n]} \cong V(S)^{[n]}.$$

And we may chase the $\mathbf{E}_{\text{fin}}$-action maps through all the above equivalences, and so extend the above isomorphism in $\mathscr{E}/N$ to an isomorphism $S \cong H(V(S))$ in $\mathscr{E}^{\mathbf{E}_{\text{fin}}{}^{\text{op}}}$. But it is clear that for any X in $\mathscr{E}$ we have $V(HX) \cong X^{[so]} \cong X$, so that H and V define the required equivalence of categories. ∎

6.33 Theorem (G. C. Wraith). *There exists an internal diagram U on $\mathbf{E}_{\text{fin}}$ such that the functor*

$$\mathfrak{Top}/\mathscr{E}(\mathscr{F}, \mathscr{E}^{\mathbf{E}_{\text{fin}}}) \longrightarrow \mathscr{F}; g \longmapsto g^*(U)$$

is an equivalence of categories for any $\mathscr{E}$-topos $\mathscr{F}$. In other words, $\mathscr{E}^{\mathbf{E}_{\text{fin}}}$ (which we henceforth denote by $\mathscr{E}[U]$) is an object classifier *for $\mathfrak{Top}/\mathscr{E}$ (cf.* 4.37(iv)). *Moreover, the generic object U is the "inclusion functor $\mathbf{E}_{\text{fin}} \longrightarrow \mathscr{E}$" defined as in* 2.38.

Proof. Let $\mathscr{F} \xrightarrow{f} \mathscr{E}$ be an $\mathscr{E}$-topos. Then we have equivalences

$$\begin{aligned} \mathfrak{Top}/\mathscr{E}(\mathscr{F}, \mathscr{E}[U]) &\simeq \text{Flat}(f^*\mathbf{E}_{\text{fin}^{\text{op}}}, \mathscr{F} && \text{by } 4.34 \\ &\simeq \text{Flat}(\mathbf{F}_{\text{fin}^{\text{op}}}, \mathscr{F}) && \text{by } 6.31 \\ &\simeq \mathscr{F} && \text{by } 6.32. \end{aligned}$$

And these equivalences are natural in $\mathscr{F}$, since it follows from 6.24 that the functor H defined in the proof of 6.32 commutes with inverse image functors. So it remains only to justify the last sentence of the theorem, which is simply a matter of chasing the Yoneda profunctor $Y(\mathbf{E}_{\text{fin}})$ through the equivalence of 6.32. In fact we find that $V(Y(\mathbf{E}_{\text{fin}}))$ is the representable diagram $R(so)$, which is easily seen to be isomorphic to the inclusion functor of 2.38. ∎

The notation $\mathscr{E}[U]$ (which is due to Wraith) is deliberately reminiscent of that used for polynomial rings; for it is illuminating to think of $\mathscr{E}[U]$ as the "free $\mathscr{E}$-topos obtained by adjoining an indeterminate object U". Indeed, there is even a sense in which the objects of $\mathscr{E}[U]$ can be regarded as "polynomials in U with coefficients in $\mathscr{E}$" (see [58]).

If X is an object of an $\mathscr{E}$-topos $\mathscr{F}$, we shall write $\mathscr{F} \xrightarrow{\bar{X}} \mathscr{E}[U]$ for the geometric morphism which classifies it; and if T is an object of $\mathscr{E}[U]$, we shall write $T \otimes X$ for the object of $\mathscr{F}$ classified by the composite

$$\mathscr{F} \xrightarrow{\bar{X}} \mathscr{E}[U] \xrightarrow{\bar{T}} \mathscr{E}[U],$$

i.e. the object $\bar{X}^*(T)$. In particular, taking $\mathscr{F} = \mathscr{E}[U]$, we have a bifunctor $\otimes : \mathscr{E}[U] \times \mathscr{E}[U] \longrightarrow \mathscr{E}[U]$; it is clear from the definition that this bifunctor is associative (up to canonical, and therefore coherent, natural isomorphism), and that the generic object U is a two-sided unit for it (again up to coherent isomorphism). So $(\mathscr{E}[U], \otimes, U)$ is a monoidal category ([CW], p. 158).

6.34 COROLLARY. *Let $\mathscr{E}$ be a topos with a natural number object. Then the following concepts are equivalent (in the sense that they define equivalent categories):*

(i) *Objects of $\mathscr{E}[U]$, i.e. internal diagrams T on $\mathbf{E}_{\mathrm{fin}}$.*
(ii) *Left flat profunctors $\mathbf{E}_{\mathrm{fin}} \overset{\hat{T}}{\dashrightarrow} \mathbf{E}_{\mathrm{fin}}$ (cf. 4.39).*
(iii) *Geometric morphisms $\mathscr{E}[U] \xrightarrow{\bar{T}} \mathscr{E}[U]$ over $\mathscr{E}$.*
(iv) *Natural endomorphisms of the forgetful functor $(\mathfrak{Top}/\mathscr{E})^{\mathrm{op}} \longrightarrow \mathfrak{Cat}$, i.e. families of functors $T_{\mathscr{F}} : \mathscr{F} \longrightarrow \mathscr{F}$ for every $\mathscr{E}$-topos $\mathscr{F}$, such that the diagram*

$$\begin{array}{ccc} \mathscr{F} & \xrightarrow{g^*} & \mathscr{G} \\ {\scriptstyle T_{\mathscr{F}}}\downarrow & & \downarrow{\scriptstyle T_{\mathscr{G}}} \\ \mathscr{F} & \xrightarrow{g^*} & \mathscr{G} \end{array}$$

commutes up to coherent natural isomorphism for every $\mathscr{G} \xrightarrow{g} \mathscr{F}$ over $\mathscr{E}$.

Proof. (i) $\simeq$ (iii) is the special case $\mathscr{F} = \mathscr{E}[U]$ of 6.33, and (ii) $\simeq$ (iii) is a special case of 4.39. For (i) $\simeq$ (iv), suppose we are given an object T of $\mathscr{E}[U]$. Then we can define $T_{\mathscr{F}}$ to be the functor $X \longmapsto T \otimes X$, and the naturality is obvious. Conversely, if we are given a natural endomorphism $(T_{\mathscr{F}})$ as in (iv), then naturality says that it is canonically isomorphic to that induced by the object $T_{\mathscr{E}[U]}(U)$ of $\mathscr{E}[U]$. ■

6.35 REMARK. Let $(\mathbf{E}_{\mathrm{fin}})^2$ be the (internal) category of morphisms of $\mathbf{E}_{\mathrm{fin}}$, constructed as in Ex. 2.3. Then arguments similar to those of 6.32 and 6.33 show that the topos $\mathscr{E}^{(\mathbf{E}_{\mathrm{fin}})^2} = \mathscr{E}[U_1 \longrightarrow U_2]$ is a *morphism classifier* for $\mathfrak{Top}/\mathscr{E}$, in the sense that we have

$$\mathfrak{Top}/\mathscr{E}(\mathscr{F}, \mathscr{E}[U_1 \longrightarrow U_2]) \simeq \mathscr{F}^{\mathbf{2}}$$

for any $\mathscr{E}$-topos $\mathscr{F}$. (Details of the proof will be found in [58], Theorem 6.5.) Moreover, it is not hard to verify that the geometric morphisms $\bar{U}_1, \bar{U}_2$: $\mathscr{E}[U_1 \longrightarrow U_2] \longrightarrow \mathscr{E}[U]$ which classify the domain and codomain of the generic morphism are those induced (as in 2.35) by the internal functors $\partial_0, \partial_1 : (\mathbf{E}_{\mathrm{fin}})^2 \longrightarrow \mathbf{E}_{\mathrm{fin}}$ constructed in Ex. 2.3.

Now if we form the pullback of the diagram

$$\begin{array}{ccc} & & \mathscr{E}[U_1 \longrightarrow U_2] \\ & & \downarrow{\scriptstyle \bar{U}_1} \\ \mathscr{E}[U_1 \longrightarrow U_2] & \xrightarrow{\bar{U}_2} & \mathscr{E}[U] \end{array}$$

in $\mathfrak{BTop}/\mathscr{E}$ (cf. 4.48), we clearly obtain a topos $\mathscr{E}[U_1 \longrightarrow U_2 \longrightarrow U_3]$ which classifies composable pairs of morphisms in $\mathscr{E}$-toposes. More generally still, it is easily seen that any finite (external) category $\mathbf{D}$ may be expressed as a colimit of a finite diagram in $\mathfrak{Cat}$ whose vertices are copies of $\mathbf{1}$, $\mathbf{2}$ and the "commutative triangle" category $\mathbf{3}$; so by forming the limit of the corresponding diagram in $\mathfrak{BTop}/\mathscr{E}$ with vertices $\mathscr{E}[U]$, $\mathscr{E}[U_1 \longrightarrow U_2]$ and $\mathscr{E}[U_1 \longrightarrow U_2 \longrightarrow U_3]$, we obtain a classifying topos $\mathscr{E}[\mathbb{D}]$ for diagrams of type $\mathbf{D}$ in $\mathscr{E}$-toposes. ∎

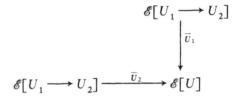

We now return to the problem of defining objects of $\mathscr{E}/N$ by recursion, which we mentioned in §6.1. As we observed there, it is essential that the "process" T which we use to construct new objects from old should be compatible with the internal structure of $\mathscr{E}$. One (very strong) way in which we can interpret this requirement is to demand that T should be an endofunctor of $\mathscr{E}$ which extends in a natural way to an endofunctor of any $\mathscr{E}$-topos; i.e. that it should be a natural endomorphism of $(\mathfrak{Top}/\mathscr{E})^{\mathrm{op}} \longrightarrow \mathfrak{Cat}$ as described in 6.34 (iv). (Note in passing that any such T is a strong functor, since it extends naturally to $\mathscr{E}$-toposes of the form $\mathscr{E}/Z$, and hence we can obtain $T_{X,Y}: Y^X \longrightarrow TY^{TX}$ by applying the functor $T_{(\mathscr{E}/Y^X)}$ to the generic morphism $(Y^X)^*X \longrightarrow (Y^X)^*Y$ in $\mathscr{E}/Y^X$, and regarding the result as a global element of $(Y^X)^*(TY^{TX})$. So the uniqueness theorem 6.19 certainly applies to such functors T.)

But if we require T to be of this form, then the existence theorem we wish to prove can be stated in terms of classifying morphisms, as follows: given any diagram $\mathscr{E} \xrightarrow{\bar{X}} \mathscr{E}[U] \xrightarrow{\bar{T}} \mathscr{E}[U]$ in $\mathfrak{Top}/\mathscr{E}$, there exists a geometric morphism $\mathscr{E}/N \xrightarrow{\bar{F}} \mathscr{E}[U]$ such that the diagram

$$\begin{array}{ccccc} \mathscr{E} & \xrightarrow{o} & \mathscr{E}/N & \xrightarrow{s} & \mathscr{E}/N \\ & \searrow^{\bar{X}} & \downarrow{\scriptstyle \bar{F}} & & \downarrow{\scriptstyle \bar{F}} \\ & & \mathscr{E}[U] & \xrightarrow{\bar{T}} & \mathscr{E}[U] \end{array}$$

commutes up to isomorphism. The similarity of this diagram with that of 6.11 is immediately apparent.

In fact we shall prove a stronger result: namely that $\mathscr{E}/N$ is a natural number object (in the sense appropriate to 2-categories) in the full subcategory $\mathfrak{DTop}/\mathscr{E}$ of $\mathfrak{Top}/\mathscr{E}$ whose objects are toposes of the form $\mathscr{E}^{\mathbf{C}}$, $\mathbf{C} \in \mathbf{cat}(\mathscr{E})$. It can actually be shown that $\mathscr{E}/N$ is a natural number object in the larger 2-category $\mathfrak{BTop}/\mathscr{E}$ of bounded $\mathscr{E}$-toposes defined in 4.48: but the proof of this fact requires a strengthening of 4.39 to give a "profunctor description" of $\mathfrak{Top}/\mathscr{E}(\mathscr{F}, \mathscr{G})$ for arbitrary bounded $\mathscr{E}$-toposes $\mathscr{F}$ and $\mathscr{G}$. The details of this strengthening are somewhat complicated, and so we do not give them here; the interested reader is referred to [58], Proposition 3.2.

Let $\mathbf{C}$ be an internal category in $\mathscr{E}$. Now the effect of 4.39 is to reduce the problem of constructing the "nth iterate" of a geometric morphism $\mathscr{E}^{\mathbf{C}} \longrightarrow \mathscr{E}^{\mathbf{C}}$ over $\mathscr{E}$ to that of constructing the "nth tensor power" of a profunctor $\mathbf{C} \dashrightarrow \mathbf{C}$. The next two propositions are devoted to this construction. First we consider the case of a discrete category $\mathbf{C} = (C \underset{1}{\overset{1}{\rightrightarrows}} C)$; in this case we have simply $\mathrm{Prof}_{\mathscr{E}}(\mathbf{C}, \mathbf{C}) \simeq \mathscr{E}/C \times C$, and $\otimes_{\mathbf{C}}$ is identified with $\times_C$ (where we adopt the convention of 2.11 for interpreting the latter symbol). Also, the Yoneda profunctor $Y(\mathbf{C})$ is identified with the diagonal $C \xrightarrow{\Delta} C \times C$.

Let p be a natural number in $\mathscr{E}$. Then the equations $sp = so + p = p + so$ give us (via 6.23(iii)) two different coproduct decompositions

$$[sp] \simeq 1 \amalg [p] \simeq [p] \amalg 1.$$

We shall (for the moment) write μ_1, μ_2 for the coproduct inclusions corresponding to the first decomposition, and ν_1, ν_2 for the second. (If we identify

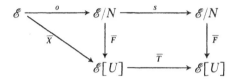

$[p]$ with the down-segment of p for the strict order-relation, then μ_1, μ_2, ν_1 and ν_2 are respectively the factorizations through $[sp] \rightarrowtail N$ of $1 \xrightarrow{o} N$, $[p] \rightarrowtail N \xrightarrow{s} N$, $[p] \rightarrowtail N$ and $1 \xrightarrow{p} N$.) Then it is easily seen from the associativity of $+$ that the diagram

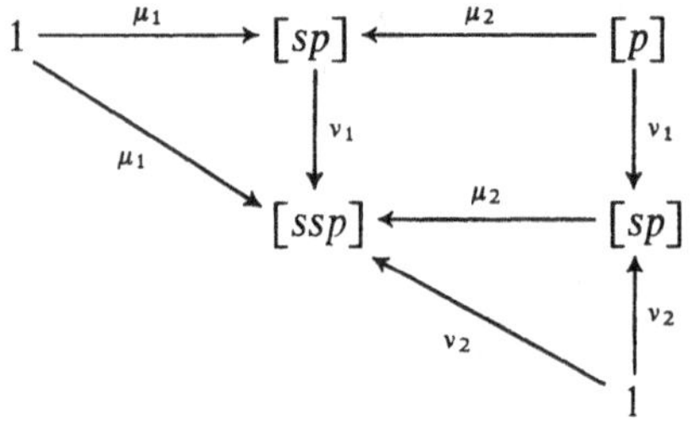

commutes for any p.

6.36 PROPOSITION. *Let $\mathscr{E}$ be a topos with a natural number object, and C an object of $\mathscr{E}$. Then there exists (uniquely up to canonical isomorphism) a functor*

$$(-)^{\langle n \rangle} : \mathscr{E}/C \times C \longrightarrow (\mathscr{E}/C \times C)/N$$

with the property that for any $X \xrightarrow{(\gamma_1, \gamma_2)} C \times C$ we have

$$o^*(X^{\langle n \rangle}) \cong (C \xrightarrow{\Delta} C \times C)$$

and

$$s^*(X^{\langle n \rangle}) \cong X^{\langle n \rangle} \times_C X \cong X \times_C X^{\langle n \rangle}.$$

Proof. Since $1 \xrightarrow{o} N \xleftarrow{s} N$ is a coproduct, we may define $X^{\langle n \rangle}$ by specifying separately the two objects $o^*(X^{\langle n \rangle})$ and $s^*(X^{\langle n \rangle})$. So we define the former to be Δ, and the latter to be the equalizer in $\mathscr{E}/N$ of the morphisms

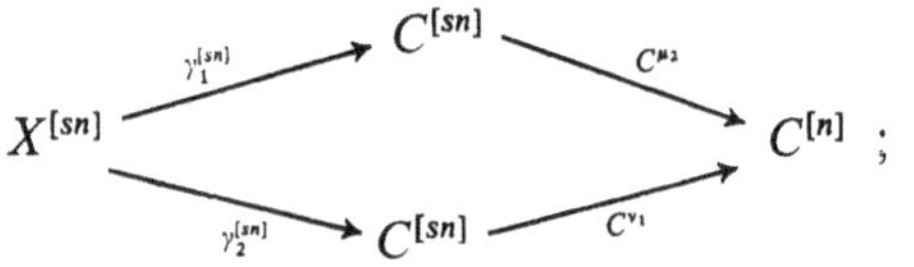

the projections $s^*(X^{\langle n \rangle}) \longrightarrow C \times C$ being induced by the composites

$$X^{[sn]} \xrightarrow{\gamma_1^{[sn]}} C^{[sn]} \xrightarrow{C^{\mu_1}} C \quad \text{and} \quad X^{[sn]} \xrightarrow{\gamma_2^{[sn]}} C^{[sn]} \xrightarrow{C^{\nu_2}} C$$

respectively.

Then the first of the isomorphisms in the statement of the proposition is trivial; to establish the second, note first that

$$(so)^*(X^{\langle n\rangle}) \cong \mathrm{eq}(X^{[so]} \rightrightarrows C^{[o]}) \cong \mathrm{eq}(X \rightrightarrows 1) \cong X \cong X \times_C \Delta \cong \Delta \times_C X,$$

and then that $(ss)^*(X^{\langle n\rangle})$ may (using the commutative diagram immediately above the proposition) be expressed as the equalizer of

$$\begin{array}{ccccc} & & C^{[sn]} \times C & & \\ & \overset{\gamma_1^{[sn]} \times \gamma_1}{\nearrow} & & \overset{C^{\mu_2} \times 1_c}{\searrow} & \\ X^{[sn]} \times X & & & & C^{[n]} \times C \\ & \underset{\gamma_2^{[sn]} \times \gamma_2}{\searrow} & & \underset{(C^{\nu_1}, C^{\nu_2})\pi_2}{\nearrow} & \\ & & C^{[sn]} \times C & & \end{array}$$

which is easily seen to be $s^*(X^{\langle n\rangle}) \times_C X$. Similarly, we may obtain an isomorphism $(ss)^*(X^{\langle n\rangle}) \cong X \times_C s^*(X^{\langle n\rangle})$.

The proof of the uniqueness of $X^{\langle n\rangle}$ is, of course, an application of the ideas of 6.19. Unfortunately, the functor $(-) \times_C X$ is not in general strong as a functor $\mathscr{E}/C \times C \longrightarrow \mathscr{E}/C \times C$; but it is enriched over $\mathscr{E}$, and so we can express $X^{\langle n\rangle}$ as the unique solution of a recursion problem in $\mathscr{E}$. The remaining details are straightforward. ■

Now let $\mathbf{C}$ be an arbitrary internal category in $\mathscr{E}$, and $\mathbf{C} \overset{X}{\dashrightarrow} \mathbf{C}$ a profunctor. In the diagram

$$\begin{array}{ccccc} X_0 \times_{C_0} C_1 \times_{C_0} X_0 \times_{C_0} C_1 \times_{C_0} X_0 & \rightrightarrows & X_0 \times_{C_0} X_0 \times_{C_0} C_1 \times_{C_0} X_0 & \longrightarrow & (X \otimes_{\mathbf{C}} X)_0 \times_{C_0} C_1 \times_{C_0} X_0 \\ \downdownarrows & & \downdownarrows & & \downdownarrows \\ X_0 \times_{C_0} C_1 \times_{C_0} X_0 \times_{C_0} X_0 & \rightrightarrows & X_0 \times_{C_0} X_0 \times_{C_0} X_0 & \longrightarrow & (X \otimes_{\mathbf{C}} X)_0 \times_{C_0} X_0 \\ \downarrow & & \downarrow & & \downarrow \\ X_0 \times_{C_0} C_1 \times_{C_0} (X \otimes_{\mathbf{C}} X)_0 & \rightrightarrows & X_0 \times_{C_0} (X \otimes_{\mathbf{C}} X)_0 & \longrightarrow & (X \otimes_{\mathbf{C}} X \otimes_{\mathbf{C}} X)_0 \end{array}$$

which we used in 2.43 to prove the associativity of $\otimes_{\mathbf{C}}$, the rows and columns are in fact reflexive coequalizers, the splittings being induced by the inclusion-of-identities of $\mathbf{C}$. So by Lemma 0.17, $(X \otimes_{\mathbf{C}} X \otimes_{\mathbf{C}} X)_0$ may be computed as the coequalizer of the single pair

$$X_0 \times_{C_0} C_1 \times_{C_0} X_0 \times_{C_0} C_1 \times_{C_0} X_0 \underset{\beta \times \beta \times 1}{\overset{1 \times \alpha \times \alpha}{\rightrightarrows}} X_0 \times_{C_0} X_0 \times_{C_0} X_0,$$

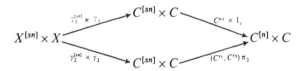

where α and β are the left and right actions of $\mathbf{C}$ on X_0. This provides the motivation for the construction used in the next proposition.

6.37 PROPOSITION. *Let $\mathscr{E}$ be a topos with a natural number object, and $\mathbf{C}$ an internal category in $\mathscr{E}$. Then there exists (uniquely up to canonical isomorphism) a functor*

$$(-)^{\otimes n}\colon \mathrm{Prof}_{\mathscr{E}}(\mathbf{C},\mathbf{C}) \longrightarrow \mathrm{Prof}_{\mathscr{E}}(\mathbf{C},\mathbf{C})/N$$

with the property that for any $\mathbf{C} \overset{X}{\dashrightarrow} \mathbf{C}$ we have

$$o^*(X^{\otimes n}) \cong Y(\mathbf{C}) \quad \textit{and} \quad s^*(X^{\otimes n}) \cong X^{\otimes n} \otimes_{\mathbf{C}} X \cong X \otimes_{\mathbf{C}} X^{\otimes n}.$$

(Note that objects of $\mathrm{Prof}_{\mathscr{E}}(\mathbf{C},\mathbf{C})/N$ may be regarded equivalently as profunctors $N^*\mathbf{C} \dashrightarrow N^*\mathbf{C}$ in $\mathscr{E}/N$, or as profunctors $\mathbf{C} \times N \dashrightarrow \mathbf{C}$ or $\mathbf{C} \dashrightarrow \mathbf{C} \times N$ in $\mathscr{E}$, where we regard N as a discrete internal category.)

Proof. As before, we define $X^{\otimes n}$ in two stages: $o^*(X^{\otimes n}) = Y(\mathbf{C})$, and $(s^*(X^{\otimes n}))_0$ is the coequalizer in $\mathscr{E}/C_0 \times C_0 \times N$ of

$$\begin{array}{ccccc}
X_0 \times_{C_0} (C_1 \times_{C_0} X_0)^{\langle n\rangle} & \xrightarrow{1 \times \alpha^{\langle n\rangle}} & X_0 \times_{C_0} X_0^{\langle n\rangle} & \searrow & \\
\downarrow \wr & & & & s^*(X_0^{\langle n\rangle}) \\
(X_0 \times_{C_0} C_1)^{\langle n\rangle} \times_{C_0} X_0 & \xrightarrow{\beta^{\langle n\rangle} \times 1} & X_0^{\langle n\rangle} \times_{C_0} X_0 & \nearrow &
\end{array}$$

where the vertical arrow on the left is constructed recursively, using the method of 6.18 to define a morphism

$$\Delta \longrightarrow \mathrm{Hom}_{C_0}(X_0 \times_{C_0} (C_1 \times_{C_0} X_0)^{\langle n\rangle}, (X_0 \times_{C_0} C_1)^{\langle n\rangle} \times_{C_0} X_0).$$

(Here $\mathrm{Hom}_{C_0}(X, -)$ denotes the right adjoint of $(-) \times_{C_0} X$, i.e. the closed structure on the monoidal category $(\mathscr{E}/C_0 \times C_0, \times_{C_0}, \Delta)$; this may be regarded as a particular instance of the biclosed structure on the bicategory $\mathfrak{Prof}_{\mathscr{E}}$, cf. 2.49.)

The left and right actions of $\mathbf{C}$ on $X^{\otimes n}$ are induced in the obvious way by α and β.

The proof that $X^{\otimes n}$ satisfies the given recursion data is similar to the argument of 6.36, using Lemma 0.17 (as already indicated) at the inductive step. And the uniqueness argument is again similar to 6.36. ■

6.38 THEOREM (Johnstone–Wraith). *The topos $\mathscr{E}/N$ is a natural number object in the 2-category $\mathfrak{D}\mathfrak{T}\mathfrak{op}/\mathscr{E}$.*

Proof. Suppose given morphisms $\mathscr{E} \xrightarrow{x} \mathscr{F} \xrightarrow{t} \mathscr{F}$ over $\mathscr{E}$, where $\mathscr{F} = \mathscr{E}^{\mathbf{C}}$. Then we can represent t by a left flat profunctor $\mathbf{C} \overset{T}{\dashrightarrow} \mathbf{C}$, by 4.39. Consider the profunctor $T^{\otimes n}$; we shall show that $T^{\otimes n}$ is left flat, and so defines a geometric morphism $\mathscr{F}/N \longrightarrow \mathscr{F}/N$ over $\mathscr{E}/N$, or equivalently $\mathscr{F}/N \longrightarrow \mathscr{F}$ over $\mathscr{E}$. But this amounts to showing that, if we have any finite diagram in $\mathscr{F}$ with vertices $(V_\alpha \mid \alpha \in A)$, then the canonical map

$$(\varprojlim_\alpha V_\alpha) \otimes_{\mathbf{C}} T^{\otimes n} \longrightarrow \varprojlim_\alpha (V_\alpha \otimes_{\mathbf{C}} T^{\otimes n})$$

is iso in $\mathscr{F}/N$. But to do this, we use the fact that $(-)\otimes_{\mathbf{C}} T$ is enriched over $\mathscr{E}$ and left exact to construct a global element of

$$\varprojlim_{\mathbf{C}} \operatorname{Iso}((\varprojlim_\alpha V_\alpha) \otimes_{\mathbf{C}} T^{\otimes n}, \varprojlim_\alpha (V_\alpha \otimes_{\mathbf{C}} T^{\otimes n}))$$

in $\mathscr{E}/N$ by the method of 6.18; and it is then easy to see that the transpose of this element must indeed be the canonical map.

Now define f to be the composite geometric morphism

$$\mathscr{E}/N \xrightarrow{x/N} \mathscr{F}/N \xrightarrow{t^{\otimes n}} \mathscr{F},$$

where $t^{\otimes n}$ is the morphism corresponding to $T^{\otimes n}$ and x/N is the pullback of x along $\mathscr{E}/N \longrightarrow \mathscr{E}$. Then it follows easily from 4.39 and 6.37 that the diagram

$$\begin{array}{ccccc}
\mathscr{E} & \xrightarrow{o} & \mathscr{E}/N & \xrightarrow{s} & \mathscr{E}/N \\
 & \searrow^{x} & \downarrow{\scriptstyle f} & & \downarrow{\scriptstyle f} \\
 & & \mathscr{F} & \xrightarrow{t} & \mathscr{F}
\end{array}$$

commutes up to isomorphism. But in fact this diagram determines f up to canonical isomorphism, by the uniqueness clause of 6.37. ■

6.39 COROLLARY (Recursion Theorem). *Let $\mathscr{E}$ be a topos with a natural number object, X an object of $\mathscr{E}$, and T a natural endomorphism of the forgetful functor $(\mathfrak{T}\mathfrak{op}/\mathscr{E})^{\mathrm{op}} \longrightarrow \mathfrak{Cat}$. Then there exists (uniquely up to canonical* *isomorphism) an object F of $\mathscr{E}/N$ satisfying the recursion data*

$$o^*F \cong X \text{ and } s^*F \cong T_{\mathscr{E}/N}(F).$$

Proof. Take $\mathscr{F} = \mathscr{E}[U]$ in 6.38. ■

6.4. ALGEBRAIC THEORIES

One of the most important uses of the axiom of infinity in classical set theory is in the construction of free algebraic structures. It is therefore not surprising to find that in topos theory, too, it is the existence of a natural number object which unlocks the door to universal algebra. We begin with a theorem, due to C. J. Mikkelsen, which is implict in a number of results already proved, but which we have refrained from proving explicitly until the present paragraph.

6.41 THEOREM. *Let $\mathscr{E}$ be a topos, and let* **mon**$(\mathscr{E})$ *denote the category of monoids in $\mathscr{E}$. Then the following are equivalent:*

(i) *$\mathscr{E}$ has a natural number object.*
(ii) *The forgetful functor U:* **mon**$(\mathscr{E}) \longrightarrow \mathscr{E}$ *has a left adjoint.*

Proof. (i) ⇒ (ii): Let X be an object of $\mathscr{E}$. Consider the object $MX = \Sigma_N(X^{[n]})$; we shall show that it is the underlying object of the free monoid generated by X. The unit of the monoid structure on MX is given by the pullback

$$\begin{array}{ccc} 1 \cong X^{[o]} & \xrightarrow{\quad e \quad} & MX \\ \downarrow & & \downarrow \\ 1 & \xrightarrow{\quad o \quad} & N \end{array},$$

and the multiplication by the pullback

$$\begin{array}{ccc} MX \times MX \cong \Sigma_{N \times N}(X^{[\pi_1^* n + \pi_2^* n]}) & \xrightarrow{\quad m \quad} & MX \\ \downarrow & & \downarrow \\ N \times N & \xrightarrow{\quad + \quad} & N \end{array}.$$

Since the above squares are pullbacks, the associativity and unit diagrams for (MX, m, e) follow immediately from those for the monoid $(N, +, o)$.

The unit of the adjunction $(M \dashv U)$ is obtained from the pullback

$$\begin{array}{ccc} X \cong X^{[so]} & \xrightarrow{\eta_X} & MX \\ \downarrow & & \downarrow \\ 1 & \xrightarrow{so} & N \end{array} .$$

To construct the counit, let (G, p, f) be a monoid in $\mathscr{E}$. We must define a morphism $MG \xrightarrow{\varepsilon_G} G$ in $\mathscr{E}$, or equivalently $G^{[n]} \longrightarrow G$ in $\mathscr{E}/N$, or equivalently a global element of $G^{(G^{[n]})}$ in $\mathscr{E}/N$. To construct the latter, we use the method of 6.18 and the data

$$1 \xrightarrow{f} G \cong G^{(G^{[o]})} \cong o^*(G^{(G^{[n]})})$$

and

$$G^{(G^{[n]})} \xrightarrow{\theta} G^{(G^{[n]}} \times G) \cong s^*(G^{(G^{[n]})}),$$

where the transpose of θ is the composite

$$G^{(G^{[n]})} \times G^{[n]} \times G \xrightarrow{(\mathrm{ev}\,.\,\pi_{12},\, \pi_3)} G \times G \xrightarrow{p} G .$$

It is now a straightforward induction to show that ε_G is a monoid homomorphism, and that η and ε satisfy the triangular identities.

(ii) $\Rightarrow$ (i): Let (M, m, e) be the free monoid generated by 1. We shall show that $M \cong 1 \amalg M$, from which the result follows by 6.15.

Let $1 \xrightarrow{g} M$ be the insertion of the generator (i.e. the unit of the adjunction), and define s to be the composite

$$M \cong M \times 1 \xrightarrow{1 \times g} M \times M \xrightarrow{m} M .$$

Now define μ to be the composite

$$(1 \amalg M) \times (1 \amalg M) \cong 1 \amalg M \amalg M \amalg (M \times M) \xrightarrow{\begin{pmatrix} \nu_1 \\ \nu_2 \\ \nu_2 \\ \nu_2 sm \end{pmatrix}} 1 \amalg M .$$

Then it is easily checked that $(1 \amalg M, \mu, \nu_1)$ is a monoid, that $\binom{e}{s}: 1 \amalg M \longrightarrow M$

is a monoid homomorphism, and that $1 \amalg M$ is a free monoid with generator $1 \xrightarrow{\nu_2 e} 1 \amalg M$. So $\binom{e}{s}$ is an isomorphism.

(In fact, of course, M is itself a natural number object, as is obvious from the construction used in the first part; but it seems hard to give a direct proof of this.) ∎

Once we have a free monoid functor, it is a straightforward task to construct free functors for other finitary algebraic theories. In this context, we shall (for the moment) find it convenient to use, not the "structure-semantics" definition of an algebraic theory introduced by F. W. Lawvere in [176], but the older "universal-algebra" definition of a theory in terms of a specific presentation by operations and equations (see [180]). (Shortly, we shall be in a position to introduce a new definition of algebraic theories, due to G. C. Wraith [58], which is much more convenient for use in the topos-theoretic context.)

A *finitary algebraic theory*, then, consists of a set of *operations*, each labelled with a natural number or "arity", together with a set of finitary *equations* between them. We say the theory is *finitely-presented* if both these sets are finite. If $\mathbb{T}$ is such a theory and $\mathscr{C}$ is a category with finite products, a *model* of $\mathbb{T}$ in $\mathscr{C}$ means an object M equipped with a morphism $\alpha_M : M^m \longrightarrow M$ for each m-ary operation α of $\mathbb{T}$, such that each equation of $\mathbb{T}$ gives rise to a commutative diagram in $\mathscr{C}$ in the obvious way. We write $\mathbb{T}(\mathscr{C})$ for the category of $\mathbb{T}$-models in $\mathscr{C}$.

6.42 LEMMA. *Let $\mathscr{E}$ be a cartesian closed category, $\mathbb{T}$ a finitary algebraic theory. Then the forgetful functor $\mathbb{T}(\mathscr{E}) \xrightarrow{U} \mathscr{E}$ creates reflexive coequalizers.*

Proof. Let $X \underset{g}{\overset{f}{\rightrightarrows}} Y$ be a reflexive pair of $\mathbb{T}$-model homomorphisms, and $Y \xrightarrow{h} Z$ their coequalizer in $\mathscr{E}$. Applying Ex. 0.1 inductively, we deduce that $X^m \rightrightarrows Y^m \longrightarrow Z^m$ is a coequalizer for any natural number m, and so there is a unique way of defining a $\mathbb{T}$-model structure on Z which makes h into a $\mathbb{T}$-model homomorphism. It is then clear that h is a coequalizer of f and g in $\mathbb{T}(\mathscr{E})$. ∎

Clearly, too, the forgetful functor $\mathbb{T}(\mathscr{E}) \longrightarrow \mathscr{E}$ reflects isomorphisms; so if we can construct a left adjoint for it (i.e. a free $\mathbb{T}$-model functor), it will follow from the Crude Tripleability Theorem (0.13) that $\mathbb{T}(\mathscr{E})$ is monadic over $\mathscr{E}$.

6.43 THEOREM (B. Lesaffre [78]). *Let $\mathscr{E}$ be a topos with a natural number object, and $\mathbb{T}$ a finitely-presented, finitary algebraic theory. Then*

(i) *There exists a free functor $\mathscr{E} \longrightarrow \mathbb{T}(\mathscr{E})$.*
(ii) *$\mathbb{T}(\mathscr{E})$ is monadic over $\mathscr{E}$.*
(iii) *$\mathbb{T}(\mathscr{E})$ has finite colimits.*

Proof. We have already remarked that (ii) follows from (i), 6.42 and 0.13; and (iii) similarly follows from (ii) and Linton's theorem (0.16).

To prove (i), consider first the case when $\mathbb{T}$ is the free theory generated by a finite set $A = \{\alpha_1, \alpha_2, \ldots, \alpha_m\}$ of finitary operations. As usual, we identify A with the m-fold coproduct of copies of 1 in $\mathscr{E}$, and write $A \xrightarrow{\gamma} N$ for the morphism indexing each operation by its arity. Now let X be any object of $\mathscr{E}$, and consider the free monoid $M(A \amalg X)$, which we can think of as the set of all words in the α_i and the elements of X. It is not hard to see that $M(A \amalg X)$ has a $\mathbb{T}$-model structure, the operations being defined in a similar manner to the monoid structure on MX; to obtain the free $\mathbb{T}$-model on X, we need to identify the subobject of $M(A \amalg X)$ consisting of the *well-formed* words, i.e. those in which each operation is applied to exactly the right number of arguments.

To do this, we define a morphism

$$\mathrm{prs}\colon M(A \amalg X) \longrightarrow \Omega \times N$$

which "parses" a word as a sequence of well-formed words, as follows: first define the predecessor map $N \xrightarrow{p} N$ to be $N \cong 1 \amalg N \xrightarrow{\binom{o}{1_N}} N$, and the subtraction map $N \times N \xrightarrow{-} N$ to be the morphism whose exponential transpose satisfies the recursion data $1 \xrightarrow{\bar{1}_N} N^N \xrightarrow{p^N} N^N$. [Note that $-(q, r) = r - q$ if $q \leqslant r$, $= o$ if $q > r$.] Let θ denote the composite

$$\Omega \times N \times A \xrightarrow{(\pi_1, \gamma\pi_3, \pi_2)} \Omega \times N \times N \xrightarrow{1 \times (\phi, -)} \Omega \times \Omega \times N \xrightarrow{\wedge \times s} \Omega \times N\ ,$$

where $\phi : N \times N \longrightarrow \Omega$ is the classifying map of the order-relation $N \times N \xrightarrow{(\pi_1, +)} N \times N$. Now define an element of the object $(\Omega \times N)^{((A \amalg X)^{[n]})}$ of $\mathscr{E}/N$, using the method of 6.18 and the data

$$1 \xrightarrow{(t, o)} \Omega \times N \cong o^*((\Omega \times N)^{((A \amalg X)^{[n]})})$$

and

$$(\Omega \times N)^{(A \amalg X)^{[n]}} \times (A \amalg X)^{[n]} \times (A \amalg X) \xrightarrow{\text{ev} \times 1} \Omega \times N \times (A \amalg X)$$

$$\cong (\Omega \times N \times A) \amalg (\Omega \times N \times X) \xrightarrow{\left(\substack{\theta \\ \pi_1,\, s\pi_2}\right)} \Omega \times N \ .$$

The transpose of this element gives us a morphism $(A \amalg X)^{[n]} \longrightarrow \Omega \times N$ in $\mathscr{E}/N$, and hence a morphism $M(A \amalg X) \xrightarrow{\text{prs}} \Omega \times N$ in $\mathscr{E}$.

Define FX by the pullback

$$\begin{array}{ccc} FX & \longrightarrow & 1 \\ \downarrow & & \downarrow{\scriptstyle (t,\, so)} \\ M(A \amalg X) & \xrightarrow{\text{prs}} & \Omega \times N \end{array} \quad ;$$

then it is straightforward to check that the $\mathbb{T}$-model structure on $M(A \amalg X)$ restricts to one on FX, that the inclusion $X \xrightarrow{\nu_2} A \amalg X \xrightarrow{\eta} M(A \amalg X)$ factors through FX, and that any morphism $X \longrightarrow Y$, where Y is a $\mathbb{T}$-model, extends uniquely to a homomorphism $FX \longrightarrow Y$.

We now prove the general case by induction on the number of equations of $\mathbb{T}$. Suppose $\mathbb{T}$ is obtained by adding a single m-ary equation $(f = g)$ to a theory $\mathbb{S}$ for which the theorem has been proved; let F denote the free $\mathbb{S}$-model functor, and let Y be any $\mathbb{S}$-model. Then we can regard f and g as a parallel pair of morphisms $Y^m \rightrightarrows Y$ in $\mathscr{E}$, which are equal iff Y is a $\mathbb{T}$-model. Now let $Y \xrightarrow{q} \hat{Y}$ be the coequalizer of the corresponding pair $F(Y^m) \rightrightarrows Y$ in $\mathbb{S}(\mathscr{E})$; then q is epi in $\mathscr{E}$ by 6.42, since we can use the existence of coproducts in $\mathbb{S}(\mathscr{E})$ to express it as the coequalizer of a reflexive pair.

Now consider the diagram

$$\begin{array}{ccc} Y^m & \xrightarrow{q^m} & \hat{Y}^m \\ {\scriptstyle f}\downarrow\downarrow{\scriptstyle g} & & {\scriptstyle \hat{f}}\downarrow\downarrow{\scriptstyle \hat{g}} \\ Y & \xrightarrow{q} & \hat{Y} \end{array} \ .$$

Since q is an $\mathbb{S}$-model homomorphism, we clearly have $\hat{f} q^m = qf = qg = \hat{g} q^m$; but q^m is epi since q is, and hence $\hat{f} = \hat{g}$, i.e. $\hat{Y}$ is a $\mathbb{T}$-model. Moreover, it is immediate from the construction that any $\mathbb{S}$-model homomorphism from Y to a $\mathbb{T}$-model factors (uniquely) through q; so the assignment $Y \longmapsto \hat{Y}$

defines a left adjoint for the inclusion $\mathbb{T}(\mathscr{E}) \longrightarrow \mathbb{S}(\mathscr{E})$. In particular, if $Y = FX$ for some object X, then $\hat{Y}$ is the free $\mathbb{T}$-model generated by X. ■

Let $\mathbb{T}$ be a finitary algebraic theory, and $\mathscr{F} \xrightarrow{f} \mathscr{E}$ a geometric morphism. Since f_* and f^* both preserve finite products, they can be lifted to an adjoint pair of functors between $\mathbb{T}(\mathscr{F})$ and $\mathbb{T}(\mathscr{E})$. Moreover, we have

6.44 Lemma. *Let* $\mathbb{T}$ *and* $\mathscr{E}$ *be as in* 6.43, *and let* $\mathscr{F} \xrightarrow{f} \mathscr{E}$ *be a geometric morphism. Then the diagram*

$$\begin{array}{ccc} \mathscr{E} & \xrightarrow{f^*} & \mathscr{F} \\ \downarrow{\scriptstyle F} & & \downarrow{\scriptstyle F} \\ \mathbb{T}(\mathscr{E}) & \xrightarrow{f^*} & \mathbb{T}(\mathscr{F}) \end{array}$$

commutes up to canonical isomorphism, where the F's *denote free* $\mathbb{T}$-*model functors.*

Proof. By definition, the corresponding diagram of right adjoints (i.e.

$$\begin{array}{ccc} \mathbb{T}(\mathscr{F}) & \xrightarrow{f_*} & \mathbb{T}(\mathscr{E}) \;) \\ \downarrow{\scriptstyle U} & & \downarrow{\scriptstyle U} \\ \mathscr{F} & \xrightarrow{f_*} & \mathscr{E} \end{array}$$

commutes, so the result follows from uniqueness of adjoints. ■

It follows from 6.44 that if $\mathbb{T}$ and $\mathscr{E}$ are as in 6.43, then the composite functor $T_{\mathscr{E}}: \mathscr{E} \xrightarrow{F} \mathbb{T}(\mathscr{E}) \xrightarrow{U} \mathscr{E}$ extends to a "natural endomorphism of $\mathscr{E}$-toposes" as described in 6.34(iv). Moreover, this functor has a (natural) monad structure arising from the adjunction $(F \dashv U)$; in terms of 6.34, this is equivalent to giving T a monoid structure in the monoidal category $(\mathscr{E}[U], \otimes, \mathrm{U})$, to giving $\bar{T}^*$ a monad structure, or (via 0.14) to giving $\bar{T}_*$ a comonad structure. We are thus led to make the following definition:

6.45 Definition (G. C. Wraith). Let $\mathscr{E}$ be a topos with a natural number object. By an (*internal*) *finitary algebraic theory* in $\mathscr{E}$, we mean a monoid in the monoidal category $(\mathscr{E}[U], \otimes, U)$. We write **alg**$(\mathscr{E})$ for the category of such monoids. If $\mathbb{T} = (T, \mu, \eta)$ is a finitary algebraic theory in $\mathscr{E}$, and $\mathscr{F} \xrightarrow{f} \mathscr{E}$

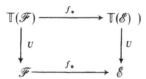

is an $\mathscr{E}$-topos, we define a $\mathbb{T}$-*model* in $\mathscr{F}$ to be an algebra for the monad $\mathbb{T}_{\mathscr{F}}$ whose functor part is $T_{\mathscr{F}}$. We write $\mathbb{T}(\mathscr{F})$for the category of $\mathbb{T}$-models in $\mathscr{F}$. ■

The advantage of this definition over the previous "external" one is that it enables us to handle theories which have "built into them" the internal structure of the topos $\mathscr{E}$. For example, the theory of rings is external, because its operations are constant throughout the domain of variation of $\mathscr{E}$; but if R is a particular (non-constant) internal ring in $\mathscr{E}$, then the theory of R-modules cannot similarly be described by external operations. However, it is an easy matter to modify 6.43 to produce a free R-module functor; and so we obtain a description of R-modules as an algebraic theory in the sense of 6.45.

6.46 LEMMA. *Let $\mathscr{E}$ be a topos with a natural number object, $\mathscr{F} \xrightarrow{f} \mathscr{E}$ an $\mathscr{E}$-topos. Then*

(i) *If T is any object of $\mathscr{E}[U]$, the functor $T_{\mathscr{F}}: \mathscr{F} \longrightarrow \mathscr{F}$ preserves reflexive coequalizers.*

(ii) *If $\mathbb{T}$ is a finitary algebraic theory in $\mathscr{E}$, $\mathbb{T}(\mathscr{F})$ has finite colimits.*

(iii) *The forgetful functor* **alg**$(\mathscr{E}) \longrightarrow \mathscr{E}[U]$ *creates reflexive coequalizers.*

Proof. (i) Suppose first that T is representable as an internal diagram on $\mathbf{E}_{\mathrm{fin}}$, i.e. $T = R(\gamma)$ for some $A \xrightarrow{\gamma} N$. Then it is easily verified from the proof of 6.32 that $T_{\mathscr{F}}$ may be described as the composite

$$\mathscr{F} \xrightarrow{N^*} \mathscr{F}/N \xrightarrow{(-)^{[n]}} \mathscr{F}/N \xrightarrow{(-)\times f^*\gamma} \mathscr{F}/N \xrightarrow{\Sigma_N} \mathscr{F} .$$

Now all the functors in the above composite, except the second, have right adjoints; and the second preserves reflexive coequalizers by 6.25. So $T_{\mathscr{F}}$ preserves reflexive coequalizers.

In general, we can form a free presentation of T as an algebra for the monad of 2.21, i.e. a coequalizer diagram $R \rightrightarrows S \longrightarrow T$ where R and S are representable. Since, for any X in $\mathscr{F}$, the functor $(T \longmapsto T_{\mathscr{F}}(X)) = \bar{X}^*$ preserves coequalizers, it is an easy diagram-chase to show that $T_{\mathscr{F}}$ preserves reflexive coequalizers, since $R_{\mathscr{F}}$ and $S_{\mathscr{F}}$ do.

(ii) Since $T_{\mathscr{F}}$ preserves reflexive coequalizers, the forgetful functor $\mathbb{T}(\mathscr{F}) \longrightarrow \mathscr{F}$ creates them; so this is a straightforward application of 0.16.

(iii) Let $R \overset{f}{\underset{g}{\rightrightarrows}} S \overset{h}{\longrightarrow} T$ be a reflexive diagram in $\mathscr{E}[U]$. Then in the diagram

$$\begin{array}{ccccc}
R\otimes R & \rightrightarrows & R\otimes S & \longrightarrow & R\otimes T \\
\downdownarrows & & \downdownarrows & & \downdownarrows \\
S\otimes R & \rightrightarrows & S\otimes S & \longrightarrow & S\otimes T \\
\downarrow & & \downarrow & & \downarrow \\
T\otimes R & \rightrightarrows & T\otimes S & \longrightarrow & T\otimes T
\end{array}$$

the rows are reflexive coequalizers by part (i), and the columns are reflexive coequalizers since $(-)\otimes R = \overline{R^*}$ preserves colimits. So by 0.17 the diagonal $R\otimes R \rightrightarrows S\otimes S \longrightarrow T\otimes T$ is a coequalizer. It now follows exactly as in 6.42 that if R and S are $\otimes$-monoids and f and g are $\otimes$-monoid homomorphisms, there is a unique $\otimes$-monoid structure on T which makes it a coequalizer in **alg** $(\mathscr{E})$. ■

6.47 THEOREM (Wraith). *Let $\mathscr{E}$ be a topos with a natural number object. Then*

(i) *The forgetful functor* **alg**$(\mathscr{E}) \longrightarrow \mathscr{E}[U]$ *has a left adjoint.*
(ii) **alg**$(\mathscr{E})$ *is monadic over* $\mathscr{E}[U]$.
(iii) **alg**$(\mathscr{E})$ *is monadic over* $\mathscr{E}/N$.
(iv) **alg**$(\mathscr{E})$ *has finite colimits.*

Proof. (i) The construction of the free $\otimes$-monoid functor is exactly analogous to 6.41, except that the functor $(-)^{\otimes n}$ of 6.37 replaces $(-)^{[n]}$. We leave the details to the reader.

(ii) and (iv) follow from (i) and 6.46(iii) in the same way as the corresponding parts of 6.43; and (iii) follows from (ii) and 2.21, since the object-of-objects of $\mathbf{E}_{\mathrm{fin}}$ is N and the hypotheses of the Crude Tripleability Theorem are stable under composition. ■

From 6.47(iii) we recover the idea of a *presentation* of a theory in terms of objects of operations and equations, each indexed over N. In particular, if $\mathbb{T}$ is the free theory generated by an object $(A \overset{\gamma}{\longrightarrow} N)$ of $\mathscr{E}/N$, and X is an object of $\mathscr{E}$, we find that a $\mathbb{T}$-model structure on X is equivalent to a morphism

$$\Sigma_N(X^{[n]} \times \gamma) \longrightarrow X$$

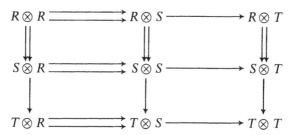

in $\mathscr{E}$, i.e. a "γ-indexed family of finitary operations on X". Yet another way of internalizing the notion of finitary algebraic theory would be to take Lawvere's definition [176] and translate it directly into the language of **cat**($\mathscr{E}$), replacing the category of finite sets by the internal category $\mathbf{E}_{\text{fin}}$. It can be shown (see [58]) that this process yields a category of algebraic theories which is equivalent to **alg**($\mathscr{E}$), as defined in 6.45.

In conclusion, we should make some remarks on the possibility of extending the ideas of this paragraph to infinitary theories. When we attempt to do this, three difficulties immediately arise: first (and perhaps most important), that models of such theories are not normally preserved by inverse image functors; second, that our arguments involving reflexive coequalizers are no longer applicable; and third, that the methods of recursion over N which we have developed are not normally sufficient for the construction of free functors. However, some progress has been made in this area; in particular R. Paré [97] has shown that if $\mathbb{T}$ is the free theory generated by a single "I-ary operation" (i.e. a morphism $X^I \longrightarrow X$, for some fixed object I of $\mathscr{E}$), then we can construct a free $\mathbb{T}$-model functor in any topos $\mathscr{E}$ with a natural number object.

6.5. GEOMETRIC THEORIES

In the last paragraph, we saw how to do universal algebra in a manner which is truly "internal" to an arbitrary topos with a natural number object. But in mathematics we frequently have to consider objects with a structure which is not purely algebraic; i.e. the structure is defined by formulae which are not simply equational. A good example is the theory of local rings, which plays a crucial rôle in algebraic geometry:

6.51 Definition. Let $\mathscr{E}$ be a topos, A an internal (commutative) ring in $\mathscr{E}$. Write o, e for the zero and one of A, and $\Upsilon A \rightarrowtail A$ for the *group of units* of A, i.e. the extension of the formula $\exists \mathbf{a}'.(\mathbf{a}\,.\,\mathbf{a}' = e)$. We say A is a *local ring* if

$$\text{(i)}\quad \mathscr{E} \models \neg(o = e)$$

and

$$\text{(ii)}\quad \mathscr{E} \models (\mathbf{a} \in \ulcorner \Upsilon A \urcorner \vee (e - \mathbf{a}) \in \ulcorner \Upsilon A \urcorner)\ .$$

A ring homomorphism $A \xrightarrow{f} B$ between local rings is said to be *local* if the square

$$\begin{array}{ccc} \Upsilon A & \xrightarrow{\Upsilon f} & \Upsilon B \\ \downarrow & & \downarrow \\ A & \xrightarrow{f} & B \end{array}$$

is a pullback. ■

Since the property of being a local ring is expressed by formulae of $L_{\mathscr{E}}$, it is clearly preserved by logical functors. But in fact the formulae of 6.51 belong to the more restricted class of "geometric formulae" whose truth is preserved by inverse image functors. (To see this, note that (i) says $0 \longrightarrow 1 \overset{o}{\underset{e}{\rightrightarrows}} A$ is an equalizer, and (ii) says that a certain morphism $\Upsilon A \amalg \Upsilon A \longrightarrow A$ is epi; but inverse image functors preserve 0, coproducts and epimorphisms as well as finite limits.) It follows easily, for example, that in the topos of sheaves on a topological space, a ring is local iff each of its stalks is a local ring in $\mathscr{S}$.

To formalize the notion of a "geometric formula" introduced above, we first describe the notion of a finitary geometric language (also called a positive first-order language), in which we shall be able to express the notion of "finitary geometric theory". The language which we describe differs from that introduced in §5.4 in that it is constructed in an abstract setting, rather than over a particular topos $\mathscr{E}$; but our experience with the Mitchell–Bénabou language will nevertheless enable us to introduce geometric languages more quickly and concisely than would otherwise have been the case.

6.52 Definition. A *finitary geometric language* $\mathbb{L}$ consists of

(i) A set of *types* $X, Y, Z, \ldots$

(ii) A set of *primitive function-symbols* $\alpha, \beta, \ldots$ Each function-symbol has a *signature* which is a pair $(\overline{X}; Y)$, where $\overline{X} = (X_1, X_2, \ldots, X_n)$ is a finite (possibly empty) string of types and Y is a type.

(iii) A set of *primitive relation-symbols* $r, s, \ldots$, each equipped with a signature $\overline{X}$ which is a finite string of types.

(iv) *Terms* of each type are defined as follows:

(a) for each type X, we have a stock of variables $\mathbf{x}, \mathbf{x}', \mathbf{x}'', \ldots$.

(b) if α is a function-symbol of signature $(\overline{X}; Y)$ and $\bar{\tau} = (\tau_1, \tau_2, \ldots, \tau_n)$

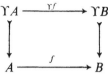

is a string of terms of types $X_1, X_2, \ldots, X_n$ (which we shall loosely refer to as a "term of type $\overline{X}$"), then $\alpha(\bar{\tau})$ is a term of type Y.

(v) *Formulae* of $\mathbb{L}$ are defined as follows:

(a) we have two *atomic sentences true, false.*

(b) if σ and τ are two terms of the same type, then $\sigma = \tau$ is a formula.

(c) if r is a relation of signature $\overline{X}$ and $\bar{\tau}$ is a term of type $\overline{X}$, then $r(\bar{\tau})$ is a formula.

(d) if ϕ and ψ are formulae, then so are $\phi \wedge \psi$ and $\phi \vee \psi$.

(e) if ϕ is a formula and $\mathbf{x}$ a variable, then $\exists \mathbf{x} \phi$ is a formula. ■

Note that the language $\mathbb{L}$ does *not* contain the symbols $\neg$, $\Rightarrow$ or $\forall$; this is because the interpretation of these symbols in $L_{\mathscr{E}}$ is not in general preserved by inverse image functors.

6.53 Definition. Let $\mathbb{L}$ be a finitary geometric language, $\mathscr{E}$ a category with finite products. By an *interpretation* of $\mathbb{L}$ in $\mathscr{E}$, we mean a function assigning to each type X of $\mathbb{L}$ an object M_X of $\mathscr{E}$, to each function-symbol α of signature $(\overline{X}; Y)$ a morphism

$$M_\alpha : M_{\overline{X}} = \prod_{i=1}^{n} M_{X_i} \rightarrowtail M_Y,$$

and to each relation-symbol r of signature $\overline{X}$ a subobject $M_r \rightarrowtail M_{\overline{X}}$. A *morphism of interpretations* $f: M \longrightarrow N$ consists of a morphism $f_X : M_X \longrightarrow N_X$ for each X, such that

$$\begin{array}{ccc} M_{\overline{X}} & \xrightarrow{f_{\overline{X}}} & N_{\overline{X}} \\ \downarrow{\scriptstyle M_\alpha} & & \downarrow{\scriptstyle N_\alpha} \\ M_Y & \xrightarrow{f_Y} & N_Y \end{array}$$

commutes for each α, and $M_r \rightarrowtail M_{\overline{X}} \xrightarrow{f_{\overline{X}}} N_{\overline{X}}$ factors through $N_r \rightarrowtail N_{\overline{X}}$ for each r. In this way we obtain a category $\mathbb{L}(\mathscr{E})$ of interpretations of $\mathbb{L}$ in $\mathscr{E}$; and if $T: \mathscr{E} \longrightarrow \mathscr{F}$ is a left exact functor, it clearly induces a functor $\mathbb{L}(\mathscr{E}) \longrightarrow \mathbb{L}(\mathscr{F})$. ■

Now suppose we are given an interpretation M of $\mathbb{L}$ in a topos $\mathscr{E}$ (or more generally, in any regular category with universal finite unions of subobjects; cf. [RM]). Then we may proceed as we did in 5.42 to interpret each term of type X in $\mathbb{L}$ as a morphism with codomain M_X, and each formula ϕ with free variables $(\mathbf{x}_1, \mathbf{x}_2, \ldots, \mathbf{x}_n)$ as a subobject

$$M_\phi \rightarrowtail \prod_{i=1}^{n} M_{X_i}.$$

(Specifically, we interpret *true*, *false* as the maximal and minimal subobjects of 1; the equality-predicate is interpreted as the diagonal subobject of $M_X \times M_X$; substitution of terms for variables is interpreted by pullback; conjunction and disjunction by intersection and union of subobjects (after first pulling back along the appropriate product projections); and existential quantification by forming images.) Moreover, if $T: \mathscr{E} \longrightarrow \mathscr{F}$ is an exact functor (in particular an inverse image functor), then it preserves all these interpretations.

6.54 Definition (Joyal–Reyes). Let $\mathbb{L}$ be a finitary geometric language. By a *sequent* of $\mathbb{L}$ we mean an expression of the form $(\phi \vdash \psi)$, where ϕ and ψ are formulae of $\mathbb{L}$. [The symbol $\vdash$ may be read as "entails"; it is not a symbol of $\mathbb{L}$.]

A *finitary geometric theory* is a pair $\mathbb{T} = (\mathbb{L}, A)$, where $\mathbb{L}$ is a finitary geometric language and A is a set of sequents of $\mathbb{L}$, called *axioms* of $\mathbb{T}$. We say $\mathbb{T}$ is *finitely-presented* if the language $\mathbb{L}$ has finitely many types, function-symbols and relation-symbols, and the set of axioms is finite.

If M is an interpretation of $\mathbb{L}$ in a topos $\mathscr{E}$, we say the sequent $(\phi \vdash \psi)$ is *satisfied* in M if (in the notation developed above) we have $M_\phi \leqslant M_\psi$ as subobjects of $M_{\bar{X}}$. (Note that if ϕ and ψ do not have the same free variables, we must first pull back M_ϕ and M_ψ along the appropriate product projections before comparing them.) We say M is a *model* of $\mathbb{T} = (\mathbb{L}, A)$ if each axiom of $\mathbb{T}$ is satisfied in M; and we write $\mathbb{T}(\mathscr{E})$ for the full subcategory of $\mathbb{L}(\mathscr{E})$ whose objects are $\mathbb{T}$-models. ■

Clearly the property of being a $\mathbb{T}$-model is preserved by any exact functor $T: \mathscr{E} \longrightarrow \mathscr{F}$; and if in addition T is faithful, then it also reflects this property.

6.55 Examples. (i) Let $\mathbb{T}$ be an (external) finitary algebraic theory. We may present $\mathbb{T}$ as a geometric theory as follows: the language $\mathbb{L}$ has one type X, one function-symbol α of signature $(X^m; X)$ for each m-ary operation α of $\mathbb{T}$, and no relations. For each equation $f(x_1, \ldots, x_n) = g(x_1, \ldots, x_n)$ of

$\mathbb{T}$, we take the sequent

$$(true \vdash f(\mathbf{x}_1, \dots, \mathbf{x}_n) = g(\mathbf{x}_1, \dots, \mathbf{x}_n))$$

as an axiom.

(ii) Similarly, we may describe many-sorted algebraic theories [58] or essentially algebraic theories [FK] as geometric theories. However, it is often convenient to introduce an operation whose domain is not simply a product of types by means of a relation rather than a function-symbol, together with axioms which say that the relation is the graph of a function. For example, the theory **cat** of categories (2.11) may be described (as in [72]) as having two types 0 and 1, three function-symbols d_0, d_1 and i of signatures (1; 0), (1; 0) and (0; 1) respectively, and a relation-symbol Γ of signature (1, 1, 1), satisfying the axioms

$$\begin{aligned}
&true \vdash d_0(i(\mathbf{u})) = \mathbf{u} \wedge d_1(i(\mathbf{u})) = \mathbf{u} \\
&d_1(\mathbf{f}_1) = d_0(\mathbf{f}_2) \vdash \exists \mathbf{f}_3 . \Gamma(\mathbf{f}_1, \mathbf{f}_2, \mathbf{f}_3) \\
&\Gamma(\mathbf{f}_1, \mathbf{f}_2, \mathbf{f}_3) \vdash d_0(\mathbf{f}_1) = d_0(\mathbf{f}_3) \wedge d_1(\mathbf{f}_1) = d_0(\mathbf{f}_2) \wedge d_1(\mathbf{f}_2) = d_1(\mathbf{f}_2) \\
&\Gamma(\mathbf{f}_1, \mathbf{f}_2, \mathbf{f}_3) \wedge \Gamma(\mathbf{f}_1, \mathbf{f}_2, \mathbf{f}_4) \vdash \mathbf{f}_3 = \mathbf{f}_4 \\
&true \vdash \Gamma(\mathbf{f}, i(d_1(\mathbf{f})), \mathbf{f}) \wedge \Gamma(i(d_0(\mathbf{f})), \mathbf{f}, \mathbf{f}) \\
&\Gamma(\mathbf{f}_1, \mathbf{f}_2, \mathbf{f}_4) \wedge \Gamma(\mathbf{f}_2, \mathbf{f}_3, \mathbf{f}_5) \wedge \Gamma(\mathbf{f}_4, \mathbf{f}_3, \mathbf{f}_6) \wedge \Gamma(\mathbf{f}_1, \mathbf{f}_5, \mathbf{f}_7) \vdash \mathbf{f}_6 = \mathbf{f}_7.
\end{aligned}$$

(Here **u**, **f** denote variables of types 0, 1 respectively.)

(iii) The theory **lann** of local rings, as defined in 6.51, clearly has a geometric presentation; we simply take the presentation of the theory **ann** of rings given by (i) above, and adjoin the axioms

$$o = e \vdash false$$

and

$$true \vdash \exists \mathbf{a}'(\mathbf{a} . \mathbf{a}' = e) \vee \exists \mathbf{a}''((e - \mathbf{a})\mathbf{a}'' = e).$$

Note, however, that morphisms in the category **lann**($\mathscr{E}$) are arbitrary ring homomorphisms between local rings, and not local homomorphisms as defined in 6.51.

(iv) The theory **filt** of filtered categories may be presented geometrically;

the three conditions of 2.51 may be expressed by the sequents

$$\begin{aligned}&\mathit{true} \vdash \exists \mathbf{u}(\mathbf{u} = \mathbf{u})\\&\mathit{true} \vdash \exists \mathbf{f}_1 \exists \mathbf{f}_2(d_0(\mathbf{f}_1) = \mathbf{u}_1 \wedge d_0(\mathbf{f}_2) = \mathbf{u}_2 \wedge d_1(\mathbf{f}_1) = d_1(\mathbf{f}_2)), \text{ and}\\&d_0(\mathbf{f}_1) = d_0(\mathbf{f}_2) \wedge d_1(\mathbf{f}_1) = d_1(\mathbf{f}_2) \vdash \exists \mathbf{f}_3 \exists \mathbf{f}_4(\Gamma(\mathbf{f}_1, \mathbf{f}_3, \mathbf{f}_4) \wedge \Gamma(\mathbf{f}_2, \mathbf{f}_3, \mathbf{f}_4))\end{aligned}$$

which we add to the axioms of the theory **cat**.

(v) Let $\mathbb{T}$ be any geometric theory. We may construct a new theory $\mathbb{T}^2$ whose models are morphisms of $\mathbb{T}$-models, as follows: for each type X of $\mathbb{T}$, we take two new types X^0, X^1 and a function-symbol γ_X of signature $(X^0; X^1)$. For each function-symbol α (resp. relation r), we take two new symbols α^0, α^1 (resp r^0, r^1) and a new axiom

$$\mathit{true} \vdash \gamma_Y(\alpha^0(\mathbf{x}_1, \mathbf{x}_2, \ldots, \mathbf{x}_n)) = \alpha^1(\gamma_{X_1}(\mathbf{x}_1), \gamma_{X_2}(\mathbf{x}_2), \ldots, \gamma_{X_n}(\mathbf{x}_n))$$

(resp.

$$r^0(\mathbf{x}_1, \mathbf{x}_2, \ldots, \mathbf{x}_n) \vdash r^1(\gamma_{X_1}(\mathbf{x}_1), \gamma_{X_2}(\mathbf{x}_2), \ldots, \gamma_{X_n}(\mathbf{x}_n))).$$

Each axiom $(\phi \dashv \psi)$ of $\mathbb{T}$ gives rise to two axioms $(\phi^0 \dashv \psi^0)$, $(\phi^1 \dashv \psi^1)$ in an obvious way. Then it is easily seen that for any topos $\mathscr{E}$ we have $\mathbb{T}^2(\mathscr{E}) \cong (\mathbb{T}(\mathscr{E}))^2$. ■

6.56 Theorem (A. Joyal, J. Bénabou [8], M. Tierney [119]). *Let $\mathbb{T}$ be a finitely-presented, finitary geometric theory, and $\mathscr{E}$ a topos with a natural number object. Then there exists an $\mathscr{E}$-topos $\mathscr{E}[\mathbb{T}]$ which is a* classifying topos *for $\mathbb{T}$, in the sense that for any $\mathscr{E}$-topos $\mathscr{F}$ we have $\mathfrak{Top}/\mathscr{E}\,(\mathscr{F}, \mathscr{E}[\mathbb{T}]) \simeq \mathbb{T}(\mathscr{F})$, the equivalence being natural in $\mathscr{F}$.*

Proof. Let $\mathbb{T} = (\mathbb{L}, A)$. First we construct a finite category $\mathbf{D}$ as follows: $\mathbf{D}$ has one object v_X (resp. v_α, v_r) for each type X (resp. function-symbol α, relation-symbol r) of $\mathbb{L}$, and (in addition to identity morphisms) one morphism $v_\alpha \longrightarrow v_X$ (resp. $v_r \longrightarrow v_X$) for each occurrence of X as a type in the signature of α (resp. r). Now each interpretation M of $\mathbb{L}$ in a topos $\mathscr{F}$ gives rise to a diagram of type $\mathbf{D}$ in $\mathscr{F}$, which sends v_X to M_X, v_r to M_r and v_α to the graph of M_α. Conversely, a diagram F of type $\mathbf{D}$ is isomorphic to one constructed as above iff

$$F(v_r) \longrightarrow \prod_{i=1}^{n} F(v_{X_i})$$

is mono for each r of signature $(X_1, \ldots, X_n)$, and

$$F(v_\alpha) \longrightarrow \prod_{i=1}^{n} F(v_{X_i})$$

is iso for each α of signature $(X_1, \ldots, X_n; Y)$. So we may identify $\mathbb{L}(\mathscr{F})$ (up to equivalence) with a full subcategory of $\mathscr{F}^{\mathbf{D}}$.

Now consider the classifying topos $\mathscr{E}[\mathbb{D}]$ for diagrams of type $\mathbf{D}$, constructed as in 6.35. Let F denote the generic diagram of type $\mathbf{D}$; then by 3.59(i) and (iv) there is a unique smallest topology j in $\mathscr{E}[\mathbb{D}]$ such that the associated j-sheaf $L(F)$ of F has each $LF(v_r) \longrightarrow \prod LF(v_{X_i})$ mono and each $LF(v_\alpha) \longrightarrow \prod LF(v_{X_i})$ iso. Moreover, by 4.19 a geometric morphism $\mathscr{F} \xrightarrow{f} \mathscr{E}[\mathbb{D}]$ factors through $\text{sh}_j(\mathscr{E}[\mathbb{D}]) \longrightarrow \mathscr{E}[\mathbb{D}]$ iff the diagram of type $\mathbf{D}$ classified by f is actually an interpretation of $\mathbb{L}$; so $\text{sh}_j(\mathscr{E}[\mathbb{D}]) = \mathscr{E}[\mathbb{L}]$ is a classifying topos for interpretations of $\mathbb{L}$. (In the suggestive terminology of Tierney, we say that the topology j *forces* F to be an interpretation of $\mathbb{L}$.)

Let $M = L(F)$ be the generic interpretation of $\mathbb{L}$ in $\mathscr{E}[\mathbb{L}]$. In general M will not satisfy the axioms of $\mathbb{T}$; but once again we can force it to do so by imposing a suitable topology on $\mathscr{E}[\mathbb{L}]$. Specifically, for each sequent $(\phi \vdash \psi)$ in A we construct the diagram

$$\begin{array}{ccc} M_\phi \cap M_\psi & \rightarrowtail & M_\psi \\ \downarrow & & \downarrow \\ M_\phi & \rightarrowtail & \prod M_{X_i} \end{array},$$

and then define k to be the smallest topology in $E[\mathbb{L}]$ for which each $M_\phi \cap M_\psi \rightarrowtail M_\phi$ is dense. Since the constructions of M_ϕ, M_ψ and their intersection are preserved by inverse image functors, it is again easy to see that $\text{sh}_k(\mathscr{E}[\mathbb{L}]) = \mathscr{E}[\mathbb{T}]$ is a classifying topos for $\mathbb{T}$-models, the generic $\mathbb{T}$-model being the associated k-sheaf of M. ∎

The reason why we required finite presentation of $\mathbb{T}$ in the proof of 6.56 was, of course, that we were working with *external* geometric theories. Clearly, the next step in our development ought to be to introduce a notion of "internal geometric theory", in the same way that we internalized algebraic theories in the last paragraph. We hope that by now the reader is willing to accept at least the theoretical possibility of such a step; it is at least clear that one may define an "$\mathscr{E}$-presented finitary geometric language", provided

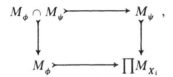

$\mathscr{E}$ has a natural number object, by specifying three objects T, F, R of types, function-symbols and relation-symbols, and morphisms

$$F \longrightarrow \Sigma_N(T^{[sn]}), \quad R \longrightarrow \Sigma_N(T^{[n]})$$

defining signatures. Then the machinery of 6.43 may be adapted to construct the object of formulae and hence the object of sequents of such a language; thus we can define an "$\mathscr{E}$-presented geometric theory" and a model of such a theory. Since the lattice of topologies in a topos is internally complete (which follows easily from 3.58 and internal completeness of Ω^Ω), the argument of 6.56 will serve to construct a classifying topos for such a theory as soon as we know how to classify diagrams on an *internal* category $\mathbf{D}$; and this question is answered in 6.60(i) below.

However, we shall not fill in all the details of the argument sketched above, essentially because the benefits gained do not seem commensurate with the work involved. Instead, we now turn our attention to a particular class of classifying toposes called spectra, which enable us to construct "global adjoints" in situations where ordinary adjoints may not exist.

Let $\mathbb{T}$ be a geometric theory. We first construct a "global" 2-category $\mathbb{T}$-$\mathfrak{Top}$ of $\mathbb{T}$-*modelled toposes*, as follows: its objects are pairs $(\mathscr{E}, M)$ where $\mathscr{E}$ is a topos and M is a $\mathbb{T}$-model in $\mathscr{E}$. A 1-arrow $(\mathscr{F}, L) \longrightarrow (\mathscr{E}, M)$ is a pair (p, f), where $\mathscr{F} \xrightarrow{p} \mathscr{E}$ is a geometric morphism and $p^*M \xrightarrow{f} L$ is a $\mathbb{T}$-model homomorphism; and a 2-arrow $(p, f) \longrightarrow (q, g)$ is a natural transformation $p \xrightarrow{\eta} q$ such that

$$\begin{array}{ccc} p^*M & \xrightarrow{\quad f \quad} & L \\ {\scriptstyle \eta_M}\big\downarrow & \nearrow{\scriptstyle g} & \\ q^*M & & \end{array}$$

commutes. We shall normally consider only $\mathbb{T}$-modelled toposes $(\mathscr{E}, M)$ such that $\mathscr{E}$ has a natural number object; we write $\mathbb{T}$-$\mathfrak{Top}_N$ for the corresponding full sub-2-category of $\mathbb{T}$-$\mathfrak{Top}$.

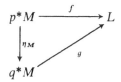

6.57 Definition. Let $\mathbb{S}$ and $\mathbb{T}$ be two geometric theories such that $\mathbb{T}$ is a *quotient theory* of $\mathbb{S}$; i.e. the theories have the same language, and the axioms of $\mathbb{S}$ are a subset of those of $\mathbb{T}$. (This clearly implies that $\mathbb{T}(\mathscr{E})$ is a full subcategory of $\mathbb{S}(\mathscr{E})$ for any $\mathscr{E}$.) Suppose further that we are given a distinguished class $\mathbb{A}$ of morphisms of $\mathbb{T}$-models; we shall say that $\mathbb{A}$ is an *admissible class* of morphisms (relative to $\mathbb{S}$) if it satisfies the following conditions:

(i) The property of being in $\mathbb{A}$ is "geometric", i.e. preserved by inverse image functors.

(ii) $\mathbb{A}$ contains all identity morphisms; and if we have $L \xrightarrow{f} M \xrightarrow{g} P$ with $g \in \mathbb{A}$, then $f \in \mathbb{A}$ iff $gf \in \mathbb{A}$.

(iii) ("Factorization lemma") Suppose given an $\mathbb{S}$-model morphism $M \xrightarrow{f} L$ where L is a $\mathbb{T}$-model. Then there exists a factorization $M \xrightarrow{q} M_f \xrightarrow{\hat{f}} L$ with M_f a $\mathbb{T}$-model and $\hat{f} \in \mathbb{A}$, which is the best possible factorization of f through an $\mathbb{A}$-morphism in the sense that, given any other such factorization $M \xrightarrow{r} P \xrightarrow{g} L$, there exists a unique $M_f \xrightarrow{h} P$ (necessarily in $\mathbb{A}$, by (ii)) such that $gh = \hat{f}$ and $hq = r$. Moreover, this factorization is preserved by inverse image functors. ■

In fact condition (i) of 6.57 is implied by the last sentence of (iii); for if we apply the factorization lemma to a morphism $M \xrightarrow{f} L$ which is already in $\mathbb{A}$, we obtain the factorization $M \xrightarrow{1} M \xrightarrow{f} L$. (i) and (ii) together imply that we may form a sub-2-category $\mathbb{A}$-$\mathfrak{Top}$ of $\mathbb{T}$-$\mathfrak{Top}$ by considering only those 1-arrows (p, f) for which $f \in \mathbb{A}$. We shall say that an $\mathbb{S}$-model morphism $M \xrightarrow{f} L$, where L is a $\mathbb{T}$-model, is *extremal* (relative to $\mathbb{A}$) if $\hat{f}$ is an isomorphism; in this situation we shall also refer to L as a $\mathbb{T}$-*quotient* of M.

6.58 Theorem (J. C. Cole [19]). *Let $\mathbb{S}$ and $\mathbb{T}$ be finitely-presented geometric theories such that $\mathbb{T}$ is a quotient of $\mathbb{S}$, and let $\mathbb{A}$ be an admissible class of morphisms of $\mathbb{T}$-models. Then the inclusion functor $\mathbb{A}$-$\mathfrak{Top}_N \longrightarrow \mathbb{S}$-$\mathfrak{Top}_N$ has a right adjoint* $\mathrm{Spec}\colon \mathbb{S}\text{-}\mathfrak{Top}_N \longrightarrow \mathbb{A}\text{-}\mathfrak{Top}_N$.

Proof. Let $(\mathscr{E}, M)$ be an $\mathbb{S}$-modelled topos. Observe first that we have morphisms $\mathscr{E}[\mathbb{S}^2] \underset{\mathrm{cod}}{\overset{\mathrm{dom}}{\rightrightarrows}} \mathscr{E}[\mathbb{S}]$ classifying the domain and codomain of the generic morphism of $\mathbb{S}$-models. (In fact $\mathscr{E}[\mathbb{S}^2] \rightrightarrows \mathscr{E}[\mathbb{S}]$ has the structure of an internal category in $\mathfrak{BTop}/\mathscr{E}$.) Now if we form the pullback

$$\begin{array}{ccccc}
\mathscr{F} & & \longrightarrow & & \mathscr{E}[\mathbb{T}] \\
\Big\downarrow p & & & & \Big\downarrow \\
 & & \mathscr{E}[\mathbb{S}^2] & \xrightarrow{\mathrm{cod}} & \mathscr{E}[\mathbb{S}] \\
 & & \Big\downarrow \mathrm{dom} & & \\
\mathscr{E} & \xrightarrow{\bar{M}} & \mathscr{E}[\mathbb{S}] & &
\end{array}$$

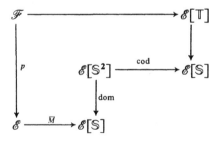

in $\mathfrak{BTop}/\mathscr{E}$, it is not hard to see that the $\mathscr{E}$-topos $\mathscr{F}$ contains a generic example of an $\mathbb{S}$-model morphism $p^*M \xrightarrow{f} L$ where L is a $\mathbb{T}$-model. Factorize this morphism as in 6.57(iii); let j be the topology in $\mathscr{F}$ which forces $\hat{f}$ to be iso, and let $\tilde{M}$ denote the associated j-sheaf of L. Then $\mathrm{sh}_j(\mathscr{F})$ is a classifying topos for the theory of $\mathbb{T}$-quotients of M, $\tilde{M}$ being the generic model of this theory.

Now let $(\mathscr{G}, P) \xrightarrow{(q,g)} (\mathscr{E}, M)$ be a 1-arrow of $\mathbb{S}$-$\mathfrak{Top}_N$, where P is a $\mathbb{T}$-model. By 6.57(iii), we may factor $q^*M \xrightarrow{g} p$ as $q^*M \xrightarrow{r} M_g \xrightarrow{\hat{g}} P$, where $\hat{g} \in \mathbb{A}$ and r is extremal. Then r is classified by a geometric morphism $\mathscr{G} \xrightarrow{\bar{r}} \mathrm{sh}_j(\mathscr{F})$ over $\mathscr{E}$ such that $\bar{r}^*(\tilde{M}) \cong M_g$; and it is easily verified that the assignment $(q, g) \longmapsto (\bar{r}, \hat{g})$ induces an equivalence of categories

$$\mathbb{S}\text{-}\mathfrak{Top}_N((\mathscr{G}, P), (\mathscr{E}, M)) \simeq \mathbb{A}\text{-}\mathfrak{Top}_N((\mathscr{G}, P), (\mathrm{sh}_j(\mathscr{F}), \tilde{M})).$$

Moreover, this equivalence is natural (up to coherent isomorphism) in all the variables involved, and so $\mathrm{Spec}(\mathscr{E}, M) = (\mathrm{sh}_j(\mathscr{F}), \tilde{M})$ is the required right adjoint. ■

Note that if, in the situation of 6.57, we had a factorization lemma in which the first factor $M \longrightarrow M_f$ depended only on M (and not on f), then M_f would be the free $\mathbb{T}$-model generated by M, i.e. $M \longmapsto M_f$ would be a left adjoint for the inclusion functor $\mathbb{A}(\mathscr{E}) \longrightarrow \mathbb{S}(\mathscr{E})$. Thus if we regard toposes as models of "generalized set theory", 6.58 says that, in situations where we may not be able to construct free $\mathbb{T}$-models while remaining within a fixed model of set theory, we can nevertheless do so if we are willing to allow our set theory as well as our models to vary. The reason why a free object appears as a *right* adjoint in 6.58 is that we defined the 1-arrows of $\mathbb{T}$-$\mathfrak{Top}$ to run in the "geometric" direction which is the opposite of the "algebraic" one.

6.59 Examples. (i) The "classical" example of a spectrum, which provided the principal motive for the proof of 6.58, is obtained by taking $\mathbb{S}$ = **ann**, $\mathbb{T}$ = **lann** and $\mathbb{A}$ to be the class of local morphisms (6.51). In this case the factorization lemma is proved as follows: given a ring homomorphism $A \xrightarrow{f} L$ where L is local, form the pullback

$$\begin{array}{ccc} S & \longrightarrow & \Upsilon L \\ \downarrow & & \downarrow \\ A & \xrightarrow{f} & L \end{array} ,$$

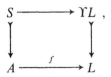

and define A_f to be the ring of fractions $A[S^{-1}]$. (Detailed accounts of the construction of rings of fractions in a topos may be found in [69] and [118]; the latter reference also contains a detailed proof of the factorization lemma.) The $\mathbb{T}$-quotients of A are thus the *localizations* of A, i.e. the rings of fractions $A[S^{-1}]$ where S is a subobject of A satisfying axioms which, in a Boolean topos, reduce to the property of being the complement of a prime ideal. If A is a commutative ring in $\mathscr{S}$, then it can be shown that Spec($\mathscr{S}$, A) is equivalent to (Shv(spec A), $\tilde{A}$), where spec A is the prime-spectrum of A with the Zariski topology, and $\tilde{A}$ is its usual structure sheaf (see [ST], §4.2); whilst it has long been known that $\tilde{A}$ is in some sense the "free local ring" generated by A, the global adjunction property of 6.58 was first observed (in this case) by M. Hakim [45], and she was also the first to construct Spec($\mathscr{E}$, A) for an arbitrary Grothendieck topos $\mathscr{E}$.

M. Hakim also pointed out that the étale topos of a ring (Ex. 0.11) is another instance of a spectrum in the sense of 6.58, in which we replace the theory of local rings by that of strictly local rings. Still further applications of spectra to quotient theories of **ann** will be found in [55].

(ii) Let $\mathbb{L}$ be a geometric language. A morphism $M \xrightarrow{f} L$ of interpretations of $\mathbb{L}$ is said to be an *embedding* if f_X is mono for each type X, and

$$\begin{array}{ccc} M_r & \longrightarrow & L_r \\ \downarrow & & \downarrow \\ \prod M_{X_i} & \longrightarrow & \prod L_{X_i} \end{array}$$

is a pullback for each relation-symbol r. An easy induction then shows that if ϕ is any formula not involving $\exists$, the square

$$\begin{array}{ccc} M_\phi & \longrightarrow & L_\phi \\ \downarrow & & \downarrow \\ \prod M_{X_i} & \longrightarrow & \prod L_{X_i} \end{array}$$

is a pullback; hence if $(\phi \vdash \psi)$ is a quantifier-free sequent of $\mathbb{L}$, M satisfies $(\phi \vdash \psi)$ whenever L does.

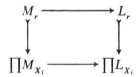

Hence if $\mathbb{S}$ and $\mathbb{T}$ are theories such that the axioms of $\mathbb{T}$ are quantifier-free we may always obtain a spectrum by taking $\mathbb{A}$ to be the class of embeddings; to prove the factorization lemma for a morphism $M \xrightarrow{f} L$, we define $(M_f)_X$ to be the image (1.52) of f_X, and $(M_f)_r$ to be the pullback of L_r. The extremal morphisms are therefore those f such that f_X is epi for each X. (This example is essentially due to J. F. Kennison [60].)

(iii) Let $\mathbb{S}$ = **cat**, $\mathbb{T}$ = **filt**, and take $\mathbb{A}$ to be the class of cofinal functors (Ex. 2.11). Then $\mathbb{A}$ has properties dual to those of 6.57; the dual of (ii) was established in Ex. 2.11, and the factorization lemma was proved by R. Street and R. F. C. Walters [115]. We may thus construct a *cospectrum* functor, i.e. a right adjoint for the inclusion

$$\mathbb{A}^{\text{op}}\text{-}\mathfrak{Top}_N \longrightarrow \mathbf{cat}^{\text{op}}\text{-}\mathfrak{Top}_N,$$

where $\mathbb{T}^{\text{op}}$-$\mathfrak{Top}$ is defined like $\mathbb{T}$-$\mathfrak{Top}$ except that its 1-arrows are pairs (p, f) with $L \xrightarrow{f} p^*M$.

It turns out that the factorization of an internal functor $\mathbf{C} \xrightarrow{f} \mathbf{D}$ has the form $\mathbf{C} \longrightarrow L(\mathbf{C}) \xrightarrow{L(f)} \mathbf{D}$, where $L(f)$ is the reflection of f in the category of discrete fibrations over $\mathbf{D}$, i.e. the left Kan extension along f of the identity discrete fibration over $\mathbf{C}$ (cf. 2.36). So the extremal morphisms with codomain $\mathbf{D}$ are simply the discrete fibrations over $\mathbf{D}$ with filtered domain, i.e. the flat presheaves on $\mathbf{D}$ (4.31). Hence 4.34 tells us that Cospec($\mathscr{E}$, $\mathbf{D}$) $\simeq$ ($\mathscr{E}^{\mathbf{D}}$, $Y(\mathbf{D})$). ■

We conclude this rather long paragraph with some applications of 6.56 to the problem of constructing exponentials in the 2-category $\mathfrak{BTop}/\mathscr{E}$.

6.60 Examples. (i) The theory **dofib** of discrete opfibrations (2.15) is clearly geometric, since it can be presented as a quotient theory of $\mathbf{cat}^2$. So if $\mathbf{C}$ is a particular category in $\mathscr{E}$, the pullback

$$\begin{array}{ccc} \mathscr{F} & \longrightarrow & \mathscr{E}[\mathbf{dofib}] \\ \big\downarrow & & \big\downarrow \\ & & \mathscr{E}[\mathbf{cat}^2] \\ & & \big\downarrow \scriptstyle\text{cod} \\ \mathscr{E} & \xrightarrow{\ \ulcorner \mathbf{C} \urcorner\ } & \mathscr{E}[\mathbf{cat}] \end{array}$$

is a classifying topos for discrete opfibrations over $\mathbf{C}$—or equivalently for internal diagrams on $\mathbf{C}$.

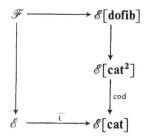

(ii) Similarly, if we are given two categories $\mathbf{C}$ and $\mathbf{D}$, we may adjoin a generic profunctor $\mathbf{D} \overset{G}{\dashrightarrow} \mathbf{C}$; and then by imposing a suitable topology, we may force G to be left flat in the sense of 4.39. Let $\mathscr{H}$ denote the topos containing this generic left flat profunctor; then for any $\mathscr{E}$-topos ($\mathscr{F} \xrightarrow{p} \mathscr{E}$) we have equivalences

$$\begin{aligned}\mathfrak{Top}/\mathscr{E}\,(\mathscr{F}, \mathscr{H}) &\simeq \mathrm{lfProf}_{\mathscr{F}}(p^*\mathbf{D}, p^*\mathbf{C}) \\ &\simeq \mathfrak{Top}/\mathscr{F}\,(\mathscr{F}^{p^*\mathbf{C}}, \mathscr{F}^{p^*\mathbf{D}}) \qquad \text{by 4.39} \\ &\simeq \mathfrak{Top}/\mathscr{E}\,(\mathscr{F} \times_{\mathscr{E}} \mathscr{E}^{\mathbf{C}}, \mathscr{E}^{\mathbf{D}}) \qquad \text{by 4.35,}\end{aligned}$$

where lfProf denotes the full subcategory of Prof consisting of left flat profunctors. So the topos $\mathscr{H}$ is (in the sense appropriate to 2-categories) the exponential $(\mathscr{E}^{\mathbf{D}})^{(\mathscr{E}^{\mathbf{C}})}$ in $\mathfrak{Top}/\mathscr{E}$.

(iii) More generally still, let $\mathscr{G} = \mathrm{sh}_j(\mathscr{E}^{\mathbf{D}})$ be any bounded $\mathscr{E}$-topos, and $\mathbf{C}$ an internal category in $\mathscr{E}$. Then by 3.47 and 4.39, we know that a geometric morphism $\mathscr{E}^{\mathbf{C}} \xrightarrow{g} \mathscr{E}^{\mathbf{D}}$ over $\mathscr{E}$ factors through $\mathscr{G}$ iff the corresponding left flat profunctor $\mathbf{D} \overset{G}{\dashrightarrow} \mathbf{C}$ has the property that $(-)\otimes_{\mathbf{D}} G$ inverts all j-dense monomorphisms. Now let $\mathscr{H}$ be the $\mathscr{E}$-topos constructed in part (ii), j' the topology in $\mathscr{H}^{\mathbf{D}}$ induced by j as in 3.59(iii), and d' the generic j'-dense monomorphism. Then if G denotes the generic left flat profunctor in $\mathscr{H}$, we can construct a topology k in $\mathscr{H}$ which forces $d' \otimes_{\mathbf{D}} G$ to be iso; and it is then easy to see that $\mathrm{sh}_k(\mathscr{H})$ is the exponential $\mathscr{G}^{(\mathscr{E}^{\mathbf{C}})}$ in $\mathfrak{Top}/\mathscr{E}$. ∎

However, it is *not* true that we can construct arbitrary exponentials in the 2-category $\mathfrak{BTop}/\mathscr{E}$. The reason is that in order to do so we would have to be able to force a given monomorphism m in $\mathscr{E}$ to be j-dense, where j is a pre-existing topology in $\mathscr{E}$; i.e. to find the unique smallest topology k in $\mathscr{E}$ such that $k + j \geqslant j_m$, where j_m is the topology which forces m to be iso. If we could do this in general, it would imply that the lattice of topologies in $\mathscr{E}$ was a co-Heyting algebra (equivalently, that the lattice of sheaf subtoposes of $\mathscr{E}$ was a Heyting algebra). But this is well known to be false; a counter-example appears in Ex. 7.1 below.

On the other hand, A. Joyal has succeeded in proving one significant extension of 6.60(iii); he has shown that if $\mathscr{F} = \mathrm{sh}_j(\mathscr{E}^{\mathbf{C}})$ where the topology j satisfies a certain "finiteness condition", and $\mathscr{G}$ is any bounded $\mathscr{E}$-topos, then we may construct the exponential $\mathscr{G}^{\mathscr{F}}$ in $\mathfrak{Top}/\mathscr{E}$. We shall sketch his argument, in the particular case $\mathscr{E} = \mathscr{S}$, in 7.49 below.

6.6. REAL NUMBER OBJECTS

As we have already observed, the "Booleanness" of the natural number object in a topos means that its arithmetic is much closer to that of standard set theory than we might at first have expected. However, when we try to carry out the construction of the real numbers from the natural numbers, we are sharply reminded of the non-classical nature of the internal logic of a topos; in particular, we find that the various (classically equivalent) constructions available yield non-isomorphic results. In this paragraph we shall be concerned with two particular constructions, which are commonly referred to as the "Dedekind reals" and the "Cauchy reals" (though in fact neither name is historically accurate); there are a number of other possibilities which have been studied (see, for example, [112]), but they appear to be less important for particular applications.

Given a natural number object N in our topos $\mathscr{E}$, we may first construct the *object of integers* Z by the pushout

$$\begin{array}{ccc} 1 & \stackrel{o}{\rightarrowtail} & N \\ {\scriptstyle o}\downarrow & & \downarrow{\scriptstyle p_1} \\ N & \stackrel{p_2}{\rightarrowtail} & Z \end{array} \quad ;$$

the arithmetic operations and order-relation on Z are defined in terms of those on N, and it is easily verified that Z is an internal ring. Then we may construct the *object of rationals* Q as the ring of fractions $Z[P^{-1}]$ (cf. 6.59(i)), where $(P \rightarrowtail Z) = (N \stackrel{s}{\rightarrowtail} N \stackrel{p_1}{\rightarrowtail} Z)$ is the subobject of positive integers. Once again Q inherits an order-relation from Z; it satisfies the "trichotomy" property (cf. 6.17(iii)), and Q is a *field* in the same sense that $\mathscr{E} \models \neg(o = e)$ and $\mathscr{E} \models (\mathbf{q} = 0) \vee (\mathbf{q} \in \ulcorner \Upsilon Q \urcorner)$ (see 6.65 below). Moreover, since the constructions used so far involve only finite limits and colimits, it follows from 6.16 that the objects Z and Q are preserved by exact functors, and in particular by inverse image functors. Thus in the topos Shv(X) of sheaves on a topological space X, Q is simply the constant sheaf $\Delta(\mathbb{Q})$ whose stalk at every point is the set $\mathbb{Q}$ of rational numbers (in $\mathscr{S}$).

6.61 DEFINITIONS. Let $\mathscr{E}$ be a topos with a natural number object. A *Dedekind real number* in $\mathscr{E}$ is an ordered pair $r = (L, U)$ of subobjects of Q satisfying the conditions

(i) $\mathscr{E} \models \forall \mathbf{q}(\mathbf{q} \in \ulcorner L \urcorner \Leftrightarrow \exists \mathbf{q}'(\mathbf{q}' \in \ulcorner L \urcorner \wedge \mathbf{q}' > \mathbf{q}))$
(ii) $\mathscr{E} \models \forall \mathbf{q}(\mathbf{q} \in \ulcorner U \urcorner \Leftrightarrow \exists \mathbf{q}'(\mathbf{q}' \in \ulcorner U \urcorner \wedge \mathbf{q}' < \mathbf{q}))$
(iii) $\mathscr{E} \models \forall \mathbf{q}\, \forall \mathbf{q}'(\mathbf{q} \in \ulcorner L \urcorner \wedge \mathbf{q}' \in \ulcorner U \urcorner \Rightarrow \mathbf{q} < \mathbf{q}')$
(iv) $\mathscr{E} \models \forall \mathbf{n}\, \exists \mathbf{q}\, \exists \mathbf{q}'(\mathbf{q} \in \ulcorner L \urcorner \wedge \mathbf{q}' \in \ulcorner U \urcorner \wedge \mathbf{n}(\mathbf{q}' - \mathbf{q}) < e)$

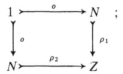

We write $R_d \xrightarrow{(l,\,u)} \Omega^Q \times \Omega^Q$ for the *object of Dedekind reals*, i.e. the extension of the formula obtained by substituting free variables **l**, **u** of type Ω^Q for the constants $\ulcorner L \urcorner$, $\ulcorner U \urcorner$ in the above.

It is easily verified that the morphism $Q \xrightarrow{(\downarrow\text{seg},\, \uparrow\text{seg})} \Omega^Q \times \Omega^Q$ is mono, and that it factors through R_d; so we have a canonical monomorphism $Q \xrightarrow{i} R_d$. ■

The object R_d, being a "higher-order" construction, is not in general preserved except by logical functors. However, the axioms (i)–(iv), if we strip them of their initial universal quantifiers, may be represented by sequents in a geometric language having N and Q as types and L, U as primitive relations of signature (Q). So the property of being a Dedekind real is preserved by inverse image functors; and indeed we can construct a classifying topos $\mathscr{E}[\mathbf{ded}]$ containing a generic real number. It is not hard to show that $\mathscr{S}[\mathbf{ded}]$ is the topos Shv($\mathbb{R}$), where $\mathbb{R}$ has its usual topology (see Ex. 7.2 below); so remembering Grothendieck's dictum that a topos is a generalized space, and a geometric morphism is a generalized continuous map, we should think of a Dedekind real in $\mathscr{E}$ as a continuous real-valued function on $\mathscr{E}$. The first of the examples which follow strengthens this intuitive idea.

6.62 EXAMPLES. (i) Let X be a topological space, and $r = (L, U)$ a Dedekind real in Shv(X). Then for each point x of X the pair (L_x, U_x) is a Dedekind section of $\mathbb{Q}$, and so it determines a real number $r(x)$ in $\mathscr{S}$. [Here F_x denotes the stalk of F at x, as usual.] Thus r determines a function $X \longrightarrow \mathbb{R}$. Now let $q \in \mathbb{Q}$, and suppose $q \in L_x$; then since L is a subsheaf of Q there exists a neighbourhood N of x such that $q \in L_y$ for all $y \in N$, i.e. the set $\{x \in X \mid q < r(x)\}$ is open. Similarly the set $\{x \in X \mid q > r(x)\}$ is open; but since the open intervals with rational endpoints form a base for the topology on $\mathbb{R}$, this is sufficient to prove that r is a continuous function $X \longrightarrow \mathbb{R}$.

Conversely, let $X \xrightarrow{r} \mathbb{R}$ be a continuous function. Then it is readily checked that the assignment

$$V \longmapsto \{q \in \Delta\mathbb{Q}(V) \mid q(x) < r(x) \text{ for all } x \in V\}$$

defines a subsheaf L of $\Delta\mathbb{Q} = Q$ [this uses the identification of $\Delta\mathbb{Q}(V)$ with the set of locally constant $\mathbb{Q}$-valued functions on V]; and that if U is defined similarly, the pair (L, U) is a Dedekind real in Shv(X).

So the Dedekind reals in Shv(X) are precisely the continuous real-valued functions on X; from which it follows easily that the object R_d is actually the sheaf $C_{\mathbb{R}}$ of continuous real-valued functions (cf. (0.22(iii)). Note also that if we modified the definition of a Dedekind real by writing down axioms involving only the lower section L, we would obtain the sheaf of upper semi-continuous real-valued functions on X.

(ii) Let (X, Σ, μ) be a measure space. P. Deligne ([GV], IV 7.4) observed that we may construct a topos $\text{Meas}(X, \Sigma, \mu)$ by taking the poset Σ of measurable subsets of X as a category, and imposing on it the Grothendieck pretopology (0.31) whose covering families are *countable* families of inclusions $\{B_i \longrightarrow B \mid i = 1, 2, \ldots\}$ such that $B - \bigcup_{i=1}^{\infty} B_i$ has measure zero. It is easily seen that in this topos Q is the sheaf of equivalence classes of almost everywhere defined, Σ-locally constant $\mathbb{Q}$-valued functions on X, under the equivalence relation

$$q \sim q' \Leftrightarrow q(x) = q'(x) \text{ almost everywhere.}$$

Now let $r = (L, U)$ be a Dedekind real in $\text{Meas}(X, \Sigma, \mu)$; then we may use 6.61(iv) to construct sequences (q_n), (q^n) of global elements of Q such that $q_n \in L$, $q^n \in U$ and $q^n - q_n < 1/n$ for all n. Then for almost all $x \in X$ the sequences $q_n(x)$ and $q^n(x)$ tend to a common limit $r(x)$; and arguments similar to those of part (i) show that the real-valued function on X thus defined must be measurable. Conversely, any measurable real-valued function on X determines a Dedekind real; so R_d is the sheaf of *random variables* on X, i.e. equivalence classes of measurable real-valued functions under the relation of almost-everywhere equality. [This example is essentially due to D. Scott [182].] ■

We have two alternative ways of defining an order relation on R_d. If $\mathbf{r}$, $\mathbf{r}'$ are variables of type R_d, we shall write $\mathbf{r} \leqslant \mathbf{r}'$ for the formula $l(\mathbf{r}) \leqslant l(\mathbf{r}')$, and $\mathbf{r} < \mathbf{r}'$ for the formula $\exists \mathbf{q}(\mathbf{q} \in u(\mathbf{r}) \wedge \mathbf{q} \in l(\mathbf{r}'))$. (Note that 6.61(i) implies that $\mathscr{E} \models \mathbf{q} \in l(\mathbf{r}) \Leftrightarrow i(\mathbf{q}) < \mathbf{r}$.) It should be pointed out that $\leqslant$ is *not* equivalent to the disjunction of $<$ and $=$; for example in $\text{Shv}(\mathbb{R})$, if we take $r(x) = 0$ and $r'(x) = x^2$, the corresponding Dedekind reals satisfy

$$\| r \leqslant r' \| = 1, \; \| r < r' \| = \mathbb{R} - \{o\}, \text{ and } \| r = r' \| = 0$$

(since the functions r and r' do not agree on any nonempty *open* set). Nor is it true that either $\leqslant$ or $<$ is a total order on R_d (again, $\text{Shv}(\mathbb{R})$ provides a counterexample if we take $r(x) = 0$, $r'(x) = x$).

To define the operation of addition on R_d is straightforward; we define

$$l(\mathbf{r} + \mathbf{r}') = \{\mathbf{q} \mid \exists \mathbf{q}' \exists \mathbf{q}''(\mathbf{q}' \in l(\mathbf{r}) \wedge \mathbf{q}'' \in l(\mathbf{r}') \wedge \mathbf{q} = \mathbf{q}' + \mathbf{q}''\},$$

with a similar definition of $u(\mathbf{r} + \mathbf{r}')$, and verify that this pair of terms of type Ω^Q satisfies the axioms of 6.61. However, the "classical" way of defining multiplication by subdivision into cases will not work, because R_d is not totally ordered; to avoid this, we use an idea due to J. H. Conway [157]. First we make R_d into a module over Q by defining

$$\begin{aligned} \mathbf{rq} &= (l(\mathbf{r}) \,.\, \mathbf{q}, u(\mathbf{r}) \,.\, \mathbf{q}) && \text{if } \mathbf{q} > 0 \\ &= i(0) && \text{if } \mathbf{q} = 0 \\ &= (u(\mathbf{r}) \,.\, \mathbf{q}, l(\mathbf{r}) \,.\, \mathbf{q}) && \text{if } \mathbf{q} < 0 \end{aligned}$$

where

$$l(\mathbf{r}) \,.\, \mathbf{q} = \{\mathbf{q}' \mid \exists \mathbf{q}''(\mathbf{q}'' \in l(\mathbf{r}) \wedge \mathbf{q}' = \mathbf{q}'' \,.\, \mathbf{q})\}.$$

(This subdivision into cases is legitimate since Q satisfies trichotomy, i.e.

$$\mathscr{E} \models (\mathbf{q} > 0) \vee (\mathbf{q} = 0) \vee (\mathbf{q} < 0).)$$

Then we define $l(\mathbf{rr}')$ to be the term

$$\{\mathbf{q} \mid (\exists \mathbf{q}' \exists \mathbf{q}'' (((\mathbf{q}' \in l(\mathbf{r}) \wedge \mathbf{q}'' \in l(\mathbf{r}')) \vee (\mathbf{q}' \in u(\mathbf{r}) \wedge \mathbf{q}'' \in u(\mathbf{r}'))) \wedge \\ \wedge \mathbf{q} \in l(\mathbf{r}'\mathbf{q}' + \mathbf{rq}'' - i(\mathbf{q}'\mathbf{q}''))\},$$

with a similar definition for $u(\mathbf{rr}')$. The proof that this definition makes R_d into a commutative ring is essentially the same as in [157].

However, R_d is not in general a field in the sense in which we used the word for Q. To obtain further information about its algebraic structure, we must investigate more closely the relationship between the two order-relations defined above.

6.63 Lemma. (i) $\mathscr{E} \models \mathbf{q} \in l(r) \Leftrightarrow \exists \mathbf{q}'(\mathbf{q}' > \mathbf{q} \wedge \neg(\mathbf{q}' \in u(\mathbf{r})))$.

(ii) $\mathscr{E} \models \mathbf{r} \leqslant \mathbf{r}' \Leftrightarrow u(\mathbf{r}) \geqslant u(\mathbf{r}')$.

(iii) $\mathscr{E} \models \mathbf{r} \leqslant \mathbf{r}' \Leftrightarrow \neg(\mathbf{r} > \mathbf{r}')$.

Proof. (i) From 6.61(iii), we immediately deduce

$$\mathscr{E} \models \mathbf{q}' \in l(\mathbf{r}) \Rightarrow \neg(\mathbf{q}' \in u(\mathbf{r})),$$

so the left-to-right implication follows from 6.61(i). Conversely, suppose we are given a Dedekind real $r = (L, U)$ and two rational numbers $1 \overset{q}{\underset{q'}{\rightrightarrows}} Q$ such that $q < q'$ and $\neg(q' \in \ulcorner U \urcorner)$. Then $q' - q$ is a positive rational, so we can express it locally as the ratio of two positive integers, and hence we can find $V \overset{\alpha}{\twoheadrightarrow} 1$ and $V \overset{n}{\longrightarrow} N$ with $(q' - q)\alpha > 1/n$. Now by 6.61(iv) we can find $W \overset{\beta}{\twoheadrightarrow} V$ and $W \overset{q''}{\underset{q'''}{\rightrightarrows}} Q$ such that $q'' \in L$, $q''' \in U$ and $q''' - q'' < 1/n\beta$. Then we have $\neg(q'\alpha\beta \geqslant q''')$ and so $q'\alpha\beta < q'''$ since Q satisfies trichotomy. Thus

$$q\alpha\beta \leqslant q'\alpha\beta - 1/n\beta < q''' - 1/n\beta < q'';$$

but $q'' \in L$, so $q\alpha\beta \in L$. And $\alpha\beta$ is epi, so $q \in L$.

(ii) Suppose $l(\mathbf{r}) \leqslant l(\mathbf{r}')$; then we have

$$\begin{aligned} \mathscr{E} \models \mathbf{q} \in u(\mathbf{r}') &\Rightarrow \exists \mathbf{q}'(\mathbf{q}' < \mathbf{q} \wedge \mathbf{q}' \in u(\mathbf{r}')) && \text{by 6.61(ii)} \\ &\Rightarrow \exists \mathbf{q}'(\mathbf{q}' < \mathbf{q} \wedge \neg(\mathbf{q}' \in l(\mathbf{r}'))) && \text{by 6.61(iii)} \\ &\Rightarrow \exists \mathbf{q}'(\mathbf{q}' < \mathbf{q} \wedge \neg(\mathbf{q}' \in l(\mathbf{r}))) && \text{by assumption.} \\ &\Rightarrow \mathbf{q} \in u(\mathbf{r}) && \text{by part (i).} \end{aligned}$$

The converse argument is dual.

(iii) From $\mathbf{r} \leqslant \mathbf{r}'$ and $\mathbf{r} > \mathbf{r}'$ we deduce $\exists \mathbf{q}(\mathbf{q} \in u(\mathbf{r}') \wedge \mathbf{q} \in l(\mathbf{r}'))$, contradicting 6.61(iii); so $\mathscr{E} \models \mathbf{r} \leqslant \mathbf{r}' \Rightarrow \neg(\mathbf{r} > \mathbf{r}')$.
Conversely, suppose $\neg(\mathbf{r} > \mathbf{r}')$. Then

$$\begin{aligned} \mathscr{E} \models \mathbf{q} \in l(\mathbf{r}) &\Rightarrow \exists \mathbf{q}'(\mathbf{q}' > \mathbf{q} \wedge \mathbf{q}' \in l(\mathbf{r})) && \text{by 6.61(i)} \\ &\Rightarrow \exists \mathbf{q}'(\mathbf{q}' > \mathbf{q} \wedge \neg(\mathbf{q}' \in u(\mathbf{r}'))) && \text{by assumption} \\ &\Rightarrow \mathbf{q} \in l(\mathbf{r}') && \text{by part (i).} \end{aligned}$$

So $\mathbf{r} \leqslant \mathbf{r}'$. ■

In order to make precise the sense in which R_d is a field, we now introduce some terminology (see also [55]):

6.64 DEFINITION. Let A be a commutative ring in a topos $\mathscr{E}$, and suppose $\mathscr{E} \models \neg(o = e)$.

(i) We saw A is a *geometric field* if $\mathscr{E} \models (\mathbf{a} = o) \vee (\mathbf{a} \in \ulcorner \Upsilon A \urcorner)$.
(ii) We say A is a *field of fractions* if $\mathscr{E} \models \neg(\mathbf{a} = o) \Rightarrow (\mathbf{a} \in \ulcorner \Upsilon A \urcorner)$.
(iii) We say A is a *residue field* if $\mathscr{E} \models \neg(\mathbf{a} \in \ulcorner \Upsilon A \urcorner \Rightarrow (\mathbf{a} = o)$.
(iv) We say A is an *integral domain* if $\mathscr{E} \models (\mathbf{a}\mathbf{a}' = o) \Rightarrow (\mathbf{a} = o) \vee (\mathbf{a}' = o)$. ■

Clearly, the first and fourth of the above axioms may be expressed as sequents in the geometric language of the theory of rings (hence the name "geometric field"). If A is an integral domain, then the subobject $N \rightarrowtail A$ of nonzero elements is multiplicatively closed, and the ring $A[N^{-1}]$ is a field of fractions; dually, if A is a local ring, then the subobject $M \rightarrowtail A$ of non-units is an ideal, and the quotient A/M is a residue field. Further discussion of the relationship between these axioms will be found in [55] and [88].

6.65 THEOREM. (i) $\mathscr{E} \models \mathbf{r} \in \ulcorner \Upsilon R_d \urcorner \Leftrightarrow (\mathbf{r} > o \vee \mathbf{r} < o)$.

(ii) R_d *is a residue field.*

(iii) R_d *is a local ring.*

Proof. (i) Suppose $\mathbf{r} > o$. Then it is easily checked that the term

$$(\{\mathbf{q} \mid \mathbf{q} \leqslant o \vee (\mathbf{q} > o \wedge \mathbf{q}^{-1} \in u(\mathbf{r}))\}, \{\mathbf{q} \mid \mathbf{q} > o \wedge \mathbf{q}^{-1} \in l(\mathbf{r})\})$$

defines a Dedekind real which is a multiplicative inverse for $\mathbf{r}$. So

$$\mathscr{E} \models \mathbf{r} > o \Rightarrow \mathbf{r} \in \ulcorner \Upsilon R_d \urcorner;$$

and similarly $E \models \mathbf{r} < o \Rightarrow \mathbf{r} \in \ulcorner \Upsilon R_d \urcorner$.

Conversely, let r be a Dedekind real, and let $V \rightarrowtail 1$ be the extension of the formula $(r > o \vee r < o)$. Then by 6.63(iii) we have $V \cong 0$ iff $r = o$. But the formula defining V is in the geometric language of the theory of Dedekind reals; hence V is preserved by inverse image functors. In particular, if i^* is the associated sheaf functor for the closed topology j_V^c, we have $i^*V \cong 0$ and so $i^*r = o$. But if r is a unit, then so is i^*r, and hence the topos $\mathrm{sh}_{j_V^c}(\mathscr{E})$ must be degenerate, i.e. $V \cong 1$. By applying this argument to generalized elements of R_d, we obtain the internal implication

$$\mathscr{E} \models \mathbf{r} \in \ulcorner \Upsilon R_d \urcorner \Rightarrow (\mathbf{r} > o \vee \mathbf{r} < o).$$

(ii) follows at once from (i) and 6.63(iii).

(iii) From 6.61(iv) we have

$$\mathscr{E} \models \exists \mathbf{q} \exists \mathbf{q}'(\mathbf{q} \in l(\mathbf{r}) \wedge \mathbf{q}' \in u(\mathbf{r}) \wedge \mathbf{q}' - \mathbf{q} < e).$$

But since Q is totally ordered, we have

$$\mathscr{E} \models \mathbf{q}' - \mathbf{q} < e \Rightarrow (\mathbf{q} > o \vee \mathbf{q}' < e);$$

so

$$\mathscr{E} \models (o \in l(\mathbf{r}) \vee e \in u(\mathbf{r})), \text{ i.e. } \mathscr{E} \models \mathbf{r} > o \vee \mathbf{r} < e.$$

Hence by (i) we have

$$\mathscr{E} \models (\mathbf{r} \in \ulcorner \Upsilon R_d \urcorner \vee (e - \mathbf{r} \in \ulcorner \Upsilon R_d \urcorner). \blacksquare$$

However, R_d is not in general a field of fractions; nor is it an integral domain. For in Shv($\mathbb{R}$) the Dedekind real defined by $r(x) = x$ satisfies $\neg(r = o)$ but not $(r \in \ulcorner \Upsilon R_d \urcorner)$. And if we define $r'(x) = \max(x, o)$, $r''(x) = \min(x, o)$, then we have $r'r'' = o$ but not $(r' = o \vee r'' = o)$.

6.66 Examples. Somewhat parenthetically to the rest of this paragraph, we give a couple of examples to illustrate the independence of the field axioms listed in 6.64.

(i) Let $\mathscr{E} = \mathrm{sh}_j(\mathscr{S}[\mathbf{ann}])$, where j is the topology which forces the generic ring to satisfy the nontriviality axiom $(o = e \vdash \mathit{false})$; and let U denote the generic nontrivial ring in $\mathscr{E}$. It is easily shown (cf. [45], [53]) that $\mathscr{S}[\mathbf{ann}]$ is the topos $\mathscr{S}^{\mathbf{C}}$, where $\mathbf{C}$ is the category of finitely-presented rings in $\mathscr{S}$, the underlying object of the generic ring being the forgetful functor $\mathbf{C} \longrightarrow \mathscr{S}$. And since the trivial ring is a strict terminal object in $\mathbf{C}$, the topology j (regarded as a Grothendieck topology on $\mathbf{C}^{\mathrm{op}}$) differs from the minimal topology only in that the empty sieve covers the trivial ring. Now consider an h^A-element $h^A \xrightarrow{a} U$ of U, where A is a finitely-presented ring; from the definition of U, this is simply an element of the ring A. Then the external interpretation of the statement $\neg(a = o)$ is clearly that the ring $A/(a)$ is trivial, where (a) is the ideal generated by a, i.e. that a is a unit. So U is a field of fractions in $\mathscr{E}$. However, U is not a residue field; for the statement $\neg(a \in \ulcorner \Upsilon U \urcorner)$ means that the ring $A[a^{-1}]$ is trivial, which merely implies that a is nilpotent, and not that $a = o$. (This example is due to A. Kock [63].)

(ii) Let $\mathscr{E} = \mathrm{Shv}(\mathbb{R})$, and let W be the sheaf of rings on $\mathbb{R}$ whose sections over an open set U are equivalence classes of real-valued continuous functions defined on dense open subsets of U, under the equivalence relation of agreement on a dense open subset. If w is a U-element of W, then the statement $\neg(w = o)$ means that the set $\{x \in U \mid w(x) = 0\}$ is nowhere dense, from which it follows that w is invertible in $W(U)$. So W is a field of fractions; but it is also a residue field, since the statement $\neg(w \in \ulcorner \Upsilon W \urcorner)$ means that there is no nonempty open set on which w takes nonzero values, and hence $w = o$. However, W is not a geometric field, nor is it a local ring; the global element of W defined by $w(x) = o$ for $x < o$, $w(x) = e$ for $x > o$, provides a counter-example to both assertions. ■

We turn now to our second definition of the real numbers, which uses the familiar notion of a Cauchy sequence of rationals.

6.67 DEFINITION. The object $S \rightarrowtail Q^N$ of *Cauchy sequences* is the extension of the formula

$$\forall \mathbf{n}\, \forall \mathbf{n}'\, (\mathbf{n}' \geqslant \mathbf{n} > o) \Rightarrow (-1/\mathbf{n} < \mathbf{f}(\mathbf{n}') - \mathbf{f}(\mathbf{n}) < 1/\mathbf{n}),$$

where $\mathbf{f}$ is a free variable of type Q^N. We define a subobject $E \overset{(a,b)}{\rightarrowtail} S \times S$ by the formula

$$\forall \mathbf{n}\, (\mathbf{n} > o) \Rightarrow (-2/\mathbf{n} \leqslant \mathbf{f}(\mathbf{n}) - \mathbf{f}'(\mathbf{n}) \leqslant 2/\mathbf{n}),$$

and we define the object R_c of *Cauchy reals* to be the coequalizer of $E \underset{b}{\overset{a}{\rightrightarrows}} S$. [Note that the above formulae differ from those normally used in defining Cauchy sequences, in that we require our sequences to converge at a "uniform" rate; this modification has the effect of reducing the number of quantifiers with which we have to wrestle in proving statements about R_c, and thereby simplifying the proofs considerably.] ■

The formulae defining S and E are clearly equivalent to ones expressible in a geometric language having Q^N as a type, since by 6.17(iii) we can rewrite $(\mathbf{n}' \geqslant \mathbf{n}) \Rightarrow \phi$ as $(\mathbf{n}' < \mathbf{n}) \vee \phi$. Hence if f^* is an inverse image factor which preserves exponentiation to the power N, it preserves these objects and hence also R_c. In particular, if $\mathscr{E} \xrightarrow{\gamma} \mathscr{S}$ is a locally connected $\mathscr{S}$-topos (for example sheaves on a locally connected space; see Exx. 0.7 and 4.9), then the object of Cauchy reals in $\mathscr{E}$ is simply the constant object $\gamma^*(\mathbb{R})$.

We may define operations of addition and multiplication on S by

$$(\mathbf{f} + \mathbf{f}')(\mathbf{n}) = \mathbf{f}(2\mathbf{n}) + \mathbf{f}'(2\mathbf{n})$$

and

$$(\mathbf{f}\mathbf{f}')(\mathbf{n}) = \mathbf{f}(k(\mathbf{f}, \mathbf{f}')\mathbf{n}) \,.\, \mathbf{f}'(k(\mathbf{f}, \mathbf{f}')\mathbf{n}),$$

where $k(\mathbf{f}, \mathbf{f}')$ is an integer greater than

$$|\mathbf{f}(so)| + |\mathbf{f}'(so)| + 2.$$

Moreover, it is not hard to check that these operations factor through the epimorphism $S \twoheadrightarrow R_c$, and give R_c the structure of a commutative ring. The transpose $Q \xrightarrow{h} Q^N$ of $Q \times N \xrightarrow{\pi_1} Q$ clearly factors through S, and so we obtain a monomorphism $Q \rightarrowtail R_c$. And we may define a monomorphism $R_c \overset{j}{\rightarrowtail} R_d$, as follows: if $\mathbf{f}$ is a free variable of type S, the term

$$(\{q \mid \exists \mathbf{n}(\mathbf{q} < \mathbf{f}(\mathbf{n}) - 1/\mathbf{n})\}, \{q \mid \exists \mathbf{n}(\mathbf{q} > \mathbf{f}(\mathbf{n}) + 1/\mathbf{n})\})$$

is easily seen to satisfy the axioms of 6.61 (for (iv), take $\mathbf{q} = \mathbf{f}(4\mathbf{n}) - 1/3\mathbf{n}$ and $\mathbf{q}' = \mathbf{f}(4\mathbf{n}) + 1/3\mathbf{n}$). Thus we obtain a morphism $S \xrightarrow{g} R_d$; to show that it factors through $S \twoheadrightarrow R_c$, suppose $(\mathbf{f}, \mathbf{f}') \in \ulcorner E \urcorner$ and $g(\mathbf{f}) > g(\mathbf{f}')$. From the latter hypothesis, we deduce

$$\exists \mathbf{q}\, \exists \mathbf{n}\, \exists \mathbf{n}'((\mathbf{q} < \mathbf{f}(\mathbf{n}) - 1/\mathbf{n}) \wedge (\mathbf{q} > \mathbf{f}'(\mathbf{n}') + 1/\mathbf{n}'));$$

then we have

$$\exists \mathbf{n}''(\mathbf{n}'' \geqslant \mathbf{n} \wedge \mathbf{n}'' \geqslant \mathbf{n}' \wedge (\mathbf{f}(\mathbf{n}) - \mathbf{f}'(\mathbf{n}') > 1/\mathbf{n} + 1/\mathbf{n}' + 2/\mathbf{n}'')).$$

whence we obtain

$$\mathbf{f}(\mathbf{n}'') - \mathbf{f}'(\mathbf{n}'') > 2/\mathbf{n}'',$$

contradicting the first hypothesis. So we have

$$\begin{aligned} \mathscr{E} \models (\mathbf{f}, \mathbf{f}) \in \ulcorner E \urcorner &\Rightarrow \neg(g(\mathbf{f}) > g(\mathbf{f}') \vee g(\mathbf{f}) < g(\mathbf{f}')) \\ &\Rightarrow g(\mathbf{f}) = g(\mathbf{f}') \qquad \text{by 6.63(iii).} \end{aligned}$$

Conversely, it follows from the definition of g that

$$\mathscr{E} \models \forall \mathbf{n}(i(\mathbf{f}(\mathbf{n}) - 1/\mathbf{n}) \leqslant g(\mathbf{f}) \leqslant i(\mathbf{f}(\mathbf{n}) + 1/\mathbf{n})),$$

whence

$$\mathscr{E} \models g(\mathbf{f}) = g(\mathbf{f}') \Rightarrow (\mathbf{f}, \mathbf{f}') \in \ulcorner E \urcorner.$$

So the factorization $R_c \xrightarrow{j} R_d$ of g through R_c is a monomorphism.

The example of sheaves on a locally connected space, mentioned above, shows that j is *not* in general an isomorphism. However, we do have the following result:

6.68 EXAMPLE. Let X be a separable, zero-dimensional topological space (cf. Ex. 5.4). Then $R_c \xrightarrow{j} R_d$ is an isomorphism in Shv(X); for if $r = (L, U)$ is a Dedekind real in Shv(X), the fact that Shv(X) satisfies (SS) allows us to choose a sequence of *global* elements (q_n, q^n) of $Q \times Q$ such that $q_n \in \ulcorner L \urcorner$, $q^n \in \ulcorner U \urcorner$ and $q^n - q_n < 1/n$. If we regard the q_n as a morphism $N \longrightarrow Q$ in Shv(X), the corresponding global element $1 \xrightarrow{f} Q^N$ is easily seen to factor through S, and to satisfy $g(f) = r$. Thus j induces an isomorphism on global elements; but by applying the same argument in Shv(V), where V is an open subset of X, we deduce that j is an isomorphism in Shv(X), as required. ■

In this paragraph, we have not been able to do more than outline the most basic properties of the Dedekind and Cauchy reals, and of the relationship between them. We therefore conclude by referring to three papers in which the reader will find much more detailed information on particular aspects of the real numbers. C. J. Mulvey [88] has studied the Dedekind reals (more particularly, modules over R_d) from an algebraic point of view; he has shown, for example, that Swan's theorem on vector bundles over a compact space X is simply the internalization to Shv(X) of Kaplansky's theorem on projective modules over local rings. L. N. Stout [112] has considered R_d as an internal topological space, and has established the sense in which familiar topological properties of $\mathbb{R}$ can be said to hold internally. M. P. Fourman [36] has taken an analytical approach; in particular he has studied the way in which the notion of "*smooth* real-valued function on $\mathscr{E}$" may be defined using a suitable subring of R_d, and also (generalizing 6.68 above) the way in which the difference between R_c and R_d can be used to give an "analytic" (as opposed to cohomological) definition of the dimension of a topos.

EXERCISES 6

1. Give a formal proof that $+$ is associative and commutative, and that . is distributive over it.

2. Let $\mathscr{E}$ be a topos with a natural number object, and $\mathbb{G}$ a left exact comonad

on $\mathscr{E}$. Define $\theta: N \longrightarrow GN$ to be the unique morphism such that

$$\begin{array}{ccccc} 1 & \xrightarrow{\;o\;} & N & \xrightarrow{\;s\;} & N \\ \big\downarrow & & \big\downarrow{\scriptstyle \theta} & & \big\downarrow{\scriptstyle \theta} \\ G1 & \xrightarrow{\;Go\;} & GN & \xrightarrow{\;Gs\;} & GN \end{array}$$

commutes. Prove that (N, θ) is a $\mathbb{G}$-coalgebra, and that it is a natural number object in $\mathscr{E}_{\mathbb{G}}$.

3. Let **end** denote the algebraic theory generated by a single unary operation e, with no equations. Prove directly (i.e. without using 6.41 or 6.43) that existence of a natural number object in $\mathscr{E}$ is equivalent to existence of a free **end**-model functor. [Observe that **end**-models are equivalent to N-objects, where we regard N as a monoid.]

†4. Let p and q be natural numbers in $\mathscr{E}$ such that $p < q$. Show that $\mathrm{Epi}([p], [q]) \cong 0$. [Hint: first show that $\mathrm{Epi}([n], [sn]) \cong 0$ in $\mathscr{E}/N$.] Deduce that if $[p] \cong [q]$ then $p = q$.

5. Use 6.18 to define the factorial map $(!)\colon N \longrightarrow N$ as an element of N in $\mathscr{E}/N$, and 6.19 to establish an isomorphism $\mathrm{Iso}([n]), [n]) \cong [n!]$ in $\mathscr{E}/N$. Using 6.29 and Ex. 4, deduce that if X and Y are finite cardinals then $\mathrm{Iso}(X, Y)$ is a complemented subobject of Y^X.

†6. Let $\mathscr{F} \xrightarrow{f} \mathscr{E}$ be a geometric morphism. Show that f^* restricts to a logical functor $\mathscr{E}_{fc} \longrightarrow \mathscr{F}_{fc}$. Is this restricted functor necessarily the inverse image of a geometric morphism?

7. Let $\mathscr{F}$ denote the equalizer of $\mathscr{S}[U_1 \longrightarrow U] \overset{U_1}{\underset{U_2}{\rightrightarrows}} \mathscr{S}[U]$ in $\mathfrak{BTop}/\mathscr{S}$. Show that $\mathscr{F}$ classifies the theory **end** defined in Ex. 3. Observe that $\mathscr{F}$ may be described as $\mathscr{S}^{\mathbf{C}}$, where $\mathbf{C}$ is the category of finitely-presented **end**-models in $\mathscr{S}$; and deduce that the functor $\mathbf{cat}(\mathscr{E}) \longrightarrow \mathfrak{Top}/\mathscr{E}$ of 2.35 does not preserve equalizers.

8. Let c denote the least upper bound of the sequence of cardinals $(\aleph_0, 2^{\aleph_0}, 2^{2^{\aleph_0}}, \ldots)$, and let $\mathscr{S}_c$ denote the category of sets of cardinality $< c$ and functions between them. Show that $\mathscr{S}_c$ is a topos, and has a natural number object, but that there exists a recursion problem (X, T)

with no solution in $\mathscr{S}_c$, even though T is a strong functor. [Take T to be the covariant power-set functor.]

†9. (B. Jonsson–A. Tarski [170], P. Freyd). Let $\mathbf{jt}_2$ denote the algebraic theory generated by two unary operations l, r and a binary operation b, subject to the equations

$$b(lx, rx) = x,\ l(b(x, y)) = x \quad \text{and} \quad r(b(x, y)) = y.$$

Show that a $\mathbf{jt}_2$-model in $\mathscr{E}$ is an object X of $\mathscr{E}$ equipped with a specified isomorphism $X \longrightarrow X \times X$, and deduce that the free models $F(1)$ and $F(1 \amalg 1)$ are isomorphic. Show also that $\mathbf{jt}_2(\mathscr{E})$ is a topos. [Hint: let M be the free monoid on two generators l and r, and show that $\mathbf{jt}_2$-models can be regarded as sheaves for a certain topology in $\mathscr{E}^{\mathbf{M}}$.]

10. Show that an internal algebraic theory $\mathbb{T}$ in $\mathscr{E}$ is equivalent to a lax diagram (4.22) $\mathbf{1} \longrightarrow \mathfrak{Top}/\mathscr{E}$ which sends the unique object of $\mathbf{1}$ to $\mathscr{E}[U]$. Show also that lax cones over this diagram with vertex $\mathscr{F}$ correspond to $\mathbb{T}$-models in $\mathscr{F}$, and deduce that the classifying topos for $\mathbb{T}$ is a lax limit of this diagram. Similarly, if $\mathbf{D}$ is any finite category, show that the classifying topos for diagrams of type $\mathbf{D}$ is a lax limit of the constant diagram $\mathbf{D} \longrightarrow \mathfrak{Top}/\mathscr{E}$ with value $\mathscr{E}[U]$.

11. Let $\mathbb{T}$ be an internal algebraic theory in a Boolean topos $\mathscr{E}$. Show that the functor $T_{\mathscr{E}}: \mathscr{E} \longrightarrow \mathscr{E}$ preserves monomorphisms, and deduce that the forgetful functor $\mathbb{T}(\mathscr{E}) \longrightarrow \mathscr{E}$ preserves injectives. [Method: let $X \overset{\sigma}{\rightarrowtail} Y$ be a monomorphism in $\mathscr{E}$. By considering the epimorphism $T_{\mathscr{E}}(X) \amalg V \twoheadrightarrow 1$, where V is the complement of the support of $T_{\mathscr{E}}(X)$, reduce to the two cases (a) $T_{\mathscr{E}}(X)$ has a global element, and (b) $T_{\mathscr{E}}(X) \cong 0$. In the former case, construct a factorization of σ through the unit map $X \longrightarrow T_{\mathscr{E}}(X)$, and deduce that $T_{\mathscr{E}}(\sigma)$ is split mono. Those who have already read the Appendix will recognize that this argument works for any locally internal monad $\mathbb{T}$ on $\mathscr{E}$ (G. C. Wraith has suggsted that such monads should be regarded as "internal infinitary algebraic theories"); but for the particular case of a finitary algebraic theory, we can dispense with the hypothesis that $\mathscr{E}$ be Boolean, by using 7.54 below. However, the result is *not* true for a locally internal monad on an arbitrary topos; see [58].]

12. Let A be a commutative ring in $\mathscr{S}$. Give a presentation of the "theory of localizations of A" as a finitary geometric theory. [Hint: take each element of A as a type.]

13. Let $\mathbb{T}$ be any representable functor $(\mathfrak{Top}/\mathscr{E})^{\text{op}} \longrightarrow \mathfrak{Cat}$. Show that $\mathbb{T}(\mathscr{E}^2) \simeq (\mathbb{T}(\mathscr{E}))^2$ [use the fact that $\mathscr{E}^2$ is a lax colimit in $\mathfrak{Top}$]. Deduce that if $\mathbb{T}$ is a finitely-presented geometric theory, then $\mathscr{E}[\mathbb{T}^2]$ is equivalent to the exponential $\mathscr{E}[\mathbb{T}]^{(\mathscr{E}^2)}$ in $\mathfrak{Top}/\mathscr{E}$.

14. Let f be a continuous real-valued function defined on an open subset of $\mathbb{R}$. Show that there exist continuous functions g, h defined on the whole of $\mathbb{R}$ such that $h(x) = o$ iff $f(x)$ is undefined and $f(x) = g(x)/h(x)$ wherever $f(x)$ is defined. Deduce that the ring W of Example 6.66(ii) is the ring of fractions $R_d[N^{-1}]$, where N is the subobject of nonzero elements of R_d. Show also that W is the associated $\neg\neg$-sheaf of R_d in $\text{Shv}(\mathbb{R})$.

15. (M. P. Fourman [36]). Let X be a topological space. Show that R_c is isomorphic to the constant sheaf $\Delta(\mathbb{R})$ in $\text{Shv}(X)$ iff, for every open $U \subseteq X$, the lattice of clopen subsets of U is countably complete (i.e. a countable union of clopen subsets is clopen). Show also that if X is second countable, this condition is equivalent to local connectedness.

16. Which, if any, of the field axioms of 6.64 are satisfied by R_c in an arbitrary topos?

Chapter 7

Theorems of Deligne and Barr

7.1. POINTS

Throughout this chapter and chapter 8, we shall be working with toposes defined and bounded over a fixed base topos $\mathscr{S}$; for convenience, we shall continue to assume that $\mathscr{S}$ is "the" category of sets, although in fact most of the results remain true for any topos $\mathscr{S}$ which satisfies (AC) and has a natural number object. We leave the details of this generalization to the interested reader.

7.11 DEFINITION. Let $\mathscr{E} \xrightarrow{\gamma} \mathscr{S}$ be an $\mathscr{S}$-topos. By a *point* of $\mathscr{E}$ we mean a geometric morphism $\mathscr{S} \xrightarrow{p} \mathscr{E}$ over $\mathscr{S}$. If K is a class of points of $\mathscr{E}$, we say that K is *sufficient* if the family of functors $(p^* \mid p \in K)$ is conservative, i.e. if any $X \xrightarrow{f} Y$ in $\mathscr{E}$ such that $p^*(f)$ is iso for all $p \in K$ is necessarily iso. We say that $\mathscr{E}$ *has enough points* if the class of all points of $\mathscr{E}$ is sufficient. ■

Note that if K is a set, then the topos $\mathscr{S}/K$ is a K-indexed copower of $\mathscr{S}$ in $\mathfrak{BTop}/\mathscr{S}$ (cf. 4.21), and so if $\mathscr{E}$ is a Grothendieck topos we can combine the points in K into a single geometric morphism $\mathscr{S}/K \xrightarrow{q} \mathscr{E}$, whose inverse image is $(X \longmapsto (p^*X \mid p \in K))$, and whose direct image is $((S_p \mid p \in K) \longmapsto \prod_{p \in K} p_*(S_p))$. In this case the statement that K is sufficient is clearly equivalent to the statement that q is a surjection (4.11(ii)).

7.12 EXAMPLES. (i) Let $\mathscr{E} = \mathrm{Shv}(X)$, X a topological space. Then every point x of X corresponds to a continuous map $P \xrightarrow{x} X$, where P is the one-point space, and so (via 0.26) to a point of $\mathscr{E}$. Moreover, the set of points of $\mathscr{E}$ which arise in this way is sufficient, as we saw in 4.18(iii). In § 7.2 we shall prove a result which enables us to characterize all the points of $\mathscr{E}$.

(ii) Let $\mathscr{E} = \mathscr{S}[U]$ (cf. 4.37(iv) and 6.33). Then we know that the category of points of $\mathscr{E}$ is equivalent to $\mathscr{S}$; so that this category, even for bounded $\mathscr{E}$, need not be equivalent to a small category. However, $\mathscr{E}$ does have a sufficient *set* of points, namely those corresponding to finite sets, since any set can be expressed as a filtered colimit of finite sets (cf. 7.17 below).

(iii) Let X be a Hausdorff space, and let $\mathscr{E} = \mathrm{sh}_{\neg\neg}(\mathrm{Shv}(X))$. We shall see in §7.2 that every point of Shv(X) is determined by a point of X as in (i); so by 3.47 the points of $\mathscr{E}$ are precisely those points $x \in X$ for which the stalk functor at x inverts all $\neg\neg$ dense monomorphisms. But it is easily seen that this condition is satisfied iff x is an isolated point of X, i.e. $\{x\}$ is open. So if X has no isolated points (e.g. if $X = \mathbb{R}$), the topos $\mathscr{E}$ has no points. But $\mathscr{E}$ is clearly non-degenerate; so it does not have enough points. ∎

The next result enables us to characterize the points of a Grothendieck topos $\mathscr{E}$ in terms of a site of definition for $\mathscr{E}$. However, we shall find it convenient to state the result in a more general form which applies to geometric morphisms between an arbitrary pair of Grothendieck toposes. Recall that, in any category $\mathscr{E}$, a family of morphisms $(Y_i \xrightarrow{f_i} X \mid i \in I)$ with common codomain is said to be (*jointly*) *epimorphic* if, for any pair $X \underset{h}{\overset{g}{\rightrightarrows}} Z$ with $g \neq h$, there exists $i \in I$ with $gf_i \neq hf_i$. (If the coproduct $\coprod_{i \in I} Y_i$ exists in $\mathscr{E}$, this is of course equivalent to saying that $\coprod Y_i \longrightarrow X$ is epi.) If $(\mathbf{C}, J)$ is a site, we shall say that a functor $\mathbf{C} \longrightarrow \mathscr{E}$ is *continuous* if it sends J-covering sieves in $\mathbf{C}$ to epimorphic families in $\mathscr{E}$. (This terminology is due to Grothendieck ([GV], III 1.2); in view of the use of the word "continuous" elsewhere in category theory ([CW], p. 112), it would perhaps have been preferable to use "cocontinuous" here.) We write $l: \mathbf{C} \longrightarrow \mathrm{Shv}(\mathbf{C}, J)$ for the composite of the Yoneda embedding $\mathbf{C} \xrightarrow{h} \mathscr{S}^{\mathbf{C}^{op}}$ and the associated sheaf functor; note that this composite is always a continuous functor.

7.13 Proposition. *Let $\mathscr{E}$ be a Grothendieck topos,* $(\mathbf{C}, J)$ *a site such that* $\mathbf{C}$ *has finite limits, and* ϕ: Shv$(\mathbf{C}, J) \longrightarrow \mathscr{E}$ *a functor. Then the following conditions are equivalent:*

(i) *ϕ is the inverse image of a geometric morphism $\mathscr{E} \longrightarrow$* Shv$(\mathbf{C}, J)$.
(ii) *ϕ is left exact and preserves $\mathscr{S}$-indexed colimits.*
(iii) *There exists a left exact, continuous functor $P: \mathbf{C} \longrightarrow \mathscr{E}$, such that ϕ is (isomorphic to) the left Kan extension of P along* $\mathbf{C} \xrightarrow{l}$ Shv$(\mathbf{C}, J)$.

Moreover, the functor P in (iii) *is determined up to isomorphism by ϕ.*

Proof. (i) $\Rightarrow$ (ii) is trivial.

(ii) $\Rightarrow$ (iii): Let P be the composite $\mathbf{C} \xrightarrow{l} \mathrm{Shv}(\mathbf{C}, J) \xrightarrow{\phi} \mathscr{E}$. Then P is certainly left exact; and it is continuous because l is continuous and ϕ preserves coproducts and epimorphisms. Let X be any object of $\mathrm{Shv}(\mathbf{C}, J)$; then by 0.12 we have an isomorphism $X \cong \varinjlim_{(h \downarrow X)} h_U$ in $\mathscr{S}^{\mathbf{C}^{\mathrm{op}}}$, and hence an isomorphism $X \cong \varinjlim_{(l \downarrow X)} l(U)$ in $\mathrm{Shv}(\mathbf{C}, J)$ since the comma category $(h \downarrow X)$ is isomorphic to $(l \downarrow X)$ and the associated sheaf functor preserves colimits. Applying the functor ϕ, we have $\phi(X) \cong \varinjlim_{(l \downarrow X)} P(U)$, which is precisely the formula defining the left Kan extension (cf. [CW], p. 236).

(iii) $\Rightarrow$ (i): We may identify P with a discrete fibration $\mathbf{F} \xrightarrow{\gamma} I(\mathbf{C})^{\mathrm{op}}$, where $I(\mathbf{C})$ denotes the internalization of $\mathbf{C}$ in the topos $\mathscr{E}$ (cf. 2.39). Since $\mathbf{C}$ has finite limits and P preserves them, the argument of 4.33 shows that γ is a flat presheaf, and so by 4.34 it determines a geometric morphism $\mathscr{E} \xrightarrow{p} \mathscr{S}^{\mathbf{C}^{\mathrm{op}}}$. And it is easy to verify that the inverse image of this morphism is given by $X \longrightarrow \varinjlim_{(h \downarrow X)} P(U)$; in particular if R is a sieve on an object U of $\mathbf{C}$, then $p^*(R)$ is the image of the corresponding family in $\mathscr{E}$ with codomain $P(U)$. So contininuity of P implies that p^* inverts J-covering sieves; hence by 3.47 p factors through $\mathrm{Shv}(\mathbf{C}, J) \longrightarrow \mathscr{S}^{\mathbf{C}^{\mathrm{op}}}$, and the inverse image of the factorization is clearly isomorphic to ϕ.

To establish the last sentence of the proposition, we must show that any left exact, continuous functor $P: \mathbf{C} \longrightarrow \mathscr{E}$ is isomorphic to the restriction of its left Kan extension along l, i.e. that the canonical map

$$P(U) \xrightarrow{\eta} \varinjlim_{(l \downarrow l(U))} P(V)$$

is an isomorphism. If the topology J is sub-canonical, this is trivial even without the assumptions on P, since then l is simply the Yoneda embedding, and therefore full and faithful; so $(U, 1_{l(U)})$ is a terminal object of the category $(l \downarrow l(U))$. In general, a morphism $l(V) \xrightarrow{f} l(U)$ in $\mathrm{Shv}(\mathbf{C}, J)$ need not derive from a morphism of $\mathbf{C}$; but for any such f, the sieve

$$\{W \xrightarrow{\alpha} V \mid f\,.\,l(\alpha) \text{is in the image of } l\}$$

is clearly J-covering, and so continuity of P implies that η is epi. Similarly, if $V \underset{\beta}{\overset{\alpha}{\rightrightarrows}} U$ are two morphisms such that $l(\alpha) = l(\beta)$, then the sieve generated by the equalizer of α and β must be J-covering, and so $P(\alpha) = P(\beta)$; from this it follows easily that η is mono. ∎

Note that the equivalence of (i) and (ii) in 7.13 may be established directly as an application of the Special Adjoint Functor Theorem ([CW], p. 125). (The existence of a set of generators for Shv($\mathbf{C}$, J) ensures that the "solution-set condition" for existence of a right adjoint to ϕ is always satisfied.) However, the explicit description of geometric morphisms in terms of continuous left-exact functors will be useful in practice.

7.14 COROLLARY. *Let $\mathscr{E}$ and $\mathscr{F}$ be Grothendieck toposes. Then the category $\mathfrak{Top}/\mathscr{S}$ $(\mathscr{F}, \mathscr{E})$ has small hom-sets and filtered $\mathscr{S}$-indexed colimits.*

Proof. Let $\mathscr{F} \underset{q}{\overset{p}{\rightrightarrows}} \mathscr{E}$ be two geometric morphisms. By 7.13, p^* is isomorphic to the left Kan extension of its restriction to a suitable small full subcategory $\mathbf{C}$ of $\mathscr{E}$. Hence any natural transformation $\eta\colon p^* \longrightarrow q^*$ is completely determined by its values at objects of $\mathbf{C}$; and since $\mathscr{F}$ has small hom-sets, there is only a set of possible choices for these values.

Now let $(p_i \mid i \in \mathbf{I})$ be a filtered system of geometric morphisms $\mathscr{F} \longrightarrow \mathscr{E}$. Then by 2.58 the (pointwise-constructed) colimit of the functors $p_i^*\colon \mathscr{E} \longrightarrow \mathscr{F}$ preserves finite limits, as well as arbitrary $\mathscr{S}$-indexed colimits. So by 7.13 it is the inverse image of a geometric morphism, which is clearly the colimit of the p_i. ■

Now let $\mathbf{C}$ be a small category with finite limits, and $P\colon \mathbf{C} \longrightarrow \mathscr{S}$ a left exact functor. If $\mathbf{F} \xrightarrow{\gamma} \mathbf{C}^{\mathrm{op}}$ is the corresponding flat presheaf (cf. 4.33), then we may reinterpret 0.12 as saying that we have a natural isomorphism

$$P \cong \varinjlim_{\mathbf{F}}(\hom_{\mathbf{C}}(\gamma_0(U), -))$$

i.e. the functor P may be expressed as a *filtered* colimit of representables. Conversely, any filtered colimit of representables is left exact, by 2.58. We therefore define a *pseudo-point* of $\mathbf{C}$ to be a filtered inverse system of objects of $\mathbf{C}$, i.e. a (small) filtered category $\mathbf{I}$ and a functor $\mathbf{I} \longrightarrow \mathbf{C}^{\mathrm{op}}$; $i \longmapsto U_i$. If $P = (U_i \mid i \in \mathbf{I})$ is a pseudo-point of $\mathbf{C}$, we shall use the letter P also to denote the left exact functor $\varinjlim_{\mathbf{I}}(\hom_{\mathbf{C}}(U_i, -))\colon \mathbf{C} \longrightarrow \mathscr{S}$.

7.15 PROPOSITION. *Let $(\mathbf{C}, J)$ be a site such that $\mathbf{C}$ has finite limits, and $P = (U_i \mid i \in \mathbf{I})$ a pseudo-point of $\mathbf{C}$. Then the point of the presheaf topos $\mathscr{S}^{\mathbf{C}^{\mathrm{op}}}$ determined by the functor P factors through* Shv($\mathbf{C}$, J) $\longrightarrow \mathscr{S}^{\mathbf{C}^{\mathrm{op}}}$ *iff the following condition is satisfied: For every object V of $\mathbf{C}$, every $R \in J(V)$, every $i \in I_0$ and every $U_i \longrightarrow V$ in $\mathbf{C}$, there exists $i \longrightarrow k$ in $\mathbf{I}$ such that the composite $U_k \longrightarrow U_i \longrightarrow V$ is in $R(U_k)$. A pseudo-point satisfying this condition will be said to be a pseudo-point of the site $(\mathbf{C}, J)$.*

Proof. The given condition is simply the statement that the functor $\varinjlim_{\mathbf{I}}(\hom_{\mathbf{C}}(U_i, -))$ sends R to an epimorphic family in $\mathscr{S}$; so this is immediate from 7.13. ∎

The next theorem provides an answer to the difficulty we encountered in Example 7.12(ii); i.e. that a Grothendieck topos may have a proper class of non-isomorphic points. If, in the spirit of § 6.5, we regard every Grothendieck topos $\mathscr{E}$ as the classifying topos of some theory, whose models are the points of $\mathscr{E}$, then the theorem will be recognizable to model-theorists as a Löwenheim–Skolem theorem for the class of theories so obtainable.

7.16 THEOREM. *Let $\mathscr{E}$ be a Grothendieck topos. Then there exists a* set K *of points of $\mathscr{E}$ such that any point of $\mathscr{E}$ is expressible as a filtered colimit of points in K.*

Proof. Let $(\mathbf{C}, J)$ be a site of definition for $\mathscr{E}$ such that $\mathbf{C}$ has finite limits, and let a be a cardinal $\geqslant \max(\aleph_0, \text{card } C_1)$. Let K denote the set of points of $\mathscr{E}$ which can be defined by pseudo-points $(U_i | i \in \mathbf{I})$ of $(\mathbf{C}, J)$ for which card $I_1 \leqslant 2^a$. Now let p be any point of E, and $(U_i | i \in \mathbf{I})$ a pseudo-point of $(\mathbf{C}\ J)$ defining it. We define operations Q, R and S on the poset of subcategories of $\mathbf{I}$, as follows:

(a) For each i $\in I_0$, each $U_i \xrightarrow{\alpha} V$ in $\mathbf{C}$ and each $R \longrightarrow h_V$ in $J(V)$, choose a morphism $v_{i,\alpha,R}: i \longrightarrow i'$ in $\mathbf{I}$ such that $U_{i'} \longrightarrow U_i \xrightarrow{\alpha} V$ is in $R(U_{i'})$; and define

$$q(i) = \{v_{i,\alpha,R} \mid U_i \xrightarrow{\alpha} V \in C_1, R \in J(V)\}.$$

(Note that card $q(i) \leqslant$ card $C_1 \times 2^{\text{card } C_1} \leqslant 2^a$.) Now for each subcategory $\mathbf{I}' \subseteq \mathbf{I}$, define $Q(\mathbf{I}')$ to be the smallest subcategory containing $\mathbf{I}'$ and $\bigcup_{i \in I'_0} q(i)$.

(b) For each finite diagram d in $\mathbf{I}$, let $r(d)$ be (the set of morphisms in) a cone under the diagram (cf. Ex. 2.9). Then for any $\mathbf{I}' \subseteq \mathbf{I}$, let D be the set of finite diagrams in $\mathbf{I}'$, and define $R(\mathbf{I}')$ to be the smallest subcategory containing $\mathbf{I}'$ and $\bigcup_{d \in D} r(d)$.

(c) Finally, for any $\mathbf{I}' \subseteq \mathbf{I}$, define

$$S(\mathbf{I}') = \bigcup_{n=0}^{\infty} (QR)^n(\mathbf{I}').$$

Since $S(\mathbf{I})'$ is the union of an increasing sequence of subcategories of $\mathbf{I}$, it is itself a subcategory; and since any finite diagram in $S(\mathbf{I}')$ must lie in $(QR)^n(\mathbf{I}')$ for some finite n, it is clear from the definition of R that $S(\mathbf{I}')$ is filtered. Similarly, it follows from the definition of Q that the restricted pseudo-point $(U_i \mid i \in S(\mathbf{I}'))$ satisfies the condition of 7.15.

But since the subcategory of $\mathbf{I}$ generated by a set of morphisms of cardinality k has cardinality $\leqslant \aleph_0 . k$, it is easily seen that the three operations Q, R and S preserve the property of having cardinality $\leqslant 2^a$; so if card $I_1' \leqslant 2^a$, then the point of E defined by $(U_i \mid i \in S(\mathbf{I}'))$ is in K. And $\mathbf{I}$ is clearly the filtered union of its finitely-generated (and hence countable) subcategories, and hence also of the subcategories $\{S(\mathbf{I}') \mid \mathbf{I}' \subseteq \mathbf{I}, \mathbf{I}' \text{ countable}\}$; so the point p is the filtered colimit of the corresponding points in K. ■

7.17 COROLLARY. *If a Grothendieck topos $\mathscr{E}$ has enough points, then it has a sufficient set of points, i.e. there exists a surjection $\mathscr{S}/K \longrightarrow \mathscr{E}$ for some K in $\mathscr{S}$.*

Proof. Let K be a set of points of $\mathscr{E}$ as in 7.16, and let f be a morphism of $\mathscr{E}$ such that $p^*(f)$ is iso for all $p \in K$. Let q be any point of $\mathscr{E}$; then by expressing q as a filtered colimit of points $p_i \in K$, we may express $q^*(f)$ as the colimit of the isomorphisms $p_i^*(f)$, and hence $q^*(f)$ must be iso. So if $\mathscr{E}$ has enough points, f must be an isomorphism in $\mathscr{E}$. ■

7.18 COROLLARY. *Let $\mathscr{E}$ be a Grothendieck topos. Then there exists a unique smallest topology j in $\mathscr{E}$ such that* $\mathrm{sh}_j(\mathscr{E})$ *has enough points.*

Proof. Let K be a set of points of $\mathscr{E}$ as in 7.16, and define j so that $\mathrm{sh}_j(\mathscr{E})$ is the image of the corresponding geometric morphism $\mathscr{S}/K \longrightarrow \mathscr{E}$ (cf. 4.14). Then if p is any point of $\mathscr{E}$, we can express p^* as a (filtered) colimit of functors each of which inverts j-dense monomorphisms; hence p^* inverts j-dense monos, i.e. p factors through $\mathrm{sh}_j(\mathscr{E})$. Hence any geometric morphism from a copower of $\mathscr{S}$ to $\mathscr{E}$ factors through $\mathrm{sh}_j(\mathscr{E})$.

Now let $\mathrm{sh}_k(\mathscr{E})$ be any sheaf subtopos of $\mathscr{E}$ with enough points, and let L be a sufficient set of points of $\mathrm{sh}_k(\mathscr{E})$. Then we can describe $\mathrm{sh}_k(\mathscr{E})$ as the image of a geometric morphism $\mathscr{S}/L \longrightarrow \mathscr{E}$; but this morphism must factor through $\mathrm{sh}_j(\mathscr{E})$, so $\mathrm{sh}_k(\mathscr{E}) \subseteq \mathrm{sh}_j(\mathscr{E})$, or equivalently $j \leqslant k$. ■

7.2. SPATIAL TOPOSES

Let $(X, \mathbf{U})$ be a topological space. In this paragraph we shall consider the extent to which $(X, \mathbf{U})$ can be recovered from the topos $\mathrm{Shv}(X, \mathbf{U})$. We begin be recalling some definitions from general topology.

7.21 DEFINITION. Let $(X, \mathbf{U})$ be a topological space.

(i) A closed set $C \subseteq X$ is said to be *irreducible* if it is not expressible as the union of two proper closed subsets. (For example, if x is a point of X, the closure of the singleton set $\{x\}$ is irreducible.)
(ii) X is said to be *sober* if each irreducible closed $C \subseteq X$ has exactly one *generic point* (i.e. a point $x \in C$ such that $C = \overline{\{x\}}$). ∎

If we regard two distinct points having the same closure as an instance of double vision (and an irreducible closed set with no generic point as a species of pink elephant!), then the reason for the term "sober space" will be apparent. Note that any Hausdorff space is sober (the irreducible closed sets being single points), as is the spectrum of any commutative ring with the Zariski topology (cf. [AG], p. 22).

7.22 PROPOSITION. *Let* **sob** *denote the full subcategory of* **esp** *whose objects are sober spaces. Then the inclusion functor* **sob** $\longrightarrow$ **esp** *has a left adjoint. Moreover, the functor* **esp** $\longrightarrow \mathfrak{Top}$ *of* 1.17(i) *factors through this left adjoint.*

Proof. Let $(X, \mathbf{U})$ be any space, and define $\hat{X}$ to be the set of irreducible closed subsets of X. If $C \subseteq X$ is closed, define $\hat{C} \subseteq \hat{X}$ to be the set $\{F \in \hat{X} \mid F \subseteq C\}$. It is easy to check that $(C_1 \cup C_2)\hat{=} \hat{C}_1 \cup \hat{C}_2$ and $(\bigcap_\alpha C_\alpha)\hat{=} \bigcap_\alpha \hat{C}_\alpha$, so the $\hat{C}$'s are the closed sets of a topology $\hat{\mathbf{U}}$ on $\hat{X}$. Moreover, as a lattice $\hat{\mathbf{U}}$ is isomorphic to $\mathbf{U}$; so the irreducible closed subsets of $\hat{X}$ are precisely $\{\hat{F} \mid F \in \hat{X}\}$. Hence $(\hat{X}, \hat{\mathbf{U}})$ is sober. Now we have a continuous map $\eta: X \longrightarrow \hat{X}$ given by $x \longmapsto \overline{\{x\}}$; and if $f: X \longrightarrow Y$ is any continuous map of X into a sober space, then we can define $\hat{f}: \hat{X} \longrightarrow Y$ by sending $F \in \hat{X}$ to the unique generic point of $\overline{f(F)}$ (which is easily seen to be irreducible in Y), and $\hat{f}$ is the unique continuous map such that $\hat{f}\eta = f$. So the adjunction is established.

Finally, we observe that η induces an isomorphism $\mathrm{Shv}(X, \mathbf{U}) \cong \mathrm{Shv}(\hat{X}, \hat{\mathbf{U}})$, since the definition of $\mathrm{Shv}(X, \mathbf{U})$ depends only on the lattice structure of $\mathbf{U}$. ∎

7.23 DEFINITION. Let $(X, \mathbf{U})$ be a sober space. We define a partial order on the points of X by

$$x \leqslant y \text{ iff } x \in \overline{\{y\}} \text{ (equivalently, } \overline{\{x\}} \subseteq \overline{\{y\}}).$$

If $x \leqslant y$, we say that x is a *specialization* of y, or that y is a *generalization* of x.

Similarly, we partially order the set of continuous maps $Y \longrightarrow X$ for any space Y, by

$$f \leqslant g \text{ iff } f(y) \leqslant g(y) \text{ for all } y \in Y. \blacksquare$$

7.24 THEOREM. *Let X and Y be spaces, X sober. Then the category $\mathfrak{Top}/\mathscr{S}$ (Shv (Y), Shv (X)) is equivalent to the poset of continuous maps $Y \longrightarrow X$, ordered as in 7.23.*

Proof. Consider first the case when Y is a single point, so that $\text{Shv}(Y) \simeq \mathscr{S}$.

Let $\mathscr{S} \xrightarrow{p} \text{Shv}(X)$ be a point of Shv(X). Let U_p denote the union of all open sets $U \subseteq X$ such that $p^*(U) = 0$; then since p^* preserves colimits we have $p^*(U_p) = 0$. Moreover, if V_1, V_2 are open sets strictly containing U_p, then $p^*(V_1) = p^*(V_2) = 1$; hence $p^*(V_1 \cap V_2) = 1$ by left exactness of p^*, so $V_1 \cap V_2 \neq U_p$. But this says that the closed set $X - U_p$ is irreducible; let x be its generic point. Then it is clear that the functors p^* and $(-)_x$ agree on all subobjects of 1 in Shv(X); but since they both preserve colimits, this means that they agree everywhere. So p is determined by a point of X, i.e. a continuous map $Y \longrightarrow X$.

Now let p, q be two points of Shv(X), and suppose we have a natural transformation $\eta: p \longrightarrow q$. Then we have a map $p^*(U_q) \longrightarrow q^*(U_q) = 0$, so $p^*(U_q) = 0$; hence $U_q \subseteq U_p$, i.e. $p \leqslant q$ in the ordering of 7.23. Conversely, if $p \leqslant q$, then for any $U \rightarrowtail 1$ we have either $p^*(U) = 0$ or $q^*(U) = 1$, so there is a unique map $p^*(U) \longrightarrow q^*(U)$ (which is clearly natural in U). But by 7.13 p^* is isomorphic to the left Kan extension of its restriction to the poset of subobjects of 1; so this map extends uniquely to a natural transformation $p^* \longrightarrow q^*$.

Now consider the case of a general space Y and a morphism $\text{Shv}(Y) \xrightarrow{f} \text{Shv}(X)$. Then each point y of Y gives rise to a point $\mathscr{S} \xrightarrow{y} \text{Shv}(Y) \xrightarrow{f} \text{Shv}(X)$ of Shv(X), i.e. to a point $g(y)$ (say) of X. So f determines a function from the set Y to the set X; and if U is any open subset of X, we have

$$g^{-1}(U) = \{y \in Y \mid (f^*U)_y = 1\} = f^*U.$$

So g is continuous, and the geometric morphism it induces is isomorphic to f. The argument for natural transformations is similar. $\blacksquare$

7.25 THEOREM. *Let $\mathscr{E}$ be a topos defined over $\mathscr{S}$, which has enough points and* *satisfies* (SG) (5.31). *Then there exists a topological space* $(X, \mathbf{U})$ *such that* $\mathscr{E}$ *is equivalent to* $\mathrm{Shv}(X, \mathbf{U})$.

Proof. By 5.37, we already know that $\mathscr{E} \simeq \mathrm{Shv}(\mathbf{H}, C)$, where $\mathbf{H}$ is a complete Heyting algebra in $\mathscr{S}$ (namely $\mathbf{H} = \gamma_*(\mathbf{\Omega}_{\mathscr{E}})$) and C is the canonical toplogy on $\mathbf{H}$. So it suffices to find a space $(X, \mathbf{U})$ for which $\mathbf{U}$ is lattice-isomorphic to $\mathbf{H}$. Let X be a sufficient set of points of $\mathscr{E}$ (which exists by 7.17, since $\mathscr{E}$ is bounded over $\mathscr{S}$); and for any $U \rightarrowtail 1$ in $\mathscr{E}$, define $\phi(U) \subseteq X$ to be the set $\{p \in X \mid p^*(U) = 1\}$. Now since the functors p^* preserve finite limits and arbitrary colimits, it is clear that the map $\phi\colon \gamma_*(\Omega_{\mathscr{E}}) \longrightarrow (\Omega_{\mathscr{S}})^X$ preserves finite intersections and arbitrary unions; so the image of ϕ is a topology ($\mathbf{U}$, say) on X. And since the set X is sufficient, ϕ is a monomorphism; for if $U_1 \neq U_2$, then (without loss of generality) $U_1 \cap U_2 \rightarrowtail U_1$ is not iso, and so there exists $p \in X$ such that $p^*(U_1 \cap U_2) = 0$, $p^*(U_1) = 1$—or in other words, $p \in \phi(U_1)$ but $p \notin \phi(U_2)$. Hence $\mathbf{U} \cong \gamma_*(\mathbf{\Omega}_{\mathscr{E}})$ as a lattice, and so as a Heyting algebra. ■

7.26 Definition. A topos satisfying the hypotheses of 7.25 is said to be *spatial*. We write $\mathfrak{STop}$ for the full sub-2-category of $\mathfrak{Top}/\mathscr{S}$ whose objects are spatial toposes. ■

7.27 Corollary. *There is an equivalence of 2-categories between* $\mathfrak{STop}$ *and* **sob**, *where the latter is made into a 2-category via the partial ordering on hom-sets defined in* 7.23.

Proof. By 7.24 the functor $\mathbf{sob} \longrightarrow \mathfrak{STop}$; $(X, \mathbf{U}) \longmapsto \mathrm{Shv}(X, \mathbf{U})$ is a full embedding; but by 7.25 and 7.22 it is essentially surjective on objects. ■

7.3. COHERENT TOPOSES

In this paragraph we introduce an important "smallness condition" for Grothendieck toposes, which (as we shall see in § 7.4) has close links with the notion of geometric theory discussed in the last chapter.

7.31 Deninition. Let $\mathscr{E}$ be a topos, X an object of $\mathscr{E}$.

(i) We say X is *compact* if any epimorphic family $(Y_i \xrightarrow{f_i} X)$ with codomain X contains a finite epimorphic subfamily.

(ii) We say X is *stable* if, whenever we have morphisms $S \longrightarrow X \longleftarrow T$ with S, T compact, the pullback $S \times_X T$ is compact.

(iii) We say X is *coherent* if it is both compact and stable. ■

7.32 EXAMPLES. Let $\mathscr{E} = \mathrm{Shv}(X)$, X a topological space. Since the image of any local homeomorphism is an open set, it is easily seen that an object of $\mathscr{E}$ is compact iff the total space $\mathscr{E}$ of the corresponding local homeomorphism $E \xrightarrow{p} X$ is compact in the topological sense. Similarly if E is a Hausdorff space, then the corresponding object of $\mathscr{E}$ is stable; but in this case the converse need not be true. For example in $\mathrm{Shv}(\mathbb{R})$ the only compact object is the initial object, and so every object is stable.

(ii) Let $\mathscr{E} = \mathscr{S}^{\mathbf{G}}$, G a group. Since any sub-G-set of a G-set M is expressible as a union of G-orbits of M, it is easily seen that M is compact iff the set $\varinjlim_{\mathbf{G}}(M)$ of G-orbits of M is finite. In particular, G itself is compact when regarded as a G-set. Now if X is any G-set, we can regard an element $x \in X$ as a G-equivariant map $G \longrightarrow X$; and then the G-orbits of the pullback $G \times_X G$ are in 1–1 correspondence with elements of the stabilizer G_x of x in G. So if X is stable, then G_x must be finite for all $x \in X$; but in fact this condition is easily seen to be sufficient as well as necessary. For example if $G = \mathbb{Z}$, a G-set is stable iff it is torsion-free, since every nonzero subgroup of G is infinite. But if $G = C_{(p^\infty)}$, then every proper subgroup of G is finite, so a G-set X is stable iff it has no G-fixed points, i.e. iff $\varprojlim_{\mathbf{G}}(X) = 0$. ■

7.33 LEMMA. *Let* $(\mathbf{C}, J)$ *be a site,* U *an object of* $\mathbf{C}$. *Then the associated sheaf* $l(U)$ *of* h_U *is compact in* $\mathrm{Shv}(\mathbf{C}, J)$ *iff every* J*-covering sieve on* U *contains a finite subfamily* $\{V_i \xrightarrow{\alpha_i} U \mid i = 1, 2, \ldots, n\}$ *such that the sieve generated by the* α_i *is still* J*-covering.*

Proof. Suppose the given condition is satisfied, and let $\{Y_i \xrightarrow{f_i} l(U) \mid i \in I\}$ be an epimorphic family. Then the sieve

$$\{V \xrightarrow{\alpha} U \mid l(\alpha) \text{ factors through } f_i \text{ for some } i\}$$

must be J-covering; let $\alpha_1, \alpha_2, \ldots, \alpha_n$ be a finite subfamily generating a covering sieve. Choose one f_i corresponding to each of the α_j; these must form an epimorphic subfamily of $\{f_i \mid i \in I\}$.

Conversely, suppose $l(U)$ is compact. Then if R is any J-covering sieve on U, the family $\{l(\alpha) \mid \alpha \in R\}$ is epimorphic in $\mathrm{Shv}(\mathbf{C}, J)$; and if $\{l(\alpha_1), \ldots, l(\alpha_n)\}$ is a finite epimorphic subfamily, the sieve generated by $\{\alpha_1, \ldots, \alpha_n\}$ must be J-covering. ■

7.34 Lemma. (i) *Let X, Y be objects of a topos $\mathscr{E}$. Then $X \amalg Y$ is compact iff both X and Y are.*

(ii) *An epimorphic image of a compact object is compact.*

(iii) *A coproduct of stable objects is stable.*

(iv) *A subobject of a stable object is stable.*

Proof. (i) Suppose $X \amalg Y$ is compact, and let $\{Z_i \xrightarrow{f_i} X\}$ be an epimorphic family. Then the family formed by the $Z_i \xrightarrow{\nu_1 f_i} X \amalg Y$ together with $Y \xrightarrow{\nu_2} X \amalg Y$ is epimorphic, so it has a finite epimorphic subfamily. On discarding ν_2 again, we are left with a finite epimorphic subfamily of the f_i.

Conversely, suppose X and Y are compact and we are given an epimorphic family $\{Z_i \xrightarrow{f_i} X \amalg Y \mid i \in I\}$. Let J, K be finite subsets of I indexing epimorphic subfamilies of $\{\nu_1^*(f_i) \mid i \in I\}$ and $\{\nu_2^*(f_i) \mid i \in I\}$; then $J \cup K$ indexes a finite epimorphic subfamily of the f_i.

(ii) Suppose we are given $X \xrightarrow{f} Y$ (epi) with X compact, and an epimorphic family $\{Z_i \xrightarrow{g_i} Y \mid i \in I\}$. Then the family $\{f^*(g_i) \mid i \in I\}$ is also epimorphic, so it has a finite epimorphic subfamily (indexed by $i_1, \ldots, i_n$, say). Now suppose $Y \underset{k}{\overset{h}{\rightrightarrows}} T$ are equalized by each of $g_{i_1}, \ldots, g_{i_n}$; then they are also equalized by f, but f is epi and so $h = k$.

(iii) Suppose X and Y are stable, and we are given $S \longrightarrow X \amalg Y \longleftarrow T$ with S, T compact. Writing S_i for $\nu_i^*(S)$, etc., we have $S \cong S_1 \amalg S_2$ and $T \cong T_1 \amalg T_2$; so by (i) the objects S_i and T_i are compact. Hence $S_1 \times_X T_1$ and $S_2 \times_Y T_2$ are compact; but $S \times_{(X \amalg Y)} T$ is isomorphic to the coproduct of these two objects.

This proves the result for finite coproducts; that for infinite coproducts follows, since if we have a morphism $S \longrightarrow \coprod_{i \in I} X_i$ with S compact, there can be only finitely many $i \in I$ such that $\nu_i^*(S)$ is nonzero.

(iv) Suppose we have

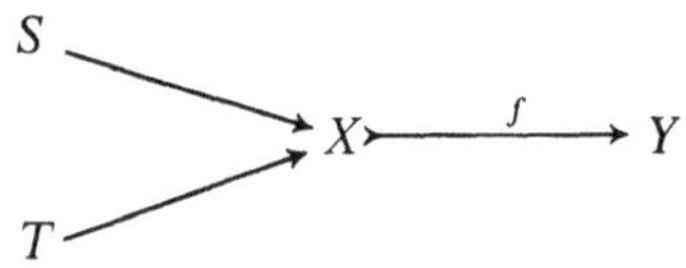

with S, T compact, Y stable. Since f is mono, we have $S \times_X T \cong S \times_Y T$ which is compact, so X is stable. ■

7.35 Theorem (Grothendieck). *Let* $(\mathbf{C}, J)$ *be a site such that* $\mathbf{C}$ *has finite limits and J is generated by a pretopology* (0.31) *in which every covering family is* *finite. Then for every $U \in C_0$, the associated sheaf $l(U)$ of h_U is coherent in the topos* Shv(**C**, *J*). *A topos which possesses a site of definition satisfying these conditions is said to be* coherent.

Proof. By 7.33 we already know $l(U)$ is compact. So suppose we have $X \xrightarrow{f} l(U) \xleftarrow{g} Y$ in Shv(**C**, J) with X, Y compact. Now X is expressible as a colimit of associated sheaves of representables, so by compactness we can find a finite epimorphic family

$$\{l(V_i) \xrightarrow{p_i} X \mid i = 1, \ldots, n\}.$$

Similarly, we can find a finite epimorphic family

$$\{l(W_j) \xrightarrow{q_j} Y \mid j = 1, \ldots, m\}.$$

If the topology J is not sub-canonical, then the composite morphisms $l(V_i) \xrightarrow{fp_i} l(U)$ and $l(W_j) \xrightarrow{gq_j} l(U)$ may not derive from morphisms $V_i \xrightarrow{\alpha_i} U$ and $W_j \xrightarrow{\beta_j} U$ in **C**; but they do so locally, i.e. at the cost of replacing each V_i and W_j by a further (finite) epimorphic family, we may assume that they do. (Compare the construction of the associated sheaf functor in § 3.3).

Now we have $l(V_i) \times_{l(U)} l(W_j) \cong l(V_i \times_U W_j)$ which is compact by 7.33; but the finite family

$$\{l(V_i) \times_{l(U)} l(W_j) \xrightarrow{p_i \times q_j} X \times_{l(U)} Y \mid i = 1, \ldots, n,\ j = 1, \ldots, m\}$$

is epimorphic, so by 7.34(i) and (ii) $X \times_{l(U)} Y$ is compact. Thus $l(U)$ is stable. ■

Even in a coherent topos, the class of coherent objects is not closed under the formation of arbitrary coequalizers (see Ex. 7.8 below). However, it is closed under the formation of coequalizers of equivalence relations, as the following result shows:

7.36 LEMMA. *Let $\mathscr{E}$ be a topos generated by compact objects, X a coherent object of $\mathscr{E}$ and $R \rightrightarrows X$ an equivalence relation with coequalizer $X \xrightarrow{f} Y$. Then Y is stable iff R is compact.*

Proof. If Y is stable, then $R \cong X \times_Y X$ is trivially compact.

Conversely, suppose R is compact and we are given $S \longrightarrow Y \longleftarrow T$ with S, T compact. Since $\mathscr{E}$ is generated by compact objects, we can find an epimorphic family $\{K_\alpha \xrightarrow{\iota_\alpha} S \times_Y X \mid \alpha \in A\}$ with each K_α compact. And since f is epi, the composites $K_\alpha \xrightarrow{\iota_\alpha} S \times_Y X \overset{\pi_1}{\twoheadrightarrow} S$ form an epimorphic family; let $\{\alpha_1, \ldots, \alpha_n\}$ index a finite epimorphic subfamily. Now $K_\alpha \times_Y X \cong K_\alpha \times_X R$ which is compact since R is compact and X is stable; and the family

$$\{K_{\alpha_i} \times_Y X \xrightarrow{\pi_1 \iota_{\alpha_i} \times 1} S \times_Y X \mid i = 1, \ldots, n\}$$

is epimorphic. Hence by 7.34(i) and (ii) $S \times_Y X$ is compact; and similarly $T \times_Y X$ is compact. Now $(S \times_Y T) \times_Y X \cong (S \times_Y X) \times_X (T \times_Y X)$ is compact, since X is stable; but since $X \overset{f}{\twoheadrightarrow} Y$ is epi, $S \times_Y T$ is an epimorphic image of $(S \times_Y T) \times_Y X$. So by 7.34(ii) $S \times_Y T$ is compact. ■

7.37 PROPOSITION. *Let $\mathscr{E}$ be a coherent topos, and let $\mathscr{E}_{\mathrm{coh}}$ denote the full subcategory of coherent objects of $\mathscr{E}$. Then $\mathscr{E}_{\mathrm{coh}}$ has the following properties:*

(i) *$\mathscr{E}_{\mathrm{coh}}$ has finite limits.*
(ii) *$\mathscr{E}_{\mathrm{coh}}$ has finite coproducts, which are disjoint and universal* (*cf.* 0.42).
(iii) *$\mathscr{E}_{\mathrm{coh}}$ has coequalizers of equivalence relations, and they are universal.*
(iv) *Every equivalence relation in $\mathscr{E}_{\mathrm{coh}}$ is effective, and every epimorphism in $\mathscr{E}_{\mathrm{coh}}$ is a coequalizer.*
(v) *$\mathscr{E}_{\mathrm{coh}}$ is* essentially small, *i.e. equivalent to a small category.*

Proof. Let $(\mathbf{C}, J)$ be a site of definition for $\mathscr{E}$ as in 7.35, and let $\mathbf{C} \xrightarrow{l} \mathscr{E}$ denote the composite of the Yoneda embedding and the associated sheaf functor, as usual.

(i) The terminal object of $\mathscr{E}$ is certainly in $\mathscr{E}_{\mathrm{coh}}$, since it is in the image of l; so it suffices to prove that $\mathscr{E}_{\mathrm{coh}}$ has pullbacks. Let

$$\begin{array}{ccc} P & \longrightarrow & X \\ \downarrow & & \downarrow \\ Y & \longrightarrow & Z \end{array}$$

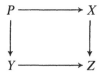

be a pullback diagram in $\mathscr{E}$, and suppose first that X, Y and Z are compact. Then we may find a finite epimorphic family $\{l(U_i) \longrightarrow Z \mid i = 1, \ldots, n\}$; and by expressing the pullbacks $l(U_i) \times_Z X$ and $l(U_i) \times_Z Y$ as colimits of representables, we may find for each i a *finite* number of commutative squares

$$\begin{array}{ccc} l(V_{ij}) & \longrightarrow & X \\ \downarrow & & \downarrow \\ l(U_i) & \longrightarrow & Z \end{array}, \qquad \begin{array}{ccc} l(W_{ik}) & \longrightarrow & Y \\ \downarrow & & \downarrow \\ l(U_i) & \longrightarrow & Z \end{array} \quad (j = 1, \ldots, m_i,\ k = 1, \ldots, p_i)$$

such that the families

$$\{l(V_{ij}) \longrightarrow X \mid i = 1, \ldots, n, j = 1, \ldots, m_i\}$$

and

$$\{l(W_{ik}) \longrightarrow Y \mid i = 1, \ldots, n, k = 1, \ldots, p_i\}$$

are epimorphic. Then the family

$$\{l(V_{ij}) \times_{l(U_i)} l(W_{ik}) \longrightarrow P \mid i = 1, \ldots, n, j = 1, \ldots, m_i,\ k = 1, \ldots, p_i\}$$

is clearly epimorphic; but the $l(V_{ij}) \times_{l(U_i)} l(W_{ik})$ are compact by 7.35, so P is compact.

Now suppose that X, Y and Z are stable, and we are given $S \longrightarrow P \longleftarrow T$ with S, T compact. Then it is easily seen that the square

$$\begin{array}{ccc} S \times_P T & \rightarrowtail & S \times_X T \\ \downarrow & & \downarrow \\ S \times_Y T & \rightarrowtail & S \times_Z T \end{array}$$

is a pullback; so by the argument just given, $S \times_P T$ is compact. Hence P is stable.

(ii) Existence of finite coproducts follows from 7.34(i) and (iii); their disjointness and universality follow from the corresponding properties in $\mathscr{E}$ (1.51 and 1.57).

(iii) Similarly, existence of coequalizers for equivalence relations follows from 7.36; their universality from 1.51.

(iv) Again, effectiveness of equivalence relations follows from 1.23, since we know their coequalizers are in $\mathscr{E}_{\text{coh}}$; the second assertion will follow from 1.53 provided we show that every epi in $\mathscr{E}_{\text{coh}}$ is epi in $\mathscr{E}$.

Let $X \xrightarrow{f} Y$ be an epi in $\mathscr{E}_{\text{coh}}$, and let $Y \rightrightarrows Q$ be its cokernel-pair in $\mathscr{E}$. Then the kernel-pair of $Y \amalg Y \twoheadrightarrow Q$ is isomorphic to $Y \amalg I \amalg I \amalg Y$, where I is the image of f in $\mathscr{E}$ (cf. 1.55); but I is compact by 7.34(ii), so Q is stable and hence coherent by 7.36. Hence the two morphisms $Y \rightrightarrows Q$ are in $\mathscr{E}_{\text{coh}}$; so they must be equal, i.e. f must be epi in $\mathscr{E}$.

(v) It follows from 7.36 that every coherent object in $\mathscr{E}$ is isomorphic to the canonical coequalizer of some compact equivalence relation on a finite canonical coproduct of objects in the image of l. But there is only a set of such canonical colimits; so we can use them to construct a small full subcategory of $\mathscr{E}_{\text{coh}}$ such that the inclusion functor is an equivalence. ■

7.38 Definition ([GV], VI 3.11). A small category satisfying the first four conditions of 7.37 is called a *pretopos*. If $\mathbf{E}$ is a pretopos, we define the *pre-canonical topology* on $\mathbf{E}$ to be that generated by the pretopology (0.31) whose covering families are all finite epimorphic families in $\mathbf{E}$. (The assumption of universality in 7.37(ii) and (iii) ensures that these families do indeed satisfy the conditions of 0.31.) ■

Note that a pretopos is an exact category in the sense of Barr [2], since we may obtain an image factorization of an arbitrary morphism by the method of 1.52. Conversely, a (small) exact category $\mathbf{E}$ is a pretopos iff $\mathbf{E}$ has finite coproducts which are disjoint and universal, and every epimorphism in $\mathbf{E}$ is a coequalizer. And since epis in a pretopos are coequalizers, it is easily verified that the precanonical topology is sub-canonical; in fact it may be characterized as the largest sub-canonical topology on $\mathbf{E}$ for which every representable presheaf is compact.

7.39 Lemma. *Let* $\mathbf{E}$ *be a pretopos,* P *the precanonical topology on* $\mathbf{E}$. *Then the Yoneda embedding* $\mathbf{E} \xrightarrow{h} \mathrm{Shv}(\mathbf{E}, P)$ *preserves finite coproducts and co-equalizers of equivalence relations.*

Proof. First note that the initial object of $\mathbf{E}$ (i.e. the empty coproduct) is P-covered by the empty sieve; and so h preserves it.

Now let $U \xrightarrow{\alpha} W \xleftarrow{\beta} V$ be a coproduct diagram in $\mathbf{E}$, and let $R \rightarrowtail h_W$ be the (clearly P-covering) sieve generated by α and β. Since

$$\begin{array}{ccc} 0 & \longrightarrow & U \\ \big\downarrow & & \big\downarrow{\scriptstyle \alpha} \\ V & \xrightarrow{\ \beta\ } & W \end{array}$$

is a pullback, it is clear that a morphism $R \longrightarrow X$ in $\mathscr{S}^{\mathbf{E}^{op}}$ is determined by an arbitrary pair of morphisms $h_U \longrightarrow X$, $h_V \longrightarrow X$ which agree with restricted to h_0, i.e.

$$\begin{array}{ccc} h_0 & \longrightarrow & h_U \\ \downarrow & & \downarrow \\ h_V & \longrightarrow & R \end{array}$$

is a pushout diagram in $\mathscr{S}^{\mathbf{E}^{op}}$. Applying the associated sheaf functor, we obtain a pushout diagram

$$\begin{array}{ccc} 0 & \longrightarrow & h_U \\ \downarrow & & \downarrow \\ h_V & \longrightarrow & h_W \end{array}$$

in Shv(**E**, P), i.e. h_W is a coproduct of h_U and h_V.

Similarly, if $U \underset{\beta}{\overset{\alpha}{\rightrightarrows}} V \overset{\gamma}{\longrightarrow} W$ is the coequalizer of an equivalence relation in **E**, let $R \rightarrowtail h_W$ be the sieve generated by γ. Then R is P-covering; and by arguments similar to those above it is the coequalizer of $h_U \rightrightarrows h_V$ in $\mathscr{S}^{\mathbf{E}^{op}}$. So $h_U \rightrightarrows h_V \longrightarrow h_W$ is a coequalizer in Shv(**E**, P). ■

7.40 Theorem (Grothendieck). *Let $\mathscr{E}$ be a Grothendieck topos. The following conditions are equivalent:*

(i) *$\mathscr{E}$ is coherent.*
(ii) *There exists a pretopos* **E** *such that $\mathscr{E} \simeq$* Shv(**E**, P), *where P is the precanonical topology on* **E**.

Moreover, the pretopos **E** *is determined up to equivalence by $\mathscr{E}$.*

Proof. (ii) $\Rightarrow$ (i) is immediate, since the site (**E**, P) satisfies the conditions of 7.35.

(i) $\Rightarrow$ (ii): Let **E** be a small full subcategory of the category $\mathscr{E}_{\text{coh}}$ as constructed in 7.37(v). Then **E** is a pretopos by 7.37, so it suffices to show that the topology J on **E** induced by $\mathscr{E}$ is the precanonical one. But it is clear that $J \subseteq P$, since J is sub-canonical and every object of **E** is compact. And if $(V_i \longrightarrow U)$ is a finite epimorphic family in **E**, it is also epimorphic in $\mathscr{E}$ by the argument of 7.37(iv), and so every object of $\mathscr{E}$ satisfies the sheaf axiom for it. Hence $P \subseteq J$.

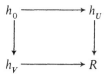

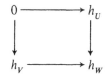

To establish the last sentence of the theorem, we must show that if **E** is a pretopos, then every coherent object in Shv(**E**, P) is isomorphic to a representable sheaf. But if X is coherent in Shv(**E**, P), we have a finite epimorphic family $\{h(U_i) \longrightarrow X \mid i = 1, \ldots, n\}$, and hence an epimorphism

$$h\left(\coprod_{i=1}^{n} U_i\right) \cong \coprod_{i=1}^{n} h(U_i) \overset{f}{\twoheadrightarrow} X.$$

Write U for $\coprod U_i$, and $R \underset{b}{\overset{a}{\rightrightarrows}} h(U)$ for the kernel-pair of f; then R is compact by 7.36, and so we can find another epimorphism $h(V) \overset{g}{\twoheadrightarrow} R$. And the kernel-pair of g is equal to the kernel-pair of the composite

$$h(V) \overset{g}{\twoheadrightarrow} R \overset{(a,\, b)}{\rightarrowtail} h(U) \times h(U) \cong h(U \times U);$$

but this composite is derived from a morphism $V \longrightarrow U \times U$ in **E**, since h is full and faithful. Thus the kernel-pair of g is the image of h; and hence so is R, and X itself. So **E** is equivalent to Shv(**E**, P)$_{\text{coh}}$, and is thus determined up to equivalence by $\mathscr{E}$. ■

7.4. DELIGNE'S THEOREM

We shall now prove the important theorem of P. Deligne ([GV], VI 9.0) which states that every coherent topos has enough points. Let (**C**, J) be a site satisfying the hypotheses of 7.35; we shall consider the class of pseudo-points of **C** (7.15) which are defined over filtered *posets*. If $P = (U_i \mid i \in \mathbf{I})$ and $Q = (V_k \mid k \in \mathbf{K})$ are two such pseudo-points, we shall say that Q is a *refinement* of P if there is an order-preserving monomorphism $I \rightarrowtail K$ such that

$$\begin{array}{ccc} \mathbf{I} & \longrightarrow & \mathbf{C}^{\mathrm{op}} \\ \downarrow & \nearrow & \\ \mathbf{K} & & \end{array}$$

commutes. Our aim is then to show that any pseudo-point of **C** has a "suitable" refinement which is actually a pseudo-point of (**C**, J). For convenience, if $P = (U_i \mid \in \mathbf{I})$ is a pseudo-point, we shall use the letter P also for the inverse image functor

$$\mathscr{S}^{\mathbf{C}^{\mathrm{op}}} \longrightarrow \mathscr{S};\ X \longmapsto \varinjlim_{\mathbf{I}} (X(U_i))$$

which it defines.

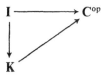

The essential inductive step of the construction is contained in

7.41 LEMMA. *Let $P = (U_i \mid i \in \mathbf{I})$ be a pseudo-point of* $\mathbf{C}$, *X a J-sheaf on* $\mathbf{C}$ *and x, y two distinct elements of $P(X)$. Let $(V_j \longrightarrow V \mid j = 1, \ldots, n)$ be a finite J-covering family in* $\mathbf{C}$, *and v an element of $P(h_V)$. Then there exists a refinement Q of P such that* (a) *the images of x and y under the natural map $P(X) \longrightarrow Q(X)$ are distinct, and* (b) *the image of v in $Q(h_V)$ is in* $\bigcup_{j=1}^{n} \operatorname{im}(Q(h_{V_j}) \longrightarrow Q(h_V))$.

Proof. By definition, v derives from a morphism $U_{i_0} \longrightarrow V$ for some $i_0 \in I$. For each $i \geqslant i_0$ and each j, form the pullback

$$\begin{array}{ccc} U_{ij} & \longrightarrow & V_j \;; \\ \downarrow & & \downarrow \\ U_i \longrightarrow U_{i_0} & \longrightarrow & V \end{array}$$

then the family $(U_{ij} \longrightarrow U_i \mid j = 1, \ldots, n)$ is J-covering, and so $X(U_i) \longrightarrow \prod_{I=1}^{n} X(U_{ij})$ is mono. But filtered colimits preserve finite products and monomorphisms, and so we have a monomorphism

$$P(X) = \varinjlim_{\mathbf{I}} X(U_i) \cong \varinjlim_{(i_0/\mathbf{I})} X(U_i) \rightarrowtail \prod_{j=1}^{n} (\varinjlim_{(i_0/\mathbf{I})} X(U_{ij})).$$

In particular, we can find k such that x and y have distinct images in $\varinjlim_{(i_0/\mathbf{I})} X(U_{ik})$. Now let K be the set $I \times \{0\} \cup (i_0/I) \times \{1\}$, with the product ordering (i.e. $(i, j) \leqslant (i', j')$ iff $i \leqslant i'$ and $j \leqslant j'$); then $\mathbf{K}$ is clearly filtered, and we may define a pseudo-point Q indexed over $\mathbf{K}$ by the assignment

$$(i, 0) \longmapsto U_i, \; (i, 1) \longmapsto U_{ik}.$$

Then for any presheaf Y on $\mathbf{C}$, we clearly have $Q(Y) = \varinjlim_{(i_0/\mathbf{I})} Y(U_{ik})$; so the pseudo-point Q has the required properties. ■

7.42 LEMMA. *Let P, X, x and y be as in 7.41. Then there exists a refinement Q of P such that* (a) *the images of x and y under the natural map $P(X) \longrightarrow Q(X)$ are distinct, and* (b) *for each object V of* $\mathbf{C}$, *each J-covering sieve R on V and each $v \in P(h_V)$, the image of v in $Q(h_V)$ is in the image of $Q(R) \rightarrowtail Q(h_V)$.*

Proof. Let Z be the set of all triples (V, R, v) where $V \in C_0$, R is a finitely-generated J-covering sieve on V, and $v \in P(h_V)$. By choosing a well-ordering of Z, we may index its members by some ordinal α. We now use transfinite induction to define a sequence of pseudo-points $(Q_\beta \mid \beta \leqslant \alpha)$, as follows:

$Q_0 = P$.

If $\beta = \gamma + 1$, then Q_β is obtained from Q_γ by applying 7.41 to a finite generating family $(V_{\gamma j} \longrightarrow V_\gamma \mid j = 1, \ldots, n_\gamma)$ for the sieve R_γ, and to the image of v_γ in $Q_\gamma(V_\gamma)$.

If β is a limit ordinal, then Q_β is the unique common refinement of the $Q_\gamma(\gamma < \beta)$ whose underlying filtered poset is the colimit of those underlying the Q_γ.

It is then easily seen that the images of x and y in $Q_\beta(X)$ are distinct for all β, and that if $\gamma < \beta$ then the image of v_γ in $Q_\beta(h_{V_\gamma})$ is in the image of $Q_\beta(R_\gamma)$. So Q_α is the required refinement of P. ∎

The pseudo-point Q constructed in 7.42 need not satisfy the condition of 7.15, since not every element of $Q(h_V)$ is necessarily in the image of $P(h_V) \longrightarrow Q(h_V)$. However, we can rectify this by a further induction of length ω, as follows:

7.43 Lemma. *Let P, X, x and y be as in 7.41. Then there exists a pseudo-point Q of $(\mathbf{C}, J)$ which is a refinement of P, such that the images of x and y in $Q(X)$ are distinct.*

Proof. Define a sequence of pseudo-points P_n by setting $P_0 = P$, and defining P_{n+1} to be the refinement of P_n constructed using 7.42. Define Q to be the colimit of the P_n; then we clearly have $Q(Y) = \varinjlim P_n(Y)$ for any presheaf Y on $\mathbf{C}$. Hence x and y have distinct images in $Q(X)$; and if $R \rightarrowtail h_V$ is a J-covering sieve, then each element of $Q(h_V)$ derives from an element of $P_n(h_V)$ for some n, and hence from an element of $P_{n+1}(R)$. So by 7.15 Q is a pseudo-point of $(\mathbf{C}, J)$. ∎

7.44 Theorem (Deligne). *A coherent topos has enough points.*

Proof. Let $\mathscr{E} = \mathbf{Shv}(\mathbf{C}, J)$, where $(\mathbf{C}, J)$ is a site satisfying the hypotheses of 7.35. It suffices to show that, given a parallel pair $Y \underset{g}{\overset{f}{\rightrightarrows}} X$ in $\mathscr{E}$ with $f \neq g$, we can find a point q of $\mathscr{E}$ with $q^*(f) \neq q^*(g)$. But we can certainly find $U \in C_0$ and $y \in Y(U)$ with $f_U(y) \neq g_U(y)$; now apply 7.43 to this pair of elements of

$X(U)$, taking P to be the pseudo-point of $\mathbf{C}$ defined by the trivial filtered system (U). Then we obtain a pseudo-point Q of $(\mathbf{C}, J)$, and hence a point q of $\mathscr{E}$, such that the image of y in $q^*(Y)$ has distinct images under $q^*(f)$ and $q^*(g)$. ■

In the same way that 7.16 is a categorical version of the Löwenheim–Skolem theorem, the proof of 7.44 should be reminiscent of the Gödel–Henkin completeness theorem for finitary first-order theories. Indeed, M. Makkai and G. E. Reyes ([MR], [RM], [82]) have shown that the connection between 7.44 and the completeness theorem may be made much more explicit; we shall now sketch the details of their argument.

Let $\mathbb{T} = (\mathbb{L}, A)$ be a finitary geometric theory. In order to talk about a completeness theorem for $\mathbb{T}$, we should first develop a notion of provability within $\mathbb{T}$; that is to say, we should specify a set L of sequents of $\mathbb{L}$, called *logical axioms*, and a set of *rules of inference* whereby sequents may be derived from other sequents. For reasons of space, we shall not give the complete list here (the interested reader will find it in [22]); but we note that the *cut-rule* (from $(\phi \vdash \psi)$ and $(\psi \vdash \chi)$, we may infer $(\phi \vdash \chi)$) is a rule of inference provided every free variable of ψ occurs free in either ϕ or χ. (This corresponds to the restriction on modus ponens which we introduced when discussing the Mitchell–Bénabou language in § 5.4.) We say that a sequent $(\phi \vdash \psi)$ of $\mathbb{L}$ is *provable* in $\mathbb{T}$ if it can be derived from those in $L \cup A$ by a finite number of applications of the rules of inference; and that two formulae ϕ and ψ with the same free variables are *provably equivalent* if the sequents $(\phi \vdash \psi)$ and $(\psi \vdash \phi)$ are both provable in $\mathbb{T}$.

We now proceed to construct a category $\mathbf{C}_{\mathbb{T}}$, called the *syntactic category* of $\mathbb{T}$, as follows: an object of $\mathbf{C}_{\mathbb{T}}$ is an equivalence class of formulae of $\mathbb{L}$ under the relation $\phi \sim \psi$ iff ψ is obtained from ϕ by a (type-preserving) renaming of the variables of ϕ. To define morphisms $\phi \longrightarrow \psi$ in $\mathbf{C}_{\mathbb{T}}$, we first suppose that ϕ and ψ have no free variables in common (say ϕ has $(\mathbf{x}_1, \ldots, \mathbf{x}_n)$ and ψ has $(\mathbf{y}_1, \ldots, \mathbf{y}_m)$), and then define a morphism $\phi \longrightarrow \psi$ to be a provable-equivalence class of formulae θ such that

(a) the free variables of θ are precisely $(\mathbf{x}_1, \ldots, \mathbf{x}_n, \mathbf{y}_1, \ldots, \mathbf{y}_m)$, and
(b) θ is "provably the graph of a function from ϕ to ψ", i.e. the sequents

$$\theta(\mathbf{x}_1, \ldots, \mathbf{x}_n, \mathbf{y}_1, \ldots, \mathbf{y}_m) \vdash \phi(\mathbf{x}_1, \ldots, \mathbf{x}_n) \wedge \psi(\mathbf{y}_1, \ldots, \mathbf{y}_m),$$

$$\phi(\mathbf{x}_1, \ldots, \mathbf{x}_n) \vdash \exists \mathbf{y}_1 \ldots \exists \mathbf{y}_m \; \theta(\mathbf{x}_1, \ldots, \mathbf{x}_n, \mathbf{y}_1, \ldots, \mathbf{y}_m), \quad \text{and}$$

$$\theta(\mathbf{x}_1, \ldots, \mathbf{x}_n, \mathbf{y}_1, \ldots, \mathbf{y}_m) \wedge \theta(\mathbf{x}_1, \ldots, \mathbf{x}_n, \mathbf{y}'_1, \ldots, \mathbf{y}'_m) \vdash \bigwedge_{i=1}^{m} (\mathbf{y}_i = \mathbf{y}'_i)$$

are provable in $\mathbb{T}$.

[To avoid confusion, we shall write $\{\phi\}$ for the equivalence class of the formula ϕ considered as an object of $\mathbf{C}_{\mathbb{T}}$, and $[\phi]$ for the equivalence class of ϕ considered as a morphism.]

We define the composite of two morphisms

$$\{\phi(\mathbf{x})\} \xrightarrow{[\theta(\mathbf{x},\mathbf{y})]} \{\psi(\mathbf{y})\} \xrightarrow{[\eta(\mathbf{y},\mathbf{z})]} \{\chi(\mathbf{z})\}$$

in $\mathbf{C}_{\mathbb{T}}$ to be $[\exists\mathbf{y}(\theta(\mathbf{x},\mathbf{y}) \wedge \eta(\mathbf{y},\mathbf{z}))]$; and the identity morphism on $\{\phi(\mathbf{x})\}$ is

$$\{\phi(\mathbf{x})\} \xrightarrow{[\mathbf{x} = \mathbf{x}' \wedge \phi(\mathbf{x})]} \{\phi(\mathbf{x}')\}$$

It is straightforward to check that these definitions make $\mathbf{C}_{\mathbb{T}}$ into a category. Moreover, $\mathbf{C}_{\mathbb{T}}$ has finite limits; the object $\{true\}$ is terminal (since for any ϕ, $[\phi]$ is easily seen to be the unique morphism from $\{\phi\}$ to $\{true\}$), and the pullback of

$$\begin{array}{ccc} & & \{\psi(\mathbf{y})\} \\ & & \downarrow{\scriptstyle [\eta(\mathbf{y},\mathbf{z})]} \\ \{\phi(\mathbf{x})\} & \xrightarrow{[\theta(\mathbf{x},\mathbf{z})]} & \{\chi(\mathbf{z})\} \end{array}$$

is the object $\{\exists\mathbf{z}(\theta(\mathbf{x},\mathbf{z}) \wedge \eta(\mathbf{y},\mathbf{z}))\}$. (In fact $\mathbf{C}_{\mathbb{T}}$ is a regular category with universal finite unions of subobjects; but we shall not need this extra information.)

We now define a family of morphisms

$$(\{\phi_i(\mathbf{x}_i)\} \xrightarrow{[\theta_i(\mathbf{x}_i,\mathbf{y})]} \{\psi(\mathbf{y})\} \,|\, i = 1, \ldots, n)$$

to be *provably epimorphic* if the sequent

$$\psi(\mathbf{y}) \vdash \bigvee_{i=1}^{n} \exists\mathbf{x}_i \theta_i(\mathbf{x}_i, \mathbf{y})$$

is provable in $\mathbb{T}$. It is again straightforward to verify that the (finite) provably epimorphic families form a pretopology on $\mathbf{C}_{\mathbb{T}}$; we write $J_{\mathbb{T}}$ for the topology which it generates. Note that $J_{\mathbb{T}}$ is sub-canonical; for if we have a compatible family of morphisms

$$\{\phi_i(\mathbf{x}_i)\} \xrightarrow{[\eta_i(\mathbf{x}_i, \mathbf{z})]} \{\chi(\mathbf{z})\}$$

relative to the probably epimorphic family $([\theta_i(\mathbf{x}_i, \mathbf{y})])$, then any factorization of the η_i through the θ_i is provably equivalent to

$$\left[\bigvee_{i=1}^{n} \exists \mathbf{x}_i(\theta_i(\mathbf{x}_i, \mathbf{y}) \wedge \eta_i(\mathbf{x}_i, \mathbf{z}))\right].$$

In fact if $\mathbf{C}_{\mathbb{T}}$ happens to be a pretopos, $J_{\mathbb{T}}$ is simply the precanonical topology as defined in 7.38.

7.45 THEOREM (Joyal–Reyes). *Let* $\mathbb{T}$ *be a finitary geometric theory. Then the topos* Shv$(\mathbf{C}_{\mathbb{T}}, J_{\mathbb{T}})$ *is a classifying topos for* $\mathbb{T}$*-models in Grothendieck toposes.*

Proof. Let $\mathscr{E}$ be a Grothendieck topos, and M a $\mathbb{T}$-model in $\mathscr{E}$. The construction which we gave in §6.5 for interpreting any formula ϕ of $\mathbb{L}$ by an object M_ϕ of $\mathscr{E}$ in fact defines a functor $\mathbf{C}_{\mathbb{T}} \longrightarrow \mathscr{E}$. (The proof of this fact requires a soundness theorem for $\mathbb{T}$, i.e. a proof that a sequent which is provable in $\mathbb{T}$ is satisfied in every model of $\mathbb{T}$; but this is a straightforward induction.) Moreover. this functor is left exact and transforms provably epimorphic families in $\mathbf{C}_{\mathbb{T}}$ to epimorphic families in $\mathscr{E}$; so by 7.13 it defines a geometric morphism $\mathscr{E} \longrightarrow$ Shv$(\mathbf{C}_{\mathbb{T}}, J_{\mathbb{T}})$. Conversely, suppose we are given a left exact continuous functor $T: \mathbf{C}_{\mathbb{T}} \longrightarrow \mathscr{E}$. Then we may define an interpretation M of $\mathbb{L}$ in $\mathscr{E}$ by setting

$$M_X = T(\{\mathbf{x} = \mathbf{x}\}),\ M_\alpha = T([\alpha(\mathbf{x}_1, \ldots, \mathbf{x}_n) = \mathbf{y}])$$
$$\text{and} \quad M_r = T(\{r(\mathbf{x}_1, \ldots, \mathbf{x}_n)\});$$

and a straightforward inductive argument then shows that $M_\phi \cong T(\{\phi\})$ for any formula ϕ. In particular, this implies that M is a model of $\mathbb{T}$; and the functor $\mathbf{C}_{\mathbb{T}} \longrightarrow \mathscr{E}$ which it induces is naturally isomorphic to T. In this way we obtain an equivalence of categories $\mathbb{T}(\mathscr{E}) \simeq \mathfrak{Top}/\mathscr{S}(\mathscr{E}, \text{Shv}(\mathbf{C}_{\mathbb{T}}, J_{\mathbb{T}}))$. ■

At first sight, 7.45 might not seem to have any significant advantage over the "semantic" construction of the classifying topos for $\mathbb{T}$ given in 6.56. True, it shows that $\mathcal{S}[\mathbb{T}]$ is a coherent topos; but in fact this information could also be obtained from a detailed analysis of the "forcing topologies" used in the proof of 6.56. The real advantage of the syntactic construction of $\mathcal{S}[\mathbb{T}]$ is that it enables us to deduce the completeness theorem for finitary geometric theories as an immediate consequence of 7.44:

7.46 COROLLARY. *Let $\mathbb{T}$ be a finitary geometric theory. Then a sequent $(\phi \vdash \psi)$ is probable in $\mathbb{T}$ iff it is satisfied in every $\mathbb{T}$-model in $\mathcal{S}$.*

Proof. Let M denote the generic $\mathbb{T}$-model in $\mathrm{Shv}(\mathbf{C}_{\mathbb{T}}, J_{\mathbb{T}})$. Since the left exact continuous functor $\mathbf{C}_{\mathbb{T}} \longrightarrow \mathrm{Shv}(\mathbf{C}_{\mathbb{T}}, J_{\mathbb{T}})$ corresponding to M is simply the Yoneda embedding, it is easy to see that a sequent $(\phi \vdash \psi)$ is satisfied in M iff it is provable in $\mathbb{T}$. But $\mathrm{Shv}(\mathbf{C}_{\mathbb{T}}, J_{\mathbb{T}})$ is coherent, and so by 7.44 it has enough points; hence a sequent is satisfied in M iff it is satisfied in every $\mathbb{T}$-model in $\mathcal{S}$. ■

It should be remarked that one can reverse the above argument, and deduce Deligne's theorem from the completeness theorem. In this direction, the most important step is the following result:

7.47 PROPOSITION. *Let $\mathcal{E}$ be a coherent topos. Then there exists a finitary geometric theory $\mathbb{T}$ such that $\mathcal{E} \simeq \mathcal{S}[\mathbb{T}]$.*

Proof. Let $\mathcal{E} \simeq \mathrm{Shv}(\mathbf{C}, J)$, where $(\mathbf{C}, J)$ is a site satisfying the hypotheses of 7.35. We shall give a geometric presentation of the "theory of left exact continuous functors on $\mathbf{C}$"; by 7.13 this will be sufficient.

For the language $\mathbb{L}$, we take each object of $\mathbf{C}$ as a type, and each morphism $U \xrightarrow{\alpha} V$ of $\mathbf{C}$ as a function-symbol of signature $(U; V)$. As axioms of $\mathbb{T}$ we take the following sequents:

(a) for each object U, the sequent $true \vdash 1_U(\mathbf{u}) = \mathbf{u}$.

(b) for each composable pair $U \xrightarrow{\alpha} V \xrightarrow{\beta} W$, the sequent

$$true \vdash \beta\alpha(\mathbf{u}) = \beta(\alpha(\mathbf{u})).$$

(c) the sequents $true \vdash \mathbf{i} = \mathbf{i}'$ and $true \vdash \exists \mathbf{i}(\mathbf{i} = \mathbf{i})$, where $\mathbf{i}, \mathbf{i}'$ are variables of type 1.

(d) for each pullback square

$$\begin{array}{ccc} P & \xrightarrow{\alpha} & U \\ {\scriptstyle\beta}\downarrow & & \downarrow{\scriptstyle\gamma} \\ V & \xrightarrow{\delta} & W \end{array},$$

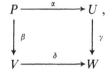

the sequents

$$\gamma(\mathbf{u}) = \delta(\mathbf{v}) \vdash \exists \mathbf{p}(\alpha(\mathbf{p}) = \mathbf{u} \wedge \beta(\mathbf{p}) = \mathbf{v})$$

and

$$\alpha(\mathbf{p}) = \alpha(\mathbf{p}') \wedge \beta(\mathbf{p}) = \beta(\mathbf{p}') \vdash \mathbf{p} = \mathbf{p}'.$$

(e) for each finite J-covering family $(V_i \xrightarrow{\alpha_i} U \mid i = 1, \ldots, n)$, the sequent

$$\mathbf{u} = \mathbf{u} \vdash \bigvee_{i=1}^{n} \exists \mathbf{v}_i(\alpha_i(\mathbf{v}_i) = \mathbf{u}),$$

where the disjunction on the right-hand side is interpreted as *false* if $n = 0$.

(Note that if $(\mathbf{C}, J)$ happens to be a pretopos with its precanonical topology, then the axioms in group (e) may be replaced by axioms referring to finite coproducts and coequalizers of equivalence relations, since the epimorphic families associated with these are sufficient to generate the topology.)

Now it is easy to see that an interpretation of $\mathbb{T}$ in a topos $\mathscr{F}$ which satisfies axioms (a) and (b) is simply a functor $\mathbf{C} \longrightarrow \mathscr{F}$; and such a functor satisfies axioms (c) and (d) iff it is left exact, and axiom (e) iff it is continuous. ■

It is evident from the proof of 7.47 that, if we generalize our definition of geometric language to allow infinite disjunctions of formulae (but with the restriction that the number of free variables in any formula must be finite), then *any* Grothendieck topos may be described as the classifying topos for a "generalized geometric theory". And indeed we have a converse result; by suitably modifying the construction of the syntactic site $(\mathbf{C}_{\mathbb{T}}, J_{\mathbb{T}})$, we may construct a classifying topos for any generalized geometric theory. Of course, the existence of toposes without points shows that we cannot hope to have a completeness theorem on the lines of 7.46 for these generalized theories; but the theorem of Barr which we prove in the next paragraph may be viewed as a "Boolean-valued completeness theorem", i.e. the assertion that provability is equivalent to satisfaction in every Boolean-valued model. We shall not pursue this idea here; for a much more detailed account of Grothendieck topos theory from this viewpoint, we refer the reader to [MR].

We conclude this paragraph with a couple of remarks concerning finitary geometric theories, the second of which fulfils a promise made in §6.5.

7.48 REMARK. Suppose a coherent topos $\mathscr{E}$ satisfies (SG). Then since $\mathscr{E}$ has enough points by 7.44, it is spatial by 7.25; i.e. there is a topological space X such that $\mathscr{E} \simeq \mathrm{Shv}(X)$. But from the description of compact objects in $\mathrm{Shv}(X)$ which we gave in 7.32(i), it is not hard to see that $\mathrm{Shv}(X)$ is coherent iff the set of compact open subsets of X is closed under finite intersections (including the empty intersection; i.e. X is compact), and forms a base for the topology. If we add the requirement that X be sober (7.21(ii)), this is previsely the definition of a *spectral space* given by M. Hochster in [169]. In view of 6.59(i), we may thus reinterpret Hochster's theorem that every spectral space is homeomorphic to the prime-spectrum of some commutative ring, as saying that every finitary geometric theory whose classifying topos satisfies (SG) is "Morita-equivalent" to the theory of localizations of some commutative ring. ■

7.49 REMARK. Let $(\mathbf{C}, J)$ be a site satisfying the hypotheses of 7.35. In 7.47 we saw how to present the "theory of points of $\mathrm{Shv}(\mathbf{C}, J)$" as a finitary geometric theory; but we may also give a presentation of the theory of sheaves on $(\mathbf{C}, J)$, as follows: once again, we take each object of $\mathbf{C}$ as a type, but this time a morphism $U \xrightarrow{\alpha} V$ of $\mathbf{C}$ becomes a function-symbol of signature $(V; U)$. For axioms, we take

(a) for each object U, the sequent $true \vdash 1_U(\mathbf{u}) = \mathbf{u}$.
(b) for each composable pair $U \xrightarrow{\alpha} V \xrightarrow{\beta} W$, the sequent

$$true \vdash \beta\alpha(\mathbf{w}) = \alpha(\beta(\mathbf{w})).$$

(c) for each finite J-covering family $(V_i \xrightarrow{\alpha_i} U \mid i = 1, \ldots, n)$, the sequents

$$\bigwedge_{i=1}^{n} (\alpha_i(\mathbf{u}) = \alpha_i(\mathbf{u}')) \vdash \mathbf{u} = \mathbf{u}'$$

and

$$\bigwedge_{i,j} (\beta_{ij}(\mathbf{v}_i) = \gamma_{ij}(\mathbf{v}_j)) \vdash \exists \mathbf{u} \left(\bigwedge_{i=1}^{n} \alpha_i(\mathbf{u}) = \mathbf{v}_i \right),$$

where

$$\begin{array}{ccc} W_{ij} & \xrightarrow{\beta_{ij}} & V_i \\ \downarrow{\scriptstyle \gamma_{ij}} & & \downarrow{\scriptstyle \alpha_i} \\ V_j & \xrightarrow{\alpha_j} & U \end{array}$$

is a pullback.

Writing $\mathbb{S}$ for the above theory, we see that an $\mathbb{S}$-model in a Grothendieck topos $\mathcal{F}$ is simply a "sheaf of $\mathcal{F}$-objects" on $(\mathbf{C}, J)$, i.e. a functor $\mathbf{C}^{\mathrm{op}} \longrightarrow \mathcal{F}$ which satisfies (the diagrammatic form of) the sheaf axiom for each J-covering family. But from 4.35 and 4.47, it is easily seen that the category of such sheaves is simply the pullback $\mathcal{F} \times_{\mathcal{S}} \mathrm{Shv}(\mathbf{C}, J)$ in $\mathfrak{BTop}/\mathcal{S}$; so we have equivalences

$$\mathfrak{Top}/\mathcal{S}(\mathcal{F}, \mathcal{S}[\mathbb{S}]) \simeq \mathcal{F} \times_{\mathcal{S}} \mathrm{Shv}(\mathbf{C}, J) \simeq \mathfrak{Top}/\mathcal{S}(\mathcal{F} \times_{\mathcal{S}} \mathrm{Shv}(\mathbf{C}, J), \mathcal{S}[U]),$$

i.e. $\mathcal{S}[\mathbb{S}]$ is the exponential $\mathcal{S}[U]^{\mathrm{Shv}(\mathbf{C}, J)}$ in $\mathfrak{Top}/\mathcal{S}$.

More generally, if $\mathbf{D}$ is any small category and $\mathbb{D}$ denotes the theory of diagrams of type $\mathbf{D}$, then we may construct the exponential $\mathcal{S}[\mathbb{D}]^{\mathrm{Shv}(\mathbf{C}, J)}$ as the classifier for the theory of diagrams of type $\mathbf{D}$ in $\mathrm{Shv}(\mathbf{C}, J)$. And then if k is any topology in $\mathcal{S}[\mathbb{D}]$, we may use forcing topologies as in 6.60(iii) to obtain the exponential $\mathrm{sh}_k(\mathcal{S}[\mathbb{D}])^{\mathrm{Shv}(\mathbf{C}, J)}$. But $\mathcal{S}^{\mathbf{D}^{\mathrm{op}}}$ is expressible as a sheaf subtopos of $\mathcal{S}[\mathbb{D}]$, via the inclusion which expresses the theory of flat presheaves on $\mathbf{D}^{\mathrm{op}}$ as a quotient of the theory of diagrams of type $\mathbf{D}$; so in this way we may obtain the exponential $\mathrm{Shv}(\mathbf{D}, L)^{\mathrm{Shv}(\mathbf{C}, J)}$ for *any* Grothendieck topos $\mathrm{Shv}(\mathbf{D}, L)$. ■

7.5. BARR'S THEOREM

As we saw in § 7.1, the existence of a sufficient set of points for an $\mathcal{S}$-topos $\mathcal{E}$ is equivalent to the existence of a surjection $\mathcal{S}/K \longrightarrow \mathcal{E}$ for some K. But for many purposes it is sufficient to assume a weaker hypothesis: namely that there exists a surjection $\mathcal{F} \xrightarrow{f} \mathcal{E}$, where $\mathcal{F}$ satisfies (AC). For example, if we wish to prove that a particular diagram in $\mathcal{E}$ commutes, or that part of it is a finite limit or colimit, it is sufficient (since f^* is exact and faithful) to prove the result for the corresponding diagram in $\mathcal{F}$; and in proving it in $\mathcal{F}$, we may avail ourselves of the fact that epis are split and monos have complements.

The existence of such a surjection, for an arbitrary Grothendieck topos $\mathcal{E}$, was first proved by M. Barr [4]; but for the first step in the argument, we use a slightly simpler method due to R. Diaconescu.

7.51 THEOREM. *Let $\mathcal{E}$ be a Grothendieck topos. Then there exists a surjection $\mathcal{F} \xrightarrow{f} \mathcal{E}$, where $\mathcal{E}$ satisfies* (SG).

Proof. Let $\mathcal{E} = \mathrm{Shv}(\mathbf{C}, J)$. We prove the result first for the presheaf topos $\mathcal{S}^{\mathbf{C}^{\mathrm{op}}}$, and then for $\mathcal{E}$ itself.

Let **P** be the poset whose objects are all finite composable strings of morphisms of **C** (i.e. n-tuples $(\alpha_1, \alpha_2, \ldots, \alpha_n)$ such that $d_1(\alpha_i) = d_0(\alpha_{i+1})$ for $1 \leqslant i < n$), ordered by setting $w_1 \leqslant w_2$ iff w_2 is a terminal segment of w_1 (i.e. there exist $\alpha_1, \ldots, \alpha_k$ such that $w_1 = (\alpha_1, \ldots, \alpha_k, w_2)$). The map $d \colon P \longrightarrow C_0$ which sends $(\alpha_1, \ldots, \alpha_n)$ to $d_0(\alpha_1)$ extends in an obvious way to a functor $\mathbf{P} \longrightarrow \mathbf{C}$; and this functor is clearly surjective on objects. It follows (cf. Ex. 4.2) that the induced geometric morphism $\mathscr{S}^{\mathbf{P}^{\mathrm{op}}} \xrightarrow{d} \mathscr{S}^{\mathbf{C}^{\mathrm{op}}}$ is a surjection; but $\mathscr{S}^{\mathbf{P}^{\mathrm{op}}}$ satisfies (SG) by 5.34(i).

We now define a Grothendieck topology K on **P**, by defining a sieve $S \rightarrowtail h_w$ to be K-covering iff, for all $w' \leqslant w$, the set

$$\{d(w'') \longrightarrow d(w') \mid w'' \leqslant w' \text{ and } w'' \in S\}$$

is a J-covering sieve on $d(w')$. The proof that this definition satisfies the first two conditions of 0.32 is straightforward; for the third, let S and T be two sieves on w such that S is K-covering and for each $w' \in S$, the sieve $\{w'' \in T \mid w'' \leqslant w'\}$ is K-covering on w'. Then for any $w_0 \leqslant w$, the sieve

$$\{d(w') \longrightarrow d(w_0) \mid w' \leqslant w_0 \text{ and } w' \in S\}$$

is J-covering; and for each such w', the sieve

$$\{d(w'') \longrightarrow d(w') \mid w'' \leqslant w' \text{ and } w'' \in T\}$$

is J-covering on $d(w')$. So by 0.32(iii) applied to J, the sieve

$$\{d(w'') \longrightarrow d(w_0) \mid w'' \leqslant w_0 \text{ and } w'' \in T\}$$

is J-covering; and hence T is K-covering.

Now we assert that a monomorphism $R \overset{\sigma}{\rightarrowtail} X$ in $\mathscr{S}^{\mathbf{C}^{\mathrm{op}}}$ is J-dense iff $d^*R \overset{d^*\sigma}{\rightarrowtail} d^*X$ is K-dense in $\mathscr{S}^{\mathbf{P}^{\mathrm{op}}}$. To prove this, it is clearly sufficient to assume that X is representable, say $X = h_U$; so we have to show that R is a J-covering sieve on U iff, for each $w \in P$ and each $d(w) \xrightarrow{\alpha} U$ in **C**, the sieve

$$\{w' \leqslant w \mid d(w') \longrightarrow d(w) \xrightarrow{\alpha} U \in R(d(w'))\}$$

is K-covering. But if R is J-covering, then so is α^*R for any such α; and hence the above sieve is K-covering. Conversely, if the above condition is satisfied, then by taking $w = (1_U)$ and $\alpha = 1_U$ we see at once that R must be J-covering.

It follows from the above and 3.47 that the composite

$$\mathrm{Shv}(\mathbf{P}, K) \longrightarrow \mathcal{S}^{\mathbf{P}^{\mathrm{op}}} \xrightarrow{d} \mathcal{S}^{\mathbf{C}^{\mathrm{op}}}$$

factors through the inclusion $\mathrm{Shv}(\mathbf{C}, J) \xrightarrow{i} \mathcal{S}^{\mathbf{C}^{\mathrm{op}}}$; and the factorization $\mathrm{Shv}(\mathbf{P}, K) \xrightarrow{\bar{d}} \mathrm{Shv}(\mathbf{C}, J)$ is a surjection, for if $\bar{d}^*(\sigma)$ is iso in $\mathrm{Shv}(\mathbf{P}, K)$, then $i_*(\sigma)$ is J-bidense in $\mathcal{S}^{\mathbf{C}^{\mathrm{op}}}$, and so σ is iso in $\mathrm{Shv}(\mathbf{C}, J)$. But $\mathrm{Shv}(\mathbf{P}, K)$ satisfies (SG) by 5.34. ■

7.52 Remark. It is not hard to verify that the square

$$\begin{array}{ccc} \mathrm{Shv}(\mathbf{P}, K) & \xrightarrow{\bar{d}} & \mathrm{Shv}(\mathbf{C}, J) \\ \downarrow & & \downarrow \\ \mathcal{S}^{\mathbf{P}^{\mathrm{op}}} & \xrightarrow{d} & \mathcal{S}^{\mathbf{C}^{\mathrm{op}}} \end{array}$$

is a pullback in $\mathfrak{Top}$; i.e. K is the "pullback topology" defined as in 3.59(iii). However, it is not true in general that if we have such a pullback square with d surjective, then $\bar{d}$ is surjective (see Ex. 7.1 below); this is why we had to give an explicit description of the topology K in 7.51. ■

For the next stage in the proof of Barr's theorem, we require a result which states that for any (elementary) topos $\mathcal{E}$ there exists a surjection $\mathcal{B} \xrightarrow{p} \mathcal{E}$ with $\mathcal{B}$ Boolean. To prove this, we follow an argument due to P. Freyd [FK].

7.53 Lemma. *Let* $X \underset{g}{\overset{f}{\rightrightarrows}} Y$ *be a parallel pair of morphisms in a topos* $\mathcal{E}$, *such that* $f \neq g$. *Then there exists a Boolean topos* $\mathcal{B}$ *and a geometric morphism* $\mathcal{B} \xrightarrow{p} \mathcal{E}$ *such that* $p^*f \neq p^*g$.

Proof. First suppose that $X = 1$, so that f and g are global elements of Y. Let $U \rightarrowtail 1$ be their equalizer; then $U \not\cong 1$, and so the topos $\mathrm{sh}_{j_U^c}(\mathcal{E})$ is non-degenerate since U is a j_U^c-sheaf. Let p be the composite inclusion

$$\mathcal{B} = \mathrm{sh}_{\neg\neg}(\mathrm{sh}_{j_U^c}(\mathcal{E})) \xrightarrow{i_1} \mathrm{sh}_{j_U^c}(\mathcal{E}) \xrightarrow{i_2} \mathcal{E};$$

now $\mathrm{eq}(i_2^*f, i_2^*g) \cong i_2^*U \cong 0$ in $\mathrm{sh}_{j_U^c}(\mathcal{E})$, and hence $\mathrm{eq}(p^*f, p^*g) \cong i_1^*0 \cong 0$ in $\mathcal{B}$. But since 0 is a $\neg\neg$-sheaf in any topos (5.18), the topos $\mathcal{B}$ is non-degenerate, and hence $p^*f \neq p^*g$.

Now consider the general case. If we apply the above argument to f and g

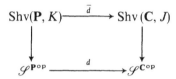

regarded as global elements $\hat{f}$ and $\hat{g}$ of X^*Y in $\mathscr{E}/X$, then we obtain $\mathscr{B} \xrightarrow{p} \mathscr{E}/X$ with $\mathscr{B}$ Boolean and $p^*\hat{f} \neq p^*\hat{g}$. But since $\hat{f}$ is the composite $1_X \xrightarrow{x} X^*X \xrightarrow{X^*f} X^*Y$, where x is the generic element of X, we must have $p^*X^*f \neq p^*X^*g$; so the composite morphism $\mathscr{B} \xrightarrow{p} \mathscr{E}/X \longrightarrow \mathscr{E}$ has the required property. ■

The composite inclusion $\mathrm{sh}_{\neg\neg}(\mathrm{sh}_{j_U^c}\,\mathscr{E})) \longrightarrow \mathrm{sh}_{j_U^c}\,\mathscr{E}) \longrightarrow \mathscr{E}$ constructed in the first part of the proof of 7.53 is of course determined by a single topology in $\mathscr{E}$, by 4.15(i). In fact it is not hard to see that this toplogy is the interpretation of the formula $((\omega \vee u) \Rightarrow u) \Rightarrow u)$ of $L_{\mathscr{E}}$, where ω is a variable of type Ω and $1 \xrightarrow{u} \Omega$ is the classifying map of $U \rightarrowtail 1$. For it follows at once from 3.53 that a subobject of an object X is j_U^c-closed iff it contains the subobject $X \times U \xrightarrow{\pi_1} X$; hence the closed-subobject classifier $\Omega_{j_U^c}$ is simply $\uparrow \mathrm{seg}(u) \rightarrowtail \Omega$, and in particular its minimal element is the factorization of $1 \xrightarrow{u} \Omega$ through $\uparrow \mathrm{seg}(u)$. We shall call this toplogy the *quasi-closed topology* determined by U, and denote it by q_U.

7.54 PROPOSITION. *Let $\mathscr{E}$ be a topos. Then there exists a surjection $\mathscr{B} \xrightarrow{p} \mathscr{E}$ where $\mathscr{B}$ is a Boolean topos. Moreover, if $\mathscr{E}$ satisfies* (SG), *then $\mathscr{B}$ may be chosen to satisfy* (SG) *also.*

Proof. Considering the generic subobject $1 \xrightarrow{t} \Omega$ as a subobject of 1_Ω in $\mathscr{E}/\Omega$, let $\mathscr{B}$ be the topos $\mathrm{sh}_{q_t}(\mathscr{E}/\Omega)$, and p the composite geometric morphism $\mathrm{sh}_{q_t}(\mathscr{E}/\Omega) \longrightarrow \mathscr{E}/\Omega \longrightarrow \mathscr{E}$. Then $\mathscr{B}$ is certainly Boolean; we shall show that p is a surjection.

Let (f, g) be a parallel pair of morphisms in $\mathscr{E}$ with $f \neq g$; then the proof of 7.53 enables us to find an $\mathscr{E}$-topos of the form $\mathrm{sh}_{q_U}(\mathscr{E}/X)$, where X is an object of $\mathscr{E}$ and U a subobject of 1_X in $\mathscr{E}/X$, in which f and g remain distinct. Let $X \xrightarrow{\phi} \Omega$ be the classifying map of U, regarded as a subobject of X in $\mathscr{E}$; then since the pullback functor ϕ^* is logical, we see easily that $\phi^*(q_t) = q_U$, and hence any q_t-dense monomorphism in $\mathscr{E}/\Omega$ is sent by ϕ^* to a q_U-dense mono in $\mathscr{E}/X$. So by 3.42 and 3.47, we have a commutative diagram

$$\begin{array}{ccccc} \mathrm{sh}_{q_U}(\mathscr{E}/X) & \longrightarrow & \mathscr{E}/X & \longrightarrow & \mathscr{E} \\ \downarrow & & \downarrow & \nearrow & \\ \mathrm{sh}_{q_t}(\mathscr{E}/\Omega) & \longrightarrow & \mathscr{E}/\Omega & & \end{array}$$

in $\mathfrak{Top}$, where the morphism $\mathscr{E}/X \longrightarrow \mathscr{E}/\Omega$ is that induced by ϕ. Hence f and g must remain distinct in $\mathrm{sh}_{q_t}(\mathscr{E}/\Omega)$, i.e. $p^*f \neq p^*g$.

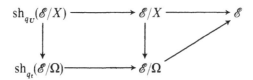

The last sentence of the proposition follows at once from 5.34, since we may regard $\mathscr{E}/\Omega$ as the topos of internal diagrams on the discrete category Ω. ■

As a by-product of 7.54, we obtain a proof of a well-known embedding theorem in lattice theory, originally due to N. Funayama [164]:

7.55 COROLLARY. *Let H be a complete Heyting algebra (in $\mathscr{S}$). Then there exists a complete Boolean algebra B and an embedding $H \rightarrowtail B$ which preserves finite meets and arbitrary joins.*

Proof. Apply 7.54 to the topos $\mathscr{E} = \mathrm{Shv}(\mathbf{H}, C)$, where C is the canonical topology. We obtain a surjection $\mathscr{B} \xrightarrow{p} \mathscr{E}$ where $\mathscr{B}$ is Boolean; and on restricting the functor p^* to subobjects of 1 in $\mathscr{E}$, we have an order-preserving map of H into the Boolean algebra of subobjects of 1 in $\mathscr{B}$. But this map is a monomorphism, since p^* reflects isomorphisms; and it preserves finite meets and arbitrary joins, since p^* preserves finite limits and arbitrary colomits. ■

7.56 REMARK. In fact we may give a purely lattice-theoretic proof of Funayama's theorem using the ideas of 7.53 and 7.54. If H is a Heyting algebra, let $H_{\neg\neg}$ denote the Boolean algebra of regular elements of H (5.16); and for each $u \in H$, consider the map

$$P(u)\colon H \longrightarrow (\uparrow \mathrm{seg}(u))_{\neg\neg}\,;\, x \longrightarrow (((x \vee u) \Rightarrow u) \Rightarrow u).$$

It is not hard to verify that $P(u)$ preserves finite meets and arbitrary joins (the latter since it is left adjoint to the inclusion $(\uparrow \mathrm{seg}(u))_{\neg\neg} \longrightarrow H$), and that the morphism $H \longrightarrow \prod_{u \in H} (\uparrow \mathrm{seg}(u))_{\neg\neg}$ whose uth component is $P(u)$ is a monomorphism.

Note also that 7.54, at least in the case when $\mathscr{E}$ is defined over $\mathscr{S}$ and satisfies (SG) (which is the case that we shall require in 7.57 below), may be deduced as a consequence of 7.55. For 5.37 implies that we can consider $\mathscr{E}$ to be $\mathrm{Shv}(\mathbf{H}, C)$, where H is a complete Heyting algebra and C is the canonical toplogy; and then the conditions on the embedding $H \xrightarrow{p} B$ of 7.55 ensure that the composite functor $\mathbf{H} \xrightarrow{p} \mathbf{B} \xrightarrow{l} \mathrm{Shv}(\mathbf{B}, C)$ is left exact and continuous for the canonical toplogy on $\mathbf{H}$ (so that it induces a geometric morphism $\mathrm{Shv}(\mathbf{B}, C) \longrightarrow \mathrm{Shv}(\mathbf{H}, C)$, by 7.13). Finally, the fact that p is a monomorphism implies that this morphism is a surjection; and $\mathrm{Shv}(\mathbf{B}, C)$ is Boolean by 5.39. ■

We are now ready to prove the main theorem of this paragraph:

7.57 THEOREM (M. Barr [4]). *Let $\mathscr{E}$ be a Grothendieck topos (and assume* (AC) *holds in $\mathscr{S}$). Then there exists a topos $\mathscr{B}$ satisfying* (AC) *and a surjection $\mathscr{B} \xrightarrow{p} \mathscr{E}$.*

Proof. By 7.51, we can find a surjection $\mathscr{F} \longrightarrow \mathscr{E}$ where $\mathscr{F}$ satisfies (SG). Then by 7.54 we can find a surjection $\mathscr{B} \longrightarrow \mathscr{F}$ where $\mathscr{B}$ is Boolean and satisfies (SG); and by 5.39 $\mathscr{B}$ satisfies (AC). ■

7.58 REMARK. In view of 4.15(ii), we may paraphrase 7.57 by saying that every Grothendieck topos has the form $\mathscr{B}_{\mathbb{G}}$, where $\mathscr{B}$ is a topos satisfying the conditions of 5.39 and $\mathbb{G}$ is a left exact comonad on $\mathscr{B}$. It is natural to ask whether every topos of the form $\mathscr{B}_{\mathbb{G}}$ is a Grothendieck topos. In fact it can be shown, by methods similar to those of Ex. 4.11, that this topos is Grothendieck iff the functor part of the commonad $\mathbb{G}$ has a rank.

EXERCISES 7

1. Let K be a set of points of a topos $\mathscr{E}$, and j a toplogy in $\mathscr{E}$. Prove that the square

$$\begin{array}{ccc} \mathscr{S}/L & \longrightarrow & \mathrm{sh}_j(\mathscr{E}) \\ \downarrow & & \downarrow \\ \mathscr{S}/K & \longrightarrow & \mathscr{E} \end{array}$$

is a pullback, where L is the set of points in K which factor through $\mathrm{sh}_j(\mathscr{E}) \longrightarrow \mathscr{E}$. Deduce, using 7.12(iii), that a pullback of a surjection in $\mathfrak{Top}$ need not be a surjection. Show also (using the same example) that the lattice of sheaf subtoposes of $\mathscr{E}$ need not be a Heyting algebra.

2. Prove that Shv($\mathbb{R}$) is a classifying topos for Dedekind real numbers in Grothendieck toposes. [Use 7.13, taking a site of definition for Shv($\mathbb{R}$) whose underlying category is the poset of open intervals with rational endpoints.]

3. (J. E. Roos [104]) Let $\mathscr{E}$ be a Grothendieck topos. Show that a functor

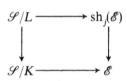

$\mathscr{E} \xrightarrow{\phi} \mathscr{S}$ has a left adjoint iff it is representable. Deduce that ϕ is the inverse image of an essential point of $\mathscr{E}$ iff it has the form $\hom_{\mathscr{E}}(P, -)$, where P is projective, connected (i.e. not decomposable as a nontrivial coproduct) and not isomorphic to 0. (We say P is an *essential object* of $\mathscr{E}$.) [Hint: if P is essential, use the ideas of Ex. 4.8 to show that $\hom_{\mathscr{E}}(P, -)$ preserves coproducts and hence all colimits; then use 7.13] If P is essential, show that a family $(X_i \xrightarrow{f_i} P \mid i \in I)$ is epimorphic iff one of the f_i is a split epimorphism, and hence show that the following conditions are equivalent:

(i) The class of essential points of $\mathscr{E}$ is sufficient.
(ii) The essential objects are a class of generators for $\mathscr{E}$.
(iii) There exists a small category $\mathbf{C}$ such that $\mathscr{E} \simeq \mathscr{S}^{\mathbf{C}^{\mathrm{op}}}$.

Show further that, if we impose the requirement that idempotents split in $\mathbf{C}$, then $\mathbf{C}$ is determined up to equivalence by $\mathscr{E}$. [M. C. Bunge [16] has proved a "relative" version of this result, characterizing $\mathscr{E}$-toposes of the form $\mathscr{E}^{\mathbf{C}}$.]

4. Let $(X, \mathbf{U})$ and $(Y, \mathbf{V})$ be two topological spaces, and let $\mathbf{B}$ be the poset of open rectangles in $X \times Y$ (i.e. subsets of the form $U \times V$, where U is open in X and V is open in Y), ordered by inclusion. [Observe that $\mathbf{B}$ is an intersection-closed base for the toplogy on $X \times Y$.] Define Grothendieck topologies J_1, J_2 on $\mathbf{B}$ such that the J_1-covering sieves have the form $(U_i \times V \longrightarrow U \times V \mid i \in I)$ where $(U_i \longrightarrow U \mid i \in I)$ is an open cover of U in X, and J_2 is similarly defined using open covers in Y. If J_3 denotes the join of J_1 and J_2, show that $\mathrm{Shv}(\mathbf{B}, J_3)$ is equivalent to $\mathrm{Shv}(X) \times_{\mathscr{S}} \mathrm{Shv}(Y)$. [Use 4.35 and 4.47.] Now suppose X is locally compact, and let R be an arbitrary sieve of open rectangles covering (set-theoretically) the rectangle $U \times V \subseteq X \times Y$. Show that R is J_3-covering. [Hint: for each $x \in U$ choose a compact neighbourhood $K_x \subseteq U$; then cover each $(\mathrm{int}\, K_x) \times V$ by open rectangles $(\mathrm{int}\, K_x) \times L_{x,y}$, each of which can be covered by finitely many rectangles in R of the form $M \times L_{x,y}$.] Deduce that if either X or Y is locally compact, then the canonical morphism

$$\mathrm{Shv}(X \times Y) \longrightarrow \mathrm{Shv}(X) \times_{\mathscr{S}} \mathrm{Shv}(Y)$$

is an equivalence.

5. A Grothendieck topos $\mathscr{E}$ is said to be an *étendue* ([GV] IV 9.8.2) if there

is an object X of $\mathscr{E}$ with global support such that $\mathscr{E}/X$ is spatial. Show that an étendue has enough points, and that its category of points is essentially small. Show that the topos $\mathrm{Shv}_G(X)$ of G-equivariant sheaves defined in Ex. 0.9 is an étendue; and that the topos $\mathscr{S}^{\mathbf{C}^{\mathrm{op}}}$ is an étendue iff every morphism of $\mathbf{C}$ is a monomorphism. [Hint: use 2.18 to show that $\mathscr{S}^{\mathbf{C}^{\mathrm{op}}}$ is an étendue iff there exists a discrete fibration $\mathbf{F} \xrightarrow{\gamma} \mathbf{C}$ such that γ_0 is epi and $\mathbf{F}$ is a poset.]

6. Let K denote the Cantor space, considered as the cartesian product (in **esp**) of countably many copies of the discrete two-point space $2 = \{l, r\}$ (so that a point of K may be regarded as an infinite string of l's and r's). Let M denote the free monoid on 2, and for each $w \in M$ define U_w to be the set of points of K which begin with the string w. Show that $B = \{U_w \mid w \in M\}$ is an intersection-closed base for the toplogy on K. Regarding B as a poset in the usual way, construct a discrete opfibration $\mathbf{B}^{\mathrm{op}} \longrightarrow \mathbf{M}$, and hence show that there is a pullback square

$$\begin{array}{ccc} \mathrm{Shv}(K) & \longrightarrow & \mathscr{S}^{\mathbf{B}^{\mathrm{op}}} \\ \downarrow & & \downarrow \\ \mathbf{jt}_2(\mathscr{S}) & \longrightarrow & \mathscr{S}^{\mathbf{M}} \end{array}$$

in $\mathfrak{Top}$, where $\mathbf{jt}_2$ is the algebraic theory defined in Ex. 6.9 and the horizontal arrows are inclusions. Deduce that $\mathbf{jt}_2(\mathscr{S})$ is an étendue.

7. Recall that an element i of a complete lattice L is said to be *inaccessible* ([LT], p. 186) if, whenever we are given a filtered sub-poset $\mathbf{P}$ of $\mathbf{L}$ with $\bigcup P = i$, we have $i \in P$. Show that an object X is compact in a topos $\mathscr{E}$ iff the maximal element of the (external) lattice of subobjects of X is inaccessible.

8. Let X be a set, R a binary relation on X (assumed reflexive and symmetric). Define R_n to be the n-fold composite of R with itself, and $R_\infty = \bigcup_{n=1}^{\infty} R_n$; show that R_∞ is the equivalence relation generated by R. Choose an infinite set X and a relation R such that the sequence (R_n) is strictly increasing; then let $\mathbf{E}$ be a small full subcategory of $\mathscr{S}$ containing X and R, and closed under finite limits and colimits. Let $\mathscr{E}$ be the topos of sheaves on $\mathbf{E}$ for the precanonical toplogy. Show that the equivalence

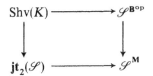

relation on $l(X)$ generated by $l(R) \rightarrowtail l(X) \times l(X)$ is the union of the $l(R_n)$ (which is *not* the same as $l(R_\infty)$), and deduce that it is not compact. Hence show that the category $\mathscr{E}_{\text{coh}}$ is not closed under the formation of coequalizers in $\mathscr{E}$. [Use 7.36.]

9. Let $\mathscr{E}$ be a coherent topos. Show that the following conditions are equivalent:

   (i) Every epimorphic image of a coherent object of $\mathscr{E}$ is coherent.
   (ii) Every subobject of a coherent object of $\mathscr{E}$ is coherent.
   (iii) Every object of $\mathscr{E}$ is stable.
   (iv) Every compact object of $\mathscr{E}$ is coherent.
   (v) If X is a coherent object of $\mathscr{E}$, then the lattice of subobjects of X satisfies the ascending chain condition.

   [Hint: for (i) $\Rightarrow$ (ii), use 7.36. For (ii) $\Leftrightarrow$ (v), use Ex. 7.]

   A coherent topos satisfying these conditions is said to be *noetherian*. If A is a noetherian commutative ring in $\mathscr{S}$, show that the topos Shv(spec A) is noetherian. Is the converse true?

10. Let $\mathscr{E}$ be a noetherian topos, j a topology in $\mathscr{E}$. Show that $\text{sh}_j(\mathscr{E})$ is noetherian.
    [Hint: if X is an object of $\mathscr{E}$ with associated sheaf LX, the lattice of subobjects of LX in $\text{Sh}_j(\mathscr{E})$ is isomorphic to the lattice of closed subobjects of X in $\mathscr{E}$.]

11. Let $T: \mathscr{S}_f \longrightarrow \mathscr{S}$ be a functor. Show that T is compact as an object of $\mathscr{S}[U]$ (cf. 4.37(iv)) iff it satisfies the following conditions:

    (a) For every finite set X, $T(X)$ if finite.
    (b) There exists a natural number n such that, for any X of cardinality $> n$ and any $x \in T(X)$, there exists $X' \xrightarrow{f} X$ with $\text{card}(X') \leqslant n$ such that x is in the image of $T(f)$.
    Hence show that $\mathscr{S}[U]$ is noetherian.

12. (M. Barr [5]) We say that a small category **C** is *atomic* if it satisfies the following conditions:

    (a) Every morphism $U \xrightarrow{\alpha} V$ of **C** is a joint coequalizer of a family of pairs of morphisms $(W_i \overset{\beta_i}{\underset{\gamma_i}{\rightrightarrows}} U \mid i \in I)$.

(*b*) Every diagram

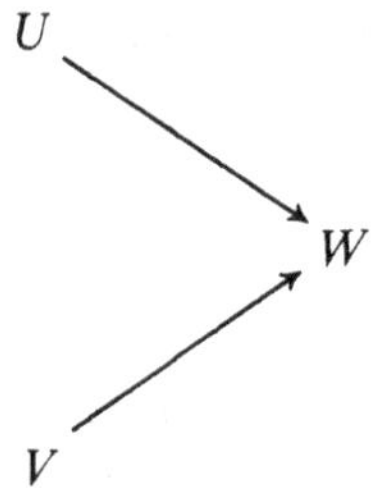

in $\mathbf{C}$ can be completed to a commutative square. If X is an infinite set, show that the monoid of epi-endomorphisms of X is an atomic category. Show also that if $\mathbf{C}$ is atomic and not equivalent to $\mathbf{1}$, then the canonical and double-negation toplogies on $\mathbf{C}$ coincide.

13. Let $\mathscr{E} \xrightarrow{\gamma} \mathscr{S}$ be an $\mathscr{S}$-topos. Show that the following conditions are equivalent:

(i) γ^* is logical.
(ii) $\mathscr{E}$ is locally connected (Ex. 4.9).and Boolean.
(iii) Every object of $\mathscr{E}$ is expressible as an $\mathscr{S}$-indexed coproduct of *atoms* (i.e. nonzero objects having no nontrivial subobjects).
(iv) if γ is bounded) There exists an atomic category $\mathbf{C}$ such that $\mathscr{E} \simeq \mathrm{sh}_{\neg\neg}(\mathscr{S}^{\mathbf{C}^{\mathrm{op}}})$.

[For (iii) $\Rightarrow$ (iv), take $\mathbf{C}$ to be (equivalent to) the full subcategory of atoms in $\mathscr{E}$.]

Chapter 8

Cohomology

8.1. BASIC DEFINITIONS

In this chapter we shall be concerned with the study of homological algebra in the category **ab**($\mathscr{E}$) of internal abelian groups in a topos $\mathscr{E}$ (usually a Grothendieck topos). Many of the results which we prove can be generalized without difficulty to the category of R-modules, where R is an internal ring in $\mathscr{E}$; but we shall not explore this generalization here. We assume that the reader is familiar with the basic techniques of homological algebra in abelian categories (see, for example, [HA]).

8.11 THEOREM. (i) *For any topos $\mathscr{E}$, the category* **ab**($\mathscr{E}$) *is abelian.*

(ii) *If $\mathscr{E}$ has a natural number object, then* **ab**($\mathscr{E}$) *is monadic over $\mathscr{E}$.*

(iii) *If $\mathscr{E}$ is a Grothendieck topos, then* **ab**($\mathscr{E}$) *satisfies Grothendieck's "Axiom* AB5" [42] *and has a generator.*

Proof. (i) Clearly **ab**($\mathscr{E}$) has finite limits, since the forgetful functor **ab**($\mathscr{E}$) $\longrightarrow \mathscr{E}$ creates them. We may now verify as in the case $\mathscr{E} = \mathscr{S}$ that the terminal object of **ab**($\mathscr{E}$) is also initial, and that the product of two objects of **ab**($\mathscr{E}$) is also their coproduct. To construct the cokernel of a morphism $A \xrightarrow{f} B$ of **ab**($\mathscr{E}$), we form the coequalizer in $\mathscr{E}$ of $A \times B \xrightarrow{f \times 1} B \times B \xrightarrow{+} B$ and $A \times B \xrightarrow{\pi_2} B$. (Note that this pair is reflexive, with splitting $B \xrightarrow{o \times 1} A \times B$— a fact which we may use, in conjunction with 6.42, to define the group structure on the coequalizer.) It is now easy to verify that any monomorphism in **ab**($\mathscr{E}$) is the kernel of its own cokernel; from this it follows that a monomorphism which is also epi in **ab**($\mathscr{E}$) is actually epi in $\mathscr{E}$, and hence (since the image in $\mathscr{E}$ of any abelian group homomorphism has an abelian group structure) that every epi in **ab**($\mathscr{E}$) is epi in $\mathscr{E}$. It now follows easily from 1.53 that every epi in $\mathbf{ab}(\mathscr{E})$ is a cokernel. We have thus verified all the axioms for an abelian category ([CW], p. 194).

(ii) is simply a particular case of 6.43(ii).

(iii) Axiom AB5 may be stated as follows: $\mathbf{ab}(\mathscr{E})$ has all $\mathscr{S}$-indexed colimits, and filtered $\mathscr{S}$-indexed colimits are universal. But existence of colimits follows from (ii) and Linton's theorem (0.16), since $\mathscr{E}$ has them; and since filtered colimits commute with finite limits in $\mathscr{E}$ (2.58), the forgetful functor $\mathbf{ab}(\mathscr{E}) \longrightarrow \mathscr{E}$ creates them. But it also creates pullbacks; and since all colimits in $\mathscr{E}$ are preserved by pullback (1.51), we deduce that filtered colimits are preserved by pullback in $\mathbf{ab}(\mathscr{E})$.

Finally, if $\{G_\alpha \mid \alpha \in A\}$ is a set of generators for $\mathscr{E}$, then the free abelian group $F(\coprod_{\alpha \in A} G_\alpha)$ is easily seen to be a generator for $\mathbf{ab}(\mathscr{E})$. ∎

From 8.11(iii) and a theorem of Grothendieck ([42], 1.10.1), we may deduce that if $\mathscr{E}$ is a Grothendieck topos then $\mathbf{ab}(\mathscr{E})$ has enough injectives. However, we can give a more direct proof of this important fact using Barr's theorem. First we require a result concerning geometric morphisms; note that, if $\mathscr{F} \xrightarrow{f} \mathscr{E}$ is a geometric morphism, then f^* and f_* can be lifted to a pair of adjoint functors between $\mathbf{ab}(\mathscr{E})$ and $\mathbf{ab}(\mathscr{F})$, since they are both left exact.

8.12 LEMMA. *Let* $\mathscr{F} \xrightarrow{f} \mathscr{E}$ *be a geometric morphism. Then*

(i) $f_*\colon \mathbf{ab}(\mathscr{F}) \longrightarrow \mathbf{ab}(\mathscr{E})$ *preserves injectives.*
(ii) *if f is a surjection and* $\mathbf{ab}(\mathscr{F})$ *has enough injectives, so has* $\mathbf{ab}(\mathscr{E})$.

Proof. (i) Let E be injective in $\mathbf{ab}(\mathscr{F})$, and suppose we are given a diagram

$$\begin{array}{ccc} A & \longrightarrow & f_*(E) \\ \downarrow & & \\ B & & \end{array}$$

in $\mathbf{ab}(\mathscr{E})$. Then the transposed diagram

$$\begin{array}{ccc} f^*(A) & \longrightarrow & E \\ \downarrow & & \\ f^*(B) & & \end{array}$$

can be completed to a commutative triangle; transposing back again, we obtain a completion of the original diagram.

(ii) Let A be an abelian group in $\mathscr{E}$; then we can find a monomorphism $f^*(A) \rightarrowtail E$ in $\mathbf{ab}(\mathscr{F})$ with E injective. But the unit of $(f^* \dashv f_*)$ is mono by 4.11(ii)(c); so the composite $A \rightarrowtail f_*f^*(A) \rightarrowtail f_*(E)$ embeds A in an injective in $\mathbf{ab}(\mathscr{E})$. ■

8.13 THEOREM. *Let $\mathscr{E}$ be a Grothendieck topos. Then* $\mathbf{ab}(\mathscr{E})$ *has enough injectives.*

Proof. From 8.12(ii) and 7.57, it is sufficient to prove the result in the case when $\mathscr{E}$ satisfies (AC), i.e. when $\mathscr{E} = \mathrm{Shv}(\mathbf{B}, C)$ where B is a complete Boolean algebra and C is the canonical topology.

But in this case we can exactly mimic the usual proof in the case $\mathscr{E} = \mathscr{S}$ (see [HA], pp. 31–33). First we prove the theorem of Baer [151] that an object E of $\mathbf{ab}(\mathscr{E})$ is injective iff it is *divisible* (i.e. iff $E(b)$ is a divisible abelian group for all $b \in B$). Then we prove that the rational circle group Q/Z (constructed either internally, as in § 6.6, or by forming the associated C-sheaf of the constant presheaf $\overline{\mathbb{Q}/\mathbb{Z}}$) is a cogenerator for $\mathbf{ab}(\mathscr{E})$; but since $\mathscr{E}$ (and therefore $\mathbf{ab}(\mathscr{E})$) has $\mathscr{S}$-indexed products, we deduce that any object of $\mathbf{ab}(\mathscr{E})$ can be embedded in a power of Q/Z. The further details of the argument are left to the reader. ■

In general, however, the category $\mathbf{ab}(\mathscr{E})$ does not have enough projectives; this is hardly suprising, since the failure of (AC) to hold in a general topos means that $\mathscr{E}$ itself does not normally have enough projectives (whereas we saw in 1.27 that it always has enough injectives).

It follows from 8.13 that we can construct right derived functors of additive functors $\mathbf{ab}(\mathscr{E}) \longrightarrow \mathscr{A}$, where $\mathscr{A}$ is any abelian category, by the usual method of forming injective coresolutions. The cohomology theory of toposes consists essentially of the study of these derived functors; we are now ready to give the fundamental definition.

8.14 DEFINITION. Let $\mathscr{E}$ be a Grothendieck topos, A an abelian group in $\mathscr{E}$. The qth *cohomology group of $\mathscr{E}$ with coefficients in A* is the group $H^q(\mathscr{E}; A) = R^q(\gamma_*)(A)$, where γ_* is the functor $\hom_{\mathscr{E}}(1, -)\colon \mathbf{ab}(\mathscr{E}) \longrightarrow \mathbf{ab}(\mathscr{S})$, i.e. the direct image of the geometric morphism $\mathscr{E} \longrightarrow \mathscr{S}$. Equivalently, we can describe $H^q(\mathscr{E}; A)$ as $\mathrm{Ext}^q_{\mathbf{ab}(\mathscr{E})}(Z, A)$, where Z is the free abelian group generated by 1 in $\mathscr{E}$.

More generally, if X is any object of $\mathscr{E}$, we write $H^q(\mathscr{E}, X; -)$ (or simply $H^q(X; -)$, if $\mathscr{E}$ is obvious from the context) for the qth right derived functor of

$\hom_{\mathscr{E}}(X, -)$. (Note that $H^q(\mathscr{E}, X; A)$ is naturally isomorphic, to $H^q(\mathscr{E}/X; X^*(A))$, since the functor $X^*: \mathbf{ab}(\mathscr{E}) \longrightarrow \mathbf{ab}(\mathscr{E}/X)$ is exact and preserves injectives.)

8.15 EXAMPLES. (i) Let $\mathscr{E} = \mathrm{Shv}(X)$, X a topological space. Then the cohomology groups $H^q(\mathscr{E}; A)$ are just the cohomology groups of X (in the usual sense) with coefficients in the sheaf A. (Cf. [ST], §5.3.)

(ii) Let $\mathscr{E} = \mathscr{S}^{\mathbf{G}}$, G a group. Then $\mathbf{ab}(\mathscr{E})$ is the category of (right) modules over the integral group-ring $\mathbb{Z}G$ of G; in particular $Z \in \mathbf{ab}(\mathscr{E})$ is the group of integers with trivial G-action. So the cohomology of $\mathscr{E}$ coincides with the Eilenberg–Mac Lane cohomology [161] of the group G.

(iii) More generally, let G be a topological group. We say that a G-set M is *continuous* if the G-action map $M \times G \longrightarrow M$ is continuous for the discrete topology on M; this is easily seen to be equivalent to the assertion that the stabilizer of each point of M is an open subgroup of G. Let $\mathscr{E}$ denote the category of continuous G-sets and G-equivariant maps; then as in Ex. 5.5 it is easy to verify that the inclusion functor $\mathscr{E} \longrightarrow \mathscr{S}^{\mathbf{G}}$ is left exact and has a right adjoint, so that $\mathscr{E}$ is a topos. The cohomology of $\mathscr{E}$ is in general different from the Eilenberg–Mac Lane cohomology of G; in particular if G is profinite (see 8.41 below), it coincides with the Galois cohomology of G (see [183]).

(iv) Suppose X is a projective object of a topos $\mathscr{E}$. Since epimorphisms in $\mathbf{ab}(\mathscr{E})$ are epi in $\mathscr{E}$, the functor $\hom_{\mathscr{E}}(X, -): \mathbf{ab}(\mathscr{E}) \longrightarrow \mathbf{ab}(\mathscr{S})$ preserves them, and is therefore exact. So $H^q(\mathscr{E}, X; A) = 0$ for all $q > 0$ and all $A \in \mathbf{ab}(\mathscr{E})$. We note in particular two important cases of this: (*a*) if $\mathscr{E}$ satisfies (AC), then $H^q(\mathscr{E}, X; A) = 0$ for all X and A, and all $q > 0$; and (*b*) if $\mathscr{E} = \mathscr{S}^{\mathbf{C}^{\mathrm{op}}}$ for some small category $\mathbf{C}$, then $H^q(\mathscr{E}, h_U; A) = 0$ for all $U \in C_0$, all $A \in \mathbf{ab}(\mathscr{E})$ and all $q > 0$. ■

We shall now investigate the functoriality of cohomology groups with respect to geometric morphisms. First we state without proof a well-known theorem of Grothendieck ([42], 2.4.1) on spectral sequences; the reader will find a detailed proof of it in [HA], VIII 9.3.

8.16 THEOREM. *Let $\mathscr{A}$, $\mathscr{B}$ and $\mathscr{C}$ be abelian categories such that $\mathscr{A}$ and $\mathscr{B}$ have enough injectives, and $\mathscr{A} \xrightarrow{S} \mathscr{B} \xrightarrow{T} \mathscr{C}$ two left exact functors. Suppose that S sends injectives in $\mathscr{A}$ to* T-acyclic *objects in $\mathscr{B}$ (i.e. objects in the kernel of R^qT for all $q > 0$). Then there is a spectral sequence $E_r^{p,q}$ of functors $\mathscr{A} \longrightarrow \mathscr{C}$*

(called the spectral sequence of the composite TS), such that $E_2^{p,q} \cong (R^pT)(R^qS)$*, and converging finitely to* $E_\infty^{p,q}$ *which is a filtration of* $R^{p+q}(TS)$. ■

8.17 PROPOSITION. *Let* $\mathscr{F} \xrightarrow{f} \mathscr{E}$ *be a geometric morphism between Grothendieck toposes. Then*

(i) *If A is an abelian group in* $\mathscr{E}$, *we have a homomorphism* $H^q(\mathscr{E}; A) \xrightarrow{f^*} H^q(\mathscr{F}; f^*A)$ *for each q, which is functorial in f and natural in A.*

(ii) *If B is an abelian group in* $\mathscr{F}$, *we have a spectral sequence (the* Leray spectral sequence) $H^p(\mathscr{E}; R^qf_*(B)) \Rightarrow H^{p+q}(\mathscr{F}; B)$, *which is natural in B.*

Proof. (i) Let $\mathscr{E} \xrightarrow{\gamma} \mathscr{S} \xleftarrow{\delta} \mathscr{F}$ be the geometric morphisms from $\mathscr{E}$ and $\mathscr{F}$ to $\mathscr{S}$, and let α denote the unit of $(f^* \dashv f_*)$. Since f^* is exact, the sequence of functors $(R^q(\delta_*) . f^*: \mathbf{ab}(\mathscr{E}) \longrightarrow \mathbf{ab}(\mathscr{S}) \mid q \geqslant 0)$ is exact and connected; hence by the universal property of derived functors, the natural transformation $\gamma_*(\alpha): \gamma_* \longrightarrow \gamma_* f_* f^* \cong \delta_* f^*$ extends uniquely to a natural transformation $R^q(\gamma_*) \longrightarrow R^q(\delta_*) . f^*$. Evaluating this transformation at A yields the desired homomorphism.

(ii) Apply 8.16 to the composite

$$(\mathbf{ab}(\mathscr{F}) \xrightarrow{f_*} \mathbf{ab}(\mathscr{E}) \xrightarrow{\gamma_*} \mathbf{ab}(\mathscr{S})) = (\mathbf{ab}(\mathscr{F}) \xrightarrow{\delta_*} \mathbf{ab}(\mathscr{S})). \quad ■$$

It follows from 8.17(i) that the group $H^q(\mathscr{E}, X; A)$ is contravariantly functorial in the variable X, since if $X \xrightarrow{f} Y$ is a morphism of $\mathscr{E}$ then $\mathscr{E}/Y \xrightarrow{f^*} \mathscr{E}/X$ is an inverse image functor. In view of 8.17(ii), we shall find it convenient to have an explicit description of the functors R^qf_*; this is provided by the following lemma.

8.18 LEMMA. *Let* $\mathscr{F} \xrightarrow{f} \mathscr{E}$ *be a geometric morphism, and suppose* $\mathscr{E} = \mathrm{Shv}(\mathbf{C}, J)$. *Then for an abelian group A in* $\mathscr{F}$ *and any q,* $R^qf_*(A)$ *is the associated J-sheaf of the presheaf*

$$U \longmapsto H^q(\mathscr{F}, f^*l(U); A)$$

on $\mathbf{C}$, *where* $\mathbf{C} \xrightarrow{l} \mathrm{Shv}(\mathbf{C}, J)$ *is the canonical functor.*

Proof. Suppose first that J is the minimal topology, so that $\mathscr{E} = \mathscr{S}^{\mathbf{C}^{\mathrm{op}}}$. Then the functor "evaluate at U": $\mathbf{ab}(\mathscr{E}) \longrightarrow \mathbf{ab}(\mathscr{S})$ is exact by 8.15(iv), so the functor $R^q f_*(-)(U)$ is the qth right derived functor of the functor

$$f_*(-)(U) \cong \hom_{\mathscr{E}}(h_U, f_*(-)) \cong \hom_{\mathscr{F}}(f^*(h_U), -) \cong H^0(\mathscr{F}, f^*(h_U), -):$$
$$\mathbf{ab}(\mathscr{F}) \longrightarrow \mathbf{ab}(\mathscr{S}).$$

This proves the result in the particular case; for the general case, let $\mathscr{E} \xrightarrow{i} \mathscr{S}^{\mathbf{C}^{\mathrm{op}}}$ be the inclusion, and write g for the composite morphism if. Then $i^*g_* = i^*i_*f_* \cong f_*$; but i^* is exact, and so $R^q f_* \cong R^q(i^*g_*) \cong i^*.R^q g_*$. Since i^* is the associated sheaf functor, the general statement of the lemma now follows from the particular case and the fact that $g^*(h_U) = f^*i^*(h_U) = f^*l(U)$. ■

The existence of enough injectives in $\mathbf{ab}(\mathscr{E})$ provides us with an existence theorem for derived functors; but for the actual computation of cohomology groups, it is not necessary to use injective coresolutions. Often it is more convenient to coresolve a given abelian group by groups which are acyclic for the functor under consideration; and in fact Barr's theorem provides us with a functorial method of doing this. We shall say that an abelian group A in a topos $\mathscr{E}$ is *flabby* if $H^q(\mathscr{E}, X; A) = 0$ for all objects X of $\mathscr{E}$ and all $q > 0$. (The word "flabby" is a translation of the French *flasque*, introduced by R. Godement [TF] for sheaves on topological spaces.)

8.19 LEMMA. *Let $\mathscr{F} \xrightarrow{f} \mathscr{E}$ be a geometric morphism, and A a flabby abelian group in $\mathscr{F}$. Then*

(i) *A is acyclic for the functor f_*:* $\mathbf{ab}(\mathscr{F}) \longrightarrow \mathbf{ab}(\mathscr{E})$.
(ii) *$f_*(A)$ is flabby in $\mathscr{E}$.*

Proof. (i) follows immediately from the description of $R^q f_*$ given in 8.18.

(ii) Consider the Leray spectral sequence of f. We have

$$\begin{aligned} E_2^{p,q} = H^p(\mathscr{E}; R^q f_*(A)) &= 0 \qquad \text{for } q > 0, \text{ by part (i)} \\ &= H^p(\mathscr{E}; f_*(A)) \qquad \text{for } q = 0. \end{aligned}$$

Now all the differentials in the sequence must vanish, and so $E_2^{p,q} = E_\infty^{p,q}$. But the latter is a filtration of $H^{p+q}(\mathscr{F}; A)$, which is zero for $p + q > 0$; hence we must have $H^p(\mathscr{E}; f_*(A)) = 0$ for $p > 0$.

So $f_*(A)$ is acyclic for the functor $\hom_{\mathscr{E}}(1, -)$; to obtain the same result for $\hom_{\mathscr{E}}(X, -)$, apply the above argument to the geometric morphism

$\mathscr{F}/f^*X \xrightarrow{f/X} \mathscr{E}/X$, using the fact that

$$X^*f_*(A) \cong (f/X)_*(f^*X)^*(A)$$

by the Beck condition for pullbacks in $\mathfrak{Top}$ (Ex. 4.7). ■

Now let $\mathscr{F} \xrightarrow{f} \mathscr{E}$ be a geometric morphism; write T for the composite functor $f_* f^*$, and α for the unit of $(f^* \dashv f_*)$. Then for any object A of $\mathscr{E}$, we have an augmented cosimplicial object of the form

$$A \xrightarrow{\alpha_A} TA \underset{T\alpha_A}{\overset{\alpha TA}{\rightrightarrows}} TTA \;\substack{\longrightarrow\\ \longrightarrow\\ \longrightarrow}\; TTTA \;\substack{\longrightarrow\\ \longrightarrow\\ \longrightarrow\\ \longrightarrow}\; \ldots;$$

and if A has an abelian group structure, we can turn this into a cochain complex $F^{\cdot}(A)$ by taking alternating sums of the co-face maps. Now the counit of $(f^* \dashv f_*)$ provides a contracting homotopy for the augmented complex

$$f^*A \longrightarrow f^*TA \longrightarrow f^*TTA \longrightarrow \ldots$$

in $\mathbf{ab}(\mathscr{F})$; hence if f is a surjection (so that f^* reflects the property of having zero cohomology) the sequence

$$0 \longrightarrow A \longrightarrow TA \longrightarrow TTA \longrightarrow \ldots$$

is exact in $\mathbf{ab}(\mathscr{E})$, i.e. $F^{\cdot}(A)$ is a coresolution of A.

8.20 THEOREM. *Let $\mathscr{E}$ be a Grothendieck topos, and let $\mathscr{F} \xrightarrow{f} \mathscr{E}$ be a surjection such that $\mathscr{F}$ satisfies* (AC). *(The existence of such a surjection was established in*7.57.*) Then the cohomology of $\mathscr{E}$ coincides with the cohomology of the comonad $\mathbb{G}$ on $\mathscr{F}$ induced by $(f^* \dashv f_*)$ (cf.* [152]).

Proof. Let X be any object of $\mathscr{E}$, A an abelian group in $\mathscr{E}$. By definition, the comonad cohomology group $H^q(\mathbb{G}, X; A)$ is the qth cohomology group of the complex $\hom_{\mathscr{E}}(X, F^{\cdot}(A))$ in $\mathbf{ab}(\mathscr{S})$. Now epimorphisms in $\mathbf{ab}(\mathscr{F})$ are split epi in $\mathscr{F}$, so the functor f_* preserves them and is therefore exact; but f^* is also exact, and so $F^{\cdot}$ is an exact functor from $\mathbf{ab}(\mathscr{E})$ to the category of cochain complexes in $\mathbf{ab}(\mathscr{E})$. Moreover, every abelian group in $\mathscr{F}$ is flabby by 8.15(iv); so by 8.19(ii) the groups in the complex $F^{\cdot}(A)$ are flabby in $\mathscr{E}$, and hence

$\hom_{\mathscr{E}}(X, F^{\cdot}(-))$ is an exact functor. Thus the functors $H^q(\mathbb{G}, X; -) = H^q(\hom_{\mathscr{E}}(X, F^{\cdot}(-)))$ form an exact connected sequence.

Now suppose A is injective in $\mathbf{ab}(\mathscr{E})$, and let $TA \overset{s}{\twoheadrightarrow} A$ be any splitting for the monomorphism $A \overset{\alpha_A}{\rightarrowtail} TA$. Then by applying successive powers of T to s, we obtain a contracting homotopy for the augmented complex $A \longrightarrow F^{\cdot}(A)$ in $\mathbf{ab}(\mathscr{E})$, and hence $H^q(\mathbb{G}, X; A) = 0$ for $q > 0$. So the functor $H^q(\mathbb{G}, X; -)$ is (isomorphic to) the qth right derived functor of

$$H^0(\mathbb{G}, X; -) = H^0(\hom_{\mathscr{E}}(X, F^{\cdot}(-))) \cong \hom_{\mathscr{E}}(X, -). \ \blacksquare$$

If $\mathscr{E} = \mathrm{Shv}(X)$ for some topological space X, then we can take $\mathscr{F} = \mathscr{S}/X$ in 8.20, by virtue of 4.18(iii). In this case, the coresolution $F^{\cdot}(A)$ is the well-known "Godement construction" [TF] for calculating sheaf cohomology. Similarly, if $\mathscr{E} = \mathscr{S}^{\mathbf{G}}$ or the topos of 8.15(iii), we can take $\mathscr{F} = \mathscr{S}$; in this case we obtain what is commonly called the "co-induced resolution". (Specifically, $F^n(A)$ is the set of all continuous (but not necessarily G-equivariant) maps $G^{n+1} \longrightarrow A$, with abelian group structure induced by that of A and G acting by translation of each factor of the domain.)

8.2. ČECH COHOMOLOGY

In this paragraph, we shall develop a technique for computing the cohomology of a Grothendieck topos in terms of a suitable site of definition for it. We shall consider a fixed topos $\mathscr{E} = \mathrm{Shv}(\mathbf{C}, J)$; we write $\mathscr{P}$ for the corresponding presheaf topos $\mathscr{S}^{\mathbf{C}^{\mathrm{op}}}$, and $i \colon \mathscr{E} \longrightarrow \mathscr{P}$ for the canonical inclusion. Let $\mathbb{H}^q \colon \mathbf{ab}(\mathscr{E}) \longrightarrow \mathbf{ab}(\mathscr{P})$ denote the qth right derived functor of i_*.

8.21 LEMMA. *Let A be an abelian group in $\mathscr{E}$. Then for all $q > 0$, we have $(\mathbb{H}^q(A))^+ = 0$, where $+ : \mathscr{P} \longrightarrow \mathscr{P}$ is the "half-sheafification" functor defined in* §3.3.

Proof. Consider the spectral sequence of the composite

$$(\mathbf{ab}(\mathscr{E}) \xrightarrow{i_*} \mathbf{ab}(\mathscr{P}) \xrightarrow{i^*} \mathbf{ab}(\mathscr{E})) = (\mathbf{ab}(\mathscr{E}) \xrightarrow{1} \mathbf{ab}(\mathscr{E})).$$

Since i^* is exact, we have $E_2^{p,q} = 0$ for $p > 0$, and $E_2^{0,q} = i^*.R^q(i_*) = i^*.\mathbb{H}^q$. But $1_{\mathbf{ab}(\mathscr{E})}$ is also exact, and so $E_\infty^{p,q} = 0$ for all $p + q > 0$. Since there cannot be any nonzero differentials in the sequence, we must therefore have $E_2^{0,q} = 0$

for $q > 0$. Thus $(\mathbb{H}^q(A))^{++} = i_* i^*(\mathbb{H}^q(A)) = 0$ for $q > 0$. But by 3.36 $(\mathbb{H}^q(A))^+$ is j-separated, and so by 3.34 the canonical map $(\mathbb{H}^q(A))^+ \longrightarrow (\mathbb{H}^q(A))^{++}$ is mono. Hence we must have $(\mathbb{H}^q(A))^+ = 0$. ■

We now suppose that **C** has pullbacks. Let $\mathscr{U} = (U_\gamma \longrightarrow U \mid \gamma \in \Gamma)$ be a family of morphisms of **C** with common codomain; and for each $\sigma = (\gamma_0, \gamma_1, \ldots, \gamma_n) \in \Gamma^{n+1}$, define U_σ to be the $(n+1)$-fold pullback $U_{\gamma_0} \times_U U_{\gamma_1} \times_U \ldots \times_U U_{\gamma_n}$. Then we have an augmented simplicial object in $\mathscr{P}$, of the form

$$\ldots \begin{smallmatrix}\longrightarrow\\ \longrightarrow\\ \longrightarrow\\ \longrightarrow\end{smallmatrix} \coprod_{\tau \in \Gamma^3} h_{U_\tau} \begin{smallmatrix}\longrightarrow\\ \longrightarrow\\ \longrightarrow\end{smallmatrix} \coprod_{\sigma \in \Gamma^2} h_{U_\sigma} \rightrightarrows \coprod_{\gamma \in \Gamma} h_{U_\gamma} \longrightarrow h_U,$$

in which the face maps are obtained by combining the canonical projections $U_\tau \longrightarrow U_\sigma$ where σ is a "face" of τ. Applying the free functor $\mathscr{P} \xrightarrow{Z} \mathbf{ab}(\mathscr{P})$, we obtain a simplicial object in $\mathbf{ab}(\mathscr{P})$, which we can convert into a chain complex by taking alternating sums of the face maps. We denote this complex by

$$N_{\bullet}(\mathscr{U}) \longrightarrow Z(h_U).$$

8.22 LEMMA. *The sequence* $\ldots \longrightarrow N_3(\mathscr{U}) \longrightarrow N_2(\mathscr{U}) \longrightarrow N_1(\mathscr{U}) \longrightarrow N_0(\mathscr{U})$ *is exact in* $\mathbf{ab}(\mathscr{P})$.

Proof. It is sufficient to show that for each $V \in C_0$, the sequence

$$\ldots \longrightarrow N_2(\mathscr{U})(V) \longrightarrow N_1(\mathscr{U})(V) \longrightarrow N_0(\mathscr{U})(V)$$

is exact in $\mathbf{ab}(\mathscr{S})$. But since the functor "evaluate at V" is an inverse image functor, it commutes with the free-abelian-group functor, and so $N_q(\mathscr{U})(V)$ is the free abelian group generated by $\coprod_{\sigma \in \Gamma^{q+1}} (\hom_{\mathbf{C}}(V, U_\sigma))$. For each morphism $V \xrightarrow{\phi} U$ in **C**, let S_ϕ denote the set $\coprod_{\gamma \in \Gamma} (\hom_{\mathbf{C}/U}(\phi, \theta_\gamma))$; then since the U_σ were defined as pullbacks, it is easily seen that

$$\coprod_{\sigma \in \Gamma^{q+1}} (\hom_{\mathbf{C}}(V, U_\sigma)) \cong \coprod_{\phi \in \hom(V, U)} (S_\phi)^{q+1}.$$

And since $\mathbf{ab}(\mathscr{S})$ satisfies axiom AB4, a coproduct of exact sequences is exact, and we are thus reduced to proving that a sequence of the form $\ldots \longrightarrow Z(S^3) \longrightarrow Z(S^2) \longrightarrow Z(S)$ is exact, where the maps are alternating sums of those induced by the product projections $S^{q+1} \longrightarrow S^q$. But if S

is empty, this is trivial; if not, we can construct a contracting homotopy for this complex by choosing an element $t \in S$, and taking the homomorphism $Z(S^q) \longrightarrow Z(S^{q+1})$ induced by $(s_1, \ldots, s_q) \longmapsto (t, s_1, \ldots, s_q)$. ∎

Now let A be any abelian group in $\mathscr{P}$, and define $C^{\cdot}(\mathscr{U}; A)$ to be the cochain complex $\hom_{\mathbf{ab}(\mathscr{P})}(N_{\cdot}(\mathscr{U}), A)$ in $\mathbf{ab}(\mathscr{S})$. Since the groups in the complex $N_{\cdot}(\mathscr{U})$ are free, this can equivalently be written in the form

$$\prod_{\gamma \in \Gamma} A(U_\gamma) \longrightarrow \prod_{\sigma \in \Gamma^2} A(U_\sigma) \longrightarrow \prod_{\tau \in \Gamma^3} A(U_\tau) \longrightarrow \cdots .$$

We write $H^q(\mathscr{U}; A)$ for the qth cohomology group of this complex, and call it the qth *Čech cohomology group of* $\mathscr{U}$ with coefficients in A. We also use the terms *Čech cochain*, *Čech cocycle*, etc., for elements of the groups $C^q(\mathscr{U}; A)$. Note in particular that

$$H^0(\mathscr{U}; A) = \text{eq}\, \Big(\prod_{\gamma \in \Gamma} A(U_\gamma) \rightrightarrows \prod_{\sigma \in \Gamma^2} A(U_\sigma)\Big) \cong \hom_{\mathscr{P}}(R, A),$$

where R is the sieve on U generated by the family $\mathscr{U}$; so $H^0(\mathscr{U}; A) \cong A(U)$ iff A "satisfies the sheaf axiom" for $\mathscr{U}$.

8.23 PROPOSITION. *$H^q(\mathscr{U}; -)$ is the qth right derived functor of $H^0(\mathscr{U}; -)$.*

Proof. Since representable presheaves are projective in $\mathscr{P}$, the groups in the complex $N_{\cdot}(U)$ are projective in $\mathbf{ab}(\mathscr{P})$, and so $A \longmapsto C^{\cdot}(\mathscr{U}; A)$ is an exact functor from $\mathbf{ab}(\mathscr{E})$ to the category of cochain complexes in $\mathbf{ab}(\mathscr{S})$. So the functors $H^q(\mathscr{U}; -)$ form an exact connected sequence. But if A is injective, then it follows from 8.22 that the complex $C^{\cdot}(\mathscr{U}; A)$ is exact in positive dimensions, and so $H^q(\mathscr{U}; A) = 0$ for $q > 0$. The result follows from the uniqueness theorem for derived functors. ∎

Now let $U = (U_\gamma \longrightarrow U \mid \gamma \in \Gamma)$ and $\mathscr{V} = (V_\delta \longrightarrow U \mid \delta \in \Delta)$ be two families of morphisms of $\mathbf{C}$ with the same codomain. By a *refinement map* $r: \mathscr{V} \longrightarrow \mathscr{U}$, we mean a function $r: \Delta \longrightarrow \Gamma$, together with a family of factorizations

$$\begin{array}{ccc} V_\delta & \xrightarrow{r_\delta} & U_{r(\delta)} \\ & \searrow & \downarrow \\ & & U \end{array}$$

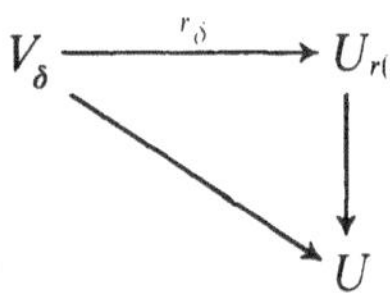

for each $\delta \in \Delta$. Clearly, each refinement map $\mathscr{V} \xrightarrow{r} \mathscr{U}$ induces a morphism $V_\sigma \longrightarrow U_{r(\sigma)}$ over U, and hence a chain map $r. : N.(\mathscr{V}) \longrightarrow N.(\mathscr{U})$. ■

8.24 Lemma. *Let r, s be two refinement maps* $\mathscr{V} \longrightarrow \mathscr{U}$. *Then the chain maps* $r.$ *and* $s.$ *are chain-homotopic.*

Proof. For each $\sigma = (\delta_0, \delta_1, \ldots, \delta_q) \in \Delta^{q+1}$, and each $i \in \{0, 1, \ldots, q\}$, define t_σ^i to be the morphism

$$(r_{\delta_0}, \ldots, r_{\delta_i}, s_{\delta_i}, \ldots, s_{\delta_q}) \colon V_\sigma \longrightarrow U_{(r(\delta_0), \ldots, r(\delta_i), s(\delta_i), \ldots, s(\delta_q))}$$

over U; and let $t_q^i \colon N_q(\mathscr{V}) \longrightarrow N_{q+1}(\mathscr{U})$ be the group homomorphism induced by the t_σ^i. Then it is straightforward to verify that the alternating sum $\sum_i (-1)^i t_q^i$ is the qth component of a chain-homotopy between $r.$ and $s.$. ■

8.25 Corollary. *Let* $\mathscr{U} = (U_\gamma \to U \mid \gamma \in \Gamma)$ *be a family of morphisms of* **C**, *and* $R \rightarrowtail h_U$ *the sieve on U generated by* $\mathscr{U}$. *Then the groups* $H^q(\mathscr{U}; A)$ *and* $H^q(R; A)$ *are canonically isomorphic for any* $A \in \mathbf{ab}(\mathscr{P})$.

Proof. Since every morphism in R factors through one of the U_γ, there exists a refinement map $R \longrightarrow \mathscr{U}$; but the inclusion map is clearly a refinement map $\mathscr{U} \longrightarrow R$. So by 8.24 the complexes $N.(\mathscr{U})$ and $N.(R)$ are chain-homotopy equivalent; hence so are $C^\cdot(\mathscr{U}; A)$ and $C^\cdot(R; A)$, and so their cohomology groups are isomorphic. Moreover, this isomorphism does not depend on the choice of refinement map $R \longrightarrow \mathscr{U}$, since two different choices induce cochain maps $C^\cdot(\mathscr{U}; A) \longrightarrow C^\cdot(R; A)$ which differ by a homotopy. ■

We are now ready to define the Čech cohomology of a site (**C**, J); the definition involves the groups $H^q(\mathscr{U}; A)$ only in the case when $\mathscr{U}$ is a J-covering sieve, but 8.25 tells us that we may replace any covering sieve by a family which generates it, if computation of the groups is thereby made easier. Note in passing that if R and S are two sieves on U, then there exists a refinement map $R \longrightarrow S$ iff $R \subseteq S$, in which case the map may be taken to be the inclusion. So the complexes $C^\cdot(R; A)$ (and *a fortiori* the groups $H^q(R; A)$) form a diagram over the poset $\mathbf{J}(U)^{\mathrm{op}}$.

8.26 Definition. Let (**C**, J) be a site such that **C** has pullbacks, U an object of **C** and A a presheaf of abelian groups on **C**. We define the qth *Čech cohomology* *group of U* with coefficients in A to be the (filtered) colimit

$$\check{H}^q(U;A) = \varinjlim_{R\in J(U)} H^q(R;A). \blacksquare$$

Since the functor $\varinjlim_{J(U)}$ is exact, the functors $\check{H}^q(U;-)$ form an exact connected sequence on $\mathbf{ab}(\mathscr{P})$, and they vanish on injectives for $q > 0$. Normally, we shall consider Čech cohomology groups with coefficients in a *sheaf*; however, the functors $\check{H}^q(U;-)$ do not in general form an exact connected sequence on $\mathbf{ab}(\mathscr{E})$, since the inclusion functor $i_*: \mathbf{ab}(\mathscr{E}) \longrightarrow \mathbf{ab}(\mathscr{P})$ is not exact. Note that

$$\check{H}^0(U;A) \cong \varinjlim_{J(U)} \hom_{\mathscr{E}}(R,A) = A^+(U)$$

(cf. 3.33); and so the functor $\mathbf{ab}(\mathscr{P}) \longrightarrow \mathbf{ab}(\mathscr{P})$ which sends A to the presheaf $U \longmapsto \check{H}^q(U;A)$ is the qth right derived functor of $+$. (We have not in fact proved that $\check{H}^q(U;A)$ is functorial in the variable U; but this is not difficult—see Ex. 8.6 below.)

8.27 THEOREM. *Let* $\mathscr{E} = \mathrm{Shv}(\mathbf{C}, J)$, *let* U *be an object of* $\mathbf{C}$ *and let* A *be an abelian group in* $\mathscr{E}$. *Then there is a homomorphism* $\check{H}^q(U;A) \longrightarrow H^q(\mathscr{E}, l(U); A)$ *which is iso for* $q = 0$ *and* 1, *and mono for* $q = 2$. (*Here* l *denotes the canonical functor* $\mathbf{C} \longrightarrow \mathrm{Shv}(\mathbf{C}, J)$.)

Proof. Consider the spectral sequence of the composite

$$(\mathbf{ab}(\mathscr{E}) \xrightarrow{i_*} \mathbf{ab}(\mathscr{P}) \xrightarrow{+} \mathbf{ab}(\mathscr{P})) = (\mathbf{ab}(\mathscr{E}) \xrightarrow{i_*} \mathbf{ab}(\mathscr{P})),$$

composed with the exact functor "evaluate at U": $\mathbf{ab}(\mathscr{P}) \longrightarrow \mathbf{ab}(\mathscr{S})$. By the remarks immediately above, we have $E_2^{p,q} = \check{H}^p(U; \mathbb{H}^q(A))$; and $E_\infty^{p,q}$ is a filtration of the $(p+q)$th right derived functor of $\hom_{\mathscr{P}}(h_U, i_*(-)) \cong \hom_{\mathscr{E}}(l(U), -)$, evaluated at A. So we have an edge homomorphism

$$\check{H}^p(U;A) = E_2^{p,0} \twoheadrightarrow E_\infty^{p,0} \rightarrowtail H^p(\mathscr{E}, l(U); A).$$

Now

$$E_2^{0,q} = \check{H}^0(U; \mathbb{H}^q(A)) = (\mathbb{H}^q(A))^+(U) = 0 \text{ for } q > 0,$$

by 8.21. So for $p = 0, 1$ and 2 the canonical map $E_2^{p,0} \twoheadrightarrow E_\infty^{p,0}$ is iso, since these groups cannot be affected by any nonzero differential; and for $p = 0$ and 1 the map $E_\infty^{p,0} \rightarrowtail H^p(E, l(U); A)$ is iso, since $E_\infty^{p,0}$ is the only nonzero term in the filtration. ■

In general, Čech and ordinary cohomology need not agree in dimensions higher than 1. However, they do agree in many of the cases most frequently encountered; the following result, due to H. Cartan, is often useful.

8.28 PROPOSITION. *Let $\mathscr{E} = \mathbf{Shv}(\mathbf{C}, J)$, and let A be an abelian group in $\mathscr{E}$. Suppose that there is a subset K of C_0 such that*

(i) *For each $V \in K$, we have $\check{H}^q(V; A) = 0$ for $q > 0$.*
(ii) *For every $U \in C_0$, there exists a J-covering family $(V_\gamma \longrightarrow U \mid \gamma \in \Gamma)$ with each V_γ in K.*
(iii) *If V and W are in K, then any pullback of the form $V \times_U W$ is in K.*

Then the canonical map $\check{H}^q(U; A) \longrightarrow H^q(\mathscr{E}, l(U); A)$ is iso for all $U \in C_0$ and all q.

Proof. Since "a cover of a cover is a cover" (0.31(iii)), it follows from (ii) that any J-covering family $\mathscr{U} = (U_\beta \longrightarrow U \mid \beta \in \beta)$ can be refined to a covering family $\mathscr{V} = (V_\gamma \longrightarrow U \mid \gamma \in \Gamma)$ with each V_γ in K; hence each element of $\check{H}^q(U; A)$ derives from an element of $H^q(\mathscr{V}; A)$ for such a family $\mathscr{V}$. Moreover, if $\mathscr{V}$ is such a family, then it follows from (iii) that the objects $V_\sigma (\sigma \in \Gamma^{q+1})$ are all in K.

We now prove the proposition by induction on q. Suppose the result has been established for all $q < n$; and consider the complex $C^\bullet(\mathscr{V}; \mathbb{H}^q(A))$ where $0 < q < n$ and $\mathscr{V}$ is a covering family of the type considered above. By 8.18, we know that $\mathbb{H}^q(A)$ is the presheaf $U \longmapsto H^q(\mathscr{E}, l(U); A)$; but $H^q(\mathscr{E}, l(U); A) \cong \check{H}^q(U; A)$ by the inductive hypothesis, and so $C^p(\mathscr{V}; \mathbb{H}^q(A)) = 0$ for all p, since the objects at which we have to evaluate $\mathbb{H}^q(A)$ are all in K. Hence $\check{H}^p(U; \mathbb{H}^q(A)) = 0$; i.e. in the spectral sequence of 8.27 we have $E_2^{p,q} = 0$ for $0 < q < n$ and all p. Since we also know that $E_2^{0,n} = 0$, it follows that the groups $E_2^{p,0}, 0 \leqslant p \leqslant n$, are the only nonzero entries on or below the diagonal $p + q = n$; and we deduce as in 8.27 that the canonical maps $E_2^{n,0} \longrightarrow E_\infty^{n,0} \longrightarrow H^n(\mathscr{E}, l(U); A)$ are iso. So the result is established for $q = n$. ■

8.29 EXAMPLE. Let X be an open subspace of Euclidean n-space, and let K be the set of convex open subsets of X. Since a (nonempty) convex set is contractible, it may be shown that the first condition of 8.28 holds for any sheaf A on X; the second condition holds since open balls are convex, and the third holds since the intersection of two convex sets is convex. So the ordinary and Čech cohomology groups of X, with coefficients in any sheaf, are equal. ■

Further examples of the coincidence of Čech and ordinary cohomology are given in Exercises 8 and 9 at the end of this chapter.

8.3. TORSORS

In this paragraph we introduce a method of representing 1-dimensional cohomology classes by torsors (cf. 4.37(ii)), and sketch how the representation may be extended to higher-dimensional classes. The advantage of this representation is not only that it gives us a fairly explicit "picture" of a cohomology class, but also that it enables us to extend the definition of cohomology to the case when the coefficient group is non-abelian.

Recall that if G is an internal group in a topos $\mathscr{E}$, a G-torsor is a flat presheaf on the internal category $\mathbf{G}$; i.e. a left G-object $(G \times T \xrightarrow{\alpha} T)$ such that $T \longrightarrow 1$ is epi and $G \times T \xrightarrow{(\alpha, \pi_2)} T \times T$ is iso. (The second condition is clearly equivalent to the existence of a *division map* $T \times T \xrightarrow{\delta} G$ such that

$$\begin{array}{ccc} T \times T & \xrightarrow{(\delta, \pi_2)} & G \times T \\ & {}_{\pi_1}\searrow & \downarrow{\scriptstyle \alpha} \\ & & T \end{array} \qquad \text{and} \qquad \begin{array}{ccc} G \times T & \xrightarrow{(\alpha, \pi_2)} & T \times T \\ & {}_{\pi_1}\searrow & \downarrow{\scriptstyle \delta} \\ & & G \end{array}$$

commute.)

8.31 LEMMA. (i) *The category* $\mathrm{Flat}(\mathbf{G}^{\mathrm{op}}, \mathscr{E})$ *of G-torsors in $\mathscr{E}$ is a groupoid.*

(ii) *A G-torsor is isomorphic to the trivial torsor* $(G \times G \xrightarrow{m} G)$ *iff it has a global element.*

Proof. (i) Let T and U be G-torsors and $T \xrightarrow{f} U$ a G-equivariant map. Then the square

$$\begin{array}{ccc} G \times T & \xrightarrow{1 \times f} & G \times U \\ \downarrow{\scriptstyle (\alpha, \pi_2)} & & \downarrow{\scriptstyle (\alpha, \pi_2)} \\ T \times T & \xrightarrow{f \times f} & U \times U \end{array}$$

commutes, and hence so does the triangle

$$\begin{array}{ccc} T\times T & \xrightarrow{f\times f} & U\times U; \\ & \searrow^{\delta} & \downarrow{\scriptstyle\delta} \\ & & G \end{array}$$

but the square

$$\begin{array}{ccc} T & \longrightarrow & 1 \\ \downarrow{\scriptstyle\Delta} & & \downarrow{\scriptstyle e} \\ T\times T & \xrightarrow{\delta} & G \end{array}$$

is a pullback, and hence

$$\begin{array}{ccc} T & \xrightarrow{f} & U \\ \downarrow{\scriptstyle\Delta} & & \downarrow{\scriptstyle\Delta} \\ T\times T & \xrightarrow{f\times f} & U\times U \end{array}$$

is a pullback, which implies that f is mono. Also, it is easy to verify that the diagram

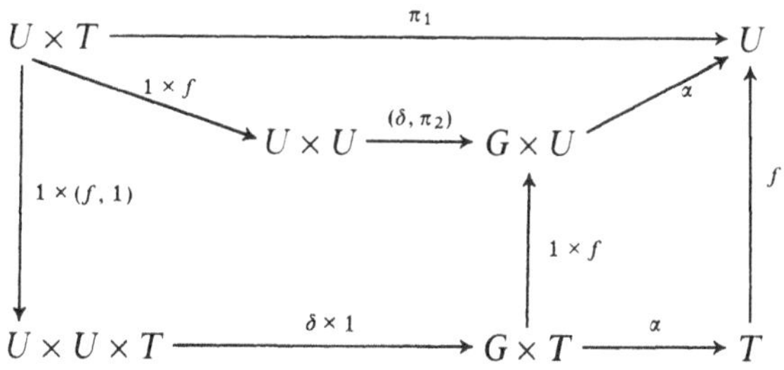

commutes; but $T \longrightarrow 1$ is epi and so $U \times T \xrightarrow{\pi_1} U$ is epi. Hence f must be epi.

(ii) Suppose T has a global element $1 \xrightarrow{t} T$. Then the morphism $G \xrightarrow{1\times t} G\times T \xrightarrow{\alpha} T$ is clearly G-equivariant; so by (i) it is an isomorphism. The converse is trivial. ■

Now suppose that G is abelian. Then if $(G \times M \xrightarrow{\alpha} M)$ is any left G-object,

$(M \times G \xrightarrow{(\pi_2, \pi_1)} G \times M \xrightarrow{\alpha} M)$ is a right G-object; and indeed the left and right G-actions on M commute, so that we can regard M as a profunctor $\mathbf{G} \dashrightarrow \mathbf{G}$. Hence there is a full subcategory of $\mathrm{Prof}_{\mathscr{E}}(\mathbf{G}, \mathbf{G})$ which is isomorphic both to $\mathscr{E}^{\mathbf{G}}$ and to $\mathscr{E}^{\mathbf{G}^{\mathrm{op}}}$; a profunctor in this subcategory will be said to be *symmetric*.

8.32 LEMMA. (i) *Let L, M be symmetric profunctors $\mathbf{G} \dashrightarrow \mathbf{G}$. Then the tensor product $L \otimes_{\mathbf{G}} M$ (2.42) is symmetric, and is isomorphic to $M \otimes_{\mathbf{G}} L$.*

(ii) *Let T be a presheaf on $\mathbf{G}$. The T is flat iff it is left flat (4.39) when regarded as a symmetric profunctor.*

Proof. (i) Consider the diagram

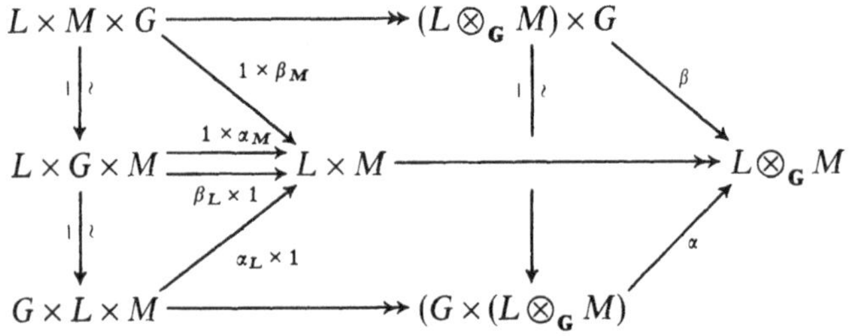

where α and β denote left and right actions of G. Since L and M are symmetric, the two left-hand triangles commute; and the middle row is a coequalizer, so the two possible composites $L \times M \times G \longrightarrow L \otimes_{\mathbf{G}} M$ are equal. But $L \times M \times G \twoheadrightarrow (L \otimes_{\mathbf{G}} M) \times G$ is epi, so the right-hand triangle commutes; i.e. $L \otimes_{\mathbf{G}} M$ is symmetric.

Now consider the diagram

$$\begin{array}{ccc}
L \times G \times M & \xrightarrow[\beta_L \times 1]{1 \times \alpha_M} L \times M & \longrightarrow L \otimes_{\mathbf{G}} M \\
\downarrow & \downarrow & \\
M \times G \times L & \xrightarrow[1 \times \alpha_L]{\beta_M \times 1} M \times L & \twoheadrightarrow M \otimes_{\mathbf{G}} L
\end{array}$$

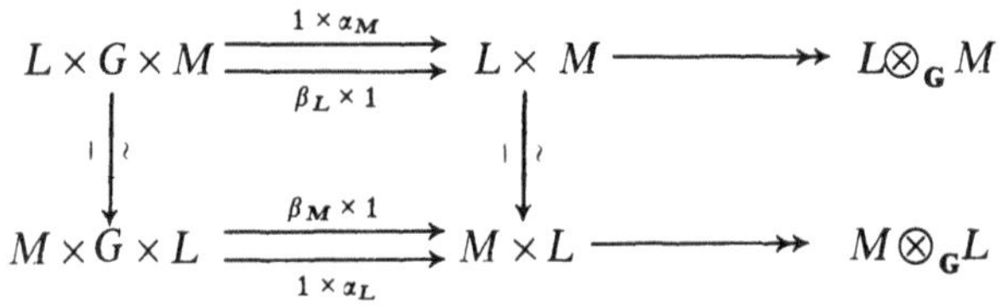

where the rows are coequalizers and the two left-hand squares commute by symmetry of L and M. We obtain an isomorphism $L \otimes_{\mathbf{G}} M \xrightarrow{\sim} M \otimes_{\mathbf{G}} L$, which is easily seen to be an isomorphism of profunctors.

(ii) follows at once from the fact that the forgetful functor $\mathscr{E}^{\mathbf{G}} \longrightarrow \mathscr{E}$ creates limits; for both conditions on T are equivalent to saying that $(-) \otimes_{\mathbf{G}} T$ preserves finite limits, the only difference being the category in which the limits are computed. ∎

Now a tensor product of left flat profunctors is clearly left flat; so it follows from 8.32 that the bifunctor $\otimes_{\mathbf{G}}$ restricts to a *symmetric* monoidal structure on Flat($\mathbf{G}^{\mathrm{op}}, \mathscr{E}$). (The unit of this structure is the trivial torsor ($G \times G \xrightarrow{m} G$), which corresponds under the isomorphism to the Yoneda profunctor $Y(\mathbf{G})$.) Moreover, any torsor ($G \times T \xrightarrow{\alpha} T$) has a $\otimes_{\mathbf{G}}$-inverse $\bar{T}$, which has the same underlying object as T and G-action map $G \times T \xrightarrow{i \times 1} G \times T \xrightarrow{\alpha} T$ (i is the inverse map $G \longrightarrow G$); for the division map $T \times T \xrightarrow{\delta} G$ induces a G-equivariant map $\bar{T} \otimes_{\mathbf{G}} T \longrightarrow G$, which must be an isomorphism by 8.31(i). So the isomorphism classes of objects (= connected components) of the groupoid Flat($\mathbf{G}^{\mathrm{op}}, \mathscr{E}$) form an abelian group, which we denote by $\mathrm{Tors}^1(\mathscr{E}; G)$.

8.33 THEOREM. *Let $\mathscr{E}$ be a Grothendieck topos, G an abelian group in $\mathscr{E}$. Then*

$$H^1(\mathscr{E}; G) \cong \mathrm{Tors}^1(\mathscr{E}; G).$$

Proof. Let Z be the free abelian group generated by 1 in $\mathscr{E}$. Then since $H^1(\mathscr{E}; G) \cong \mathrm{Ext}^1_{\mathbf{ab}(\mathscr{E})}(Z, G)$, we know that elements of $H^1(\mathscr{E}; G)$ correspond to isomorphism classes of short exact sequences $0 \longrightarrow G \longrightarrow E \longrightarrow Z \longrightarrow 0$ in $\mathbf{ab}(\mathscr{E})$. Suppose we are given such a sequence; form the pullback

$$\begin{array}{ccc} T & \longrightarrow & 1 \\ \downarrow & & \downarrow{\scriptstyle u} \\ E & \longrightarrow & Z \end{array}$$

in $\mathscr{E}$, where u is the insertion of the generator of Z. Since

$$\begin{array}{ccc} G & \longrightarrow & 1 \\ \downarrow & & \downarrow{\scriptstyle 0} \\ E & \longrightarrow & Z \end{array}$$

is also a pullback, the addition in E restricts to a G-action on T. Now $T \longrightarrow 1$ is epi, since $E \longrightarrow Z$ is; and it is easily verified that the composite $E \times E \xrightarrow{i \times 1} E \times E \xrightarrow{+} E$ restricts to a division map $T \times T \longrightarrow G$, so that T is a G-torsor.

Conversely, suppose we are given a G-torsor T. Define E to be the coproduct $\coprod_{p \in \mathbb{Z}} T^{\otimes p}$, where $T^{\otimes p}$ denotes the pth power of T for the product $\otimes_{\mathbf{G}}$. Then the canonical maps

$$T^{\otimes p} \times T^{\otimes q} \twoheadrightarrow T^{\otimes p} \otimes_{\mathbf{G}} T^{\otimes q} \cong T^{\otimes (p+q)} \ (p, q \in \mathbb{Z})$$

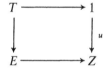

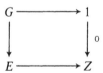

combine to give a morphism $E \times E \longrightarrow E$, which is easily seen to be an abelian group structure on E. Moreover, the canonical map $\coprod_{p \in \mathbb{Z}} T^{\otimes p} \longrightarrow \coprod_{p \in \mathbb{Z}} 1$ is an abelian group homorphism $E \longrightarrow Z$, which is epi (since each $T^{\otimes p} \longrightarrow 1$ is), and whose kernel is $T^{\otimes 0} \cong G$. We have thus constructed a short exact sequence $0 \longrightarrow G \longrightarrow E \longrightarrow Z \longrightarrow 0$ in $\mathbf{ab}(\mathscr{E})$.

To show that the two constructions given above are inverse up to natural isomorphism, we need to show that if $0 \longrightarrow G \longrightarrow E \longrightarrow Z \longrightarrow 0$ is any short exact sequence and T_p is the pullback of $E \longrightarrow Z$ along the pth power of the generator of Z, then $T_p \cong T^{\otimes p}$. But this is an easy induction on p, since the addition in E restricts to a map $T_p \times T_q \longrightarrow T_{p+q}$, and hence induces an isomorphism $T_p \otimes_{\mathbf{G}} T_q \cong T_{p+q}$, for each pair (p, q).

We have thus established a bijection between the sets $H^1(\mathscr{E}; G)$ and $\mathrm{Tors}^1(\mathscr{E}; G)$. The proof that the additive structure on $\mathrm{Ext}^1_{\mathbf{ab}(\mathscr{E})}(Z, G)$ agrees with that induced by tensor product of torsors is tedious but straightforward. ■

In fact the equivalence between G-torsors and extensions of Z by G is valid in any topos with a natural number object, since we can use the internal nth-tensor-power construction of 6.37 to take the place of the external $\mathbb{Z}$-indexed coproduct which we used above. However, in the general case we have no guarantee that the isomorphism classes of G-torsors in $\mathscr{E}$ can be parametrized by a set.

8.34 Corollary. *Let G be an abelian group in a Grothendieck topos $\mathscr{E}$. Then the topos $\mathscr{E}^{\mathbf{G}}$ is a classifying topos for 1-dimensional cohomology with coefficients in G; i.e. a representing object for the functor*

$$H^1(-; G): (\mathfrak{BTop}/\mathscr{E})^{\mathrm{op}} \longrightarrow \mathbf{ab}(\mathscr{S}). \blacksquare$$

Now let us return to the case of a nonabelian group G. In view of 8.33, it seems reasonable to *define* $H^1(\mathscr{E}; G)$ as the set of isomorphism classes of G-torsors in $\mathscr{E}$. This definition of course assumes that these isomorphism classes may indeed by parametrized by a set—an assumption which is valid for any Grothendieck topos, as the following lemma shows:

8.35 Lemma. *Let $\mathscr{E}$ be a Grothendieck topos, X an object of $\mathscr{E}$. Then the $\mathscr{E}$-isomorphism classes of objects of objects which are locally isomorphic to X (i.e. isomorphic to V^*X in $\mathscr{E}/V$, for some V having global support) may be parametrized by a set.*

Proof. Let Y be an object of $\mathscr{E}$ such that we have an isomorphism $V^*Y \xrightarrow[\simeq]{\alpha} V^*X$ in $\mathscr{E}/V$. Since V has global support, the diagram

$$Y \times V \times V \underset{\pi_{13}}{\overset{\pi_{12}}{\rightrightarrows}} Y \times V \xrightarrow{\pi_1} Y$$

is a coequalizer, and hence so is

$$X \times V \times V \underset{\pi_{13}.\beta}{\overset{\pi_{12}}{\rightrightarrows}} X \times V \xrightarrow{\pi_1.\alpha^{-1}} Y,$$

where β is the composite isomorphism

$$(V \times V)^*(X) \xrightarrow{\pi_1^*(\alpha^{-1})} (V \times V)^*(Y) \xrightarrow{\pi_2^*(\alpha)} (V \times V)^*(X).$$

But since $\mathscr{E}$ is defined over $\mathscr{S}$, there is only a set of possible choices for β, by 4.41; and Y is determined up to isomorphism by β.

We must now consider the possible choices for V. Let $(\mathbf{C}, J)$ be a site of definition for $\mathscr{E}$ such that $\mathbf{C}$ has a terminal object; then saying that V has global support is clearly equivalent to saying that $\{U \in C_0 \mid V(U) \text{ is nonempty}\}$ is a J-covering sieve on 1 in $\mathbf{C}$. Let R denote this sieve, and let $W = \coprod_{U \in R} l(U)$; then W has global support, and there exists a morphism $W \longrightarrow V$ in $\mathscr{E}$. So it is sufficient to restrict our attention to objects V of the form $\coprod_{U \in R} l(U)$, for some $R \in J(1)$; but there is only a set of such objects. ■

Now (the underlying object of) a G-torsor T is locally isomorphic to G, since $T^*T \cong T^*G$ in $\mathscr{E}/T$; so we may now define $H^1(\mathscr{E}; G)$ (for a Grothendieck topos $\mathscr{E}$) as the set of isomorphism classes of G-torsors in $\mathscr{E}$. Of course, we no longer have a group structure on this set, since the arguments of 8.32 are valid only when G is abelian; but the isomorphism class of the trivial G-torsor gives us a distinguished element of $H^1(\mathscr{E}; G)$, and so we can regard $H^1(\mathscr{E}; -)$ as a functor from $\mathbf{gp}(\mathscr{E})$ to the category $\mathbf{pt}(\mathscr{S})$ of pointed sets. (To show that it is functorial, let $G \xrightarrow{f} H$ be a group homomorphism and T a G-torsor; then we define $f_*(T)$ to be $f^\# \otimes_{\mathbf{G}} T$, where $f^\#: \mathbf{H} \dashrightarrow \mathbf{G}$ is the (left flat) profunctor defined in 2.44.) Similarly, we define $H^0(\mathscr{E}; G)$ to be the group of global elements of G in $\mathscr{E}$, viewed as a pointed set; and then we have

8.36 PROPOSITION. *Let G be a group in a topos $\mathscr{E}$, and K a normal subgroup of G.*

Then we have an exact sequence of pointed sets:

$$1 \longrightarrow H^0(\mathscr{E}; K) \longrightarrow H^0(\mathscr{E}; G) \longrightarrow H^0(\mathscr{E}; G/K) \xrightarrow{\partial} H^1(\mathscr{E}; K) \longrightarrow$$
$$H^1(\mathscr{E}; G) \longrightarrow H^1(\mathscr{E}; G/K).$$

Proof. Let $K \xrightarrow{k} G \xrightarrow{q} G/K$ denote the canonical homomorphisms.

The connecting map ∂ is constructed as in 8.33; i.e. given an element $1 \xrightarrow{u} G/K$, we form the pullback

$$\begin{array}{ccc} T & \longrightarrow & 1 \\ \downarrow{\scriptstyle h} & & \downarrow{\scriptstyle u} \\ G & \xrightarrow{q} & G/K \end{array}$$

and let K act on T by means of the multiplication in G. The proof that T is a K-torsor is as in 8.33.

Now exactness of the sequence at $H^0(\mathscr{E}; K)$ and $H^0(\mathscr{E}; G)$ follows from left exactness of the functor $\hom_{\mathscr{E}}(1, -)$; for exactness at $H^0(\mathscr{E}; G/K)$, observe that the K-torsor $T = \partial(u)$ has a global element iff u factors through $G \xrightarrow{q} G/K$.

Now let $T = \partial(u)$ be a K-torsor in the image of ∂, and consider the diagram

$$\begin{array}{ccccc} K \times T & \overset{\alpha}{\underset{\pi_2}{\rightrightarrows}} & T & \twoheadrightarrow & 1 \\ \downarrow{\scriptstyle (im(k\times h),\, 1)} & & \downarrow{\scriptstyle (ih,\, 1)} & & \\ G \times K \times T & \overset{1\times\alpha}{\underset{m(1\times k)\times 1}{\rightrightarrows}} & G \times T & \twoheadrightarrow & k_*(T) \end{array}$$

where i, m denote the inverse and multiplication maps of G. Here the first row is a coequalizer since T is a K-torsor, and the second is a coequalizer by the definition of $k_*(T)$. So $k_*(T)$ has a global element, and is thus trivial. Conversely, let T be any K-torsor such that $k_*(T)$ is trivial; then it is not hard to verify that the composite $T \xrightarrow{e\times 1} G \times T \twoheadrightarrow k_*(T) \cong G$ is K-equivariant (where K acts on G by left multiplication), and hence we have a commutative diagram of the form

$$\begin{array}{ccc} T & \longrightarrow & 1 \\ \downarrow & & \downarrow{\scriptstyle u} \\ G & \longrightarrow & G/K \end{array}$$

(since 1 and G/K are the quotients of T and G by their respective K-actions). So we have a morphism of K-torsors $T \longrightarrow \partial(u)$, which must be an isomorphism by 8.31(i). Thus the sequence is exact at $H^1(\mathscr{E}; K)$.

Since $H^1(\mathscr{E}; -)$ is functorial, the composite $H^1(\mathscr{E}; K) \longrightarrow H^1(\mathscr{E}; G) \longrightarrow H^1(\mathscr{E}; G/K)$ is clearly the zero map. Conversely, let V be a G-torsor: then it is easily verified that the composite $V \xrightarrow{e \times 1} G/K \times V \twoheadrightarrow q_*(V)$ is epi, and hence that

$$K \times V \underset{\pi_2}{\overset{\alpha(k \times 1)}{\rightrightarrows}} V \twoheadrightarrow q_*(V)$$

is a coequalizer. So if $q_*(V)$ is trivial, the pullback

$$\begin{array}{ccc} T & \longrightarrow & 1 \\ \downarrow & & \downarrow \\ V & \longrightarrow & q_*(V) \end{array}$$

is a sub-K-object of V, and is easily checked to be a K-torsor; and the morphism

$$G \times T \rightarrowtail G \times V \xrightarrow{\alpha} V$$

induces a G-equivariant map $k_*(T) \longrightarrow V$, which must be an isomorphism. So the sequence is exact at $H^1(\mathscr{E}; G)$. ■

It is natural to ask whether we may define the cohomology of a topos, with nonabelian coefficients, in dimensions greater than 1, in such a way as to extend the exact sequence of 8.36. The first significant answer to this question was given by J. Giraud [38], who produced a workable, though somewhat unwieldy, definition of $H^2(\mathscr{E}; G)$ in terms of equivalence classes of "extensions of $\mathscr{E}$ by G". (See also [WB], chapter 10.) More recently, J. W. Duskin [32] has introduced a notion of "$K(G, n)$-torsor", which appears to solve the problem of defining $H^n(\mathscr{E}; G)$ for all n.

Duskin's definition relies heavily on the theory of simplicial objects; he first defines, for each $n \geqslant 0$, a simplicial object $K(G, n)$, which plays roughly the same rôle in simplical theory that Eilenberg-Mac Lane spaces do in homotopy theory (in particular, $K(G, 1)$ is the category $\mathbf{G} = (G \rightrightarrows 1)$ considered as a simplicial object (2.13) and $K(G, 0)$ is the discrete category with object-of-objects G). He then defines a $K(G, n)$-torsor to be simplicial object satisfying certain exactness conditions and "fibred over $K(G, n)$" in a suitable sense; in particular for $n = 0$ or 1, a $K(G, n)$-torsor is simply a flat presheaf on $K(G, n)$ (thus agreeing with our previous definitions of H^0 and H^1). For $n > 1$, the category of $K(G, n)$-torsors is not in general a groupoid (cf. the Yoneda representation [187] for Ext-groups), and so we must define $H^n(\mathscr{E}; G)$ to be the set of connected components (rather than isomorphism classes) of the category of $K(G, n)$-torsors. The distinguished element of $H^n(\mathscr{E}; G)$ is that containing a certain simplicial object $L(G, n)$, which is fibred over $K(G, n)$ with fibre $K(G, n - 1)$.

To give a complete account of Duskin's generalized torsors would involve us in an extensive discussion of the theory of simplicial objects, and would thus occupy more space than we can reasonably afford to devote to it. However, in view of its potential importance, we shall give a brief description (equivalent but not identical to Duskin's) of the notion of a $K(G, 2)$-torsor, together with a few examples. (In fact what we describe is the fibre of a $K(G, 2)$-torsor, rather than the torsor itself.)

Let $\mathbf{T} = (T_1 \rightrightarrows T_0)$ be an internal groupoid in $\mathscr{E}$. We say $\mathbf{T}$ is *connected* if $T_1 \xrightarrow{(d_0, d_1)} T_0 \times T_0$ is epi. Form the pullback

$$\begin{array}{ccc} V & \longrightarrow & T_1 \\ {\scriptstyle v}\downarrow & & \downarrow{\scriptstyle (d_0, d_1)} \\ T_0 & \xrightarrow{\ \Delta\ } & T_0 \times T_0 \end{array}$$

then $(V \xrightarrow{v} T_0)$ is a group in the topos $\mathscr{E}/T_0$. Even if $\mathbf{T}$ is connected, V need not be a constant group (i.e. isomorphic to $T_0^*(G)$ for some $G \in \mathbf{gp}(\mathscr{E})$); but if we do have such an isomorphism, we shall call G the *vertex group* of $\mathbf{T}$. Suppose $\mathbf{T}$ has vertex group G, and let $1 \xrightarrow{t} T_1$ be a global element of T_1. Since t is invertible in $\mathbf{T}$, the operation of conjugating by t defines a morphism from the fibre of $V \xrightarrow{v} T_0$ over $d_0 t$ to the fibre over $d_1 t$, i.e. a morphism $G \longrightarrow G$ in $\mathscr{E}$. And this morphism is easily seen to be an automorphism of G; so by the method of generic elements, we may construct a morphism $T_1 \longrightarrow \mathrm{aut}(G)$ in $\mathscr{E}$, where $\mathrm{aut}(G)$ denotes the (internal) group of automorphisms of G. We say that T *acts properly* on its vertex group if this morphism factors through the subgroup $\mathrm{inn}(G)$ of inner automorphisms, i.e. the image of the homomorphism $G \longrightarrow \mathrm{aut}(G)$ defined by conjugation. (Note in particular $\mathbf{G}$ acts properly on its vertex group.)

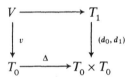

It should be emphasized that the requirement of proper action is really a restriction, not on the groupoid $\mathbf{T}$ itself, but on the "descent data" for V, i.e. the isomorphism whereby we identify V with $G \times T_0$. If G is abelian, then there is always a *unique* way of choosing this ismorphism so that the action of $\mathbf{T}$ on G is proper; but for non-abelian G there may be several possible choices, and so we have to regard the isomorphism $V \cong G \times T_0$ as part of the structure of a torsor in the following definition.

8.37 DEFINITION. Let G be a group in a topos $\mathscr{E}$. A $K(G, 2)$-*torsor* in $\mathscr{E}$ is an internal groupoid $\mathbf{T}$ which is connected and has global support, and acts properly on its vertex group which is isomorphic to G. ■

This definition is not functorial on arbitrary group homomorphisms $G \xrightarrow{f} H$; but it is so on *central* homomorphisms, i.e. those which map the centre of G into the centre of H (and hence induce a homomorphism $\mathrm{inn}(G) \longrightarrow \mathrm{inn}(H)$). In particular, it is functorial on epimorphisms, and on arbitrary homomorphisms between abelian groups. Note that if the epimorphisms $T_0 \twoheadrightarrow 1$ and $T_1 \twoheadrightarrow T_0 \times T_0$ are split, then $\mathbf{T}$ is actually equivalent to $\mathbf{G}$ as a category; so provided the latter epi is locally split (5.24), $\mathbf{T}$ will be locally equivalent to $\mathbf{G}$. This will always happen if the cohomology class represented by $\mathbf{T}$ is a Čech cohomology class.

8.38 EXAMPLES. (i) Let $K \overset{k}{\rightarrowtail} G \overset{q}{\twoheadrightarrow} G/K$ be a short exact sequence of groups as in 8.36, and let T be a (G/K)-torsor. Then $G \times T \underset{\pi_2}{\overset{\alpha(q \times 1)}{\rightrightarrows}} T$ is a connected groupoid in $\mathscr{E}$, with global support and vertex group K; and it is a $K(K, 2)$-torsor iff G acts on K by inner automorphisms, i.e. iff G is equal to the product of K and its centralizer in G. (This condition implies that the homomorphism k is central; and if K is abelian the converse is true.) If T is of the form $q_*(V)$ for some G-torsor V, let $V \overset{r}{\twoheadrightarrow} T$ be the canonical epimorphism; then the diagram

$$\begin{array}{ccc}
G \times T & \underset{\pi_2}{\overset{\alpha_T(q \times 1)}{\rightrightarrows}} & T \\
\uparrow{\scriptstyle (\pi_2,\, r\pi_3)} & & \uparrow{\scriptstyle r} \\
K \times G \times V & \underset{\pi_3}{\overset{\alpha_V \cdot \pi_{23}}{\rightrightarrows}} & V \\
\downarrow{\scriptstyle \pi_1} & & \downarrow \\
K & \rightrightarrows & 1
\end{array}$$

connects $\mathbf{T}$ to $\mathbf{K}$ in the category of $K(K, 2)$-torsors. So the composite

$$H^1(\mathscr{E}; G) \longrightarrow H^1(\mathscr{E}; G/K) \longrightarrow H^2(\mathscr{E}; K)$$

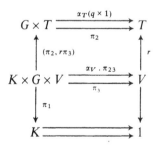

is the zero map; and J. Duskin and P. Glenn have shown that this sequence is in fact exact.

(ii) Let G be an abelian group in $\mathscr{S}$, X a topological space, and V a G-torsor in Shv(X). We may define a $K(G, 2)$-torsor $\mathbf{T} = \sigma(V)$ in Shv(ΣX), where ΣX denotes the suspension of X ([AT], §1.6), as follows: T_0 is (the sheaf of sections of) a covering of ΣX by two overlapping cones on X. If we denote the intersection of these two cones by $X \times U$, then T_1 is the coproduct of $T_0 \times G$ (the vertex group) and two copies of $V \times U$, which constitute the morphisms of T between the two sheets of the covering over $X \times U$. The picture below illustrates the case $X = S^1$, $G = \mathbb{Z}/2\mathbb{Z}$.

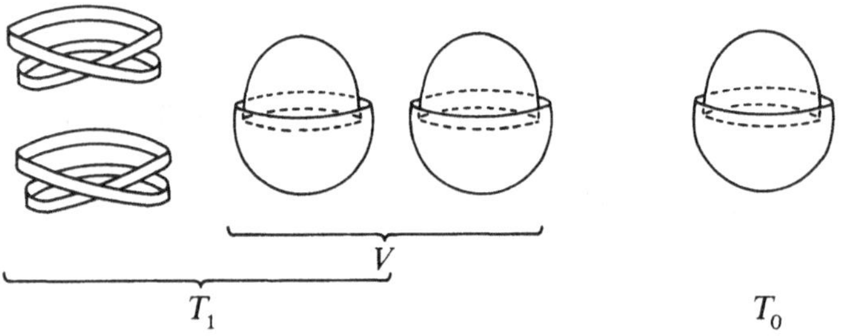

(iii) Let G be an abelian group in $\mathscr{E}$. Then $\mathbf{G}$ is an internal abelian group in **cat**($\mathscr{E}$), and hence $\mathscr{E}^{\mathbf{G}}$ is an abelian group in $\mathfrak{Top}/\mathscr{E}$ (cf. 4.37(ii)). Let $\mathbf{T}$ be a $K(G, 2)$-torsor; then left multiplication by elements of the vertex group induces an internal functor $\mathbf{G} \times \mathbf{T} \xrightarrow{\alpha} \mathbf{T}$, which is an action of $\mathbf{G}$ on $\mathbf{T}$ in **cat**($\mathscr{E}$). So the $\mathscr{E}$-topos $\mathscr{E}^{\mathbf{T}}$ has an $\mathscr{E}^{\mathbf{G}}$-action. Moreover, there exists a geometric morphism $\mathscr{E}/X \longrightarrow \mathscr{E}^{\mathbf{T}}$ over $\mathscr{E}$, where X is an object of $\mathscr{E}$ with global support (take $X = T_0$), and the morphism $\mathscr{E}^{\mathbf{G}} \times_{\mathscr{E}} \mathscr{E}^{\mathbf{T}} \xrightarrow{(\alpha, \pi_2)} \mathscr{E}^{\mathbf{T}} \times_{\mathscr{E}} \mathscr{E}^{\mathbf{T}}$ is an equivalence, since $\mathbf{G} \times \mathbf{T} \xrightarrow{(\alpha, \pi_2)} \mathbf{T} \times \mathbf{T}$ is full, faithful and essentially epimorphic on objects (cf. Ex. 4.2). It seems reasonable to call such an $\mathscr{E}$-topos an $\mathscr{E}^{\mathbf{G}}$-*torsor* in $\mathfrak{Top}/\mathscr{E}$. Conversely, G. C. Wraith has shown that any $\mathscr{E}^{\mathbf{G}}$-torsor $\mathscr{F}$ in $\mathfrak{Top}/\mathscr{E}$ determines a $K(G, 2)$-torsor in $\mathscr{E}$, as follows: let X be an object with global support such that there exists a geometric morphsim $\mathscr{E}/X \xrightarrow{q} \mathscr{F}$ over $\mathscr{E}$, and consider the composite

$$\mathscr{E}/X \times X \simeq \mathscr{E}/X \times_{\mathscr{E}} \mathscr{E}/X \xrightarrow{q \times q} \mathscr{F} \times_{\mathscr{E}} \mathscr{F} \xrightarrow{\sim} \mathscr{E}^{\mathbf{G}} \times_{\mathscr{E}} \mathscr{F} \xrightarrow{\pi_1} \mathscr{E}^{\mathbf{G}}.$$

By 4.37(ii), this determines a G-torsor T in $\mathscr{E}/X \times X$; but since the action of $\mathscr{E}^{\mathbf{G}}$ on $\mathscr{F}$ is associative and unitary, it is easily verified that we have isomorphisms $\pi_{12}^*(T) \otimes_{\mathbf{G}} \pi_{23}^*(T) \cong \pi_{13}^*(T)$ in $\mathscr{E}/X \times X \times X$ and $\Delta^*(T) \cong X^*(G)$ in $\mathscr{E}/X$. This means that we can define a groupoid in $\mathscr{E}$ whose object-of-objects is X and whose object-of-morphisms is $\Sigma_{X \times X}(T)$; and this groupoid is a $K(G, 2)$-torsor. ∎

8.4. PROFINITE FUNDAMENTAL GROUPS

In this paragraph we shall introduce a definition of the fundamental group of a topos, which is convenient for many applications. The starting-point of this definition is the close connection, which we observed in Exercise 4.5, between locally constant sheaves on a space X and the fundamental group of X. However, our general definition of fundamental group will not agree precisely with that used by algebraic topologists; the reason is that, for a general topos, the most we can hope to recover using locally constant objects is the "best approximation" to the fundamental group by means of its finite quotients. To make precise what we mean by this, we begin by recalling the definition of a profinite group.

8.41 THEOREM. *Let G be a group. The following conditions are equivalent:*

(i) *There exists a set $\{H_i \mid i \in I\}$ of normal subgroups of finite index in G, closed under finite intersections, such that G is (canonically) isomorphic to the limit of the diagram whose vertices are the groups G/H_i, $i \in I$, and whose arrows are the quotient maps $G/H_i \longrightarrow G/H_j$ induced by in- inclusions $H_i \subseteq H_j$.*
(ii) *G is isomorphic to the limit of a diagram whose vertices are finite groups.*
(iii) *There exists a topology $\mathbf{U}$ on (the underlying set of) G, making G into a topological group, and such that $(G, \mathbf{U})$ is a* Stone space *(i.e. a totally disconnected compact space, or equivalently a Hausdorff spectral space; cf. 7.48).*

If these conditions are satisfied, we say that G is profinite.

Proof. (i) ⇒ (ii) is trivial.

(ii) ⇒ (iii): Topologize the finite groups in the diagram as discrete spaces, and take the limit in the category of topological groups. Since the class of Stone spaces contains finite discrete spaces, and is closed under the formation of limits in **esp**, we obtain a Stone-space topology on G making it a topological group.

(iii) $\Rightarrow$ (i): Let $\{H_i \mid i \in I\}$ be the set of open normal subgroups of G. Now if K is any open subgroup of G, the coset space $G{:}K$ is discrete and compact, and therefore finite; so the H_i all have finite index in G. We shall show that the canonical map $G \longrightarrow \varprojlim_I (G/H_i)$ is iso, i.e. that the diagram

$$G \longrightarrow \prod_i (G/H_i) \rightrightarrows \prod_{i,j} (G/H_i \, . \, H_j)$$

is an equalizer. Let $(x_i \mid i \in I)$ be an element of the equalizer; i.e. $x_i \in G/H_i$ for each i, and x_i is congruent to x modulo $H_i . H_j$ for each (i, j). Now a basic open neighbourhood of (x_i) in $\prod G/H_i$ has the form

$$\{(y_i \mid i \in I) \mid y_{i_k} = x_{i_k} \text{ for } k = 1, \ldots, n\}$$

for some finite subset $\{i_1, \ldots, i_n\}$ of I; but if we write H_{i_0} for $\bigcap_{k=1}^{n} H_{i_k}$, then an element of the equalizer is in the above neighbourhood iff $y_{i_0} = x_{i_0}$. So if $t \in G$ is any representative of the coset x_{i_0}, the element $(tH_i \mid i \in I)$ is in the given neighbourhood. So the image of G is dense in the equalizer; but it is compact and hence closed, and so G maps surjectively onto the equalizer.

To show that $G \longrightarrow \prod G/H_i$ is one-to-one, let x be a non-identity element of G. Since G is totally disconnected, we can find a clopen (and therefore compact) neighbourhood U of e not containing x. We shall show that U contains an open normal subgroup of G, which is clearly sufficient. Consider the set $F = (G - U) \cap U^2$; since U is compact, F is closed and $U \cap F$ is empty, we can find an open neighbourhood V of e such that $UV \cap F$ is empty. And we may clearly assume $V = V^{-1}$ and $V \subseteq U$; so $UV \subseteq U^2$ and hence $UV \cap (G - U)$ is empty, i.e. $UV \subseteq U$. It now follows by induction that $UV^n \subseteq U$ for all n, and hence $\bigcup_{n=1}^{\infty} V^n$ is an open subgroup of G contained in U. This subgroup need not be normal; but it has finite index by the argument already given, and hence has only finitely many conjugates in G. On taking the intersection of these conjugates, we obtain an open normal subgroup as required. ■

If G is any (discrete) group, we define the *profinite completion* $\hat{G}$ of G to be the limit of the diagram whose vertices are all finite quotients of G. It is not true in general that any subgroup of finite index in a profinite group is open (and so a profinite group need not be isomorphic to its profinite completion!), but it is true for profinite groups of the form $\hat{G}$.

Now let $\mathscr{E}$ by any topos with a natural number object. By a *locally constant finite object* in $\mathscr{E}$, we mean an object X such that V^*X is isomorphic to a finite cardinal in $\mathscr{E}/V$, for some object V of $\mathscr{E}$ having global support. We write $\mathscr{E}_{lcf}$ for the full subcategory of locally constant finite objects of $\mathscr{E}$.

8.42 PROPOSITION. (i) *$\mathscr{E}_{lcf}$ is a Boolean topos, and the inclusion functor $\mathscr{E}_{lcf} \longrightarrow \mathscr{E}$ is logical iff $\mathscr{E}$ is Boolean.*

(ii) *If $\mathscr{F} \xrightarrow{p} \mathscr{E}$ is a geometric morphism, then p^* restricts to a logical functor $\mathscr{E}_{lcf} \longrightarrow \mathscr{F}_{lcf}$.*

(iii) *If $\mathscr{E}$ is a connected $\mathscr{S}$-topos (cf. Ex. 4.8) and $\mathscr{S} \xrightarrow{p} \mathscr{E}$ is a point of $\mathscr{E}$, then the functor $\mathscr{E}_{lcf} \longrightarrow \mathscr{S}_{lcf} = \mathscr{S}_f$ induced by p^* reflects isomorphisms.*

(iv) *If $\mathscr{E}$ is Grothendieck topos, then $\mathscr{E}_{lcf}$ is equivalent to a small category.*

Proof. (i) follows directly from 6.29 and the fact that pullback functors are logical. For example, to show that $\mathscr{E}_{lcf}$ is closed under the formation of exponentials in $\mathscr{E}$, let X and Y be two objects of $\mathscr{E}_{lcf}$ and V, W two objects with global support such that V^*X and W^*Y are isomorphic to finite cardinals. Then $(V \times W)^*(Y^X)$ is isomorphic to a finite cardinal in $\mathscr{E}/V \times W$, by 6.23(v).

(ii) similarly follows from the corresponding result for finite cardinals (Ex. 6.6) and the fact that p^* preserves the property of having global support.

(iii) Let $X \xrightarrow{f} Y$ be a morphism of $\mathscr{E}_{lcf}$, and let $U \rightarrowtail 1$ be the extension of the sentence

$$\bar{f} \in \ulcorner \mathrm{Iso}(X, Y) \urcorner$$

of $L_{\mathscr{E}_{lcf}}$. Since $\mathscr{E}_{lcf}$ is Boolean, U is a complemented subobject of 1 in $\mathscr{E}_{lcf}$ and hence in $\mathscr{E}$; so it must be either 0 or 1. But the restriction of p^* to $\mathscr{E}_{lcf}$ preserves extensions of formulae; hence if $p^*(f)$ is iso, we must have $U \cong 1$, i.e. f is iso.

(iv) follows at once from 8.35, since there is only a set of natural numbers (and hence of finite cardinals) in $\mathscr{E}$. ■

8.43 DEFINITION (A. Grothendieck [GV], V 5.1). By a *Galois category* we mean a pair $(\mathscr{G}, F)$, where $\mathscr{G}$ is a small Boolean (pre)topos and $F: \mathscr{G} \longrightarrow \mathscr{S}_f$ is an exact, isomorphism reflecting functor. (It follows from the structure theorem, to be proved below, that every Galois category is in fact a topos, though we shall use only the pretopos structure in proving the theorem. The structure theorem also implies that F is determined up to (non-canonical) isomorphism by $\mathscr{G}$, so we shall feel free to write "$\mathscr{G}$ is a Galois category" without mentioning F.) ■

Recall that an object A of a Boolean topos is said to be an atom (Ex. 7.13) if it is nonzero but cannot be decomposed as a nontrivial coproduct (equivalently, has no nontrivial subobjects).

8.44 LEMMA. *Let $\mathscr{G}$ be a Galois category. Then*

(i) *Any object of $\mathscr{G}$ is (uniquely) expressible as a finite coproduct of atoms.*
(ii) *Any endomorphism of an atom of $\mathscr{G}$ is an automorphism.*

Proof. (i) Let X be an object of $\mathscr{G}$. If $X \cong 0$, it is expressible as an empty coproduct; so assume $X \not\cong 0$. Then $F(X)$ is nonempty, since F preserves 0 and reflects isomorphisms. Now if X is not an atom, we can decompose it as a nontrivial coproduct $X \cong Y_1 \amalg Y_2$; and this corresponds to a nontrivial decomposition $F(X) \cong F(Y_1) \amalg F(Y_2)$ in $\mathscr{S}_f$. But $F(X)$ is a finite set, so the process of breaking down X as a coproduct must terminate after finitely many steps.

(ii) Let $A \xrightarrow{f} A$ be an endomorphism of an atom of $\mathscr{G}$. Since A is nonzero, so is the image of f, and hence it must be the whole of A; i.e. f is epi. So $F(f)$ is an epimorphism from the finite set $F(A)$ to itself, and must therefore be an isomorphism. Hence f is an isomorphism. ∎

8.45 PROPOSITION. *Let $(\mathscr{G}, F)$ be a Galois category. Then the functor F is "pro-representable"; i.e. there exists a filtered inverse system $(A_i \mid i \in \mathbf{I})$ of objects of $\mathscr{G}$ and a natural isomorphism*

$$F(X) \cong \varinjlim_{\mathbf{I}} \hom_{\mathscr{G}}(A_i, X).$$

Proof. Consider the category $\mathbf{I}$ whose objects are pairs (A, a), where A is an atom of $\mathscr{G}$ and $a \in F(A)$, and whose morphisms $(A, a) \xrightarrow{f} (B, b)$ are morphisms $A \xrightarrow{f} B$ of $\mathscr{G}$ such that $F(f)(a) = b$.

Now $\mathbf{I}$ is in fact a poset, since if two morphisms $A \rightrightarrows B$ have the same effect on $a \in F(A)$, their equalizer is nonzero and so must be the whole of A. And $\mathbf{I}^{\mathrm{op}}$ is filtered, since both (A, a) and (B, b) are preceded in $\mathbf{I}$ by $(C, \langle a, b \rangle)$, where C is the component of $A \times B$ (in the decomposition of 8.44(i)) containing the element $\langle a, b \rangle$ of $F(A) \times F(B) \cong F(A \times B)$. (Nonemptiness of I follows from the fact that $F(1) \cong 1 \not\cong F(0)$ and hence $1 \not\cong 0$ in $\mathscr{G}$.)

Let (A, a) be an object of $\mathbf{I}$; then by 0.11 we can regard a as a natural transformation $\hom_{\mathscr{G}}(A, -) \longrightarrow F$. And from the definition of morphisms in $\mathbf{I}$, these transformations combine to induce a transformation

$$\theta\colon G = \varinjlim_{\mathbf{I}} (\hom_{\mathscr{G}}(A, -)) \longrightarrow F.$$

Now θ is epi by 8.44(i), since any element of $F(X)$ lies in (the image of) $F(A)$ for some atom $A \rightarrowtail X$; and θ is mono, since if we are given $x, y \in G(X)$ such that $\theta(x) = \theta(y)$, then we can represent both x and y by morphisms $A \underset{y}{\overset{x}{\rightrightarrows}} X$ for some $(A, a) \in I$, such that $F(x)(a) = F(y)(a)$. And then it follows as before that the equalizer of x and y is the whole of A; i.e. $x = y$ in $\hom_{\mathscr{G}}(A, X)$ and hence in $G(X)$. ■

Now if $(A, a) \in I$, we have a natural map $\text{aut}_{\mathscr{G}}(A) = \hom_{\mathscr{G}}(A, A) \longrightarrow F(A)$ which is clearly mono, since the transition maps in the inverse system are epimorphisms. We say (A, a) is a *normal object* if this map is also epi. (Note: since this is equivalent to saying that the group $\text{aut}(A)$ acts transitively on $F(A)$, the definition of normality is in fact independent of the element a.)

8.46 PROPOSITION. *For any object X of $\mathscr{G}$, there exists a normal object (A, a) such that the canonical map* $\hom(A, X) \longrightarrow F(X)$ *is iso. Hence in particular the normal objects form a cofinal subcategory of* $\mathbf{I}$ (*cf.* Ex. 2.11).

Proof. We can certainly find $(B, b) \in I$ such that each element x of $F(X)$ derives from a morphism $B \xrightarrow{\bar{x}} X$, since $F(X)$ is finite and $\mathbf{I}^{\text{op}}$ is filtered. Now consider the image A of the canonical map $B \longrightarrow \prod_{F(X)} X$ in $\mathscr{G}$ whose xth component is $\bar{x}$. Since A is a quotient of B, it is an atom; so if a denotes the image of b in $F(A)$, the pair (A, a) is an element of I, and we again have an isomorphism $\hom(A, X) \xrightarrow{\cong} F(X)$.

Now let (C, c) be another element of I such that each element of $F(A)$ derives from a morphism $C \longrightarrow A$; in particular let $C \xrightarrow{\phi} A$ be the morphism such that $F(\phi)(c) = a$. To show that (A, a) is normal, we have to show that for any $C \xrightarrow{\psi} A$ there exists $A \xrightarrow{v} A$ such that $v\phi = \psi$. But for each $x \in F(X)$, the composite $C \xrightarrow{\psi} A \xrightarrow{\bar{x}} X$ determines an element x' of $F(X)$ such that $\bar{x}\psi = \overline{x'}\phi$; and the map $F(X) \longrightarrow F(X)$ defined by $x \longmapsto x'$ is clearly mono (since ψ is epi) and hence iso.

Now consider the diagram

$$\begin{array}{ccccc} C & \overset{\phi}{\twoheadrightarrow} & A & \rightarrowtail & \prod_{F(X)} X \\ \downarrow{\scriptstyle 1} & & & & \downarrow{\scriptstyle u} \\ C & \overset{\psi}{\twoheadrightarrow} & A & \rightarrowtail & \prod_{F(X)} X \end{array}$$

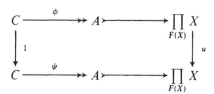

where u is the isomorphism obtained by permuting factors; by functoriality of image factorization, we obtain an isomorphism $A \xrightarrow{v} A$ such that $v\phi = \psi$, as required. ■

If G is any topological group, we shall write $\mathscr{C}(G)$ for the topos of continuous G-sets (8.15(iii)), and $\mathscr{C}_f(G)$ for the full subcategory of continuous finite G-sets (which is clearly a logical subtopos of $\mathscr{C}(G)$, cf. 1.14).

8.47 THEOREM (Grothendieck). *Let $\mathscr{G}$ be a small category, and $\mathscr{G} \xrightarrow{F} \mathscr{S}_f$ a functor. The following conditions are equivalent:*

(i) *$(\mathscr{G}, F)$ is a Galois category.*
(ii) *There exists a topological group G and an equivalence $\mathscr{G} \simeq \mathscr{C}_f(G)$ which identifies F with the forgetful functor.*

Moreover, if we demand that the group G in (ii) *be profinite, it is determined up to isomorphism by $(\mathscr{G}, F)$; we call the group so determined the* fundamental group *of $(\mathscr{G}, F)$.*

Proof. (ii) ⇒ (i) is trivial, since $\mathscr{C}_f(G)$ is a Boolean topos and the forgetful functor $\mathscr{C}_f(\mathscr{G}) \longrightarrow \mathscr{S}_f$ is logical.

(i) ⇒ (ii): Let $\mathbf{N}$ denote the full subcategory of $\mathbf{I}$ whose objects are normal. If $(A, a) \xrightarrow{f} (B, b)$ is a morphism of $\mathbf{N}$, we can regard $F(f)\colon F(A) \longrightarrow F(B)$ as a map $\operatorname{aut}(A) \xrightarrow{\phi} \operatorname{aut}(B)$; and in fact this map sends an automorphism $A \xrightarrow{v} A$ to the unique $B \xrightarrow{w} B$ such that

$$\begin{array}{ccc} A & \xrightarrow{f} & B \\ \downarrow{\scriptstyle v} & & \downarrow{\scriptstyle w} \\ A & \xrightarrow{f} & B \end{array}$$

commutes. It follows easily from this description that ϕ is in fact a group homomorphism; and so the groups $\operatorname{aut}(A)$ $((A, a) \in N)$ form a diagram over $\mathbf{N}$ in the category of finite groups. Define G to be the limit of this diagram; then G is clearly a profinite group.

Now for any object X of $\mathscr{G}$, we obtain a continuous G-action on the finite set $F(X)$, by choosing a normal object (A, a) as in 8.46 and letting $\operatorname{aut}(A)$ (which is clearly a discrete quotient of G) act on $\hom(A, X) \cong F(X)$. (It is easy to verify that the G-action thus obtained is independent of the choice of A.) So we have a factorization of F through the forgetful functor $\mathscr{C}_f(G) \longrightarrow \mathscr{S}_f$.

To construct an inverse $\mathscr{C}_f(G) \xrightarrow{T} \mathscr{G}$ for this factorization, it suffices in view of 8.44(i) to define T on transitive G-sets (i.e. atoms of $\mathscr{C}_f(G)$), and then extend the definition by requiring that T preserve coproducts. But any atom of $\mathscr{C}_f(\mathscr{G})$ has the form $G:H$, where H is an open subgroup of G; and then since H is a neighbourhood of the identity in G, it must contain the kernel of the projection $G \longrightarrow \text{aut}(A)$ for some normal object (A, a). Let $\bar{H} \subseteq \text{aut}(A)$ be the image of H under this projection; then since $\mathscr{G}$ has finite coproducts, images and coequalizers of equalizers of equivalence relations, we can form the quotient of A by this finite group of automorphisms, i.e. the joint co-equalizer of the pairs $(v, 1_A)$ for each $v \in \bar{H}$. Define $T(G: H)$ to be this quotient; then it is again straightforward to show that $T(G: H)$ is independent (up to canonical isomorphism) of the choice of normal object (A, a), that T is functorial on maps between transitive G-sets, and that it is inverse to the functor $\mathscr{G} \longrightarrow \mathscr{C}_f(G)$ already constructed.

The last sentence of the theorem follows from 8.41; for if G is profinite, then it is determined up to isomorphism by its discrete quotients, and these are precisely the normal objects in $\mathscr{C}_f(G)$. ■

8.48 COROLLARY. *Let $(\mathscr{G}, F)$ be a Galois category, and $F' : \mathscr{G} \longrightarrow \mathscr{S}_f$ another functor such that $(\mathscr{G}, F')$ is also a Galois category. Then F and F' are naturally isomorphic.*

Proof. By 8.47, we may assume that $\mathscr{G} = \mathscr{C}_f(G)$ for some profinite group G, and that F is the forgetful functor. Since F' preserves coproducts, it is sufficient to establish the isomorphism $F \cong F'$ on atoms of $\mathscr{G}$. Now let $\mathbf{I}'$ be the poset whose objects are pairs (A, a'), where A is an atom of $\mathscr{G}$ and $a' \in F'(A)$ (cf. 8.45), and consider the limit

$$X = \varprojlim_{(A, a') \in \mathbf{I}'} F(A),$$

computed in $\mathscr{S}$. Since the sets $F(A)$ are finite and nonempty, $\mathbf{I}'^{\text{op}}$ is filtered and the transition maps in the inverse system are epimorphisms, an easy Zorn's Lemma argument shows that X is nonempty. Let x be an element of X, and let $x(a')$ denote the image of x in the factor corresponding to the element (A, a') of I'. Then the assignment $a' \longmapsto x(a')$ defines a map $F'(A) \xrightarrow{\xi} F(A)$, which is clearly natural in A. Moreover, this map is mono, since if $x(a') = x(a'')$, then we can find $(B, b') \in I'$ and $B \underset{g}{\overset{f}{\rightrightarrows}} A$ in $\mathscr{G}$ such that $F'(f)(b') = a'$, $F'(g)(b') = a''$; and then f and g agree on the element $x(b')$ of $F(B)$, so they must be equal.

Now F' preserves 1 and finite coproducts; hence ξ is an isomorphism for any G-set with trivial action. But any object of $\mathscr{C}_f(\mathscr{G})$ is locally isomorphic to one with trivial action; for if H is an open normal subgroup of G which stabilizes every element of A, then the map $(a, Hg) \longmapsto (ag^{-1}, Hg)$ is a well-defined isomorphism $A \times (G/H) \xrightarrow{\cong} A_0 \times (G/H)$, where A_0 denotes the underlying set of A with trivial G-action. It follows easily that $F'(A)$ has the same number of elements as $F'(A_0)$, and hence the same as $F(A)$; so ξ must be an isomorphism for all A. ■

Note that the isomorphism $F' \cong F$ of 8.48 is *not* canonical; that is, it does depend on the choice of the element x.

8.49 DEFINITION. Let $\mathscr{E}$ be a connected Grothendieck topos, p a point of $\mathscr{E}$. We define the *profinite fundamental group* $\Pi_1(\mathscr{E}, p)$ to be fundamental group of the Galois category $(\mathscr{E}_{lcf}, p^*)$. ■

8.50 EXAMPLES. (i) Let X be a connected topological space having a universal covering space $\tilde{X}$, and let x be a point of X. Now a finite cardinal in Shv(X) is simply a constant sheaf on X with finite stalks; so the locally constant finite objects are (the sheaves of sections of) finite covering spaces over X, which correspond (via the argument of Ex. 4.5) to finite sets equipped with a $\Pi_1(X, x)$-action. So $\Pi_1(\mathrm{Shv}(X), x)$ is isomorphic to the profinite completion of the usual fundamental group.

(ii) Let $\mathscr{E} = \mathscr{S}^{\mathbf{G}}$, where G is a (discrete) group. By the argument used in the last part of the proof of 8.48, every finite G-set is locally isomorphic to a finite cardinal in $\mathscr{E}$; so $\mathscr{E}_{lcf} = (\mathscr{S}_f)^{\mathbf{G}}$. Hence the profinite fundamental group of $\mathscr{E}$ (based at its unique point) is the profinite completion of G.

(iii) Let $\mathscr{E} = \mathscr{C}(G)$, where G is a compact topological group. As in (ii), we find that $\mathscr{E}_{lcf} = \mathscr{C}_f(G) \cong \mathscr{C}_f(G/N)$, where N is the component of the identity in G; so G/N is the profinite fundamental group of $\mathscr{E}$. ■

EXERCISES 8

1. Let $\mathscr{F} \xrightarrow{f} \mathscr{E}$ be an essential geometric morphism. Use 0.15 to define a left adjoint $f_\#$ for $f^*\colon \mathbf{ab}(\mathscr{E}) \longrightarrow \mathbf{ab}(\mathscr{F})$. Show that $f_\#$ preserves projectives; and deduce that if f is surjective and $\mathbf{ab}(\mathscr{F})$ has enough projectives, then $\mathbf{ab}(\mathscr{E})$ has enough projectives.

2. Let j be a topology in a topos $\mathscr{E}$, and suppose that $\mathbf{ab}(\mathscr{E})$ has *injective envelopes*; i.e. for each $A \in \mathbf{ab}(\mathscr{E})$ we can find $A \rightarrowtail E$ such that E is injective and A intersects each nonzero subobject of E nontrivially. Prove that $\mathbf{ab}(\mathrm{sh}_j(\mathscr{E}))$ has injective envelopes. [Hint: if $A \rightarrowtail E$ is an injective envelope in $\mathbf{ab}(\mathscr{E})$ of a j-sheaf A, prove that $E \longrightarrow L(E)$ is mono, and deduce that E is a sheaf.]

3. Let $\mathscr{E}$ be a topos, A an abelian group in $\mathscr{E}$. Show that the following conditions are equivalent:

   (i) For every object X of $\mathscr{E}$, $H^1(X; A) = 0$.
   (ii) For every short exact sequence $0 \longrightarrow A \longrightarrow B \longrightarrow C \longrightarrow 0$ in $\mathbf{ab}(\mathscr{E})$, the epimorphism $B \twoheadrightarrow C$ has a *transversal* (i.e. a splitting in $\mathscr{E}$).

   [Hint for (i) $\Rightarrow$ (ii): take X to be (the underlying object of) C.]

   Are these conditions sufficient for flabbiness of A? [Consider the topos $\mathscr{S}^{\mathbf{C}}$, where C is a cyclic group of order 2, and take $A = \mathbb{Z}$ with trivial C-action.]

4. Let G be a (not necessarily abelian) group in a topos $\mathscr{E}$, such that $H^1(X; G) = 0$ for all $X \in \mathscr{E}$. Prove that G is injective as an object of $\mathscr{E}$. [Hint: given a diagram

$$\begin{array}{ccc} Y & \xrightarrow{\ f\ } & G, \\ {\scriptstyle\sigma}\big\downarrow & & \\ X & & \end{array}$$

   let $X \rightrightarrows Q$ be the cokernel-pair of σ; and use f to construct a G-torsor in $\mathscr{E}/Q$, in such a way that any global element of this torsor determines an extension of f along σ.] Show, however, that the converse is false even if G is abelian. [Take $\mathscr{E}$ to be the same topos as in Ex. 3, and let $G = \mathbb{Z}/2\mathbb{Z}$ with trivial C-action.]

5. Let X be a topological space, A a sheaf of abelian groups on X. Prove that the following conditions are equivalent:

   (i) A is flabby.
   (ii) A satisfies the equivalent conditions of Ex. 3.

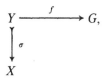

(iii) A is injective as an object of Shv(X).
(iv) For each open $U \subseteq X$, the restriction map $\rho^X_U\colon A(X) \longrightarrow A(U)$ is surjective.

[Hint for (iv) $\Rightarrow$ (i): suppose we are given a short exact sequence $0 \longrightarrow A \longrightarrow B \longrightarrow C \longrightarrow 0$, and use Zorn's Lemma to construct a transversal for $B \twoheadrightarrow C$. Now suppose that B is flabby, and show that C also satisfies (iv).]

6. Let $\mathbf{C}$ be a small category with pullbacks, $\mathscr{U} = (U_\gamma \xrightarrow{\theta_\gamma} U \mid \gamma \in \Gamma)$ a family of morphisms of $\mathbf{C}$ with common codomain, and $V \xrightarrow{\phi} U$ a morphism of $\mathbf{C}$. Writing $\phi^*\mathscr{U}$ for the family $(U_\gamma \times_U V \xrightarrow{\pi_2} V \mid \gamma \in \Gamma)$, construct a chain map $N.(\phi^*\mathscr{U}) \longrightarrow N.(\mathscr{U})$, and hence a map $C^{\cdot}(\mathscr{U}; A) \longrightarrow C^{\cdot}(\phi^*\mathscr{U}; A)$ for any presheaf of abelian groups A on $\mathbf{C}$. Deduce that the Čech cohomology groups $\check{H}^q(U : A)$ (for a Grothendieck topology on $\mathbf{C}$) are contravariantly functorial in U. Show also that if θ_γ is split epi for some $\gamma \in \Gamma$, then the complex $N.(\mathscr{U})$ is contractible; and deduce that for any $\mathscr{U}$, we have $H^q(\theta_\gamma^*\mathscr{U}; A) = 0$ for all $\gamma \in \Gamma$ and all $q > 0$.

7. Let $(\mathbf{C}, J)$ be a site satisfying the hypotheses of 7.35 (so that $\mathscr{E} = \mathrm{Shv}(\mathbf{C}, J)$ is a coherent topos). Show that the inclusion $\mathrm{Shv}(\mathbf{C}, J) \longrightarrow \mathscr{S}^{\mathbf{C}^{\mathrm{op}}}$ preserves filtered colimits [i.e. that the colimit in $\mathscr{S}^{\mathbf{C}^{\mathrm{op}}}$ of a filtered diagram of sheaves is a sheaf], and hence show that the functor $\check{H}^q(U; -)$ preserves filtered colimits for any $U \in C_0$. Deduce that a filtered colimit of flabby sheaves is acyclic for the functor $\hom_{\mathscr{E}}(1, -)$ [use 8.28, taking $K = C_0$], and hence prove that $H^q(\mathscr{E}; -)$ preserves filtered colimits.

8. Let A be a commutative ring, and M an A-module. Show that the assignment $D(f) \longmapsto M \otimes_A A[f^{-1}]$ $(f \in A)$ defines a sheaf $\tilde{M}$ on spec(A), and that $\check{H}^q(\mathrm{spec}(A); \tilde{M}) = 0$ for all $q > 0$. [Method: given a finite open covering $\mathscr{U} = (D(f_i) \mid i \in I)$ of spec(A) by basic Zariski-open sets, express the augmented cochain complex $M = \tilde{M}(\mathrm{spec}\, A) \longrightarrow \mathbf{C}^{\cdot}(\mathscr{U}; \tilde{M})$ as a filtered colimit of augmented complexes

$$\mathbf{C}^{\cdot}_n(\mathscr{U}; M)\colon M \longrightarrow \prod_{i \in I} M \longrightarrow \prod_{(i,j) \in I \times I} M \longrightarrow \dots \qquad (n \in \mathbb{N}),$$

in which the coboundary maps are defined by

$$m \longmapsto (f_i^n . m \mid i \in I),$$

$$(m_i \mid i \in I) \longmapsto (f_j^n . m_i - f_i^n . m_j \mid (i, j) \in I \times I),$$

etc. Then show that each $C_n^{\bullet}(\mathcal{U} : M)$ is contractible, using the fact that there exist elements $g_{n,i} \in A$ such that $\sum_i g_{n,i} \cdot f_i^n = 1$.] Hence show that if $(X, \mathcal{O}_X)$ is a scheme and F is a quasi-coherent $\mathcal{O}_X$-module, then the Čech and ordinary cohomology of X with coefficients in F coincide. [Use 8.28; a quasi-coherent module is one whose restriction to each affine open subset of X is a module of the form $\tilde{M}$.]

9. Let X be a paracompact (Hausdorff) topological space, and A a presheaf of abelian groups on X whose stalk at every point is 0. Show that $\check{H}^q(X; A) = 0$ for all q. [Method: given a cochain $\xi \in C^q(\mathcal{U}; A)$, where $\mathcal{U} = (U_\gamma \mid \gamma \in \Gamma)$ is a locally finite open cover of X, let $\mathcal{V} = (V_\gamma \mid \gamma \in \Gamma)$ be a shrinking ([GT], p. 104) of $\mathcal{U}$, and define a further refinement $\mathcal{W}$ of $\mathcal{V}$ by choosing for each $x \in X$ a neighbourhood W_x such that

   (a) $x \in U_\gamma \Rightarrow W_x \subseteq U_\gamma$.

   (b) $x \in V_\gamma \Rightarrow W_x \subseteq V_\gamma$.

   (c) $x \notin U_\gamma \Rightarrow W_x \cap V_\gamma = \varnothing$.

   (d) If $x \in U_\sigma$ for some $\sigma \in \Gamma^{q+1}$ (so that $W_x \subseteq U_\sigma$, by (a)), then the element ξ_σ of $A(U_\sigma)$ restricts to 0 in $A(W_x)$.

   Now show that ξ restricts to the zero cochain in $C^q(W; A)$.]
   Deduce that if F is any sheaf of abelian groups on X, then the Čech and ordinary cohomology of X with coefficients in F coincide.

†10. Let $\mathscr{E}$ be a topos, and Φ a filter of open objects of $\mathscr{E}$ (i.e. a set of subobjects of 1 such that $1 \in \Phi$,

$$(U_1 \in \Phi \text{ and } U_2 \in \Phi) \Rightarrow U_1 \cap U_2 \in \Phi,$$

and

$$(U_1 \in \Phi \text{ and } U_1 \subseteq U_2) \Rightarrow U_2 \in \Phi).$$

Let A be an abelian group in $\mathscr{E}$, and $1 \xrightarrow{a} A$ a global element of A; we define the *cosupport* of a to be the equalizer of a and the zero element $1 \xrightarrow{0} A$. Let $\gamma_\Phi(A)$ denote the set of global elements of A with cosupport in Φ; prove that $\gamma_\Phi(A)$ is a group, and that the functor $\gamma_\Phi : \mathbf{ab}(\mathscr{E}) \longrightarrow \mathbf{ab}(\mathscr{S})$ is left exact. Show that flabby abelian groups are acyclic for γ_Φ [use Ex. 3 plus induction]. (The qth right derived functor of γ_Φ is denoted $H^q_\Phi(\mathscr{E}; -)$, and called the qth *cohomology group of $\mathscr{E}$ with cosupports in* Φ.)

11. (R. Diaconescu) Let V be an object with global support in a topos $\mathscr{E}$; let $G = \mathrm{Iso}(V, V)$ be the group of permutations of V in $\mathscr{E}$, and let $G \times V \xrightarrow{\alpha} V$ be the canonical action (i.e. the evaluation map). Show that α is not in general transitive [consider the object $(V_0 \xrightarrow{f} V_1)$ of $\mathscr{S}^{\mathbf{2}}$, where V_0 has three elements, V_1 has two and f is surjective]. Show, however, that if $\mathscr{E}$ is Boolean then $G \times V \xrightarrow{(\alpha, \pi_2)} V \times V$ is split epi [use generalized elements]; let $V \times V \xrightarrow{(\sigma, \pi_2)} G \times V$ be a splitting for it. Suppose there exists a group H in $\mathscr{E}$ such that the groupoid $(G \times V \overset{\alpha}{\underset{\pi_2}{\rightrightarrows}} V)$ is a *trivial* $K(H, 2)$-torsor; then show that σ can be chosen to satisfy the coherence conditions

$$\begin{array}{ccccc} V \times V \times V & \xrightarrow{\pi_{13}} & V \times V & \xleftarrow{\Delta} & V. \\ \Big\downarrow{\scriptstyle (\sigma\pi_{12}, \sigma\pi_{23})} & & \Big\downarrow{\scriptstyle \sigma} & & \Big\downarrow \\ G \times G & \xrightarrow{m} & G & \xleftarrow{e} & 1 \end{array}$$

If these conditions are satisfied, show that the coequalizer $G \times V \xrightarrow{q} T$ of

$$G \times V \times V \overset{\pi_{12}}{\underset{(m(1 \times \sigma), \pi_3)}{\rightrightarrows}} G \times V$$

is a G-torsor in $\mathscr{E}$, and that $G \times V \xrightarrow{\alpha} V$ factors through q. Deduce that if T is also a trivial torsor, then V has a global element.

12. Let G and H be profinite groups, and $C_f(G) \xrightarrow{T} \mathscr{C}_f(H)$ a functor commuting with the underlying-set functors. (Observe that T is automatically exact and reflects isomorphisms.) Prove that there is a (unique) continuous homomorphism $H \xrightarrow{f} G$ such that T is induced by "change of operators" along f. [Hint: for each open normal subgroup K of G,

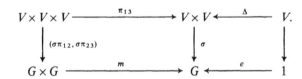

show that

$$H \xrightarrow{1 \times T(e)} H \times T(G/K) \xrightarrow{\alpha} T(G/K)$$

is a group homomorphism.] Deduce that the fundamental group $\Pi_1(\mathscr{E}, p)$ defined in 8.49 is a (covariant) functor from the category of pointed, connected Grothendieck toposes to the category of profinite groups.

13. Let K be a field, and let $\mathscr{E}$ be the étale topos K_{et} of Ex. 0.11. Show that the locally constant finite objects in $\mathscr{E}$ are precisely the étale sheaves represented by K-algebras of the form $\prod_{i=1}^{n} L_i$, where n is finite and each L_i is a finite separable field extension of K. Deduce that the normal objects in $\mathscr{E}_{lcf}$ are the finite Galois extensions of K. Hence show that if K^s is a separable closure of K, then the inclusion $K \rightarrowtail K^s$ induces a geometric morphism $\mathscr{S} \simeq (K^s)_{et} \xrightarrow{p} \mathscr{E}$, and $\Pi_1(\mathscr{E}, p)$ is the profinite Galois group $\mathrm{Gal}(K^s/K)$.

14. (Hilbert's Theorem 90) Let L be a finite Galois extension of a field K, with Galois group G, and consider the multiplicative group ΥL of nonzero elements of L as a G-module (i.e. an object of $\mathbf{ab}(\mathscr{S}^{\mathbf{G}})$). Show that $H^1(\mathscr{S}^{\mathbf{G}}; \Upsilon L) = 0$. [Hint: given a 1-cocycle $G \xrightarrow{\xi} \Upsilon L$, show that $f(x) = \sum_{\sigma \in G} \xi(\sigma) \cdot \sigma(x)$ is a nonzero for some $x \in \Upsilon L$, and deduce that ξ is the coboundary of the 0-cochain $1 \xrightarrow{f(x)^{-1}} \Upsilon L$] Hence show that $H^1(K_{et}; \Upsilon) = 0$, where Υ is the étale sheaf on K which sends a K-algebra to its group of units. Show also that the morphism $\Upsilon \xrightarrow{\phi_n} \Upsilon$ which sends a unit of L to its nth power is epi in K_{et}, and deduce that if M_n denotes the kernel of ϕ_n (i.e. the sheaf of nth roots of 1), then $H^1(K_{et}; M_n) \cong \Upsilon K/(\Upsilon K)^n$.

Chapter 9

Topos Theory and Set Theory

9.1. KURATOWSKI-FINITENESS

The first paragraph of this chapter is perhaps less directly related to the overall theme of "topos theory and set theory" than those which follow; but it resembles them in that the concept we consider here is motivated by set-theoretic considerations, and that it is logical rather than geometric functors which play a dominant role. Specifically, we are concerned here with an alternative definition of "finite object in a topos", which differs from that of a finite cardinal (6.21) in that we do not need to assume that the topos has a natural number object. In set theory, this definition was first introduced by C. Kuratowski [174]; its application to topos theory is largely the work of A. Kock, P. Lecouturier and C. J. Mikkelsen [65].

Let $\mathbb{T}$ be a finitely-presented algebraic theory. If X is a $\mathbb{T}$-model in a topos $\mathscr{E}$ and $Y \rightarrowtail X$ is a subobject of X, we may (even without 6.43) construct the sub-$\mathbb{T}$-model of X generated by Y, by applying the internal intersection operator (5.34) to "the object of sub-$\mathbb{T}$-models of X which contain Y". In particular, this is true for the theory **slat** of *semilattices* ([LT], p. 9). For any object X, we define $K(X) \rightarrowtail \Omega^X$ to be the sub-$\vee$-semilattice of Ω^X generated by $X \xrightarrow{\{\}} \Omega^X$.

9.11 Definition. We say X is *Kuratowski-finite* (abbreviated K-finite) if the maximal element $1 \xrightarrow{\ulcorner 1_X \urcorner} \Omega^X$ of Ω^X factors through $K(X) \rightarrowtail \Omega^X$. We write $\mathscr{E}_{kf}$ for the full subcategory of K-finite objects of $\mathscr{E}$. ■

Note that the assertion "X is K-finite" may be expressed by the sentence

$$\forall \mathbf{z}(\ulcorner 0 \rightarrowtail X \urcorner \in \mathbf{z} \wedge \forall \mathbf{x}(\{\mathbf{x}\} \in \mathbf{z}) \wedge \forall \mathbf{y}\, \forall \mathbf{y}'(\mathbf{y} \in \mathbf{z} \wedge \mathbf{y}' \in \mathbf{z} \Rightarrow (\mathbf{y} \vee \mathbf{y}') \in \mathbf{z}) \Rightarrow \ulcorner 1_X \urcorner \in \mathbf{z})$$

of $L_{\mathscr{E}}$, where $\mathbf{y}$ and $\mathbf{y}'$ are variables of type Ω^X, and $\mathbf{z}$ is of type Ω^{Ω^X}. Hence the property of being K-finite is preserved by logical functors, and reflected by faithful logical functors; in particular if X is locally K-finite (i.e. V^*X is K-finite in $\mathscr{E}/V$, for some V with global support), then X is globally K-finite. Note also that X is K-finite iff the poset $K(X)$ has a maximal element; for such an element must be an (internal) "upper bound for the singletons" in Ω^X, and hence must be $\ulcorner 1_X \urcorner$.

Now let $X \xrightarrow{f} Y$ be a morphism of $\mathscr{E}$. Then the diagram

$$\begin{array}{ccc} X & \xrightarrow{\quad f \quad} & Y \\ \Big\downarrow{\scriptstyle \{\}} & & \Big\downarrow{\scriptstyle \{\}} \\ \Omega^X & \xrightarrow{\quad \exists f \quad} & \Omega^Y \end{array}$$

commutes, and $\exists f$ (being an internal left adjoint) is a $\vee$-semilattice homomorphism; so it restricts to a morphism $K(f)$: $K(X) \longrightarrow K(Y)$. So K is actually a subfunctor of the covariant power-set functor.

9.17 Lemma. (i) *If X is K-finite and $X \overset{f}{\twoheadrightarrow} Y$ is epi, then Y is K-finite.*

(ii) *$X \amalg Y$ is K-finite iff both X and Y are.*

Proof. (i) Since $\ulcorner 1_X \urcorner$ factors through $K(X)$, it follows from the remarks above that $\exists f . \ulcorner 1_X \urcorner = \ulcorner \mathrm{im} f \urcorner = \ulcorner 1_Y \urcorner$ factors through $K(Y)$.

(ii) Suppose X and Y are K-finite. Then as in (i) the elements $\exists v_1 . \ulcorner 1_X \urcorner = \ulcorner v_1 \urcorner$ and $\exists v_2 . \ulcorner 1_Y \urcorner = \ulcorner v_2 \urcorner$ both factor through $K(X \amalg Y)$; but $K(X \amalg Y)$ is a semilattice, so $\ulcorner 1_{X \amalg Y} \urcorner = \vee (\ulcorner v_1 \urcorner, \ulcorner v_2 \urcorner)$ factors through $K(X \amalg Y)$.

Conversely, suppose $X \amalg Y$ is K-finite. Now it is easily seen that the square

$$\begin{array}{ccc} X \amalg Y & \xrightarrow{\quad 1_X \amalg Y \quad} & X \amalg 1 \\ \Big\downarrow{\scriptstyle \{\}} & & \Big\downarrow{\scriptstyle \binom{\{\}}{\ulcorner 0 \urcorner}} \\ \Omega^{(X \amalg Y)} & \xrightarrow{\quad \Omega^{v_1} \quad} & \Omega^X \end{array}$$

commutes, since both ways round correspond to the subobject $X \xrightarrow{(v_1, 1)} (X \amalg Y) \times X$. But Ω^{v_1} is a $\vee$-semilattice homomorphism since pullback preserves unions of subobjects; and since $K(X)$ is a semilattice it contains the element $\ulcorner 0 \urcorner$. Hence Ω^{v_1} maps $K(X \amalg Y)$ into $K(X)$, and so $\ulcorner 1_X \urcorner = \Omega^{v_1} . \ulcorner 1_{X \amalg Y} \urcorner$ factors through $K(X)$. ■

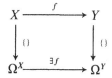

9.13 Corollary. (i) *$K(X)$ is "the object of Kuatowski-finite subobjects of X", i.e. given a subobject $Y \overset{f}{\rightarrowtail} X$, we have*

$$\mathscr{E} \models (Y \text{ is } K\text{-finite}) \Leftrightarrow (\ulcorner f \urcorner \in \ulcorner K(X) \urcorner).$$

(ii) *For any two objects X and Y, we have $K(X \amalg Y) \cong K(X) \times K(Y)$.*

Proof. (i) If Y is K-finite, then clearly $\ulcorner f \urcorner = \exists f . \ulcorner 1_Y \urcorner$ factors through $K(X)$.

Conversely, let $K'(X)$ denote the object of K-finite subobjects of X (which may be constructed as the truth-value of the sentence "$\in_X$ is K-finite" in $L_{(\mathscr{E}/\Omega^X)}$); then it follows easily from 9.12 that $K'(X)$ is a sub-$\vee$-semilattice of Ω^X. But $(1 \xrightarrow{\{\}} \Omega) = (1 \xrightarrow{t} \Omega) = (1 \xrightarrow{\ulcorner 1 \urcorner} \Omega)$, from which it follows that 1 is K-finite; and so $X \overset{\{\}}{\rightarrowtail} \Omega^X$ factors through $K'(X)$. Hence $K'(X) \geqslant K(X)$, and thus $(\ulcorner f \urcorner \in \ulcorner K(X) \urcorner)$ implies that Y is K-finite.

(ii) By 9.12(ii), a subobject $S \rightarrowtail X \amalg Y$ is K-finite iff $\nu_1^*(S)$ and $\nu_2^*(S)$ are; hence by (i) we have a natural bijection between the generalized elements of $K(X \amalg Y)$ and those of $K(K) \times K(Y)$. ■

Kuratowski-finiteness turns out to be equivalent to another concept of finiteness involving filtered posets and cofinality. Recall that if **P** and **Q** are internal posets, an order-preserving map $\mathbf{P} \xrightarrow{f} \mathbf{Q}$ is said to be cofinal if $Q_1 \times_Q P \xrightarrow{d_0 \pi_1} Q$ is epi. (The second condition for cofinality given in Ex. 2.11 is redundant for posets.)

9.14 Theorem (Kock–Lecouturier–Mikkelsen). *Let X be an object of a topos $\mathscr{E}$. Then X is K-finite iff, for every filtered poset* **P** *in $\mathscr{E}$, the canonical morphism $\mathbf{P} \xrightarrow{h} \mathbf{P}^X$ induced by $X \longrightarrow 1$ is cofinal.*

Proof. Suppose X is K-finite, and let **P** be a filtered poset. Let $A \rightarrowtail \Omega^\phi \times P$ be the extension of the formula $\forall \mathbf{p}'(\mathbf{p}' \in \mathbf{w} \Rightarrow \mathbf{p}' \leqslant \mathbf{p})$, where **w** is a variable of type Ω^P, and let $B \rightarrowtail \Omega^P$ be the extension of

$$\exists \mathbf{p}\, \forall \mathbf{p}'(\mathbf{p}' \in \mathbf{w} \Rightarrow \mathbf{p}' \leqslant \mathbf{p}),$$

i.e. the image of $A \rightarrowtail \Omega^P \times P \xrightarrow{\pi_1} \Omega^P$. Thus B is "the object of subobjects of P which are bounded above". Since **P** is filtered, it is easy to see that B is a sub-$\vee$-semilattice of Ω^P (i.e. the empty subobject is bounded, and the union of two bounded subobjects is bounded). Moreover, $P \overset{\{\}}{\rightarrowtail} \Omega^P$ factors through

B, since $P \xrightarrow{(\{\},1)} \Omega^P \times P$ factors through A; so $B \geqslant K(P)$. Now suppose we are given a U-element $U \xrightarrow{\gamma} P^X$; then we can regard it as a morphism $U^*X \xrightarrow{\bar{\gamma}} U^*P$ in the topos $\mathscr{E}/U$. But the image I of this morphism is K-finite by 9.12(i), so by 9.13(i) it determines a U-element of $K(P)$ and hence of B. Now $A \longrightarrow B$ is epi, so by pulling back along it we obtain an epi $V \xrightarrow{\varepsilon} U$ and a V-element $V \xrightarrow{p} P$ which is an upper bound for $\varepsilon^*(I \rightarrowtail U^*P)$. But this says that the pair $(\gamma\varepsilon, hp)$ factors through $P_1^X \rightarrowtail P^X \times P^X$; so $\mathbf{P} \xrightarrow{h} \mathbf{P}^X$ is cofinal.

Conversely, suppose the given condition is satisfied. Since $K(X)$ is a $\vee$-semilattice, it is a filtered poset (cf. the proof that $\mathbf{J}^{\mathrm{op}}$ is filtered in 3.31), and so the map $K(X) \xrightarrow{h} K(X)^X$ is cofinal. Consider the global element $1 \xrightarrow{s} K(X)^X$ corresponding to the inclusion $X \rightarrowtail K(X)$; by cofinality we can find $V \twoheadrightarrow 1$ and a V-element $V \xrightarrow{u} K(X)$ such that the map $V \xrightarrow{s \times hu} K(X)^X \times K(X)^X$ factors through the order-relation. But then the K-finite subobject $U \rightarrowtail V^*X$ corresponding to u must be an upper bound for the singletons of V^*X; hence it is the whole of V^*X. So V^*X is K-finite, and hence X is K-finite by the remarks after 9.11. ∎

It is interesting to note that the definition of filteredness (2.51(a) and (b)) for a poset $\mathbf{P}$ is equivalent to saying that the unique map $\mathbf{P} \longrightarrow \mathbf{1}$ $(= \mathbf{P}^0)$ and the diagonal map $\mathbf{P} \xrightarrow{\Delta} \mathbf{P} \times \mathbf{P}$ $(= \mathbf{P}^{1 \amalg 1})$ are cofinal. So 9.14 implies that the class of K-finite objects is the closure of the two-element class $\{0, 1 \amalg 1\}$ under the the Galois connection between classes of objects and classes of posets in $\mathscr{E}$ defined by the relation

$$D(X, \mathbf{P}) \Leftrightarrow \mathbf{P} \xrightarrow{h} \mathbf{P}^X \text{ is cofinal.}$$

9.10 Proposition. (i) *Let X be a K-finite object of $\mathscr{E}$, and $\mathbf{P}$ a filtered poset. Then $\mathbf{P}^X$ is filtered.*

(ii) *Let X and Y be K-finite objects of $\mathscr{E}$. Then $X \times Y$ is K-finite.*

Proof. (i) Consider the diagram

$$\begin{array}{ccc} \mathbf{P} & \xrightarrow{h_1} & \mathbf{P}^X \\ \downarrow{\scriptstyle h_2} & & \downarrow{\scriptstyle \Delta} \\ \mathbf{P}^{X \amalg X} & \xrightarrow{\cong} & \mathbf{P}^X \times \mathbf{P}^X \end{array}$$

Since $X \amalg X$ is K-finite by 9.12(ii), h_1 and h_2 are cofinal by 9.14. Hence by Ex. 2.11 Δ is cofinal. A similar argument shows that $\mathbf{P}^X \longrightarrow \mathbf{1}$ is cofinal; hence $\mathbf{P}^X$ is filtered.

(ii) Let $\mathbf{P}$ be a filtered poset, and consider the diagram

$$\begin{array}{ccc} \mathbf{P} & \xrightarrow{h_1} & \mathbf{P}^X \\ {\scriptstyle h_3}\downarrow & & \downarrow{\scriptstyle h_2} \\ \mathbf{P}^{X \times Y} & \xrightarrow{\cong} & (\mathbf{P}^X)^Y . \end{array}$$

Since $\mathbf{P}^X$ is filtered by (i), h_1 and h_2 are cofinal by 9.14. Hence h_3 is cofinal by Ex. 2.11; so by 9.14 $X \times Y$ is K-finite. ∎

In fact a similar proof of the coproduct theorem (9.12(ii)) may be given using 9.14 as a definition of finiteness; we leave this as an exercise for the reader.

We have already remarked that $X \longmapsto K(X)$ is a covariant functor $\mathscr{E} \longrightarrow \mathscr{F}$, and in fact it can be regarded as taking values in **slat**($\mathscr{E}$). Somewhat surprisingly, we have

9.16 THEOREM (Mikkelsen). *K is the free functor for the theory* **slat**; *hence in particular* **slat**($\mathscr{E}$) *is monadic over $\mathscr{E}$ even if $\mathscr{E}$ does not have a natural number object.*

Proof. Let L be a semilattice. Defining the order-relation $L_1 \rightarrowtail L \times L$ as the equalizer of $L \times L \underset{\pi_2}{\overset{\vee}{\rightrightarrows}} L$, we make $\mathbf{L}$ into a filtered poset; so we could proceed as in 9.14 to define the object of bounded subobjects of L and to prove that it contains $K(L)$. But in fact we can do better than this; let $C \rightarrowtail \Omega \quad \times L$ be the extension of the formula

$$\forall \mathbf{l}'(\mathbf{l}' \in \mathbf{w} \Rightarrow \mathbf{l}' \leqslant \mathbf{l}) \wedge \forall \mathbf{l}'(\forall \mathbf{l}''(\mathbf{l}'' \in \mathbf{w} \Rightarrow \mathbf{l}'' \leqslant \mathbf{l}') \Rightarrow \mathbf{l} \leqslant \mathbf{l}')$$

(i.e. the formula "$\mathbf{l}$ is a least upper bound for $\mathbf{w}$"), and $D \rightarrowtail \Omega^L$ the image of $C \rightarrowtail \Omega^L \times L \xrightarrow{\pi_1} \Omega^L$. Since the order-relation on $\mathbf{L}$ is by definition antisymmetric, a least upper bound is unique if it exists, and so $C \twoheadrightarrow D$ is actually an isomorphism.

Now it is easy to show that C is a sub-semilattice of $\Omega^L \times L$, and that $L \xrightarrow{(\{\}, 1)} \Omega^L \times L$ factors through it; hence $K(L) \leqslant D$ as a subobject of Ω^L.

Let ε_L denote the composite semilattice homorphism

$$K(L) \rightarrowtail D \xrightarrow{\cong} C \rightarrowtail \Omega^L \times L \xrightarrow{\pi_2} L;$$

then it is easily seen that the composite $L \rightarrowtail K(L) \xrightarrow{\varepsilon_L} L$ is the identity. Hence if we are given a morphism $X \xrightarrow{f} L$ from any object X to L, we may factor it through $X \rightarrowtail K(X)$ by the semilattice homomorphism

$$K(X) \xrightarrow{K(f)} K(L) \xrightarrow{\varepsilon_L} L.$$

But this factorization is unique; for if g and g' are two such, then their equalizer is a sub-$\vee$-semilattice of Ω^X containing $X \xrightarrow{\{\}} \Omega^X$, and so must be the whole of $K(X)$. So $K(X)$ is the free semilattice generated by X. The fact that **slat**($\mathscr{E}$) is monadic over $\mathscr{E}$ follows as in 6.43, since 6.42 does not require a natural number object. ■

9.16 enables us to use known properties of finitary algebraic theories in establishing further properties of Kuratowski-finiteness: for example,

9.17. COROLLARY. *Inverse image functors preserve K-finiteness.*

Proof. Let $\mathscr{F} \xrightarrow{p} \mathscr{E}$ be a geometric morphism, X an object of $\mathscr{E}$. By 9.16 and 6.44, $p^*(K(X))$ and $K(p^*(X))$ are isomorphic as semilattices in $\mathscr{F}$, and hence as posets. But the property of having a maximal element is preserved by p^*; so K-finiteness of X implies K-finiteness of $p^*(X)$. ■

9.18 PROPOSITION. *Let X be an object of a topos $\mathscr{E}$. Then X is K-finite iff $K(X)$ is.*

Proof. First suppose $K(X)$ is K-finite. From 9.16, we have a semilattice homomorphism $K(K(X)) \xrightarrow{\varepsilon_{K(X)}} K(X)$ which is a split epimorphism; hence if $1 \xrightarrow{x} K(K(X))$ is a maximal element of $(K(X))$, it is easy to see that $\varepsilon_{K(X)} \cdot x$ is a maximal element of $K(X)$.

Conversely, suppose X is K-finite. Let $Q \rightarrowtail \Omega^X$ be the extension of the formula "$K(\mathbf{w})$ is K-finite", where $\mathbf{w}$ is a variable of type Ω^X. We shall show that Q is a $\vee$-semilattice and contains $X \xrightarrow{\{\}} \Omega^X$. In fact it is clear that Q contains $1 \xrightarrow{\ulcorner 0 \urcorner} \Omega^X$ and $X \xrightarrow{\{\}} \Omega^X$, since $K(0) \cong 1$ and $K(1) \cong 1 \sqcup 1$ are K-finite; so we need only show that Q is closed under $\vee$, or equivalently that if $Y_1 \rightarrowtail X$ and $Y_2 \rightarrowtail X$ are subobjects with $K(Y_1)$ and $K(Y_2)$ K-finite, then

$K(Y_1 \cup Y_2)$ is K-finite. But K, being the free functor for a finitary algebraic theory, preserves reflexive coequalizers (and hence all epimorphisms in $\mathscr{E}$) by 6.42; so the epi $Y_1 \amalg Y_2 \twoheadrightarrow Y_1 \cup Y_2$ induces an epi $K(Y_1 \amalg Y_2) \twoheadrightarrow K(Y_1 \cup Y_2)$. But $K(Y_1 \amalg Y_2) \cong K(Y_1) \times K(Y_2)$ is K-finite by 9.13(ii) and 9.15; so $K(Y_1 \cup Y_2)$ is K-finite by 9.12(i). Hence $Q \geqslant K(X)$, and so K-finiteness of X implies that $1 \xrightarrow{\ulcorner 1_X \urcorner} \Omega^X$ factors through Q, i.e. $K(X)$ is K-finite. ■

9.19 THEOREM. *Let $\mathscr{E}$ be a Boolean topos. The $\mathscr{E}_{kf}$ is a topos, and the inclusion functor $\mathscr{E}_{kf} \longrightarrow \mathscr{E}$ is logical.*

Proof From 9.12(ii), we know that any subobject of a K-finite object is K-finite; hence if X is K-finite, 9.13(i) implies that $K(X)$ is the whole of Ω^X. So by 9.18 $\mathscr{E}_{kf}$ is closed under the formation of power-objects. But it is also closed under the formation of finite products (and hence of all finite limits) by 9.15(ii); and by Ex. 1.1 this is sufficient to ensure that $\mathscr{E}_{kf}$ has all the structure of a topos, and that the inclusion functor preserves it. ■

For non-Boolean $\mathscr{E}$, $\mathscr{E}_{kf}$ need not be a topos; indeed, it need not even be closed under the formation of equalizers (see Ex. 9.2 below).

Finally, we prove a result due to F. W. Lawvere [LC], which makes precise the relationship between Kuratowski-finiteness and cardinal-finiteness in a topos with a natural number object.

9.20 THEOREM. *Let $\mathscr{E}$ be a topos with a natural number object, X an object of $\mathscr{E}$. Then X is K-finite iff it is locally a quotient of a finite cardinal, i.e. iff there exists $V \twoheadrightarrow 1$ in $\mathscr{E}$, a natural number p in $\mathscr{E}/V$ and an epimorphism $[p] \twoheadrightarrow V^*X$.*

Proof. First we shall show that cardinals are K-finite; more specifically, that the generic cardinal $[n]$ is K-finite in $\mathscr{E}/N$. To do this, we have to construct a factorization of $1 \xrightarrow{\ulcorner 1_{[n]} \urcorner} \Omega^{[n]}$ through $K([n])$; but we can do this by induction, using 6.18 and the data

$$1 \xrightarrow{1} 1 \cong K([o]) \cong o^*(K([n]))$$

and

$$K([n]) \xrightarrow{1_{K([n])} \times \ulcorner 1_1 \urcorner} K([n]) \times K(1) \cong K([n] \amalg 1) \cong s^*(K([n])).$$

It now follows from 9.12(i) and the remarks after 9.11 that every object satisfying the given conditions is K-finite.

Conversely, suppose X is K-finite. Since the theory **slat** is a quotient of the theory **mon** of monoids (obtained by adding the equations $m(x, y) = m(y, x)$ and $m(x, x) = x$), we have an epimorphism from the free monoid $M(X) = \Sigma_N(X^{[n]})$ to the free semilattice $K(X)$. Since this map is a monoid homomorphism, it is easily seen that its "elementwise" description must be as follows: a word $[p] \longrightarrow X$ is sent to the image of $[p] \longrightarrow X$ considered as a K-finite subobject of X. Now form the pullback

$$\begin{array}{ccc} V & \twoheadrightarrow & 1\,, \\ {\scriptstyle w}\downarrow & & \downarrow{\scriptstyle \ulcorner 1_X \urcorner} \\ M(X) & \twoheadrightarrow & K(X) \end{array}$$

and let p be the composite $V \xrightarrow{w} M(X) \longrightarrow N$. Then w corresponds to a word of length p in the elements of V^*X whose image is the whole of V^*X, i.e. to an epimorphism $[p] \twoheadrightarrow V^*X$ in $\mathscr{E}/V$. ■

The three concepts studied in §6.2, §8.4 and the present paragraph have by no means exhausted the possible definitions of finiteness in a topos. In particular, it should be mentioned that T. G. Brook [14] has studied the definition "X is finite iff there exists a partial order on X such that $\mathbf{X}$ and $\mathbf{X}^{\mathrm{op}}$ are both well-ordered", and H. Volger [127] has studied the definition "X is finite iff every ultrafilter on X is principal". In a general topos, Brook's definition seems to be inconveniently restrictive (it is not always satisfied by finite cardinals), and Volger's seems excessively lax (it is always satisfied by Ω). The reason in both cases is essentially the same; namely that in a topos where (AC) does not hold, the supply of well-orderings and non-principal ultrafilters is not as plentiful as one might wish.

9.2. TRANSITIVE OBJECTS

In this and the following paragraph, our programme is to prove that (elementary) topos theory, with the addition of certain "set-like" axioms, is *logically equivalent* to a certain "weak" version of Zermelo–Fraenkel set theory. This programme was first carried out (independently) by J. C. Cole [18] and W. Mitchell [85], both of whom used the idea that the membership-relation on a set can be characterized as a tree having certain properties. We shall follow a later proof of the same result by G. Osius [92], whose starting-point is a familiar result of ZF set theory known as the Mostowski Isomorphism Theorem.

Recall that a set S is said to be *transitive* if $(x \in y \wedge y \in S) \Rightarrow x \in S$.

9.21 THEOREM (Mostowski [181]). *Let A be a set and R a binary relation on A. The following statements are equivalent:*

(i) R *is* extensional (*i.e.* $\{a \in A \mid a\, R\, x\} = \{a \in A \mid a\, R\, y\} \Rightarrow x = y$) *and* well-founded (*i.e.* $\forall B \subseteq (B \neq \varnothing \Rightarrow \exists y \in B\, \forall b \in B \neg(b\, R\, y))$).

(ii) *There exists a transitive set S such that (A, R) is isomorphic to $(S, \{\langle x, y\rangle \mid x \in y \in S\})$ in the category of sets-equipped-with-a-binary-relation.*

Proof. (i) $\Rightarrow$ (ii): we define a function f with domain A by recursion on the well-founded relation R, by the formula

$$f(a) = \{f(b) \mid b \in A \wedge b\, R\, a\}.$$

Extensionality of R ensures that f is one-to-one, and so the set $\{f(a) \mid a \in A\}$ has the required properties.

(ii) $\Rightarrow$ (i) is immediate from the axioms of extensionality and foundation. ■

We now proceed to develop, within topos theory, the basic properties of objects equipped with binary relations similar to those in 9.21.

9.22 DEFINITION. Let X be an object of a topos $\mathscr{E}$. By a *relation on* X we shall (for the moment) mean a morphism $X \xrightarrow{r} \Omega^X$. We say that r is *extensional* if it is a monomorphism, and r is *inductive* if any subobject $Y \overset{m}{\rightarrowtail} X$ satisfying $r^{-1}(\Omega^Y \overset{\exists m}{\rightarrowtail} \Omega^X) \leqslant m$ is in fact the whole of X.

By a *transitive object* of $\mathscr{E}$, we mean a pair (X, r), where r is an extensional, inductive relation on X.

Inductiveness means that we can define morphisms with domain X "by recursion on r", as is shown by the following theorem, due to C. J. Mikkelsen [84]. (Compare also 6.14.) ■

9.23 THEOREM. *Let X be an object of a topos $\mathscr{E}$, r an extensional relation on X. Then r is inductive iff (X, r) has the following "universal recursion property": for any object Y and morphism $Q^Y \xrightarrow{g} Y$, there exists a unique $f\colon X \longrightarrow Y$*

such that

$$\begin{array}{ccc} X & \xrightarrow{\;f\;} & Y \\ {\scriptstyle r}\downarrow & & \uparrow{\scriptstyle g} \\ \Omega^X & \xrightarrow{\;\exists f\;} & \Omega^Y \end{array}$$

commutes. (*We say f is r*-recursively defined *by g.*)

Proof. (i) Suppose r is inductive, and we are given $\Omega^Y \xrightarrow{g} Y$. If $\mathbf{f}$ is a variable of type $\Omega^{X \times Y}$, we shall write "$\mathbf{f}\colon X \longrightarrow Y$" as an abbreviation for the formula

$$\forall \mathbf{x}\, \forall \mathbf{y}\, \forall \mathbf{y}'\, (\langle \mathbf{x}, \mathbf{y}\rangle \in \mathbf{f} \wedge \langle \mathbf{x}, \mathbf{y}'\rangle \in \mathbf{f}) \Rightarrow \mathbf{y} = \mathbf{y}'$$

of $L_{\mathscr{E}}$. (The extension of this formula is clearly (isomorphic to) $\tilde{Y}^X \rightarrowtail (\Omega^Y)^X \cong \Omega^{X \times Y}$.) Now let $M \rightarrowtail \Omega^{X \times Y}$ be the extension of the formula

$$\forall \mathbf{f}'\, \forall \mathbf{x} (\mathbf{f}' \leqslant \mathbf{f} \wedge \mathbf{f}'\colon X \longrightarrow Y \wedge \exists\pi_1(\mathbf{f}') = r(\mathbf{x})) \Rightarrow \langle \mathbf{x}, g(\exists\pi_2(\mathbf{f}'))\rangle \in \mathbf{f},$$

and let $F \rightarrowtail X \times Y$ be the internal intersection of M. We shall show that F is the graph of a morphism $X \xrightarrow{f} Y$ with the required property.

First note that

$$\mathscr{E} \models \mathbf{f}' \leqslant \ulcorner F \urcorner \Rightarrow \forall \mathbf{f}(\mathbf{f} \in \ulcorner M \urcorner \Rightarrow \mathbf{f}' \leqslant \mathbf{f}),$$

from which it follows easily that $\ulcorner F \urcorner \in \ulcorner M \urcorner$. If we define $G \rightarrowtail X \times Y$ to be the extension of

$$\exists \mathbf{f}(\mathbf{f} \leqslant \ulcorner F \urcorner \wedge \mathbf{f}\colon X \longrightarrow Y \wedge \exists\pi_1(\mathbf{f}) = r(\mathbf{x}) \wedge g(\exists\pi_2(\mathbf{f})) = \mathbf{y}),$$

then we have $\mathscr{E} \models \langle \mathbf{x}, \mathbf{y}\rangle \in \ulcorner G \urcorner \Rightarrow \langle \mathbf{x}, \mathbf{y}\rangle \in \ulcorner F \urcorner$, i.e. $G \leqslant F$. But this in turn implies that $\ulcorner G \urcorner \in \ulcorner M \urcorner$ and so $F \leqslant G$; hence $F = G$.

Now consider the subobject

$$X' \overset{m}{\rightarrowtail} X = \| \exists! \mathbf{y} \langle \mathbf{x}, \mathbf{y}\rangle \in \ulcorner F \urcorner \|;$$

we shall show that $r^{-1}(\exists m) \leqslant m$, and so $m \cong 1_X$, i.e. F is the graph of a morphism. Let $1 \xrightarrow{x} X$ be a global element such that $rx \leqslant \ulcorner m \urcorner$; define $F_x \rightarrowtail X \times Y$ to be the extension of $\langle \mathbf{x}, \mathbf{y}\rangle \in \ulcorner F \urcorner \wedge \mathbf{x} \in rx$. Then it is easy to see that F_x is the graph of a partial map $X \longrightarrow Y$ with domain rx, and so the pair $\langle x, g(\exists\pi_2(\ulcorner F_x \urcorner))\rangle$ is in $\ulcorner G \urcorner$ and hence in $\ulcorner F \urcorner$. But since F_x is the unique such partial map contained in F, we deduce that

$$\mathscr{E} \models \langle \mathbf{x}, \mathbf{y} \rangle \in \ulcorner F \urcorner \Rightarrow \mathbf{y} = g(\exists\pi_2(\ulcorner F_x \urcorner))$$

and hence $x \in \ulcorner m \urcorner$. And since this argument works equally well for generalized elements of X, we have the required result.

To show that the morphism f is unique, suppose we have two morphisms f_1, f_2 both r-recursively defined by g. Then it is not hard to show that their equalizer $X' \stackrel{m}{\rightarrowtail} X$ satisfies $r^{-1}(\exists m) \leqslant m$, and so $f_1 = f_2$.

(ii) Conversely, suppose r is extensional and has the universal recursion property. Let

$$\hat{r}: \Omega^{\tilde{X}} \longrightarrow \tilde{X}$$

represent the partial map

$$\begin{array}{ccc} X & \xrightarrow{\quad 1 \quad} & X\ ; \\ \downarrow{\scriptstyle r} & & \\ \Omega^X & & \\ \downarrow{\scriptstyle \exists\eta_X} & & \\ \Omega^{\tilde{X}} & & \end{array}$$

then we observe that $X \stackrel{\eta_X}{\rightarrowtail} \tilde{X}$ is r-recursively defined by $\hat{r}$. Now given $Y \stackrel{m}{\rightarrowtail} X$ such that $r^{-1}(\exists m) \leqslant m$, form the pullback

$$\begin{array}{ccc} Z & \stackrel{s}{\rightarrowtail} & \Omega^Y , \\ \downarrow & & \downarrow{\scriptstyle \exists m} \\ X & \stackrel{r}{\rightarrowtail} & \Omega^X \end{array}$$

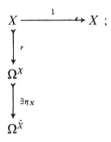

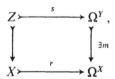

and let $\Omega^{\dot{Y}} \xrightarrow{g} \tilde{Y}$ represent the partial map

$$\begin{array}{ccc} Z & \rightarrowtail & Y\,. \\ \downarrow{\scriptstyle s} & & \\ \Omega^{Y} & & \\ \downarrow{\scriptstyle \exists \eta_Y} & & \\ \Omega^{\tilde{Y}} & & \end{array}$$

Let $X \xrightarrow{f} \tilde{Y}$ be r-recursively defined by g.

Now the square

$$\begin{array}{ccc} \Omega^{\tilde{Y}} & \xrightarrow{\exists \tilde{m}} & \Omega^{\tilde{X}} \\ \downarrow{\scriptstyle g} & & \downarrow{\scriptstyle \hat{r}} \\ \tilde{Y} & \xrightarrow{\tilde{m}} & \tilde{X} \end{array}$$

commutes, since both ways round represent the partial map

$$\begin{array}{ccc} Z & \rightarrowtail & X\,; \\ \downarrow{\scriptstyle s} & & \\ \Omega^{Y} & & \\ \downarrow{\scriptstyle \exists \eta_Y} & & \\ \Omega^{\dot{\tilde{Y}}} & & \end{array}$$

hence $\tilde{m}f = \eta_X \colon X \longrightarrow \tilde{X}$ since both morphisms are r-recursively defined by $\hat{r}$. Now η_X is mono, so that $\eta_X^*(\eta_X) \cong 1_X$; and since $\eta_X^*(\tilde{m}) \cong m$, applying η_X^* to the equation $\tilde{m}f = \eta_X$ gives us a factorization of 1_X through m. Hence m is epi. ■

9.24 Proposition. *Let (X, r) and (Y, s) be transitive objects in $\mathscr{E}$. Then there exists at most one $f \colon X \longrightarrow Y$ such that*

$$\begin{array}{ccc} X & \xrightarrow{f} & Y \\ \downarrow{\scriptstyle r} & & \downarrow{\scriptstyle s} \\ \Omega^{X} & \xrightarrow{\exists f} & \Omega^{Y} \end{array}$$

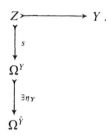

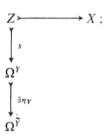

commutes, and if such an f exists, then it is mono. We say f is an inclusion *of* (X, r) *in* (Y, s); *and we write* $\mathscr{E}_{\text{tr}}$ *for the category of transitive objects and inclusions in* $\mathscr{E}$.

Proof. Let $\hat{s}: \Omega^{\tilde{Y}} \longrightarrow \tilde{Y}$ be defined as in 9.23(ii). Now the diagram

$$\begin{array}{ccccc} X & \xrightarrow{f} & Y & \xrightarrow{\eta_Y} & \tilde{Y} \\ \downarrow{\scriptstyle r} & & \downarrow{\scriptstyle s} & & \uparrow{\scriptstyle \hat{s}} \\ \Omega^X & \xrightarrow{\exists f} & \Omega^Y & \xrightarrow{\exists \eta_Y} & \Omega^{\tilde{Y}} \end{array}$$

commutes; i.e. $\eta_Y f$ is r-recursively defined by $\hat{s}$. Hence r and s determine the composite $\eta_Y f$ uniquely; but η_Y is mono, so f is uniquely determined.

To show that f is mono, let $Y \xrightarrow{g} \tilde{X}$ be the morphism s-recursively defined by $\hat{r}$. Then gf is r-recursively defined by $\hat{r}$, so $gf = \eta_X$. But η_X is mono, so f is mono. ■

The next result may be thought of as a partial converse of 9.24.

9.25 LEMMA. *Suppose we have a commutative diagram*

$$\begin{array}{ccc} X & \overset{f}{\rightarrowtail} & Y \\ \downarrow{\scriptstyle r} & & \downarrow{\scriptstyle s} \\ \Omega^X & \xrightarrow{\exists f} & \Omega^Y \end{array}$$

with f mono. Then if s is extensional (resp. inductive), so is r.

Proof. (i) Trivially s mono $\Rightarrow sf = \exists f \,.\, r$ mono $\Rightarrow r$ mono.

(ii) Let $Z \overset{m}{\rightarrowtail} X$ satisfy $r^{-1}(\exists m) \leqslant m$. Define $T \overset{n}{\rightarrowtail} Y = \forall_f(Z \overset{m}{\rightarrowtail} X)$; then from the adjunction $(f^{-1} \dashv \forall_f)$, we have $f^{-1}(n) \leqslant m$ as subobjects of X. But by Ex. 1.10 the functor $f \longmapsto \exists f$ preserves pullbacks of monos; hence $(\exists f)^{-1}(\exists n) \leqslant \exists m$ as subobjects of Ω^X. So $f^{-1}(s^{-1}(\exists n)) = r^{-1}((\exists f)^{-1}(\exists n)) \leqslant r^{-1}(\exists m) \leqslant m$, and hence $s^{-1}(\exists n) \leqslant \forall_f(m) = n$. But s is inductive, so $n \cong 1_Y$; hence $m \geqslant f^{-1}(1_Y) = 1_X$. ■

9.26 PROPOSITION. *The poset $\mathscr{E}_{\text{tr}}$ has binary intersections and unions.*

Proof. (i) Let (X, r) and (Y, s) be two transitive objects. To form their intersection in $\mathscr{E}_{\mathrm{tr}}$ (which we denote informally by $(X \cap Y, r \cap s)$), construct the pullback

$$\begin{array}{ccc} X \cap Y & \xrightarrow{p} & Y, \\ \Big\downarrow q & & \Big\downarrow \eta_Y \\ X & \xrightarrow{f} & \tilde{Y} \end{array}$$

where f is r-recursively defined by $\hat{s}$. Now we have $\hat{s}\,.\,\exists f\,.\,rq = fq = \eta_Y p$; so $\exists f\,.\,rq$ factors through

$$(\hat{s})^{-1}(Y \rightarrowtail^{\eta_Y} \tilde{Y}) = (Y \rightarrowtail^{s} \Omega^Y \rightarrowtail^{\exists\eta_Y} \Omega^{\tilde{Y}}),$$

and hence in particular through $\exists\eta_Y$. But

$$\begin{array}{ccc} \Omega^{X \cap Y} & \xrightarrow{\exists p} & \Omega^Y \\ \Big\downarrow \exists q & & \Big\downarrow \exists\eta_Y \\ \Omega^X & \xrightarrow{\exists f} & \Omega^{\tilde{Y}} \end{array}$$

is a pullback by Ex. 1.10, and so rq factors through $\exists q$, i.e. $X \xrightarrow{r} \Omega^X$ restricts to a relation $X \cap Y \xrightarrow{r \cap s} \Omega^{X \cap Y}$ on $X \cap Y$. Now $(X \cap Y, r \cap s)$ is a transitive object (and q is an inclusion) by 9.25; but p is also an inclusion, since the two ways round the square

$$\begin{array}{ccc} X \cap Y & \xrightarrow{p} & Y \\ \Big\downarrow r \cap s & & \Big\downarrow s \\ \Omega^{X \cap Y} & \xrightarrow{\exists p} & \Omega^Y \end{array}$$

are coequalized by $\Omega^Y \rightarrowtail^{\exists\eta_Y} \Omega^{\tilde{Y}}$.

To establish the universal property of $(X \cap Y, r \cap s)$, let (Z, u) be any transitive object, and $(Z, u) \xrightarrow{i} (X, r)$, $(Z, u) \xrightarrow{j} (Y, s)$ two inclusions. Then the composites fi and $\eta_Y j$ are both u-recursively defined by $\hat{s}$, so they are equal: hence (i, j) factors through the pullback $X \cap Y$. And the factorization is clearly an inclusion of transitive objects.

(ii) To form the union $(X \cup Y, r \cup s)$, construct the pushout

$$\begin{array}{ccc} X \cap Y & \overset{p}{\rightarrowtail} & Y, \\ {\scriptstyle q}\downarrow & & \downarrow{\scriptstyle b} \\ X & \xrightarrow{a} & X \cup Y \end{array}$$

and define $r \cup s$ to be the unique morphism such that

$$\begin{array}{ccccc} X & \xrightarrow{a} & X \cup Y & \xleftarrow{b} & Y \\ {\scriptstyle r}\downarrow & & \downarrow{\scriptstyle r \cup s} & & \downarrow{\scriptstyle s} \\ \Omega^X & \xrightarrow{\exists a} & \Omega^{X \cup Y} & \xleftarrow{\exists b} & \Omega^Y \end{array}$$

commutes. It is immediate from the universal property of a pushout that $r \cup s$ satisfies the universal recursion property of 9.23; so we need only prove that it is mono.

Let $X \cup Y \xrightarrow{w'} \tilde{Y}$ represent the partial map

$$\begin{array}{ccc} Y & \xrightarrow{1} & Y; \\ {\scriptstyle b}\downarrow & & \\ X \cup Y & & \end{array}$$

then since the pushout square defining $X \cup Y$ is also a pullback by 1.28, we have $wa = f$ since both represent the same partial map $X \longrightarrow Y$. But we also have $wb = \eta_Y$; and since the pair (a, b) is jointly epimorphic, we deduce that w is $(r \cup s)$-recursively defined by $\hat{s}$.

Now suppose $U \overset{\gamma_1}{\underset{\gamma_2}{\rightrightarrows}} X \cup Y$ are coequalized by $r \cup s$; then they are also coequalized by $w = \hat{s} \,.\, \exists w \,.\, (r \cup s)$. Form the pullback

$$\begin{array}{ccc} V & \longrightarrow & Y\ ; \\ {\scriptstyle v}\downarrow & & \downarrow{\scriptstyle \eta_Y} \\ U & \xrightarrow{w'\gamma_1} & \tilde{Y} \end{array} \qquad \text{since} \qquad \begin{array}{ccc} Y & \xrightarrow{1} & Y \\ {\scriptstyle b}\downarrow & & \downarrow{\scriptstyle \eta_Y} \\ X \cup Y & \xrightarrow{w'} & \tilde{Y} \end{array}$$

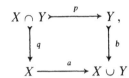

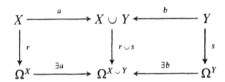

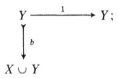

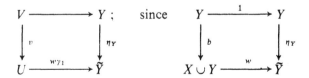

is a pullback by the definition of w, the squares

$$\begin{array}{ccc} V & \longrightarrow & Y \\ \downarrow{\scriptstyle v} & & \downarrow{\scriptstyle b} \\ U & \xrightarrow{\gamma_i} & X \cup Y \end{array}$$

are also pullbacks (and in particular they commute) for both $i = 1$ and $i = 2$. Hence γ_1 and γ_2 are equalized by $v = \gamma_1^*(b)$. A similar argument shows that they are equalized by $\gamma_1^*(a)$; but (a, b) is jointly epi, so we must have $\gamma_1 = \gamma_2$.

Finally, the universal property of $(X \cup Y, r \cup s)$ is immediate from that of a pushout in $\mathscr{E}$, and the fact that $\mathscr{E}_{\text{tr}}$ is a poset. ■

9.27 PROPOSITION. (i) $(0, 0 \longrightarrow \Omega^0 \cong 1)$ *is a transitive object.*

(ii) *If* (X, r) *is a transitive object, so is* $(\Omega^X, \exists r)$, *and* $(X, r) \xrightarrow{r} (\Omega^X, \exists r)$ *is an inclusion.*

Proof. (i) is trivial.

(ii) From 1.33, we know $\exists$ preserves monos, so $\exists r$ is extensional. Suppose we are given $\Omega^Y \xrightarrow{g} Y$. Then if $X \xrightarrow{f} Y$ is r-recursively defined by g, it is easily checked that $\Omega^X \xrightarrow{\exists f} \Omega^Y \xrightarrow{g} Y$ is $\exists r$-recursively defined by g. Moreover, if h is also $\exists r$-recursively defined by g, then we must have $hr = f$, and so $h = g . \exists h . \exists r = g . \exists f$. Thus $\exists r$ has the universal recursion property; the last part of the statement is trivial. ■

9.28 DEFINITION. An object X of $\mathscr{E}$ is said to be *partially transitive* if there exists a transitive object (Y, s) and a monomorphism $X \rightarrowtail Y$ in $\mathscr{E}$. We write $\mathscr{E}_{\text{ptr}}$ for the full subcategory of partially transitive objects in $\mathscr{E}$. ■

9.29. THEOREM (Osius). $\mathscr{E}_{\text{ptr}}$ *is a topos, and the inclusion functor* $\mathscr{E}_{\text{ptr}} \longrightarrow \mathscr{E}$ *is logical. Moreover, any logical functor* $\mathscr{E} \longrightarrow \mathscr{F}$ *restricts to a logical functor* $\mathscr{E}_{\text{ptr}} \longrightarrow \mathscr{F}_{\text{ptr}}$.

Proof. For the first sentence, it suffices by Ex. 1.1 to show that $\mathscr{E}_{\text{ptr}}$ is closed under the formation (in $\mathscr{E}$) of finite limits and power-objects, and that every subobject of a partially transitive object is partially transitive.

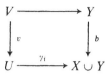

But if we have a mono $X \xrightarrow{m} Y$ where (Y, s) is transitive, then $\Omega^X \overset{\exists m}{\rightarrowtail} \Omega^Y$ expresses Ω^X as a subobject of a transitive object, by 9.27(ii). The third statement above is trivial from the definition; it follows that $\mathscr{E}_{\text{ptr}}$ is closed under formation of equalizers. And since $(1, 1 \xrightarrow{f} \Omega)$ is transitive by 9.27(i) and (ii), it remains to consider products of pairs.

Suppose X and Y are both partially transitive. Using 9.26(ii), we can find a single transitive object (Z, u) containing both of them as subobjects; but then by Ex. 5.10 we have a monomorphism

$$X \times Y \rightarrowtail Z \times Z \xrightarrow{\lambda pr} \Omega^{\Omega^Z}.$$

So $X \times Y$ is partially transitive.

To prove the second sentence of the theorem, it suffices to show that both extensionality and inductiveness of relations are preserved by logical functors. For extensionality this is trivial; for inductiveness, note that the definition is equivalent to saying that $1 \xrightarrow{\ulcorner 1_x \urcorner} \Omega^X$ is the unique fixed point of the order-preserving map

$$\Omega^X \xrightarrow{\downarrow seg} \Omega^{\Omega^X} \xrightarrow{\Omega^r} \Omega^X.$$

Since the construction of the Tarski fixed-point theorem (5.50) can be used to construct the unique *smallest* fixed point of such an order-preserving map, this property is clearly preserved by logical functors. ■

9.3. THE EQUICONSISTENCY THEOREM

We are now ready to embark on the process of constructing a model of set theory from a given topos. The set theory which we use is *not* standard ZF set theory; this is because the latter is well known not to be finitely axiomatizable, and so cannot be logically equivalent to any finitely-presented extension of elementary topos theory.

9.31 Definition. We define the *weak Zermelo set theory* $\mathbf{zer}_0$ to be a theory expressed in the (classical) first-order predicate calculus with equality, having a single type (whose terms are called "sets"), a constant (= nullary operation) $\varnothing$, a binary operation $\langle -, - \rangle$, and a binary predicate $\in$, subject to the axioms

(i) (Extensionality) $\forall \mathbf{x}(\mathbf{x} \in \mathbf{s} \Leftrightarrow \mathbf{x} \in \mathbf{t}) \Rightarrow \mathbf{s} = \mathbf{t}$.
(ii) (Empty set) $\neg(\mathbf{x} \in \varnothing)$.
(iii) (Unordered pair) $\exists \mathbf{z}\, \forall \mathbf{t}(\mathbf{t} \in \mathbf{z} \Leftrightarrow (\mathbf{t} = \mathbf{x} \vee \mathbf{t} = \mathbf{y}))$.
(iv) (Ordered pair) $\langle \mathbf{x}, \mathbf{y} \rangle = \langle \mathbf{z}, \mathbf{t} \rangle \Leftrightarrow (\mathbf{x} = \mathbf{z} \wedge \mathbf{y} = \mathbf{t})$.

(v) (Power-set) $\exists \mathbf{y}\, \forall \mathbf{z}(\mathbf{z} \in \mathbf{y} \Leftrightarrow \forall \mathbf{t}(\mathbf{t} \in \mathbf{z} \Rightarrow \mathbf{t} \in \mathbf{x}))$.

(vi) (Relative complement) $\exists \mathbf{z}\, \forall \mathbf{t}(\mathbf{t} \in \mathbf{z} \Leftrightarrow (\mathbf{t} \in \mathbf{x} \wedge \neg(\mathbf{t} \in \mathbf{y})))$.

(vii) (Cartesian product) $\exists \mathbf{z}\, \forall \mathbf{t}(\mathbf{t} \in \mathbf{z} \Leftrightarrow \exists \mathbf{u}\, \exists \mathbf{v}(\mathbf{u} \in \mathbf{x} \wedge \mathbf{v} \in \mathbf{y} \wedge \mathbf{t} = \langle \mathbf{u}, \mathbf{v} \rangle))$.

(viii) (Membership relation) $\exists \mathbf{y}\, \forall \mathbf{z}(\mathbf{z} \in \mathbf{y} \Leftrightarrow \exists \mathbf{s}\, \exists \mathbf{t}(\mathbf{s} \in \mathbf{t} \wedge \mathbf{t} \in \mathbf{x} \wedge \mathbf{z} = \langle \mathbf{s}, \mathbf{t} \rangle))$.

(ix) (Domain) $\exists \mathbf{y}\, \forall \mathbf{z}(\mathbf{z} \in \mathbf{y} \Leftrightarrow \exists \mathbf{t}(\langle \mathbf{z}, \mathbf{t} \rangle \in \mathbf{x}))$.

(x) (Permutations) $\exists \mathbf{y}\, \forall \mathbf{z}(\mathbf{z} \in \mathbf{y} \Leftrightarrow \exists \mathbf{s}\, \exists \mathbf{t}(\langle \mathbf{s}, \mathbf{t} \rangle \in \mathbf{x} \wedge \mathbf{z} = \langle \mathbf{t}, \mathbf{s} \rangle))$
$\exists \mathbf{y}\, \forall \mathbf{z}(\mathbf{z} \in \mathbf{y} \Leftrightarrow \exists \mathbf{s}\, \exists \mathbf{t}\, \exists \mathbf{u}\langle\langle \mathbf{s}, \mathbf{t} \rangle, \mathbf{u} \rangle \in \mathbf{x} \wedge \mathbf{z} = \langle\langle \mathbf{u}, \mathbf{s} \rangle, \mathbf{t} \rangle))$.

(xi) (Foundation) $\neg(\mathbf{x} = \varnothing) \Rightarrow \exists \mathbf{y}(\mathbf{y} \in \mathbf{x} \wedge \forall \mathbf{t}(\mathbf{t} \in \mathbf{y} \Rightarrow \neg(\mathbf{t} \in \mathbf{x})))$. ■

The form of 9.31 makes it clear that $\mathbf{zer}_0$ is finitely-presented. However, it is convenient to note that axioms (vi)–(x) may equivalently be replaced by the axiom that union-sets exist, and the *restricted comprehension axiom* that, for any formula $\phi(\mathbf{x})$ having only restricted quantifiers (i.e. quantifiers of the form $\forall \mathbf{s}(\mathbf{s} \in \mathbf{t} \Rightarrow \ldots)$ and $\exists \mathbf{s}(\mathbf{s} \in \mathbf{t} \wedge \ldots)$), and any set $\mathbf{y}$, the set $\{\mathbf{x} \in \mathbf{y} \mid \phi(\mathbf{x})\}$ exists. (This fact is due to E. J. Thiele [186].)

It will be noted that the Mostowski Isomorphism Theorem (9.21) is not provvble in $\mathbf{zer}_0$, since its proof involved the axiom of replacement; nor is it provable in $\mathbf{zer}_0$ that any set is embeddable in a transitive set. Since we wish both these statements to hold in our model, we proceed to add them as axioms:

9.32 Definition. The set theory **zer** is defined by adding the following two axioms to those of $\mathbf{zer}_0$:

(xii) (Transitivity) $\exists \mathbf{y}(\mathbf{x} \subseteq \mathbf{y} \wedge \mathbf{y}$ is transitive).

(xiii) (Mostowski)

$$(\mathbf{r} \text{ is an extensional, well-founded relation on } \mathbf{x}) \Rightarrow$$
$$\Rightarrow \exists \mathbf{y}\, \exists \mathbf{f}(\mathbf{y} \text{ is transitive}) \wedge (\mathbf{f} \text{ is a bijection from } \mathbf{x} \text{ to } \mathbf{y}) \wedge$$
$$\wedge\, \forall \mathbf{s}\, \forall \mathbf{t}(\langle \mathbf{s}, \mathbf{t} \rangle \in \mathbf{r} \Leftrightarrow f(\mathbf{s}) \in f(\mathbf{t})). \ ■$$

It is easy to prove (cf. 9.21) that the set **y** and the bijection **f** of 9.32(xiii) are uniquely determined by **x** and **r**. Similarly, we may deduce from 9.32(xii) that every set **x** has a transitive closure; using the restricted comprehension axiom, we may define the latter as

$$\{\mathbf{t} \in \mathbf{y} \mid \forall \mathbf{z} \subseteq \mathbf{y}(\mathbf{x} \subseteq \mathbf{z} \wedge \mathbf{z} \text{ is transitive}) \Rightarrow \mathbf{t} \in \mathbf{z}\}$$

where **y** is any transitive set containing **x**, and then prove that the definition is independent of the choice of **y**.

Now it is clear that for any model of $\mathbf{zer}_0$, the corresponding category of sets and functions is an elementary topos; but this topos enjoys certain special properties which are not shared by all toposes. It is therefore necessary to impose additional conditions on a topos in order to establish a logical equivalence.

9.33 PROPOSITION (P. Freyd [FK]). *Let $\mathscr{E}$ be a topos. The following conditions are equivalent:*

(i) 1 *is a generator for* $\mathscr{E}$, *and* $\mathscr{E}$ *is a non-degenerate* (*i.e.* $0 \longrightarrow 1$ *is not iso*).
(ii) $\mathscr{E}$ *is Boolean and satisfies* (SS) (5.21) *and the* two-valuedness *axiom* (TV): *There are exactly two morphisms from* 1 *to* Ω.

Proof. (i) $\Rightarrow$ (ii): If $X \not\cong 0$, then the coproduct inclusions $X \overset{\nu_1}{\underset{\nu_2}{\rightrightarrows}} X \amalg X$ are not equal, and so there exists a morphism $1 \xrightarrow{x} X$ with $\nu_1 x \neq \nu_2 x$. In particular $X \longrightarrow 1$ is split epi, and so $\mathscr{E}$ satisfies (SS). Moreover, if U is a nonzero subobject of 1, then $U \rightarrowtail 1$ is iso, and so $\mathscr{E}$ satisfies (TV). To show that $\mathscr{E}$ is Boolean, let $X' \overset{m}{\rightarrowtail} X$ be a subobject in $\mathscr{E}$. Then for every $1 \xrightarrow{x} X$, we have either $x \in \ulcorner m \urcorner$ or $x \cap m = 0$, i.e. $x \in \neg \ulcorner m \urcorner$. Thus every global element of X factors through $X' \amalg (\neg X') \rightarrowtail X$, so the latter map is iso, i.e. $\neg X'$ is a complement for X'.

(ii) $\Rightarrow$ (i): By 5.32, $\mathscr{E}$ satisfies (SG); but the only subobjects of 1 are 0 and 1, and 0 is clearly redundant as a member of any set of generators. So 1 is a generator; and non-degenerateness follows from (TV). ■

9.34 DEFINITION. A topos satisfying the equivalent conditions of 9.33 is said to be *well-pointed.* We write **wpt** for the theory of well-pointed toposes, i.e. the elementary theory of categories [72], plus (elementary forms of) the three topos axioms (1.11), plus either of the conditions of 9.33. We define the theory **wtt** to consist of the theory **wpt**, plus the *axiom of partial transitivity* (PT): Every object of $\mathscr{E}$ is partially transitive (cf. 9.28). ■

It is clear that if $\mathscr{E}$ is a model of **wpt**, then $\mathscr{E}_{\text{ptr}}$ is a model of **wtt**; and indeed 9.29 can be interpreted as a relative consistency theorem for (PT) relative to **wpt**. In order to avoid using the axiom of choice in our metatheory, we shall interpret (PT) in the same "constructive" way as we intepret the statement that a category has limits; i.e. as saying that we are given an operation which assigns to each object X of $\mathscr{E}$ a monomorphism $X \rightarrowtail T(X)$ and an extensional, inductive relation r_X on $T(X)$. For the same reason, we shall want to assume that we are given a canonical representative of each isomorphism class of transitive objects in $\mathscr{E}$, i.e. that we are given a functor $\tau: \mathscr{E}_{\text{tr}} \longrightarrow \mathscr{E}_{\text{tr}}$, naturally isomorphic to the identity, such that $(X, r) \cong (Y, s)$ implies $\tau(X, r) = \tau(Y, s)$. We shall denote this assumption by (TR).

9.35 LEMMA. *Let M be a model of* $\mathbf{zer}_0$. *Then the category $\mathscr{S}(M)$ of sets and functions in M is a model of* **wpt**. *If moreover M is a model of* **zer**, *then $\mathscr{S}(M)$ satisfies* (PT) *and* (TR).

Proof. It is easily seen that if x and y are sets in M, then we can construct the cartesian product $x \times y$, the equalizer of two functions $x \rightrightarrows y$, and the set of all functions $x \longrightarrow y$, within M, and that they have the required categorical properties. Moreover $1 = \{\varnothing\}$ is a terminal object and a generator for $\mathscr{S}(M)$, and $2 = \{\varnothing, 1\}$ is a subobject classifier. Now if x is a transitive set in M, then the membership relation on x is extensional and inductive by 9.31(i) and (xi), so 9.32(xii) (plus the remark about transitive closures following 9.32) implies that $\mathscr{S}(M)$ satisfies (PT). Similarly, 9.32(xiii) implies (TR), since we can take the transitive sets as canonical representatives for the objects of $\mathscr{S}(M)_{\text{tr}}$. ■

We now embark on the construction of a model $S(\mathscr{E})$ of **zer** within an arbitrary well-pointed topos $\mathscr{E}$. We consider triples of the form (X, r, m), where (X, r) is a transitive object of $\mathscr{E}$ and m is a global element of Ω^X.

9.36 LEMMA. *Let (X, r, m) and (Y, s, n) be triples as above. Then the following are equivalent:*

(i) *There exists a transitive object (Z, u) and inclusions $(X, r) \xrightarrow{a} (Z, u)$, $(Y, s) \xrightarrow{b} (Z, u)$ such that $\exists a \,.\, m = \exists b \,.\, n$.*
(ii) *For all transitive objects (Z, u) with inclusions $(X, r) \xrightarrow{a} (Z, u)$ and $(Y, s) \xrightarrow{b} (Z, u)$, we have $\exists a \,.\, m = \exists b \,.\, n$.*
(iii) *If $(Z, u) = (X \cup Y, r \cup s)$ as defined in* 9.26 *and a, b are the canonical inclusions, then $\exists a \,.\, m = \exists b \,.\, n$.*

Moreover, the relation on triples defined by these three conditions is an equivalence relation.

Proof. Trivial from 9.26 and the fact that inclusions are monomorphisms in $\mathscr{E}$. ■

We now define the sets in $S(\mathscr{E})$ to be triples as above, with equality defined by the equivalent conditions of 9.36. To define the membership-relation, say that a set (X, r, m) is an *r-element* if $1 \xrightarrow{m} \Omega^X$ factors through $X \xrightarrow{r} \Omega^X$ by a morphism $1 \xrightarrow{\hat{m}} X$ (which is of course unique, since r is mono); we then say that $(X, r, m) \in_r (X, r, n)$ if m is an r-element and $\hat{m}$ factors through the subobject $N \rightarrowtail X$ classified by n. Now if $(X, r) \xrightarrow{i} (Y, s)$ is an inclusion, it is easily verified that $(X, r, m) \in_r (X, r, n)$ iff $(Y, s, \exists i\,.\,m) \in_s (Y, s, \exists i\,.\,n)$; so we can define the global membership relation $\in$ by

$$(X, r, m) \in (Y, s, n) \quad \text{if} \quad (X \cup Y, r \cup s, \exists a\,.\,m) \in_{r \cup s} (X \cup Y, r \cup s, \exists b\,.\,n).$$

(As in 9.36, this is equivalent to saying that "$\exists a\,.\,m \in_u \exists b\,.\,n$" becomes true in some transitive object (Z, u) containing both (X, r) and $Y, s)$, or in every such transitive object.)

The empty set in $S(\mathscr{E})$ is of course the triple $(0, 0 \longrightarrow 1, 1 \longrightarrow 1)$; and the ordered-pair operation is defined by $\langle (X, r, m), (Y, s, n) \rangle = (\Omega^{\Omega^{X \cup Y}}, \exists\exists(r \cup s)$, kpr $(\exists a.m, \exists b.n))$, where kpr is the Kuratowskian ordered-pair map defined in Ex. 5.10. It is straightforward to check that this definition is independent of the choice of representatives for (X, r, m) and (Y, s, n).

We are now ready to state the fundamental theorem:

9.37 THEOREM (Cole–Mitchell–Osius). (i) *Let $\mathscr{E}$ be a model of* **wpt**. *Then $S(\mathscr{E})$ is a model of* **zer**.

(ii) *If in addition $\mathscr{E}$ satisfies* (TR), *then $\mathscr{S}(S(\mathscr{E}))$ is equivalent to $\mathscr{E}_{\text{ptr}}$.*

(iii) *Let M be a model of* **zer**. *Then $S(\mathscr{S}(M))$ is isomorphic to M.*

Proof. (i) Since a subobject of X in $\mathscr{E}$ is determined (up to isomorphism) by the global elements of X which factor through it, the axiom of extensionality holds in $S(\mathscr{E})$. Similarly, since there are no morphisms $1 \longrightarrow 0$ in $\mathscr{E}$, the empty-set axiom holds. The unordered pair $\{(X. r, m), (Y, s, n)\}$ is $(\Omega^{X \cup Y}, \exists(r \cup s)$, pr$(\exists a.m, \exists b.n))$, where pr is the map defined in Ex. 5.10; and the ordered pair has already been defined. Again, the power-set of (X, r, m) is $(\Omega\ , \exists r$, $\downarrow$ seg.$m)$ where the down-segment map is defined relative to the partial order $\Omega_1^X \rightarrowtail \Omega^X \times \Omega^X$; in each case it is easy to verify that these definitions are compatible with our definition of equality in $S(\mathscr{E})$). and that they satisfy the appropriate axioms. The remaining axioms of set-existence (9.31(vi)+ (x)) are similarly straightforward but tedious to verify.

Now let (X, r, m) be a nonempty set in $S(\mathscr{E})$, and $M \rightarrowtail X$ the subobject classified by m. Since the relation r is inductive, and since the complement of

M is not the whole of X, we can find a global element x of X such that x factors through M, but the subobject classified by $1 \xrightarrow{x} X \xrightarrow{r} \Omega^X$ is contained in the complement of M. Then (X, r, rx) is an $\in$-minimal element of (X, r, m); so the axiom of foundation holds in $S(\mathscr{E})$.

For any transitive object (X, r) in $\mathscr{E}$, the set $(X, r, \ulcorner 1_X \urcorner)$ is transitive in $S(\mathscr{E})$, since its elements are precisely the sets of the form (X, r, rx) for some $1 \xrightarrow{x} X$, and they are thus also subsets of $(X, r, \ulcorner 1_X \urcorner)$. The transitivity axiom (9.32(xii)) is therefore trivial. For the Mostowski axiom, we observe that an extensional and well-founded relation on a set (X, r, m) defines an extensional and inductive relation $M \xrightarrow{s} \Omega^M$, where M is the subobject of X classified by m; and then the identity morphism $M \xrightarrow{1} M$ defines the required bijection $(X, r, m) \longrightarrow (M, s, \ulcorner 1_M \urcorner)$.

(ii) We define a triple (X, r, m) to be *minimal* if $\ulcorner 1_X \urcorner$ is the unique fixed point of the (order-preserving) map

$$\Omega^X \xrightarrow{\exists r} \Omega^{\Omega^X} \xrightarrow{\bigcup \times m} \Omega^X \times \Omega^X \xrightarrow{\vee X} \Omega^X.$$

For any triple (X, r, m), let $1 \xrightarrow{h} \Omega^X$ be the minimal fixed point of this map (which may be constructed using 5.50); then if $X' \rightarrowtail^{i} X$ is the corresponding subobject of X, it is not hard to show that r restricts to a relation r' on X', that m factors through $\exists i$ (say by $1 \xrightarrow{m'} \Omega^{X'}$), and that (X', r', m') is a minimal triple which is equal to (X, r, m) in $S(\mathscr{E})$. Moreover, if (X, r, m) and (Y, s, n) are two minimal triples which are equal in $S(\mathscr{E})$, then (X, r) and (Y, s) are isomorphic in $\mathscr{E}_{\mathrm{tr}}$ (since if $X \cap Y \rightarrowtail^{q} X$ is the canonical inclusion of 9.26(i), then $\ulcorner q \urcorner$ is a fixed point of the given map). So we have a well-defined map from the sets in $S(\mathscr{E})$ to the objects of $\mathscr{E}_{\mathrm{ptr}}$, which sends (X, r, m) to the (canonical) pullback of the diagram

$$\begin{array}{ccccc} & & & & 1\,. \\ & & & & \downarrow t \\ \tau X' & \xrightarrow{\;\cong\;} & X' & \xrightarrow{\overline{m'}} & \Omega \end{array}$$

And this map clearly extends to a functor $\Phi\colon \mathscr{S}(S(\mathscr{E})) \longrightarrow \mathscr{E}_{\mathrm{ptr}}$, since the property of "being the graph of a function" in $\mathscr{E}$ may be determined by considering global elements.

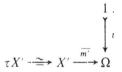

To define an inverse (up to natural isomorphism) for Φ, we simply send an object $(X \rightarrowtail^{m} Y,\ Y \xrightarrow{r} \Omega^Y)$ of $\mathscr{E}_{\mathrm{ptr}}$ to the set $(Y, r, \ulcorner m \urcorner)$. (Note that we may define this inverse functor, and prove that it is full, faithful and logical, even without assuming the axiom (TR).)

(iii) Let (X, r, m) be a set in $S(\mathscr{S}(M))$. Using the Mostowski axiom, we may choose a representative of this set for which X is actually a transitive set in M and r is the membership relation on X. Then we may define $T(X, r, m)$ to be the subset of X in M corresponding to $1 \xrightarrow{m} \Omega^X$ in $\mathscr{S}(M)$; it is straightforward to check that T is well-defined, and that it induces a bijection between sets in M and sets in $S(\mathscr{S}(M))$ which preserves and reflects the membership relation. So M and $S(\mathscr{S}(M))$ are isomorphic. ∎

9.38 COROLLARY. *The theories* **zer** *and* **wtt** + (TR) *are logically equivalent, in the sense that from a model of either theory we may construct a model of the other, from which the original model may be reconstructed. Moreover, the theories* **wtt**, **wpt** *and* $\mathbf{zer}_0$ *are equiconsistent with those above, in the sense that existence of a non-degenerate model for any one implies existence of a model for each of the others.* ∎

9.39 REMARK. It is of course possible to obtain further logical equivalences from 9.38, by adding further set-theoretic axioms to **zer** and their topos-theoretic equivalents to **wtt**. For example, the axiom of infinity is equivalent to the topos-theoretic axiom (NN): There exists a natural number object in $\mathscr{E}$, and the set-theoretic axiom of choice is equivalent to (AC) as defined in 5.21. (Note in passing that the addition of (AC) makes the axiom (PT) redundant, since it implies that every object has a well-ordering: see Ex. 9.8 below.) Similarly, we may formulate a categorical version of the axiom of replacement (this is done in [92]), and so obtain a logical equivalence for full Zermelo–Fraenkel set theory. This last step, however, is less obviously worthwhile than those before it, since it appears that those logical questions which make essential use of replacement (for example, problems involving large cardinals), are always more "naturally" handled by set-theoretic than by topos-theoretic methods. For problems not involving replacement (for example, the independence of the continuum hypothesis), the topos-theoretic viewpoint can often be a most illuminating one, as we hope to show in the next two paragraphs. ∎

9.40 REMARK. In the opposite direction, we may attempt to remove axioms from each side of the equivalence of 9.38, and thus give a "set-theoretic" characterization of a more general class of toposes than the well-pointed ones. In this direction, a definitive result has been obtained independently by A.

Boileau [11], M. Coste [26] and M. P. Fourman [35], each of whom has constructed a "higher-order intuitionistic type theory" which is logically equivalent to topos theory itself. ■

9.4. THE FILTERPOWER CONSTRUCTION

In this paragraph we introduce a method of constructing new toposes from old, which will be of central importance in the independence proofs discussed in the next paragraph.

9.41 DEFINITION. Let $\mathscr{E}$, $\mathscr{F}$ be toposes and $L: \mathscr{E} \longrightarrow \mathscr{F}$ a left exact functor. By a *filter on L* we mean a $\wedge$-semilattice homomorphism $\Phi: L(\Omega_{\mathscr{E}}) \longrightarrow \Omega_{\mathscr{F}}$ in $\mathscr{F}$. We call Φ an *ultrafilter* if it is actually a Heyting algebra homomorphism. [In the case when $\mathscr{F}$ is "the" category of sets, these are equivalent to the usual definitions of filter and ultrafilter on the Heyting algebra $L(\Omega_{\mathscr{E}})$.] ■

Now any filter on L determines an (external) filter of subobjects of 1 in $\mathscr{E}$, namely those $U \rightarrowtail 1$ whose classifying maps $1 \xrightarrow{u} \Omega_{\mathscr{E}}$ satisfy the commutative diagram

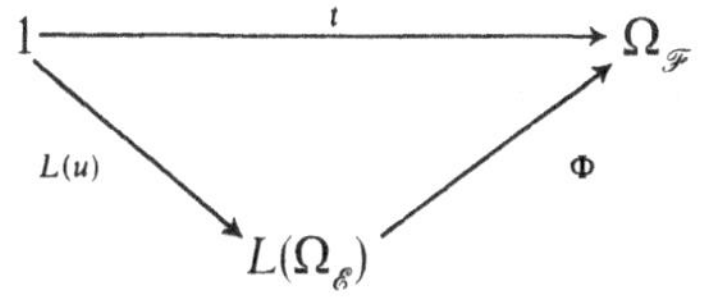

in F. We call such subobjects Φ-*dense*; note that if the topos $\mathscr{F}$ is well-pointed (so that the subobject of $L(\Omega_{\mathscr{E}})$ classified by Φ is determined by its global elements) and L is the direct image of a geometric morphism, then any filter of subobjects of 1 in $\mathscr{E}$ determines a filter on L. Since we shall normally be interested in this particular case, we shall commit the abuse of identifying Φ with the filter of Φ-dense subobjects of 1; thus we shall simply write "$U \in \Phi$" for "U is Φ-dense".

We say that a morphism $X \xrightarrow{f} Y$ in $\mathscr{E}$ is Φ-*invertible* if there exists a Φ-dense U such that $U^*(f)$ is iso in $\mathscr{E}/U$; and we write $\boldsymbol{\Phi}$ for the class of Φ-invertible morphisms in $\mathscr{E}$. (Note that $\boldsymbol{\Phi} = \bigcup_{U \in \Phi} \Xi_{J_U^o}$, by 3.54(i).)

9.42 LEMMA. $\boldsymbol{\Phi}$ *admits a (saturated) two-sided calculus of fractions on $\mathscr{E}$, and is closed under exponentiation (i.e. $X \xrightarrow{f} Y \in \boldsymbol{\Phi}$ implies $X^Z \xrightarrow{f^Z} Y^Z \in \boldsymbol{\Phi}$ for any Z).*

Proof. Suppose we have a composite $X \xrightarrow{f} Y \xrightarrow{g} Z$ with $f, g \in \Phi$. Then we can find $U, V \in \Phi$ such that $U^*(f)$ and $V^*(g)$ are iso; but since Φ is a filter, we have $U \cap V \in \Phi$ and $(U \cap V)^*(gf)$ is clearly iso. Similarly, if gf and f (resp. gf and g) are Φ-invertible, then so is g (resp. f). It follows at once from exactness of the functors U^* that pullbacks and pushouts of Φ-invertible morphisms are Φ-invertible, and that a parallel pair $X \rightrightarrows Y$ can be equalized by a Φ-invertible morphism iff they can be coequalized by one. Similarly, closure under exponentiation follows from the fact that U^* preserves exponentials. ■

9.43 Definition. Let $\mathscr{E} \xrightarrow{L} \mathscr{F}$ be a left exact functor, Φ a filter on L. We define the *filterpower of $\mathscr{E}$ modulo* Φ to be the category of fractions $\mathscr{E}_\Phi = \mathscr{E}[\Phi^{-1}]$; and we write $P_\Phi \colon \mathscr{E} \longrightarrow \mathscr{E}_\Phi$ for the canonical projection functor. If Φ is an ultrafilter, we commonly use the word *ultrapower* in place of filterpower. ■

9.44 Theorem (Lawvere–Tierney). *Let $\mathscr{E} \xrightarrow{L} \mathscr{F}$ be a left functor, Φ a filter on L. Then $\mathscr{E}_\Phi$ is a topos, and P_Φ is logical.*

Proof. (i) Existence of finite limits in $\mathscr{E}_\Phi$, and the fact that P_Φ preserves them, are immediate from 9.42 and 0.19.

(ii) By 9.42, an $\mathscr{E}_\Phi$-morphism $X \times Y \longrightarrow Z$ can be represented by a diagram

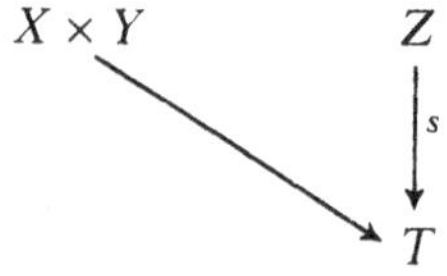

in $\mathscr{E}$ with $s \in \Phi$; and then the transposed diagram

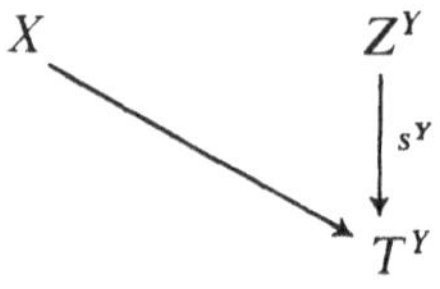

represents a morphism $X \longrightarrow Z^Y$ in $\mathscr{E}_\Phi$. Conversely, if we represent $X \longrightarrow Z^Y$

in $\mathscr{E}_\Phi$ by a diagram

$$\begin{array}{ccc} U & \searrow & \\ \downarrow{\scriptstyle u} & & \\ X & & Z^Y \end{array}$$

with $u \in \mathbf{\Phi}$, then its transpose

$$\begin{array}{ccc} U \times Y & \searrow & \\ \downarrow{\scriptstyle u \times 1} & & \\ X \times Y & & Z \end{array}$$

represents a morphism $X \times Y \longrightarrow Z$ in $\mathscr{E}_\Phi$. A straightforward diagram-chase shows that these two constructions preserve the equivalence relation defining morphisms in $\mathscr{E}_\Phi$ (0.19), and that they are mutually inverse. So $\mathscr{E}_\Phi$ has exponentials, and P_Φ preserves them.

(iii) Let $Y \xrightarrow{m} X$ be a monomorphism in $\mathscr{E}_\Phi$, and let

$$\begin{array}{ccc} U & \overset{f}{\searrow} & \\ \downarrow{\scriptstyle s} & & \\ Y & & X \end{array}$$

be a diagram in $\mathscr{E}$ representing it. Let $U \overset{q}{\twoheadrightarrow} I \overset{i}{\rightarrowtail} X$ be the image factorization of f; then q must be Φ-invertible, since P_Φ preserves image factorizations and $\mathbf{\Phi}$ is saturated. So m is isomorphic (as a subobject of X in $\mathscr{E}$) to the "honest" subobject $P_\Phi(i)$. But any morphism $X \longrightarrow \Omega$ in $\mathscr{E}_\Phi$ is actually in the image of P_Φ; for Ω is injective in $\mathscr{E}$, and it is clear from the construction of $\mathbf{\Phi}$ that we may represent a morphism $X \longrightarrow \Omega$ in $\mathscr{E}_\Phi$ by a diagram of the form

$$\begin{array}{ccc} X \times U & \searrow & \\ \downarrow{\scriptstyle \pi_1} & & \\ X & & \Omega \end{array}$$

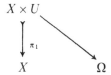

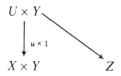

with $U \in \Phi$. And two morphisms $X \rightrightarrows \Omega$ in $\mathscr{E}$ become equal in $\mathscr{E}_\Phi$ iff their equalizer is Φ-invertible, which implies that the subobjects they classify become isomorphic in $\mathscr{E}_\Phi$. So $P_\Phi(\Omega)$ is a subobject classifier for $\mathscr{E}_\Phi$. ■

9.45 EXAMPLE. Let $\mathscr{E}$ be a topos, X an object of $\mathscr{E}$ with global support, and $\Phi : \Omega^X \longrightarrow \Omega$ an ultrafilter on $\Pi_X : \mathscr{E}/X \longrightarrow \mathscr{E}$. If $\mathscr{E}$ is well-pointed, then so is $(\mathscr{E}/X)_\Phi$ (see 9.48 below); and if we think of $\mathscr{E}$ as a model of set theory (via the equiconsistency theorem 9.38), then the construction of $(\mathscr{E}/X)_\Phi$ is precisely that used to obtain non-standard models for analysis in the sense of Robinson. Note that the composite $\mathscr{E} \xrightarrow{X^*} \mathscr{E}/X \xrightarrow{P_\Phi} (\mathscr{E}/X)_\Phi$ is faithful as well as logical; for if U is any subobject of 1 in $\mathscr{E}$, then $X \longrightarrow 1$ epi implies $U^X \cong U$, and so $P_\Phi . X^*(U) \cong 1$ implies $U \cong 1$—but since $P_\Phi . X^*$ preserves equalizers and the functors $\forall_f$, this is sufficient. However, $P_\Phi . X^*$ is not in general an equivalence of categories; cf. Ex. 9.12 below. (This example is discussed at greater length in [67] by A. Kock and C. J. Mikkelsen, who give an alternative construction of $(\mathscr{E}/X)_\Phi$ via a factorization theorem for functors preserving first-order logic.) ■

In proving the independence of the continuum hypothesis in the next paragraph, we shall take as our starting-point an axiom-system **tvm**, which is a strengthening of the system **wtt** of §9.3. We now introduce this system, and prove a result which shows how the ultrapower construction may be used to obtain new models of **tvm**.

9.46 DEFINITION. By a *boolean-valued model* (of the category of sets), we mean a topos $\mathscr{E}$ which satisfies (AC) and has a natural number object. If in addition $\mathscr{E}$ satisfies the axiom (TV) of 9.33, we call it a *two-valued model*. We write **bvm**, **tvm** respectively for the theories of boolean-valued and two-valued models. ■

It follows from 9.33 and Ex. 9.8 below that **tvm** is equivalent to **wtt** + (AC) + (NN); so we can use 9.38 to interpret independence proofs in **tvm** in terms of a suitable subsystem of ZFC set theory.

9.47 REMARK. Let $\mathscr{S}$ be a two-valued model. Then by 5.39, we know that boolean-valued models defined over $\mathscr{S}$ correspond to complete Boolean algebras in $\mathscr{S}$. So any two-valued model which is defined over $\mathscr{S}$ is actually equivalent to $\mathscr{S}$ as a category; as M. Tierney [117] has pointed out, this is the precise form of Lawvere's dictum [71] that "if you add the (non-elementary) axiom of completeness, you have characterized set theory". Hence it is essential that, in constructing *new* models of **tvm**, we should make use of a construction which takes us outside the realm of geometric morphisms. This is the rôle played by the ultrapower construction.

9.48 THEOREM. *Let $\mathscr{S}$ be a two-valued model, and $\mathscr{E} \xrightarrow{\gamma} \mathscr{S}$ a boolean-valued model defined over $\mathscr{S}$. Let Φ be an ultrafilter on $\gamma_*: \mathscr{E} \longrightarrow \mathscr{S}$ (which exists by Zorn's Lemma in $\mathscr{S}$). Then the ultrapower $\mathscr{E}_\Phi$ is a two-valued model.*

Proof. By 6.16, $\mathscr{E}_\Phi$ has a natural number object; so it remains to verify (AC) and (TV). Let $X \xrightarrow{f} Y$ be an epimorphism in $\mathscr{E}_\Phi$, and

$$\begin{array}{ccc} U & & \\ {\scriptstyle s}\downarrow & \searrow^{g} & \\ X & & Y \end{array}$$

a diagram in $\mathscr{E}$ representing it. Let $U \xrightarrow{q} I \rightarrowtail^{i} Y$ be the image factorization of g; then as in 9.44(iii) we deduce that i must be Φ-invertible. Let $I \rightarrowtail^{d} U$ be a splitting for q in $\mathscr{E}$; then

$$\begin{array}{ccc} I & & \\ {\scriptstyle i}\downarrow & \searrow^{sd} & \\ Y & & X \end{array}$$

is easily seen to be a splitting for f in $\mathscr{E}_\Phi$.

Now let $U \rightarrowtail 1$ be any subobject of 1 in $\mathscr{E}$, with classifying map $1 \xrightarrow{u} \Omega$. Since $\mathscr{S}$ satisfies (TV), we must have either $\Phi . \gamma_*(u) = t$ (in which case U is Φ-dense, so that $P_\Phi(U) \cong 1$), or $\Phi . \gamma_*(u) = f$ (in which case the complement of U is Φ-dense, so that $0 \rightarrowtail U$ is Φ-invertible and $P_\Phi(U) \cong 0$). But since any subobject of 1 in $\mathscr{E}_\Phi$ is in the image of P_Φ (and since Φ preserves the minimal element of $\gamma_*(\Omega_{\mathscr{E}})$, so that $0 \rightarrowtail 1$ is *not* Φ-dense), we deduce that there are exactly two such subobjects. ■

9.5. INDEPENDENCE OF THE CONTINUUM HYPOTHESIS

One of the most striking early applications of elementary topos theory was to give a categorical proof that the Continuum Hypothesis is independent of the axioms of set theory. This fact was first proved in 1963 by P. J. Cohen [156]; whilst categorical ideas are implicit in Cohen's work, the language of elementary topos theory was required to make the connection explicit—and indeed the desire to do this was one of the main driving forces behind the development of elementary topos theory by Lawvere and Tierney. In this paragraph we shall follow Tierney's account of the independence proof, given in [117].

Let $\mathscr{E}$ be a topos with a natural number object. We shall say that an object X *negates the continuum hypothesis* in $\mathscr{E}$ if there exist monomorphisms $N \rightarrowtail X \rightarrowtail \Omega^N$, but the objects of epimorphisms $\mathrm{Epi}(N, X)$ and $\mathrm{Epi}(X, \Omega^N)$ are both 0. (Note that the truth of this statement is preserved by logical functors.) For the rest of this paragraph, we shall suppose that we are given a two-valued model $\mathscr{S}$, and we shall seek to construct a boolean-valued model $\mathscr{E}$ defined over $\mathscr{S}$ and containing an object which negates (CH). Then by applying the ultrapower construction to $\mathscr{E}$, we shall obtain a new two-valued model $\mathscr{S}'$ in which (CH) is false.

We shall feel free to talk about objects of $\mathscr{S}$ as if they were "actual" sets; in view of 9.38, this is a reasonable simplification, although the reader who has followed all our arguments thus far should find little difficulty in translating everything we say into more rigorous topos-theoretic language. In particular, we shall denote the subobject classifier of $\mathscr{S}$ by 2 rather than Ω. Since $\mathscr{S}$ satisfies (AC), it follows easily from 9.20 that the definitions of cardinal-finiteness and Kuratowski-finiteness coincide, and so we shall not bother to specify which one we mean by the word "finite"; note also that, by virtue of Ex. 6.4, we can talk without ambiguity of the *cardinality* of a finite object (i.e. the natural number indexing its elements).

Let I be an object of $\mathscr{S}$ such that there is a monomorphism $2^N \rightarrowtail I$, but $\mathrm{Epi}(2^N, I) = 0$. (For example, we can take $I = 2^{2^N}$, by Ex. 5.7.) Our aim is now to construct an $\mathscr{S}$-topos $\mathscr{E} \xrightarrow{\gamma} \mathscr{S}$ in which γ^*I becomes a subobject of Ω^N, whilst we still have $\mathrm{Epi}(\gamma^*2^N, \gamma^*I) = 0$. Now let $\mathbf{P}$ be the poset (in $\mathscr{S}$) of partial maps $I \times N \longrightarrow 2$ with finite domain, ordered by the relation $f \leqslant g \Leftrightarrow g$ extends f, i.e. there exists a commutative diagram

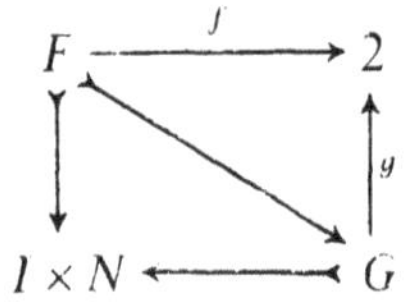

We think of an element of P as a finite set of "forcing conditions", each of

which says that a particular element n of N either is or is not in the particular element i of I.

9.51 LEMMA. *Let X be a subobject of P such that no two elements of X have a common extension in* **P** *(i.e. such that*

$$\mathscr{S} \models \forall \mathbf{p}, \mathbf{p}', \mathbf{p}''(\mathbf{p} \in \ulcorner X \urcorner \wedge \mathbf{p}' \in \ulcorner X \urcorner \wedge \mathbf{p} < \mathbf{p}'' \wedge \mathbf{p}' < \mathbf{p}'') \Rightarrow \mathbf{p} = \mathbf{p}').$$

Then X is countable, *i.e. $X \cong 0$ or there exists an epi $N \twoheadrightarrow X$. (We say that* **P** *satisfies the* countable chain condition.)

Proof. Suppose first that there is a natural number m which is an upper bound for the cardinalities of the domains of the partial maps in X. In this case, we shall show by induction on m that X is actually finite, and its cardinality is at most $2^m . m!$.

But this is trivial for $m = o$; so we can assume that $m > o$ and that the assertion above has been proved for $m - 1$. We may assume that X is non-empty, and so pick a global element $1 \xrightarrow{x} X$. Let q denote the cardinality of the domain of the corresponding partial map. Now if y is any other partial map in X, there must exist (i, n) in $\mathrm{dom}(x) \cap \mathrm{dom}(y)$ such that $x(i, n) \neq y(i, n)$, since otherwise x and y would have a common extension in **P**. So for each $(i, n) \in \mathrm{dom}(x)$, let $Y(i, n)$ be the set of partial maps obtained by taking a partial map in X which disagrees with x at (i, n), and deleting (i, n) from its domain. Then it is clear that no two partial maps in $Y(i, n)$ can have a common extension in **P**, and their domains all have cardinality $\leqslant m - 1$, so by induction

$$\mathrm{card}(Y(i, n)) \leqslant 2^{m-1} . (m - 1)!.$$

But we have a morphism $\coprod_{(i,n) \in \mathrm{dom}(x)} Y(i, n) \longrightarrow X$ whose image contains every element of X except x itself, and so

$$\mathrm{card}(X) \leqslant q . 2^{m-1} . (m - 1)! + 1 \leqslant 2^{m-1} . m! + 1 \leqslant 2^m . m!.$$

In the general case, let X_m be the subobject of X consisting of those partial maps whose domains have cardinality $\leqslant m$. Then X_m is finite by the above argument, and so $X = \bigcup_{m \in N} X_m$ is countable. ■

We now consider the topos $\mathscr{E} = \mathrm{sh}_{\neg\neg}(\mathscr{S}^{\mathbf{P}})$. By 5.17 and 5.34, $\mathscr{E}$ is Boolean and satisfies (SG); so by 5.39 it satisfies (AC) and is thus a boolean-valued model. As usual, we write γ^* for the inverse image of the geometric morphism $\mathscr{E} \longrightarrow \mathscr{S}$. We shall require the following technical lemma:

9.52 LEMMA. *Let S be an object of $\mathscr{S}$, and suppose we are given an element x of $\gamma^*S(p)$, where p is an element of P. Then there exists $q \geqslant p$ in P such that the image of x in $\gamma^*S(q)$ is in the image of the canonical map $S \xrightarrow{\sigma} \gamma^*S(q)$.*

Proof. Since $\mathscr{S}$ is Boolean, the constant functor $\bar{S}$ is decidable (cf. Ex. 5.3) and hence $\neg\neg$-separated in $\mathscr{S}^{\mathbf{P}}$, and so γ^*S (being the associated sheaf of $\bar{S}$) is equal to $(\bar{S})^+$ in the notation of §3.3. So any element of $\gamma^*S(p)$ derives from a morphism $R \longrightarrow \bar{S}$ in $\mathscr{S}^{\mathbf{P}}$, where R is a $\neg\neg$-covering sieve on p in $\mathbf{P}^{\mathrm{op}}$. And since any $\neg\neg$-covering sieve must be nonempty, we have simply to choose an element $(p \longrightarrow q)$ of R to obtain the desired result. ∎

We shall say that an object S of $\mathscr{S}$ is *infinite* if there is an isomorphism $S \times N \cong S$. We shall not discuss the relationship between this definition and the various definitions of finiteness which we have considered; we merely pause to note that N is infinite in any topos (since the assignment

$$(p, q) \longmapsto \tfrac{1}{2}(p + q)(p + q + 1) + q$$

defines an isomorphism $N \times N \longrightarrow N$), and Ω^N is infinite in any Boolean topos (since we have monos

$$\Omega^N \xrightarrow{1 \times o} \Omega^N \times N \xrightarrow{1 \times \{\}} \Omega^N \times \Omega^N \cong \Omega^{N \cup N} \cong \Omega^N,$$

to which we can apply the Schröder–Bernstein theorem). (See also Ex. 9.15 below.)

9.53 PROPOSITION. *Let $\mathbf{P}$ be a poset in $\mathscr{S}$ satisfying the countable chain condition, and let $\mathscr{E} = \mathrm{sh}_{\neg\neg}(\mathscr{S}^{\mathbf{P}})$. Then the functor $\gamma^*\colon \mathscr{S} \longrightarrow \mathscr{E}$ "preserves cardinals", in the sense that if S is infinite and* $\mathrm{Epi}(S, T) = 0$ *in $\mathscr{S}$, then* $\mathrm{Epi}(\gamma^*S, \gamma^*T) = 0$ *in $\mathscr{E}$.*

Proof. Suppose $\mathrm{Epi}(\gamma^*S, \gamma^*T) \neq 0$. Then there exists $p \in P$ and an epimorphism $l(p)^*\gamma^*S \xrightarrow{f} l(p)^*\gamma^*T$ in $\mathscr{E}/l(p)$. We may clearly assume that S and T are nonzero; so γ^*T has a global element and is therefore injective in $\mathscr{E}$. And since $l(p)$ is a subobject of 1 in $\mathscr{E}$, we may extend f to a morphism $\bar{f}\colon \gamma^*S \longrightarrow \gamma^*T$

(not necessarily epi) such that $l(p)^*(\bar{f}) = f$. Now define $X \rightarrowtail P \times S \times T$ to be the extension of the formula

$$\bar{f}_{\mathbf{p}}(\sigma_{\mathbf{p}}(\mathbf{s})) = \tau_{\mathbf{p}}(\mathbf{t})$$

of $L_{\mathcal{P}}$, where $S \xrightarrow{\sigma_p} \gamma^*S(p)$ and $T \xrightarrow{\tau_p} \gamma^*T(p)$ are the canonical monomorphisms.

We claim first that the composite $X \rightarrowtail P \times S \times T \xrightarrow{\pi_3} T$ is epi. To prove this, let t be an element of T; since $l(p)^*(\bar{f}) = f$ is (split) epi, we can find $x \in \gamma^*S(p)$ such that $\bar{f}_p(x) = \tau_p(t)$. Then by 9.52 we can find $q \geqslant p$ such that the image x' of x in $\gamma^*S(q)$ is in the image of $S \xrightarrow{\sigma_q} \gamma^*S(q)$. Let s be the (unique) element of S such that $\sigma_q(s) = x'$; then $\bar{f}_q(\sigma_q(s)) = \bar{f}_q(x') = \tau_q(t)$, so the triple (q, s, t) is in X.

Now let $T \overset{h}{\rightarrowtail} X$ be a splitting for this epimorphism, and consider the composite

$$T \overset{h}{\rightarrowtail} X \rightarrowtail P \times S \times T \xrightarrow{\pi_2} S.$$

For each element s of S, let T_s denote the fibre of this map over $1 \xrightarrow{s} S$; i.e. T_s is the set of those $t \in T$ such that $h(t) = (p, s, t)$ for some p. Let g_s be the composite

$$T_s \rightarrowtail T \overset{h}{\rightarrowtail} X \rightarrowtail P \times S \times T \xrightarrow{\pi_1} P;$$

we shall show that g_s is mono, and that no two elements of its image have a common extension in $\mathbf{P}$.

For suppose that t and t' are elements of T_s such that $g_s(t) \leqslant q \geqslant g_s(t')$ for some $q \in P$. Then since the triples $(g_s(t), s, t)\,(= h(t))$ and $(g_s(t'), s, t')$ are in X, it follows easily from the definition of X that (q, s, t) and $q, s, t')$ are in X. So $\tau_q(t) = \bar{f}_q(\sigma_q(s)) = \tau_q(t')$; but τ_q is mono (since T is $\neg\neg$-separated), and so $t = t'$.

So by the countable chain condition on $\mathbf{P}$, either $T_s = 0$ or we have an epimorphism $N \twoheadrightarrow T_s$. But we are assuming that T is nonzero; so we can construct an epimorphism $S \times N \twoheadrightarrow \coprod_{s \in S} T_s = T$. But $S \times N \cong S$, so we have constructed an epimorphism $S \twoheadrightarrow T$ in S, i.e. an element of Epi(S, T). ■

We now return to the particular poset $\mathbf{P}$ of partial maps $I \times N \longrightarrow 2$, which we defined earlier. The crucial feature of this poset is contained in the following result:

9.54 PROPOSITION. *In the topos* $\mathscr{E} = \mathrm{sh}_{\neg\neg}(\mathscr{S}^{\mathbf{P}})$, *there is a monomorphism* $\gamma^* I \overset{m}{\rightarrowtail} \Omega^N$.

Proof. Define a subfunctor R of $\overline{I \times N}$ in $\mathscr{S}^{\mathbf{P}}$ by

$$R(p) = \{(i, n) \mid p(i, n) = true\}.$$

We claim first that R is a $\neg\neg$-closed subobject of $\overline{I \times N}$. To prove this, we have to show that if i, n and p are such that for all $q \geqslant p$ there exists $r \geqslant q$ with $r(i, n) = true$ (so that (i, n) is in the $\neg\neg$-closure of $R(p)$), then $(i, n) \in R(p)$. But if this condition is satisfied, we must have $(i, n) \in \mathrm{dom}(p)$, since otherwise we could find an extension q of p with $q(i, n) = false$; and for the same reason we must have $p(i, n) = true$, i.e. $(i, n) \in R(p)$.

So R is a $\neg\neg$-closed subobject; hence it is classified by a morphism $\overline{I \times N} \longrightarrow \Omega_{\neg\neg}$ in $\mathscr{S}^{\mathbf{P}}$, or equivalently by a morphism $\bar{I} \overset{\phi}{\longrightarrow} (\Omega_{\neg\neg})^{\bar{N}}$. Now for any $p \in P$, $(\Omega_{\neg\neg})^{\bar{N}}(p)$ may be identified with the set of $\neg\neg$-closed subobjects of $\bar{N} \times h^p$ (cf. 1.12); from this, it is easy to see that an element i of $I = \bar{I}(p)$ is sent by ϕ_p to the functor

$$\begin{aligned} q &\longmapsto \{n \in N \mid q(i, n) = true\} && (q \geqslant p) \\ &\longmapsto 0 && (q \ngeqslant p). \end{aligned}$$

Suppose i and j are two distinct elements of I; then since dom (p) is finite, we can find a natural number n such that neither (i, n) nor (j, n) is in $\mathrm{dom}(p)$, and then define an extension q of p such that $q(i, n) = true$ but $q(j, n) = false$. Hence $\phi_p(i) \neq \phi_p(j)$; so ϕ is a monomorphism in $\mathscr{S}^{\mathbf{P}}$.

But $(\Omega_{\neg\neg})^{\bar{N}}$ is a $\neg\neg$-sheaf by 3.24 and 3.27; so on applying the associated sheaf functor to ϕ, we obtain a monomorphism

$$\gamma^* I \rightarrowtail (\Omega_{\neg\neg})^{\bar{N}} \cong \Omega^{\gamma^* N}$$

in $\mathrm{sh}_{\neg\neg}(\mathscr{S}^{\mathbf{P}})$. (The latter isomorphism was established in Ex. 3.6.) And this is the monomorphism we require. ■

9.55 COROLLARY. *In the topos $\mathscr{E}$, the object $\gamma^*(2^N)$ negates the continuum hypothesis.*

Proof. Since γ^* preserves monomorphisms, we have monos

$$N \cong \gamma^*N \rightarrowtail \gamma^*(2^N) \rightarrowtail \gamma^*I \overset{m}{\rightarrowtail} \Omega^N$$

in $\mathscr{E}$; and since $\text{Epi}(N, 2^N) = \text{Epi}(2^N, I) = 0$ in $\mathscr{S}$, we have

$$\text{Epi}(N, \gamma^*(2^N)) = \text{Epi}(\gamma^*(2^N), \gamma^*I) = 0$$

in $\mathscr{E}$ by 9.51 and 9.53. But since γ^*I has a global element, it is injective in the Boolean topos $\mathscr{E}$, and so m is split—say by $\Omega^N \overset{q}{\twoheadrightarrow} \gamma^*I$. Then composition with q induces a morphism $\text{Epi}(\gamma^*(2^N), \Omega^N) \longrightarrow \text{Epi}(\gamma^*(2^N), \gamma^*I)$ in $\mathscr{E}$; so by 1.56 $\text{Epi}(\gamma^*(2^N), \Omega^N) = 0$. ■

Combining 9.55 with the results of the last paragraph, we have proved

9.56 THEOREM (Cohen–Lawvere–Tierney). *The Continuum Hypothesis is independent of the axiom-system* **tvm**.

Proof. Let $\mathscr{S}$ be a model of **tvm**. Construct a boolean-valued model $\mathscr{E}$ over $\mathscr{S}$ as above, and let Φ be an ultrafilter on $\gamma^*: \mathscr{E} \longrightarrow \mathscr{S}$. Define $\mathscr{S}' = \mathscr{E}_\Phi$; then $\mathscr{S}'$ is a two-valued model by 9.48, and since $\gamma^*(2^N)$ negates the continuum hypothesis in $\mathscr{E}$, the object $P_\Phi\gamma^*(2^N)$ negates it in $\mathscr{S}'$. ■

9.57 REMARK. The arguments used in this paragraph are merely a particular case of a general method for proving independence theorems within **tvm**, similar to the method of "Cohen forcing" in set theory. For example, M. C. Bunge [15] has used the method to prove the independence of Souslin's Hypothesis. The general features of the method are as follows: first one expresses the statement whose consistency one is trying to prove (e.g. "X negates the continuum hypothesis" or "X is a Souslin tree") in the internal language of an arbitrary topos) or at least of an arbitrary Boolean topos). Then, given a two-valued model $\mathscr{S}$, one chooses a poset $\mathbf{P}$ in $\mathscr{S}$ (normally the poset of "finite partial models" of the thing one is trying to construct), such that the appropriate statement becomes true in the boolean-valued model $\mathscr{E} = \text{sh}_{\neg\neg}(\mathscr{S}^{\mathbf{P}})$. Finally, one uses the ultrapower construction to "collapse" $\mathscr{E}$ to a new two-valued model $\mathscr{S}'$. ■

EXERCISES 9

1. Let $\mathscr{E}$ be a topos. Show that an object $X = (X_0 \xrightarrow{f} X_1)$ of the Sierpinski topos $\mathscr{E}^{\mathbf{2}}$ is K-finite iff X_0 is K-finite in $\mathscr{E}$ and f is epi. [Hint: using the fact that "evaluate at 0" and "evaluate at 1" are inverse image functors, show that

$$K(X_0 \xrightarrow{f} X_1) \cong (K(X_0) \xrightarrow{K(f)} K(X_1)).]$$

More generally, if $\mathbf{C}$ is any internal category in $\mathscr{E}$, show that an internal diagram $\mathbf{F} \xrightarrow{\gamma} \mathbf{C}$ is K-finite in $\mathscr{E}^{\mathbf{C}}$ iff $F_0 \xrightarrow{\gamma_0} C_0$ is K-finite in $\mathscr{E}/C_0$ and the action of $\mathbf{C}$ on F_0 (regarded as map

$$(C_1 \xrightarrow{(d_0\ d_1)} C_0 \times C_0) \longrightarrow \pi_2^*(\gamma_0)^{\pi_1^*(\gamma_0)}$$

in $\mathscr{E}/C_0 \times C_0$) factors through the object of epimorphisms $\mathrm{Epi}(\pi_1^*(\gamma_0), \pi_2^*(\gamma_0))$.

2. Using the topos $\mathscr{E} = \mathscr{S}^{\mathbf{2}}$, give examples of

(i) A parallel pair $X \rightrightarrows Y$ in $\mathscr{E}_{kf}$ whose equalizer (in $\mathscr{E}$) is not K-finite.

(ii) A surjection $\mathscr{F} \xrightarrow{p} \mathscr{E}$ and an object X of $\mathscr{E}$ such that p^*X is K-finite but X is not.

(iii) An object X such that Ω^X is K-finite but X is not.

(iv) A K-finite object of $\mathscr{E}$ which is not internally projective (cf. 6.25).

(v) An exponential of the form Y^X, with X K-finite, which is not preserved by the inverse image of a geometric morphism $\mathscr{S} \longrightarrow \mathscr{E}$ (cf. 6.24).

3. Let X be a K-finite object in a topos $\mathscr{E}$ with a natural number object, and p a natural number in $\mathscr{E}$. Show that the object $\mathrm{Epi}([p], X)$ is preserved by inverse image functors. [Hint: construct a pullback diagram

$$\begin{array}{ccc} \mathrm{Epi}([p], X) & \longrightarrow & X^{[p]} \\ \downarrow & & \downarrow \\ 1 & \xrightarrow{\ulcorner 1_X \urcorner} & K(X) \end{array}$$

.]

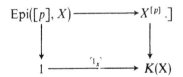

Show by an example in $\mathscr{S}^2$ that the hypothesis of K-finiteness of X cannot be omitted.

4. Let $\mathscr{E}$ be a topos with a natural number object, and define $\mathbf{E}_{\text{finepi}}$ to be the internal category of finite cardinals and epimorphisms in $\mathscr{E}$; i.e. its object-of-objects is N, and its object-of-morphisms is the object $\text{Epi}(\pi_1^*[n], \pi_2^*[n])$ of $\mathscr{E}/N \times N$. If X is any object of $\mathscr{E}$, show that the object $\text{Epi}([n], X)$ of $\mathscr{E}/N$ has the structure of an internal presheaf $H(X)$ on $\mathbf{E}_{\text{finepi}}$; and that this presheaf is flat iff X is K-finite. [For the first condition of flatness, use 9.20; for the second, use the fact that finite cardinals are internally projective.] Deduce that $\mathscr{E}^{\mathbf{E}_{\text{finepi}}}$ is a classifying topos for the theory whose models are K-finite objects. and whose morphisms are epimorphisms.

5. Let p be a natural number in a topos $\mathscr{E}$, and $[p] \xrightarrow{f} X$ an epimorphism in $\mathscr{E}$. Show that X is isomorphic to a finite cardinal in $\mathscr{E}$ iff the diagonal subobject $X \xrightarrow{\Delta} X \times X$ has a complement. [Hint: consider the kernel-pair of f.] Deduce that if $\mathscr{E}$ is a Boolean topos with a natural number object, then the subcategories $\mathscr{E}_{kf}$ and $\mathscr{E}_{lcf}$ coincide.

6. Define the notion of *inaccessible element* of an internally complete poset in a topos, and show that X is K-finite iff the maximal element of $\mathbf{\Omega}^X$ is inaccessible. [Compare Ex. 7.7.]

7. Let $\mathscr{E}$ be a topos with a natural object N. Prove that (N, card) is a transitive object, where $N \xrightarrow{\text{card}} \Omega^N$ is the down-segment map for the strict order-relation of 6.17.

8. Let $\mathbf{X} = (X_1 \rightrightarrows X)$ be a well-ordered object (5.49) in a Boolean topos. Let $X \xrightarrow{r} \Omega^X$ be the down-segment map for the strict order relation corresponding to X_1 (i.e. the complement of $X \xrightarrow{\Delta} X \times X$ in $X_1 \rightarrowtail X \times X$). Prove that (X, r) is a transitive object.

9. Let $\mathfrak{Log}$ denote the 2-category of toposes, logical functors and natural transformations. Show that the forgetful functor $\mathfrak{Log} \longrightarrow \mathfrak{Cat}$ creates filtered (pseudo-)colimits. Show that the topos $\mathscr{E}_\Phi$ of 9.43 may be described as a filtered colimit of a diagram in $\mathfrak{Log}$ whose vertices have the form $\mathscr{E}/U$ for some $U \in \Phi$.

10. (Freyd's proof of independence of (AC)) Let $\mathscr{E}$ be the topos of Ex. 5.5,

and let P be the set of positive integers, ordered by the relation of divisibility (i.e. $p \leqslant q$ iff p divides q). Show that the functor $\mathscr{E} \longrightarrow \mathscr{E}$ which sends a $\mathbb{Z}$-set (M, α) to (M, α^k) is logical, and hence construct a diagram over $\mathbf{P}$ in $\mathfrak{Log}$ whose vertices are copies of $\mathscr{E}$. Let $\mathscr{F}$ be the colimit of this diagram; show that objects of $\mathscr{F}$ may be described as triples (M, α, k), where (M, α) is an object of $\mathscr{E}$ and k is a positive integer, and that morphisms $(M, \alpha, k) \xrightarrow{f} (Q, \beta, l)$ are functions $M \xrightarrow{f} Q$ in $\mathscr{S}$ such that

$$\begin{array}{ccc} M & \xrightarrow{f} & Q \\ \downarrow{\scriptstyle \alpha^{lm}} & & \downarrow{\scriptstyle \beta^{km}} \\ M & \xrightarrow{f} & Q \end{array}$$

commutes for some $m \geqslant 1$. Hence show that $\mathscr{F}$ is well-pointed, but that it does not satisfy (AC).

11. (P. Freyd) Let B be a (not necessarily complete) Boolean algebra in $\mathscr{S}$. Show that there exists a boolean-valued model whose lattice of subobjects of 1 is isomorphic to B. [Express B as the (filtered) union of its finite subalgebras; then observe that any finite Boolean algebra B_0 is complete and atomic (so that $\mathrm{Shv}(\mathbf{B}_0, C) \simeq \mathscr{S}/X$ for some X), and use Ex. 9.]

12. Let Φ be a non-principal ultrafilter on the set of natural numbers, and let $\mathscr{E} = (\mathscr{S}/N)_\Phi$. Show that there are natural numbers in $\mathscr{E}$ which are not in the image of the functor $\mathscr{S} \xrightarrow{N^*} \mathscr{S}/N \xrightarrow{P_\Phi} (\mathscr{S}/N)_\Phi$.

13. (P. Freyd [FK]) Let $\mathscr{E}$ be a well-pointed topos, and $\mathscr{F}$ any topos. Show that any exact functor $\mathscr{E} \xrightarrow{T} \mathscr{F}$ is faithful. [Hint: T preserves images of morphisms $1 \amalg 1 \longrightarrow X$ in $\mathscr{E}$.] Deduce that there does not exist an exact functor $\mathscr{E} \longrightarrow \mathscr{S}$, where $\mathscr{E}$ is the topos of Ex. 12.

14. Let $\mathscr{E}$ be a Grothendieck topos, and Φ a filter on $\gamma_*: \mathscr{E} \longrightarrow \mathscr{S}$. Show that the functor $P_\Phi: \mathbf{ab}(\mathscr{E}) \longrightarrow \mathbf{ab}(\mathscr{E}_\Phi)$ preserves injectives, and deduce that $\mathbf{ab}(\mathscr{E}_\Phi)$ has enought injectives. If A is an abelian group in $\mathscr{E}$, show that we have an exact sequence

$$0 \longrightarrow \gamma_\Phi(A) \longrightarrow \hom_{\mathscr{E}}(1, A) \longrightarrow \hom_{\mathscr{E}_\Phi}(1, P_\Phi(A)),$$

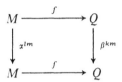

where γ_Φ is the functor defined in Ex. 8.10, and that the last morphism in this sequence is epi if A is injective. Hence show that we have a long exact sequence of cohomology groups

$$\ldots \longrightarrow H^q_\Phi(\mathscr{E} : A) \longrightarrow H^q(\mathscr{E} : A) \longrightarrow H^q(\mathscr{E}_\Phi : P_\Phi(A)) \longrightarrow H^{q+1}_\Phi(\mathscr{E} : A)$$
$$\longrightarrow \ldots .$$

15. Let $(X_0 \longrightarrow X_1)$ denote the object Ω^N of $\mathscr{S}^2$. Describe the sets X_0 and X_1, and hence show that Ω^N is not infinite in the sense used in §9.5. [Hint: there is exactly one element of X_1 having a unique pre-image in X_0.]

Appendix

Locally Internal Categories

If an elementary topos is to be regarded as a generalized set theory, we should clearly try to develop a theory of "large" mathematical notions relative to a given topos $\mathscr{E}$, as well as "small" mathematical notions within $\mathscr{E}$. In particular, we need to do this for the most commonly studied "large" mathematical notion, namely that of a category. Whilst the machinery developed in chapter 2 enables us to handle "small categories" in a topos with some confidence, the need for some comparable machinery for large categories is emphasized by the numerous examples we have met (notably toposes defined over $\mathscr{E}$, and categories of models of algebraic and other theories) of categories which clearly have such a structure. To take a typical example of the questions which arise, how should we formulate and prove an analogue of 7.14 when the particular topos $\mathscr{S}$ is replaced by an arbitrary base topos?

In the development of category theory, there have been two main lines of approach to the study of large categories $\mathscr{C}$ "relative" to a base category $\mathscr{E}$. One involves the theory of fibrations and pseudofunctors [44], [166], and is designed to allow us to talk about families of objects of $\mathscr{C}$ indexed by an object of $\mathscr{E}$; the other, stemming from the theory of enriched categories [154], [160], allows us to index families of morphisms of $\mathscr{C}$ by objects of $\mathscr{E}$. When the base category is a topos, it seems desirable to adopt a viewpoint which unites these two lines of development; the first detailed investigation of this viewpoint was given by F. W. Lawvere in [76], and others who have considered the same idea include J. Bénabou [9], J. Celeyrette [17], J. Penon [100], R. Paré and D. Schumacher [97]. In this brief account, we shall generally follow the terminology introduced by Penon.

Recall that if $(\mathscr{E}, \otimes, I)$ is a (symmetric) monoidal category, then an $\mathscr{E}$-*category* $\mathscr{C}$ is defined by specifying a collection $\mathscr{C}_0$ of objects and, for each pair of objects (A, B), an object $\mathscr{C}(A, B)$ of $\mathscr{E}$ (to be thought of as "the object of $\mathscr{C}$-morphisms from A to B"), together with morphisms $I \xrightarrow{i} \mathscr{C}(A, A)$ and $\mathscr{C}(A, B) \otimes \mathscr{C}(B, C) \xrightarrow{m} \mathscr{C}(A, C)$ defining the inclusion-of-identities and multiplication of $\mathscr{C}$; these latter being required to satisfy the appropriate commutative diagrams in $\mathscr{E}$. [In all the examples we consider, $\mathscr{E}$ will be a category with finite products, and the monoidal structure will be that induced by cartesian product.] If $|\mathscr{C}|$ is an ordinary category, we say $|\mathscr{C}|$ is *enriched over* $\mathscr{E}$ (or has an $\mathscr{E}$-category structure) if there exists an $\mathscr{E}$-category $\mathscr{C}$ whose objects are those of $|\mathscr{C}|$, together with a functorial bijection between $|\mathscr{C}|$-morphisms from A to B and I-elements of $\mathscr{C}(A, B)$. Note in particular that a closed monoidal category (e.g. a cartesian closed category) is enriched over itself. We may similarly define $\mathscr{E}$-functors, $\mathscr{E}$-natural transformations, etc.; we write $\mathscr{E}$-$\mathfrak{Cat}$ for the 2-category of $\mathscr{E}$-categories.

Now if $T: \mathscr{E} \longrightarrow \mathscr{F}$ is a monoidal functor between monoidal categories and $\mathscr{C}$ is an $\mathscr{E}$-category, we may define an $\mathscr{F}$-category $T(\mathscr{C})$ by $T(\mathscr{C})_0 = \mathscr{C}_0$ and $T(\mathscr{C})(A, B) = T(\mathscr{C}(A, B))$. (In fact T induces a functor $\mathscr{E}$-$\mathfrak{Cat} \longrightarrow \mathscr{F}$-$\mathfrak{Cat}$.) This leads to the following definition:

A.1 DEFINITION (J. Penon). Let $\mathscr{E}$ be a topos. A *locally internal category* $\mathscr{C}$ over $\mathscr{E}$ is defined by specifying, for each object X of $\mathscr{E}$, an $(\mathscr{E}/X)$-category $\mathscr{C}_X$ [to be thought of as "the category of X-indexed families of objects of $\mathscr{C}$"], and for each morphism $X \xrightarrow{f} Y$ of $\mathscr{E}$, an $(\mathscr{E}/X)$-full embedding $\theta_f: f^*(\mathscr{C}_Y) \longrightarrow \mathscr{C}_X$, such that θ_f is functorial in f up to coherent natural isomorphism.

If $\mathscr{C}$ and $\mathscr{D}$ are locally internal categories over $\mathscr{E}$, a *locally internal functor* $T: \mathscr{C} \longrightarrow \mathscr{D}$ consists of an $(\mathscr{E}/X)$-functor $T_X: \mathscr{C}_X \longrightarrow \mathscr{D}_X$ for each X, such that the diagram

$$\begin{array}{ccc} f^*(\mathscr{C}_Y) & \xrightarrow{f^*(T_Y)} & f^*(\mathscr{D}_Y) \\ \downarrow{\scriptstyle \theta_f} & & \downarrow{\scriptstyle \theta_f} \\ \mathscr{C}_X & \xrightarrow{T_X} & \mathscr{D}_X \end{array}$$

commutes up to (specified, coherent) natural isomorphism for each f. Locally internal natural transformations may be similarly defined; we thus have a 2-category of locally internal categories over $\mathscr{E}$, which we denote by $\mathfrak{Cat}(\mathscr{E})$. ■

Let $\mathscr{C}$ be a locally internal category over $\mathscr{E}$. If we "forget" the enriched structure on the categories $\mathscr{C}_X$, then we have simply a pseudofunctor

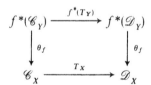

$\mathscr{E}^{\mathrm{op}} \longrightarrow \mathfrak{Cat}$; and by applying the "Grothendieck construction" ([44], VI 8), we obtain a cloven fibration $\mathscr{C} \xrightarrow{p} \mathscr{E}$. Conversely, if we are given a cloven fibration over $\mathscr{E}$, it is natural to ask when the corresponding pseudofunctor $\mathscr{E}^{\mathrm{op}} \longrightarrow \mathfrak{Cat}$ has the structure of a locally internal category. The following lemma provides the answer.

A.2 LEMMA. *Let $\mathscr{C} \xrightarrow{p} \mathscr{E}$ be a fibration over $\mathscr{E}$, equipped with a cleavage θ. Then specifying a locally internal category structure on the corresponding pseudofunctor $\mathscr{E}^{\mathrm{op}} \longrightarrow \mathfrak{Cat}$ is equivalent to specifying, for each pair (A, B) of objects of $\mathscr{C}$, an object $\mathscr{C}(A, B)$ of $\mathscr{E}/p(A) \times p(B)$, together with, for each object $Z \xrightarrow{(f,g)} p(A) \times p(B)$ of $\mathscr{E}/p(A) \times p(B)$, a bijection (functorial in Z) between morphisms $(f, g) \longrightarrow \mathscr{C}(A, B)$ in $\mathscr{E}/p(A) \times p(B)$ and morphisms $\theta_f(A) \longrightarrow \theta_g(B)$ in $\mathscr{C}_Z$.*

Proof. Suppose $\mathscr{C}$ is a locally internal category; write X, Y for $p(A)$, $p(B)$ respectively. Define $\mathscr{C}(A, B) = \mathscr{C}_{X \times Y}(\theta_{\pi_1}(A), \theta_{\pi_2}(B))$; now morphisms $(f, g) \longrightarrow \mathscr{C}(A, B)$ in $\mathscr{E}/X \times Y$ correspond to global elements of $(f, g)^*\mathscr{C}(A, B)$ in $\mathscr{E}/Z$. But since $\theta_{(f,g)}$ is an $(\mathscr{E}/Z)$-full embedding, we have an isomorphism

$$(f, g)^*(\mathscr{C}_{X \times Y}(\theta_{\pi_1}(A), \theta_{\pi_2}(B))) \cong \mathscr{C}_Z(\theta_f(A), \theta_g(B)),$$

so we have the required bijection.

Conversely, suppose the condition is satisfied; let X be an object of $\mathscr{E}$, and A, B two objects of $\mathscr{C}_X$. Define $\mathscr{C}_X(A, B) = \Delta^*(\mathscr{C}(A, B))$, where $X \xrightarrow{\Delta} X \times X$ is the diagonal map. We now make $\mathscr{C}_X$ into an $(\mathscr{E}/X)$-category using the method of generic elements; for example, to define the multiplication $\mathscr{C}_X(A, B) \times \mathscr{C}_X(B, C) \longrightarrow \mathscr{C}_X(A, C)$, write $(Z \xrightarrow{f} X)$ for $\mathscr{C}_X(A, B) \times \mathscr{C}_X(B, C)$, and then compose the morphisms $\theta_f(A) \longrightarrow \theta_f(B) \longrightarrow \theta_f(C)$ in $\mathscr{C}_Z$. Now let $Y \xrightarrow{g} X$ be any morphism with codomain X; then for any $(T \xrightarrow{h} Y)$ we have a bijection between h-elements of $\mathscr{C}_Y(\theta_g(A), \theta_g(B))$ and gh-elements of $\mathscr{C}_X(A, B)$, since both correspond to morphisms $\theta_{gh}(A) \longrightarrow \theta_{gh}(B)$ in $\mathscr{C}_T$. So $\mathscr{C}_Y(\theta_g(A), \theta_g(B)) \cong g^*(\mathscr{C}_X(A, B))$ in $\mathscr{E}/Y$; i.e. θ_g is an $(\mathscr{E}/Y)$-full embedding. ∎

A.3 EXAMPLES. (i) $\mathscr{E}$ itself becomes a locally internal category, if we set

$$\mathscr{E}_X = \mathscr{E}/X \qquad \text{and} \qquad \theta_f = f^*.$$

The fact that θ_f is an $\mathscr{E}/X$-full embedding is simply another way of saying that pullback functors preserve exponentials.

(ii) More generally, let $\mathscr{F} \xrightarrow{p} \mathscr{E}$ be a geometric morphism. Then for any object X of $\mathscr{E}$, we have a geometric morphism $\mathscr{F}/p^*X \xrightarrow{p/X} \mathscr{E}/X$, whose inverse image is simply p^* applied to objects over X, and whose direct image is the composite

$$\mathscr{F}/p^*X \xrightarrow{p^*} \mathscr{E}/p_*p^*X \xrightarrow{\alpha_X^*} \mathscr{E}/X,$$

where α is the unit of $(p^* \dashv p_*)$. We may thus make $\mathscr{F}$ into a locally internal category over $\mathscr{E}$, by setting

$$\mathscr{F}_X = (p/X)_*(\mathscr{F}/p^*X) \qquad \text{and} \qquad \theta_f = (p/X)_*\,((p^*f)^*).$$

The fact that θ_f is well-defined depends on the Beck condition for pullbacks in $\mathfrak{Top}$ (Ex. 4.7), i.e. the fact that the square

$$\begin{array}{ccc} \mathscr{F}/p^*Y & \xrightarrow{(p^*f)^*} & \mathscr{F}/p^*X \\ \downarrow{\scriptstyle (p/Y)_*} & & \downarrow{\scriptstyle (p/X)_*} \\ \mathscr{E}/Y & \xrightarrow{f^*} & \mathscr{E}/X \end{array}$$

commutes up to isomorphism.

And if $\mathscr{G} \xrightarrow{q} \mathscr{F}$ is any geometric morphism over $\mathscr{E}$, it is not hard to verify that both q_* and q^* induce locally internal functors over $\mathscr{E}$; so we have functors $(\mathfrak{Top}/\mathscr{E})^{\mathrm{co}} \longrightarrow \mathfrak{Cat}(\mathscr{E})$ and $(\mathfrak{Top}/\mathscr{E})^{\mathrm{op}} \longrightarrow \mathfrak{Cat}(\mathscr{E})$. (The superscript "co" denotes the "second dual" of a 2-category, i.e. the 2-category obtained by reversing 2-arrows.)

(iii) More generally still, let $\mathscr{F} \xrightarrow{p} \mathscr{E}$ be a geometric morphism and $\mathscr{C}$ a locally internal category over $\mathscr{F}$. Then as in (ii), it is easy to verify that the assignment

$$X \longmapsto (p/X)_*(\mathscr{C}_{p^*X})$$

defines a locally internal category $p_*(\mathscr{C})$ over $\mathscr{E}$. So p_* becomes a functor $\mathfrak{Cat}(\mathscr{F}) \longrightarrow \mathfrak{Cat}(\mathscr{E})$; and in fact the assignment

$$\mathscr{E} \longmapsto \mathfrak{Cat}(\mathscr{E})$$

is itself functorial on $\mathfrak{Top}^{\mathrm{co}}$, for if we are given a natural transformation $\eta\colon p \longrightarrow q$ between geometric morphisms $\mathscr{F} \rightrightarrows \mathscr{E}$, then the evident natural transformation $\Sigma_{\eta_X} \cdot (p/X)^* \longrightarrow (q/X)^*$ transposes to give a transformation

$$(q/X)_* \xrightarrow{\bar{\eta}} (p/X)_* \cdot \eta_X^*,$$

and hence we have an $\mathscr{E}/X$-functor

$$(q_*(\mathscr{C}))_X = (q/X)_*(\mathscr{C}_{q*X}) \xrightarrow{\bar{\eta}} (p/X)_*(\eta_X^*(\mathscr{C}_{q*X})) \xrightarrow{(p/X)_*(\theta_{\eta_*})}$$

$$(p/X)_*(\mathscr{C}_{p*X}) = (p_*(\mathscr{C}))_X.$$

(iv) Let $\mathscr{C}$ be a locally internal category over $\mathscr{E}$, and $\mathbb{T}$ a locally internal monad on $\mathscr{C}$ (i.e. a lax functor $\mathbf{1} \longrightarrow \mathfrak{Cat}(\mathscr{E})$ sending the unique object of $\mathbf{1}$ to $\mathscr{C}$, cf. 4.23). Then for each object X of $\mathscr{E}$, $\mathbb{T}_X$ is a strong monad on $\mathscr{C}_X$, and so the category of $\mathbb{T}_X$-algebras has an $(\mathscr{E}/X)$-category structure. And pullback functors in $\mathscr{E}$ preserve the equalizers used to define the objects of morphisms in $(\mathscr{C}_X)^{\mathbb{T}_X}$; so the assignment

$$X \longmapsto (\mathscr{C}_X)^{\mathbb{T}_X}$$

defines a locally internal category $\mathscr{C}^{\mathbb{T}}$ of $\mathbb{T}$-algebras in $\mathscr{C}$. In particular, any internal finitary algebraic theory in $\mathscr{E}$ (6.45) defines a locally internal monad on $\mathscr{E}$ itself, since the free $\mathbb{T}$-model functor commutes with inverse image functors and hence with pullback functors. So if $\mathbb{T}$ is such a theory, the assignment

$$X \longmapsto \mathbb{T}(\mathscr{E}/X)$$

defines a locally internal category over $\mathscr{E}$. ■

The above examples show that the concept of locally internal category includes most of the examples of "large" categories which we wish to study. We next investigate its relationship with the concept of "small" (i.e. internal) category.

A.4 Proposition. *Let* $\mathfrak{cat}(\mathscr{E})$ *denote the 2-category of internal categories in* $\mathscr{E}$. *(The 2-category structure was defined in Ex. 2.1.) Then there is a full embedding of 2-categories* $L\colon \mathfrak{cat}(\mathscr{E}) \longrightarrow \mathfrak{Cat}(\mathscr{E})$.

Proof. Let $\mathbf{C}$ be an internal category in $\mathscr{E}$. We define the locally internal category $L\mathbf{C}$ as follows: the objects of $L\mathbf{C}_X$ are the X-elements of C_0 in $\mathscr{E}$, and

$LC_X(c_1, c_2)$ is the pullback

$$\begin{array}{ccc} LC_X(c_1, c_2) & \longrightarrow & C_1 \\ \downarrow & & \downarrow{\scriptstyle (d_0, d_1)} \\ X & \xrightarrow{(c_1, c_2)} & C_0 \times C_0 \end{array} .$$

If $X \xrightarrow{f} Y$ is a morphism of $\mathscr{E}$ and c is a Y-element of C_0, we define $\theta_f(c)$ to be the composite cf; since a composite of pullback squares is a pullback, θ_f is an $(\mathscr{E}/X)$-full embedding. The definition of L on 1-arrows and 2-arrows of $\mathfrak{cat}(\mathscr{E})$ is straightforward.

Now suppose we are given a locally internal functor $T: L\mathbf{C} \longrightarrow L\mathbf{D}$. Let $f_0: C_0 \longrightarrow D_0$ denote the object $T_{C_0}(1_{C_0})$ of $L\mathbf{D}_{C_0}$; then since $T_{C_0 \times C_0}$ is a strong functor, it defines a morphism

$$(C_1 \longrightarrow C_0 \times C_0) \longrightarrow (f_0 \times f_0)^*(D_1 \longrightarrow D_0 \times D_0)$$

in $\mathscr{E}/C_0 \times C_0$, which in turn defines a morphism $f_1: C_1 \longrightarrow D_1$ such that

$$\begin{array}{ccc} C_1 & \xrightarrow{f_1} & D_1 \\ \downarrow{\scriptstyle (d_0, d_1)} & & \downarrow{\scriptstyle (d_0, d_1)} \\ C_0 \times C_0 & \xrightarrow{f_0 \times f_0} & D_0 \times D_0 \end{array}$$

commutes. It is now easy to verify that (f_0, f_1) is an internal functor $\mathbf{C} \longrightarrow \mathbf{D}$, and that $L(f)$ is canonically isomorphic to T; so L induces an equivalence of categories $\mathfrak{cat}(\mathscr{E})(\mathbf{C}, \mathbf{D}) \simeq \mathfrak{Cat}(\mathscr{E})(L\mathbf{C}, L\mathbf{D})$. ■

A.5 Remark. The equivalence $\mathfrak{cat}(\mathscr{E})(\mathbf{C}, \mathbf{D}) \simeq \mathfrak{Cat}(\mathscr{E})(L\mathbf{C}, L\mathbf{D})$ which we have just established is *not* in general an isomorphism of categories, as is shown by the following example, due to R. Paré:

Let X and Y be two objects of a topos $\mathscr{E}$ such that we have monomorphisms $X \rightarrowtail Y \rightarrowtail X$ but no isomorphism $X \cong Y$ (cf. Ex. 5.8), and let $\mathbf{C}$ and $\mathbf{D}$ be the indiscrete categories (trivial connected groupoids) whose objects are X and Y. Then it is easily seen that the given monos induce an isomorphism $L\mathbf{C} \cong L\mathbf{D}$ in $\mathfrak{Cat}(\mathscr{E})$, whereas $\mathbf{C}$ and $\mathbf{D}$ are merely equivalent in $\mathfrak{cat}(\mathscr{E})$.

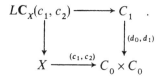

We do, therefore, lose a certain amount of "1-categorical" information by passing from internal to locally internal categories. One way to avoid this difficulty would be to work with *strict* functors $\mathscr{E}^{\mathrm{op}} \longrightarrow \mathfrak{Cat}$ instead of pseudo-functors, i.e. with split fibrations instead of cloven ones. However, the "large" categories considered in A.3 do not normally have such a "strict" structure (though it is possible to "strictify" them e.g. by Freyd's technique for canonizing finite limits [37]), and so the correct attitude to the problem would seem to be that we should work in $\mathfrak{cat}(\mathscr{E})$ when considering 1-categorical questions (i.e. those which depend on isomorphism rather than equivalence), and in $\mathfrak{Cat}(\mathscr{E})$ when considering 2-categorical ones. ■

An important advantage of locally internal categories is that they enable us to generalize the "internal full subcategory" construction of 2.38 to any of the examples considered above. Specifically, let $\mathscr{C}$ be a locally internal category over $\mathscr{E}$, and A an object of $\mathscr{C}_X$. Then we define the internal category $\mathrm{Full}_{\mathscr{C}}(A)$ in $\mathscr{E}$ to have object-of-objects X, and object-of-morphisms given by the object $\mathscr{C}_{X\times X}(\theta_{\pi_1}(A), \theta_{\pi_2}(A))$ of $\mathscr{E}/X \times X$. The multiplication of $\mathrm{Full}_{\mathscr{C}}(A)$ is given by the composition map

$$\mathscr{C}_{X\times X\times X}(\theta_{\pi_1}(A), \theta_{\pi_2}(A)) \times \mathscr{C}_{X\times X\times X}(\theta_{\pi_2}(A), \theta_{\pi_3}(A)) \xrightarrow{m} \mathscr{C}_{X\times X\times X}(\theta_{\pi_1}(A), \theta_{\pi_3}(A))$$

in $\mathscr{E}/X \times X \times X$, and the inclusion-of-identities is similarly defined. The "inclusion functor" of 2.38 is replaced by a locally internal full embedding $L(\mathrm{Full}_{\mathscr{C}}(A)) \longrightarrow \mathscr{C}$, which is defined on objects by $(Y \xrightarrow{f} X) \longmapsto \theta_f(A)$.

A.6 Proposition. *Let* $\mathbf{A}$ *be an internal category in* $\mathscr{E}$, $\mathscr{C}$ *a locally internal category. Then any locally internal functor* $L\mathbf{A} \xrightarrow{f} \mathscr{C}$ *can be factored as* $L\mathbf{A} \xrightarrow{Lg} L\mathbf{B} \xrightarrow{h} \mathscr{C}$, *where* g *is an internal functor which is the identity on objects, and* h *is a full embedding; and this factorization is unique up to canonical isomorphism.*

Proof. Define $\mathbf{B} = \mathrm{Full}_{\mathscr{C}}(f_{A_0}(1_{A_0}))$. Then h is simply the inclusion functor defined above, and g is the internal functor defined on morphisms by

$$(A_1 \longrightarrow A_0 \times A_0) \cong L\mathbf{A}_{A_0\times A_0}(\pi_1, \pi_2) \xrightarrow{f} \mathscr{C}_{A_0\times A_0}(f(\pi_1), f(\pi_2))$$
$$\cong (B_1 \longrightarrow B_0 \times B_0).$$

It is now easy to construct a natural isomorphism $h.Lg \cong f$; moreover if f is a full embedding, then g is an isomorphism of internal categories, which proves the uniqueness clause of the statement. ■

A.4 and A.6 together allow us to pass freely back and forth between internal and locally internal categories. Another useful result in this area allows us to exponentiate a locally internal category to the power of an internal category:

A.7 PROPOSITION. *Let* $\mathbf{A}$ *be an internal category in* $\mathscr{E}$, $\mathscr{C}$ *a locally internal category. Then there exists a locally internal category* $\mathscr{C}^{\mathbf{A}}$, *which has the universal property of an exponential* $\mathscr{C}^{L\mathbf{A}}$ *in* $\mathfrak{Cat}(\mathscr{E})$.

Proof. We define $\mathscr{C}^{\mathbf{A}}$ as follows: an object of $(\mathscr{C}^{\mathbf{A}})_X$ is a pair (F, α) where F is an object of $\mathscr{C}_{A_0 \times X}$ and α is a morphism

$$\theta_{d_0 \times 1_X}(F) \longrightarrow \theta_{d_1 \times 1_X}(F)$$

in $\mathscr{C}_{A_1 \times X}$, such that $\theta_{i \times 1}(\alpha)$ is the identity morphism on F, and the diagram

$$\begin{array}{ccc} \theta_{\pi_1 \times 1}(F) & \xrightarrow{\theta_{\pi_1 \times 1}(\alpha)} & \theta_{\pi_2 \times 1}(F) \\ & \searrow^{\theta_{m \times 1}(\alpha)} & \downarrow{\scriptstyle \theta_{\pi_2 \times 1}(\alpha)} \\ & & \theta_{\pi_3 \times 1}(F) \end{array}$$

commutes in $\mathscr{C}_{A_2 \times X}$. To define the objects of morphisms of $(\mathscr{C}^{\mathbf{A}})_X$, we shall assume $X = 1$ for notational convenience; so let (F, α) and (G, β) be two objects of $(\mathscr{C}^{\mathbf{A}})_1$. Then $(\mathscr{C}^{\mathbf{A}})_1((F, \alpha), (G, \beta))$ is the equalizer in $\mathscr{E}$ of a pair of morphisms

$$\Pi_{A_0}(\mathscr{C}_{A_0}(F, G)) \rightrightarrows \Pi_{A_1}(\mathscr{C}_{A_1}(\theta_{d_0}(F), \theta_{d_1}(G))),$$

of which the first is the transpose of

$$A_1^*\Pi_{A_0}(\mathscr{C}_{A_0}(F, G)) \cong d_1^* A_0^* \Pi_{A_0}(\mathscr{C}_{A_0}(F, G)) \xrightarrow{d_1^*(\varepsilon)} d_1^*(\mathscr{C}_{A_0}(F, G))$$

$$\cong \mathscr{C}_{A_1}(\theta_{d_1}(F), \theta_{d_1}(G)) \xrightarrow{\bar{\alpha}} \mathscr{C}_{A_1}(\theta_{d_0}(F), \theta_{d_1}(G))$$

(here ε is the counit of $(A_0^* \dashv \Pi_{A_0})$, and $\bar{\alpha}$ is induced by composition with α), and the second is similarly induced by composition with β.

The verification that $\mathscr{C}^{\mathbf{A}}$ forms a locally internal category is long but straightforward.

Now suppose we are given another locally internal category $\mathscr{D}$. Then to any locally internal functor $\mathscr{D} \times L\mathbf{A} \xrightarrow{f} \mathscr{C}$, we can associate $\mathscr{D} \xrightarrow{\bar{f}} \mathscr{C}^{\mathbf{A}}$, which sends an object V of $\mathscr{D}_X$ to the object $f_{A_0 \times X}(\theta_{\pi_2}(V), \pi_1)$ of $\mathscr{C}_{A_0 \times X}$, equipped with a structure map obtained by applying $f_{A_1 \times X}$ to the generic morphism $d_0\pi_1 \longrightarrow d_1\pi_1$ in $L\mathbf{A}_{A_1 \times X}$. Conversely, given $\mathcal{D} \xrightarrow{g} \mathcal{C}^{\mathbf{A}}$, we may define $\mathcal{D} \times L\mathbf{A} \xrightarrow{\bar{g}} \mathcal{C}$ by

$$\bar{g}_X(V, X \xrightarrow{\gamma} A_0) = \theta_{(\gamma, 1_X)}(g_X(V)).$$

The proof that $\bar{f}$ and $\bar{g}$ are indeed locally internal functors, and that the assignments $f \longmapsto \bar{f}$ and $g \longmapsto \bar{g}$ induce an equivalence of categories $\mathfrak{Cat}(\mathcal{E})(\mathcal{D} \times L\mathbf{A}, \mathcal{C}) \simeq \mathfrak{Cat}(\mathcal{E})(\mathcal{D}, \mathcal{C}^{\mathbf{A}})$, is again straightforward. ■

The interested reader may verify that the functor L: $\mathfrak{cat}(\mathcal{E}) \longrightarrow \mathfrak{Cat}(\mathcal{E})$ preserves exponentials (the cartesian closed structure on $\mathfrak{cat}(\mathcal{E})$ was defined in Ex. 2.4); and that if we take $\mathcal{C} = \mathcal{E}$ in A.7, the locally internal category $\mathcal{E}^{\mathbf{A}}$ which we construct is equivalent to that defined (as in A.3(ii)) by the $\mathcal{E}$-topos $\mathcal{E}^{\mathbf{A}}$. Thus by the use of locally internal categories, we are able to regard internal diagrams on an internal category $\mathbf{A}$ as "actual" functors from $\mathbf{A}$ to $\mathcal{E}$.

We next investigate the notions of completeness and cocompleteness for locally internal categories.

A.8 Definition. Let $\mathcal{C}$ be a locally internal category over $\mathcal{E}$. We say $\mathcal{C}$ is *tensored over* $\mathcal{E}$ if, for every $X \xrightarrow{f} Y$ in $\mathcal{E}$, the functor $\theta_f : |\mathcal{C}_Y| \longrightarrow |\mathcal{C}_X|$ has a (non-enriched) left adjoint σ_f. (Note that σ_f is automatically pseudofunctorial in f.) If in addition, for each pullback square

$$\begin{array}{ccc} P & \xrightarrow{f} & X \\ {\scriptstyle g}\downarrow & & \downarrow{\scriptstyle h} \\ Y & \xrightarrow{k} & Z \end{array}$$

in $\mathcal{E}$, the canonical natural transformation $\sigma_g\theta_f \longrightarrow \theta_k\sigma_h$ is an isomorphism (i.e. σ satisfies the "Beck condition"), we say $\mathcal{C}$ is *strongly tensored* over $\mathcal{E}$. Dually, we say $\mathcal{C}$ is (*strongly*) *cotensored* over $\mathcal{E}$ if each θ_f has a right adjoint τ_f (and the Beck condition holds). ■

Note that $\mathcal{E}$ itself is strongly tensored and cotensored over $\mathcal{E}$, as is any topos defined over $\mathcal{E}$. If $\mathcal{F} \xrightarrow{p} \mathcal{E}$ is a geometric morphism, then $p_* : \mathfrak{Cat}(\mathcal{F}) \longrightarrow \mathfrak{Cat}(\mathcal{E})$ preserves the property of being (strongly) tensored; and if $\mathcal{C}$ is (strongly) cotensored over $\mathcal{E}$ and $\mathbb{T}$ is a locally internal monad on $\mathcal{C}$, then $\mathcal{C}^{\mathbb{T}}$ is (strongly) cotensored. The reader may verify that if $\mathbf{P}$ is an internal poset in $\mathcal{E}$, then $\mathbf{P}$ is internally complete (5.34) iff $L\mathbf{P}$ is strongly tensored and cotensored over $\mathcal{E}$.

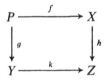

A.9 Remark. Let $\mathscr{E}$ and $\mathscr{F}$ be toposes, and $T : \mathscr{E} \longrightarrow \mathscr{F}$ a functor. Then it is easily seen that the assignment

$$X \longmapsto \mathscr{F}/TX$$

makes $\mathscr{F}$ into a locally internal category over $\mathscr{E}$ iff T has a right adjoint; for if $\mathscr{F}/T(1)$ is enriched over $\mathscr{E}$, we may define the right adjoint by

$$Y \longmapsto \mathscr{F}_1(1_{T(1)}, T(1)^* Y),$$

and conversely we may use the argument of A.3(ii). If this condition is satisfied, then $\mathscr{F}$ is certainly tensored over $\mathscr{E}$; but J. Bénabou has pointed out that it is *strongly* tensored iff T preserves pullbacks, since the Beck condition $\Sigma_g . f^* \cong k^* . \Sigma_h$ holds for a commutative square

$$\begin{array}{ccc} P & \xrightarrow{f} & X \\ {\scriptstyle g}\downarrow & & \downarrow{\scriptstyle h} \\ Y & \xrightarrow{k} & Z \end{array}$$

if $\mathscr{F}$ iff the square is a pullback. (Consider the effect of each composite on the object 1_X of $\mathscr{F}/X$.) This is undoubtedly part of the explanation why the notion of geometric morphism, which may at first sight have appeared somewhat arbitrary, is the "right" notion of morphism of toposes. ∎

A.10 Lemma. *Suppose $\mathscr{C}$ is strongly tensored over $\mathscr{E}$. Then for each $X \xrightarrow{f} Y$ in $\mathscr{E}$, σ_f is in fact an $(\mathscr{E}/Y)$-functor $\Pi_f(\mathscr{C}_X) \longrightarrow \mathscr{C}_Y$; and the adjunction $(\sigma_f \dashv \theta_f)$ is enriched over $\mathscr{E}/Y$, in the sense that we have a natural isomorphism $\Pi_f(\mathscr{C}_X(A, \theta_f(B))) \cong \mathscr{C}_Y(\sigma_f(A), B)$ for objects A, B of $\mathscr{C}_X$, $\mathscr{C}_Y$ respectively.*

Proof. We prove the second statement first. Let α, β denote the unit and counit of $(\sigma_f \dashv \theta_f)$; then the composite

$$f^*(\mathscr{C}_Y(\sigma_f(A), B)) \cong \mathscr{C}_X(\theta_f \sigma_f(A), \theta_f(B)) \xrightarrow{\bar{\alpha}} \mathscr{C}_X(A, \theta_f(B))$$

transposes to give a natural map

$$\mathscr{C}_Y(\sigma_f(A), B) \longrightarrow \Pi_f(\mathscr{C}_X(A, \theta_f(B))).$$

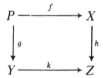

To define an inverse for this map, denote the object $\Pi_f(\mathscr{C}_X(A, \theta_f(B)))$ of $\mathscr{E}/Y$ by $(V \xrightarrow{g} Y)$, and form the pullback

$$\begin{array}{ccc} P & \xrightarrow{k} & V \\ \big\downarrow{\scriptstyle h} & & \big\downarrow{\scriptstyle g} \\ X & \xrightarrow{f} & Y \end{array} .$$

Then the counit map

$$h = f^*\Pi_f(\mathscr{C}_X(A, \theta_f(B))) \longrightarrow \mathscr{C}_X(A, \theta_f(B))$$

induces a morphism

$$\theta_h(A) \longrightarrow \theta_h\theta_f(B) \cong \theta_k\theta_g(B)$$

in $\mathscr{C}_P$. Transposing and applying the Beck condition, we obtain a morphism

$$\theta_g\sigma_f(A) \cong \sigma_k\theta_h(A) \longrightarrow \theta_g(B)$$

in $\mathscr{C}_V$, which in turn induces

$$g \longrightarrow \mathscr{C}_Y(\sigma_f(A), B)$$

in $\mathscr{E}/Y$. The proof that this map is inverse to that constructed above is straightforward verification.

To show that σ_f is an $(\mathscr{E}/Y)$-functor, substitute $B = \sigma_f(C)$ in the above; then we obtain a morphism

$$\Pi_f\, \mathscr{C}_X(A, C)) \xrightarrow{\Pi_f(\bar{\alpha})} \Pi_f\, \mathscr{C}_X(A, \theta_f\sigma_f(C)) \cong \mathscr{C}_Y(\sigma_f(A), \sigma_f(C)).$$

Once again, it is not hard to check that this morphism has the required properties. ■

Now let $\mathbf{A}$ be an internal category in $\mathscr{E}$, and $\mathscr{C}$ a locally internal category. Then we have a locally internal functor $\mathbf{A}^*: \mathscr{C} \longrightarrow \mathscr{C}^{\mathbf{A}}$, which sends an object V of $\mathscr{C}_X$ to the object $\theta_{\pi_2}(V)$ of $\mathscr{C}_{A_0 \times X}$, equipped with the identity structure map. (This is clearly the analogue of the functor $\mathbf{A}^*: \mathscr{E} \longrightarrow \mathscr{E}^{\mathbf{A}}$ defined in 2.25(ii).)

A.11 THEOREM (J. Celeyrette). *Suppose $\mathscr{C}$ is strongly tensored over $\mathscr{E}$ and has locally internal coequalizers of reflexive pairs (i.e. each $\mathscr{C}_X$ has $\mathscr{E}/X$-enriched coequalizers, and the functors θ_f preserve them). Then for each internal category $\mathbf{A}$ in $\mathscr{E}$, the functor $\mathbf{A}^*\colon \mathscr{C} \longrightarrow \mathscr{C}^{\mathbf{A}}$ has a locally internal left adjoint $\varinjlim_{\mathbf{A}}$. (We say that $\mathscr{C}$ is $\mathscr{E}$-*cocomplete.)

Proof. Consider first the case when $\mathbf{A}$ is a discrete category ($A \overset{1}{\underset{1}{\rightrightarrows}} A$); in this case it is natural to write $\mathscr{C}/A$ instead of $\mathscr{C}^{\mathbf{A}}$. Now $(\mathscr{C}/A)_X$ is simply $\Pi_{\pi_2}(\mathscr{C}_{A\times X})$, and $(A^*)_X$ is the transpose of $\theta_{\pi_2}\colon \pi_2^*(\mathscr{C}_X) \longrightarrow \mathscr{C}_{A\times X}$; so it follows at once from A.10 that σ_{π_2} is the required left adjoint. (The fact that it defines a locally internal functor uses the Beck condition for pullback squares of the form

$$\begin{array}{ccc} A\times X & \xrightarrow{1\times f} & A\times Y.) \\ \downarrow{\scriptstyle \pi_2} & & \downarrow{\scriptstyle \pi_2} \\ X & \xrightarrow{f} & Y \end{array}$$

Now consider the general case. Let (F, α) be an object of $(\mathscr{C}^{\mathbf{A}})_X$; since $\mathscr{C}$ is tensored over $\mathscr{E}$, we may regard the structure map α as a morphism

$$\sigma_{d_1\times 1}\theta_{d_0\times 1}(F) \longrightarrow F$$

in $\mathscr{C}_{A_0\times X}$; so using the argument of 2.21, we deduce that $(\mathscr{C}^{\mathbf{A}})_X$ is monadic over $(\mathscr{C}/A_0)_X$. And since $\mathscr{C}$ is strongly tensored, it is not hard to verify that $\mathscr{C}^{\mathbf{A}}$ is in fact locally internally monadic over $\mathscr{C}/A_0$. Moreover, $\mathscr{C}^{\mathbf{A}}$ has locally internal reflexive coequalizers, since $\mathscr{C}/A_0$ has them and the functor part of the appropriate monad preserves them. The construction of the functor $\varinjlim_{\mathbf{A}}$ is now a straightforward application of the adjoint lifting theorem (0.15), similar to 2.35. ■

More generally, of $\mathscr{C}$ satisfies the hypotheses of A.11, we may construct left Kan extension functors $\varinjlim_f\colon \mathscr{C}^{\mathbf{A}} \longrightarrow \mathscr{C}^{\mathbf{B}}$ for morphisms $\mathbf{A} \xrightarrow{f} \mathbf{B}$ of $\mathbf{cat}(\mathscr{E})$; and if $\mathscr{C}$ is strongly cotensored and has locally internal (coreflexive) equalizers, then we may similarly construct right Kan extensions.

With this theorem, we bring our survey of locally internal categories to a close. There are a number of other interesting results which we could have mentioned—notably the proof by R. Paré and D. Schumacher [97] of a version of the Adjoint Functor Theorem for locally internal categories. But there remain vast areas of the subject which are virtually unexplored; any attempt to explore them would require another book at least as long as this one! However, we hope that this brief account has convinced the reader that locally internal categories are a worthwhile subject of study.

Bibliography

Note: **MR** p, p. q at the end of an entry ($p \leqslant 19$) means that the article is reviewed on page q of volume p of *Mathematical Reviews*. **MR** p/q ($p \geqslant 20$) similarly denotes review number q in volume p.

A: STANDARD REFERENCES

[AC] P. Freyd, "Abelian Categories". Harper and Row, 1964. **MR** 29/3517.

[AG] I. G. Macdonald, "Algebraic Geometry: Introduction to Schemes". W. A. Benjamin Inc., 1968. **MR** 39/205.

[AT] E. H. Spanier, "Algebraic Topology". McGraw-Hill Inc., 1966. **MR** 35/1007.

[BA] R. Sikorski, "Boolean Algebras". Ergebnisse der Mathematik, Band 25, Springer-Verlag (second edition 1964). **MR** 31/2178.

[CF] P. Gabriel and M. Zisman, "Calculus of Fractions and Homotopy Theory". Ergebnisse der Mathematik, Band 35, Springer-Verlag, 1967. **MR** 35/1019.

[CW] S. Mac Lane, "Categories for the Working Mathematician". Graduate Texts in Mathematics no. 5, Springer-Verlag, 1971. **MR** 50/7275.

[GT] S. Willard, "General Topology". Addison-Wesley Publishing Co., 1970. **MR** 41/9173.

[HA] P. Hilton and U. Stammbach, "A Course in Homological Algebra". Graduate Texts in Mathematics no. 4, Springer-Verlag, 1971. **MR** 49/10751.

[LT] G. Birkhoff, "Lattice Theory". A.M.S. Colloquium Publications vol. XXV, American Mathematical Society (third edition 1967). **MR** 37/2638.

[MM] H. Rasiowa and R. Sikorski, "The Mathematics of Metamathematics". Monografie Matematyczne, tom 41, PWN (Polish Scientific Publishers), 1963. **MR** 29/1149.

[ST] B. R. Tennison, "Sheaf Theory". L.M.S. Lecture Note Series no. 20, Cambridge University Press, 1975. **MR** 53/8192.

[TF] R. Godement, "Topologie Algébrique et Théorie des Faisceaux". Publ. Inst. Math. Univ. Strasbourg, XIII, Hermann, 1958. **MR** 21/1583.

B: GENERAL AND INTRODUCTORY ACCOUNTS OF TOPOS THEORY

[AH] M. Artin, "Grothendieck Topologies". Lecture Notes, Harvard University 1962.

[BC] J. Bénabou and J. Celeyrette, "Généralites sur les topos de Lawvere et Tierney". Seminaire Bénabou, Université Paris-Nord 1971.

[FK] P. Freyd, Aspects of Topoi. *Bull. Austral. Math. Soc.* **7** (1972), 1–76 and 467–480. **MR** 53/576.

[GB] J. Giraud, Analysis Situs. Seminaire Bourbaki, exposé **256** (1963). **MR** 33/1343.

[GV] A. Grothendieck and J. L. Verdier, "Théorie des Topos" (SGA 4, exposés I–VI). Second edition, Springer Lecture Notes in Math. **269–270** (1972). **MR** 50/7130–1.

[KS] G. M. Kelly and R. Street, Elementary Topoi. Abstracts of Sydney Category Seminar, 1972.

[KW] A. Kock and G. C. Wraith, "Elementary Toposes". Aarhus Universitet Lecture Notes Series, no. **30**, 1971. **MR** 49/7324.

[LB] F. W. Lawvere, Continuously Variable Sets: Algebraic Geometry = Geometric Logic. Proc. ASL Logic Colloquium, Bristol 1973 (ed. H. E. Rose and J. C. Shepherdson), North-Holland 1975, 135–156. **MR** 52/13384.

[LC] F. W. Lawvere, "Variable Sets, Etendu and Variable Structures in Topoi" (notes by S. E. Landsburg). Lecture Notes, University of Chicago 1976.

[LE] F. W. Lawvere, Variable quantities and variable structures in topoi "Algebra, Topology and Category Theory: a collection of papers in honor of Samuel Eilenberg" (ed. A. Heller and M. Tierney), Academic Press 1976, 101–131.

[LH] F. W. Lawvere, Introduction. "Toposes, Algebraic Geometry and Logic", Springer Lecture Notes in Math. **274** (1972), 1–12. **MR** 51/12973.

[LM] F. W. Lawvere, Introduction. "Model Theory and Topoi", Springer Lecture Notes in Math. **445** (1975), 3–14. **MR** 51/12982.

[LN] F. W. Lawvere, Quantifiers and Sheaves. Actes du Congrès International des Mathématiciens, Nice 1970, 329–334.

[MB] S. Mac Lane, Sets, Topoi and Internal Logic in Categories. Proc. ASL Logic Colloquium, Bristol 1973 (ed. H. E. Rose and J. C. Shepherdson), North-Holland 1975, 119–134. **MR** 52/5419.

[MR] M. Makkai and G. E. Reyes, "Coherent Logic". To appear.

[RM] G. E. Reyes, From Sheaves to Logic. "Studies in Algebraic Logic" (ed. A. Daigneault), MAA Studies in Math. vol. 9, Mathematical Association of America 1974, 143–204. **MR** 50/13182.

[RW] K. A. Rowe, "Topoidal Set Theory". Lecture Notes, University of Waterloo 1974.

[SR] D. Schlomiuk, Topos di Grothendieck e topos di Lawvere e Tierney. *Rend. Mat.* (VI) **7** (1974), 513–553. **MR** 51/3252.

[TV] M. Tierney, Axiomatic Sheaf Theory: some constructions and applications. Proc. CIME conference on Categories and Commutative Algebra, Varenna 1971, Edizioni Cremonese 1973, 249–326. **MR** 50/7277.

[WB] G. C. Wraith, Lectures on Elementary Topoi. "Model Theory and Topoi", Springer Lecture Notes in Math. **445** (1975), 114–206. **MR** 52/13989.

[WI] G. C. Wraith, Logic from Topology: a survey of Topoi. *Bull. Inst. Math. and its Applications* **12** (1976), 115–119.

C: OTHER PAPERS ON TOPOS THEORY AND RELATED SUBJECTS

[1] C. Anghel and P. Lecouturier, Généralisation d'un résultat sur le triple de la réunion. *Ann. Fac. Sci. de Kinshasa (Zaire), Section Math.-Phys.*, **1** (1975), 65–94. **MR** 53/582.

[2] M. Barr, Exact categories. "Exact categories and categories of sheaves", Springer Lecture Notes in Math. **236** (1971), 1–120.
[3] M. Barr, The point of the empty set. *Cahiers top. et géom. diff.* **XIII** (1972), 357–368. **MR** 48/2216.
[4] M. Barr, Toposes without points. *J. Pure and Applied* **5** (1974), 265–280.
[5] M. Barr, Atomic Sites. *J. Pure and Applied Algebra*, to appear.
[6] J. Bénabou, "Les distributeurs". Univ. Cath. de Louvain, Inst. de Math. Pure et Appliquée, Rapport no. 33 (1973).
[7] J. Bénabou, "Problèmes dans les topos". Ibid., Rapport no. 34 (1973).
[8] J. Bénabou, Théories relatives à un corpus. *C.R. Acad. Sci. Paris* **281** (1975), A831–834. **MR** 52/13990.
[9] J. Bénabou, Fibrations petites et localement petites. *Ibid.*, A897–900.
[10] A. Blass, Exact functors and measurable cardinals. *Pacific J. Math.* **63** (1976), 335–346.
[11] A. Boileau, "Types vs. topos". Preprint, Université de Montreal 1975.
[12] F. Borceux, When is Ω a cogenerator in a topos? *Cahiers top. et géom. diff.* **XVI** (1975), 3–15. **MR** 52/3277.
[13] D. Bourn, Ditopos. *C.R. Acad. Sci. Paris* **279** (1974), A731–732 and A911–913. **MR** 51/639.
[14] T. G. Brook, "Order and recursion in topoi". A.N.U. Notes in Pure Math. (to appear).
[15] M. C. Bunge, Topos theory and Souslin's hypothesis. *J. Pure and Applied Algebra* **4** (1974), 159–187. **MR** 51/2908.
[16] M. C. Bunge, "Topoi of internal presheaves". Universidad Nacional Autonoma de Mexico, Comunicaciones tecnicas, serie naranja no. 132 (1976).
[17] J. Celeyrette, "Fibrations et extensions de Kan". Thèse de 3e cycle, Université Paris-Nord 1974.
[18] J. C. Cole, Categories of sets and models of set theory. Proc. Bertrand Russell Memorial Logic Conference, Uldum 1971 (ed. J. Bell and A. Slomson), Leeds 1973, 351–399. **MR** 50/9584.
[19] J. C. Cole, The bicategory of topoi, and spectra. *J. Pure and Applied Algebra* (to appear).
[20] F. Conduché, Au sujet de l'existence d'adjoints à droite aux foncteurs "image réciproque" dans la catégorie des catégories. *C.R. Acad. Sci. Paris* **275** (1972), A891–894. **MR** 46/9136.
[21] M. F. Coste, Construction d'un modèle booléen de la théorie des ensembles à partir d'un topos booléen. *C.R. Acad. Sci. Paris* **278** (1974), A1073–1076. **MR** 49/5114.
[22] M. F. Coste and M. Coste, "Théories cohérentes et topos cohérents". Seminaire Bénabou, Université Paris-Nord 1975.
[23] M. F. Coste, M. Coste and J. Parent, "Algèbres de Heyting dans les topos". Seminaire Bénabou, Université Paris-Nord 1974.
[24] M. Coste, "Langage interne d'un topos". Seminaire Bénabou, Université Paris-Nord 1972.
[25] M. Coste, "Logique du premier ordre dans les topos élémentaires". Seminaire Bénabou, Université Paris-Nord 1973.
[26] M. Coste, "Logique d'ordre supérieur dans les topos élémentaires". Seminaire Bénabou, Université Paris-Nord 1974.
[27] M. Coste, "Une approche logique des théories definissables par limites projectives finies". Seminaire Bénabou, Université Paris-Nord 1976.

[28] B. J. Day, An adjoint-functor theorem over topoi. *Bull. Austral. Math. Soc.* **15** (1976), 381–394.

[29] J. P. Delale, Ensemble sous-jacent dans un topos. *C.R. Acad. Sci. Paris* **277** (1973), A153–156. **MR** 51/8197.

[30] R. Diaconescu, Change of base for toposes with generators. *J. Pure and Applied Algebra* **6** (1975), 191–218. **MR** 52/532.

[31] R. Diaconescu, Axiom of choice and complementation. *Proc. Amer. Math. Soc.* **51** (1975), 176–178. **MR** 51/10093.

[32] J. W. Duskin, "Simplicial methods and the interpretation of "triple" cohomology". *Memoirs Amer. Math. Soc.* **163** (1975). **MR** 52/14006.

[33] H. Engenes, Subobject classifiers and classes of subfunctors. *Math. Scand.* **34** (1974), 145–152. **MR** 51/8200.

[34] H. Engenes, "Uniform spaces in topoi". Oslo Universitet Math. Preprint Series no. 5 (1976).

[35] M. P. Fourman, The logic of topoi. "Handbook in Logic" (ed. J. Barwise), North-Holland (to appear).

[36] M. P. Fourman, Comparaison des réelles d'un topos; structures lisses sur un topos élémentaire. *Cahiers top. et géom. diff.* **XVI** (1976), 233–239.

[37] P. Freyd, "On canonizing category theory, or, On functorializing model theory". Pamphlet, University of Pennsylvania 1974.

[38] J. Giraud, "Cohomologie non abélienne". Die Grundlehren der math. Wissenschaften, Band 179, Springer-Verlag 1971, **MR** 49/8992.

[39] J. Giraud, Classifying topos. "Toposes, Algebraic Geometry and Logic" (ed. F. W. Lawvere), Springer Lecture Notes in Math. **274** (1972), 43–56. **MR** 50/2300.

[40] P. A. Grillet, Directed colimits and sheaves in some non-abelian categories. "Reports of the Midwest Category Seminar, V" (ed. J. W. Gray), Springer Lecture Notes in Math. **195** (1971), 36–69. **MR** 44/6787.

[41] P. A. Grillet, Regular categories. "Exact categories and categories of sheaves", Springer Lecture Notes in Math. **236** (1971), 121–222.

[42] A. Grothendieck, Sur quelques points d'algèbre homologique. *Tohoku Math. J.* **9** (1957), 119–221. **MR** 21/1328.

[43] A. Grothendieck, Technique de descente et théorèmes d'existence en Géometrie Algébrique. Seminaire Bourbaki, exposés **190**, **195**, **212**, **221**, **232**, **236** (1959–1962). **MR** 26/3566.

[44] A. Grothendieck, "Revêtements étales et groupe fondamental" (SGA 1). Second edition, Springer Lecture Notes in Math. **224** (1971). **MR** 50/7129.

[45] M. Hakim, "Topos annelés et schémas relatifs". Ergebnisse der Mathematik, Band 64, Springer-Verlag, 1972. **MR** 51/500.

[46] A. Heller and K. A. Rowe, On the category of sheaves. *Amer. J. Math.* **84** (1962), 205–216. **MR** 26/1887.

[47] D. Higgs, "A category approach to boolean-valued set theory". Lecture Notes, University of Waterloo 1973.

[48] H. L. Hiller, Fibrations and Grothendieck topologies. *Bull. Austral. Math. Soc.* **14** (1976), 111–128.

[49] K. H. Hofmann, Representations of algebras by continuous sections. *Bull. Amer. Math. Soc.* **78** (1972), 291–373. **MR** 50/415.

[50] L. Illusie, "Complexe cotangent et déformations". Springer Lecture Notes in Math. **239** and **283** (1971–2).

[51] J. R. Isbell, Atomless parts of spaces. *Math. Scand.* **31** (1972), 5–32. **MR** 50/11184.

[52] P. T. Johnstone, The associated sheaf functor in an elementary topos. *J. Pure and Applied Algebra* **4** (1974), 231–242. **MR** 50/10002.
[53] P. T. Johnstone, Internal categories and classification theorems. "Model Theory and Topoi", Springer Lecture Notes in Math. **445** (1975), 103–113. **MR** 53/581.
[54] P. T. Johnstone, Adjoint lifting theorems for categories of algebras. *Bull. Lond. Math. Soc.* **7** (1975), 294–297. **MR** 52/10845.
[55] P. T. Johnstone, Rings, Fields and Spectra. *J. Algebra* (to appear).
[56] P. T. Johnstone, Automorphisms of Ω. To appear.
[57] P. T. Johnstone, On a topological topos. To appear.
[58] P. T. Johnstone and G. C. Wraith, Algebraic theories in toposes. To appear.
[59] O. Keane, Abstract Horn Theories. "Model Theory and Topoi", Springer Lecture Notes in Math. **445** (1975), 15–50. **MR** 52/2871.
[60] J. F. Kennison, Integral domain type representations in sheaves and other topoi. *Math. Zeitschrift* **151** (1976), 35–56.
[61] A. Kock, "Linear algebra and projective geometry in the Zariski topos". Aarhus Universitet Preprint Series 1974/75, no. 4.
[62] A. Kock, Linear algebra in a local ringed site. *Commun. Algebra* **3** (1975), 545–561. **MR** 52/5690.
[63] A. Kock, Universal projective geometry via topos theory. *J. Pure and Applied Algebra* **9** (1976), 1–24.
[64] A. Kock, A simple axiomatics for differentiation. *Math. Scand.* (to appear).
[65] A. Kock, P. Lecouturier and C. J. Mikkelsen, Some topos-theoretic concepts of finiteness. "Model Theory and Topoi", Springer Lecture Notes in Math. **445** (1975), 209–283. **MR** 52/2882.
[66] A. Kock and C. J. Mikkelsen, "Non-standard extensions in the theory of toposes". Aarhus Universitet Preprint Series 1971/72, no. 25.
[67] A. Kock and C. J. Mikkelsen, Topos-theoretic factorization of non-standard extensions. Proc. Victoria Symposium on Nonstandard Analysis (ed. A. Hurd and P. Loeb), Springer Lecture Notes in Math. **369** (1974), 122–143.
[68] A. Kock and G. E. Reyes, Doctrines in categorical logic. "Handbook in Logic" (ed. J. Barwise), North-Holland (to appear).
[69] A. Labella, Costruzione del monoide dei quozienti in un topos elementare. *Rend. Mat.* (VI) **7** (1974), 151–168. **MR** 50/10009.
[70] J. Lambek and B. A. Rattray, Localization and sheaf reflectors. *Trans. Amer. Math. Soc.* **210** (1975), 279–293.
[71] F. W. Lawvere, An elementary theory of the category of sets. *Proc. Nat. Acad. Sci.* **52** (1964), 1506–1511. **MR** 30/3025.
[72] F. W. Lawvere, The category of categories as a foundation for mathematics. Proc. La Jolla conference on Categorical Algebra, Springer-Verlag 1966, 1–20. **MR** 34/7332.
[73] F. W. Lawvere, Diagonal arguments and cartesian closed categories. "Category Theory, Homology Theory and their Applications, II", Springer Lecture Notes in Math. **92** (1969), 134–145. **MR** 39/4075.
[74] F. W. Lawvere, Adjointness in foundations. *Dialectica* **23** (1969), 281–296.
[75] F. W. Lawvere, Equality in Hyperdoctrines and the Comprehension Schema as an adjoint functor. Proc. New York Symposium on Applications of Categorical Algebra (ed. A. Heller), American Mathematical Society 1970, 1–14. **MR** 41/1829.

[76] F. W. Lawvere, "Teoria delle categorie sopra un topos di base". Lecture Notes, University of Perugia 1973.

[77] P. Lecouturier, "Quantificateurs dans les topos élémentaires". Preprint, Université Nationale du Zaire 1972.

[78] B. Lesaffre, Structures algébriques dans les topos élémentaires. *C.R. Acad. Sci Paris* **277** (1973), A663–666. **MR** 48/8597.

[79] G. Loullis, "Some aspects of the model theory in a topos". Ph.D. thesis, Yale University 1976.

[80] L. Mahé, Topos infinitesimal. *C.R. Acad. Sci. Paris* **277** (1973), A497–500. **MR** 49/8986.

[81] L. Mahé, Théorèmes, exemples et contre-examples pour certaines propriétés des entiers naturels dans les topos. Ibid. **282** (1976), A1273–1276.

[82] M. Makkai and G. E. Reyes, Model-theoretic methods in the theory of topoi and related categories. *Bull. Acad. Pol. Sci.* **XXIV** (1976), 379–392.

[83] C. Maurer, Universes in topoi. "Model Theory and Topoi", Springer Lecture Notes in Math. **445** (1975), 284–296. **MR** 52/95.

[84] C. J. Mikkelsen, "Lattice-theoretic and logical aspects of elementary topoi". Aarhus Universitet Various Publications Series **25** (1976).

[85] W. Mitchell, Boolean topoi and the theory of sets. *J. Pure and Applied Algebra* **2** (1972), 261–274. **MR** 47/8299.

[86] W. Mitchell, On topoi as closed categories. Ibid. **3** (1973), 133–139. **MR** 50/427.

[87] W. Mitchell, Categories of Boolean topoi. Ibid. **3** (1973), 193–201. **MR** 50/9585.

[88] C. J. Mulvey, Intuitionistic algebra and representations of rings. *Memoirs Amer. Math. Soc.* **148** (1974), 3–57. **MR** 53/2650.

[89] C. J. Mulvey, A generalization of Swan's theorem. *Math. Zeitschrift* **151** (1976), 57–70.

[90] C. J. Mulvey, Compact ringed spaces. To appear.

[91] C. J. Mulvey, The real numbers in a topos. To appear.

[92] G. Osius, Categorical Set Theory: a characterization of the category of sets. *J. Pure and Applied Algebra* **4** (1974), 79–119. **MR** 51/643.

[93] G. Osius, The internal and external aspects of logic and set theory in elementary topoi. *Cahiers top. et géom. diff.* **XV** (1974), 157–180. **MR** 52/7896.

[94] G. Osius, Logical and set-theoretical tools in elementary topoi. "Model Theory and Topoi", Springer Lecture Notes in Math. **445** (1975), 297–346.

[95] G. Osius, A note on Kripke–Joyal semantics for the internal language of topoi. Ibid., 349–354. **MR** 52/7898.

[96] R. Paré, Colimits in topoi. *Bull. Amer. Math. Soc.* **80** (1974), 556–561. **MR** 48/11245.

[97] R. Paré and D. Schumacher, Abstract families and the Adjoint Functor Theorems. To appear.

[98] A. M. Penk, Two forms of the axiom of choice for an elementary topos. *J. Symbolic Logic* **40** (1975), 197–212. **MR** 51/5698.

[99] J. Penon, Quasitopos. *C.R. Acad. Sci. Paris* **276** (1973), A237–240. **MR** 48/11246.

[100] J. Penon, Catégories localement internes. Ibid. **278** (1974), A1577–1580. **MR** 51/10435.

[101] R. S. Pierce, "Modules over commutative regular rings". *Memoirs Amer. Math. Soc.* **70** (1967). **MR** 36/151.

[102] G. E. Reyes, "Sheaves and Concepts: a model-theoretic introduction to Grothendieck topoi". Aarhus Universitet Preprint Series 1975/76, no. 2.

[103] G. E. Reyes, "Théorie des modèles et faisceaux". Univ. Cath. de Louvain, Inst. de Math. Pure et Appliquée, Rapport no. 63 (1976).

[104] J. E. Roos, Sur la distributivité des foncteurs $\varprojlim$ par rapport aux $\varinjlim$ dans les catégories de faisceaux (topos). *C.R. Acad. Sci. Paris* **259** (1964), 969–972, 1605–1608 and 1801–1804. **MR** 30/4816, 32/5714–5.

[105] D. Schlomiuk, An elementary theory of the category of topological spaces. *Trans. Amer. Math. Soc.* **149** (1970), 259–278. **MR** 41/3559.

[106] D. S. Scott and M. P. Fourman, Logic and Sheaves. To appear.

[107] J. P. Serre, Faisceaux algébriques cohérents. *Ann. Math.* (2) **61** (1955), 197–278. **MR** 16, p. 953.

[108] I. Sols, Bon ordre dans l'objet des nombres naturels d'un topos booléen. *C.R. Acad. Sci. Paris* **281** (1975), A601–603. **MR** 52/96.

[109] I. Sols, Programming in topoi. *Cahiers top. et géom. diff.* **XVI** (1976), 312–319.

[110] L. N. Stout, Topological space objects in a topos I: Variable spaces for variable sets. To appear.

[111] L. N. Stout, Topological space objects in a topos II: $\mathscr{E}$-completeness and cocompleteness. *Manuscripta Math.* **17** (1975), 1–14.

[112] L. N. Stout, Topological properties of the real numbers object in a topos. *Cahiers top. et géom. diff.* **XVII** (1976), 295–326.

[113] R. Street, Elementary cosmoi I. Proc. Sydney Category Seminar 1973, Springer Lecture Notes in Math. **420** (1974), 134–180. **MR** 50/7290.

[114] R. Street, Cosmoi of internal categories. *Trans. Amer. Math. Soc.* (to appear).

[115] R. Street and R. F. C. Walters, The comprehensive factorization of a functor. *Bull. Amer. Math. Soc.* **79** (1973), 936–941. **MR** 49/10753.

[116] R. Street and R. F. C. Walters, Yoneda structures on 2-categories. *J. Algebra* (to appear).

[117] M. Tierney, Sheaf Theory and the Continuum Hypothesis. "Toposes, Algebraic Geometry and Logic" (ed. F. W. Lawvere), Springer Lecture Notes in Math. **274** (1972), 13–42. **MR** 51/10088.

[118] M. Tierney, On the spectrum of a ringed topos. "Algebra, Topology and Category Theory: a collection of papers in honor of Samuel Eilenberg" (ed. A. Heller and M. Tierney), Academic Press 1976, 189–210.

[119] M. Tierney, Forcing topologies and classifying topoi. Ibid., 211–219.

[120] F. Ulmer, On the existence and exactness of the associated sheaf functor. *J. Pure and Applied Algebra* **3** (1973), 295–306. **MR** 51/10431.

[121] G. Van de Wauw-de Kinder, Arithmetique de premier ordre dans les topos. *C.R. Acad. Sci. Paris* **280** (1975), A1579–1582.

[122] D. H. Van Osdol, Sheaves in regular categories. "Exact categories and categories of sheaves", Springer Lecture Notes in Math. **236** (1971), 223–239.

[123] D. H. Van Osdol, Coalgebras, Sheaves and Cohomology. *Proc. Amer. Math. Soc.* **33** (1972), 257–263. **MR** 45/3517.

[124] D. H. Van Osdol, Homological algebra in topoi. Ibid. **50** (1975), 52–54. **MR** 51/5714.

[125] H. Volger, Completeness theorem for logical categories. "Model Theory and Topoi", Springer Lecture Notes in Math. **445** (1975), 51–86. **MR** 51/12983.

[126] H. Volger, Logical categories, semantical categories and topoi. Ibid., 87–100. **MR** 51/12984.

[127] H. Volger, Ultrafilters, ultrapowers and finiteness in a topos. *J. Pure and Applied Algebra* **6** (1975), 345–356. **MR** 52/84.

[128] H. Wolff, Flat epimorphisms in Cat. *Math. Nachrichten* **69** (1975), 243–252.

[129] G. C. Wraith, Artin glueing. *J. Pure and Applied Algebra* **4** (1974), 345–348. **MR** 49/9040.

[130] O. Wyler, Are there topoi in topology? Proc. Mannheim conference on Categorical Topology, Springer Lecture Notes in Math. **540** (1976), 699–719.

[131] A. Carboni and G. C. Meloni, Costruzione algebrica dello spazio étalé. *Boll. Un. Mat. Ital.* (4) **12** (1975), 192–197. **MR** 52/13998.

[132] P. Deligne (rédigé par J. F. Boutot), Cohomologie étale: les points de départ. "Cohomologie étale" (SGA 4½), Springer Lecture Notes in Math **569** (1977), 4–75.

[133] R. Diaconescu, Grothendieck toposes have Boolean points—a new proof. *Commun. Algebra* **4** (1976), 723–729.

[134] M. P. Fourman, "Sheaves over cHa and their logic". Lecture notes, Rijksuniversitet Utrecht 1977.

[135] R. Guitart, Calcul des relations inverses. *Cahiers top. et géom. diff.* **XVIII** (1977), 67–100.

[136] A. Radu, Sur les topos de Giraud-Grothendieck. *Ann. Fac. Sci. de Kinshasa (Zaire), Section Math.-Phys.* **2** (1976), 108–118.

[137] C. Rousseau, "Topos theory and complex analysis". Ph.D. thesis, Université de Montréal 1977.

[138] D. Schumacher, Absolutely free algebras in a topos containing an infinite object. *Canad. Math. Bull.* **19** (1976), 323–328.

[139] G. van den Bossche, "Relèvements conservatifs de morphismes conservatifs de topos". Univ. Cath. de Louvain, Inst. de Math. Pure et Appliquée, Rapport no. 65 (1977).

D: MISCELLANEOUS PAPERS

[151] R. Baer, Abelian groups that are direct summands of every containing abelian group. *Bull. Amer. Math. Soc.* **46** (1940), 800–806. **MR** 2, p. 126.

[152] M. Barr and J. Beck, Homology and standard constructions. "Seminar on Triples and Categorical Homology Theory", Springer Lecture Notes in Math. **80** (1969), 245–335. **MR** 41/3562.

[153] J. Beck, "Triples, algebras and cohomology". Ph.D. thesis, Columbia University 1967.

[154] J. Bénabou, Catégories avec multiplication. *C.R. Acad. Sci. Paris* **256** (1963), 1887–1890. **MR** 26/6225.

[155] J. Bénabou, Introduction to bicategories. "Reports of the Midwest Category Seminar, I", Springer Lecture Notes in Math. **47** (1967), 1–77. **MR** 36/3841.

[156] P. J. Cohen, "Set Theory and the Continuum Hypothesis". W. A. Benjamin Inc., 1966. **MR** 38/999.

[157] J. H. Conway, "On Numbers and Games". L.M.S. Mathematical Monographs no. 6, Academic Press 1976.

[158] B. J. Day, Limit spaces and closed span categories. Proc. Sydney Category Seminar 1973, Springer Lecture Notes in Math. **420** (1974), 65–74. **MR** 51/10440.

[159] P. Deligne, La conjecture de Weil, I. I.H.E.S. Publ. Math. **43** (1974), 273–307. **MR** 49/5013.

[160] S. Eilenberg and G. M. Kelly, Closed categories. Proc. La Jolla conference on Categorical Algebra, Springer-Verlag 1966, 421–562. **MR** 37/1432.

[161] S. Eilenberg and S. MacLane, Cohomology theory in abstract groups. *Ann. Math.* (2) **48** (1947), 51–78 and 326–341. **MR** 8, p. 367 and 9, p. 7.

[162] S. Eilenberg and J. C. Moore, Adjoint functors and triples. *Illinois J. Math.* **9** (1965), 381–398. **MR** 32/2455.

[163] S. Fakir, Monade idempotente associée à une monade. *C.R. Acad. Sci. Paris* **270** (1970), A99–101. **MR** 41/1828.

[164] N. Funayama, On imbedding infinitely distributive lattices completely isomorphically into Boolean algebras. *Nagoya Math. J.* **15** (1959), 71–81. **MR** 21/6341.

[165] P. Gabriel and F. Ulmer, "Lokal präsentierbare Kategorien". Springer Lecture Notes in Math. **221** (1971). **MR** 48/6205.

[166] J. W. Gray, Fibred and cofibred categories. Proc. La Jolla conference on Categorical Algebra, Springer-Verlag 1966, 21–83. **MR** 35/4277.

[167] J. W. Gray, "Formal Category Theory I: Adjointness for 2-categories". Springer Lecture Notes in Math. **391** (1974). **MR** 51/8207.

[168] J. W. Gray, "Formal Category Theory II". To appear.

[169] M. Hochster, Prime ideal structure in commutative rings. *Trans. Amer. Math. Soc.* **142** (1969), 43–60. **MR** 40/4257.

[170] B. Jonsson and A. Tarski, On two properties of free algebras. *Math. Scand.* **9** (1961), 95–101. **MR** 23/A3695.

[171] D. M. Kan, Adjoint functors. *Trans. Amer. Math. Soc.* **87** (1958), 295–329. **MR** 24/A1301.

[172] G. M. Kelly and R. Street, Review of the elements of 2-categories. Proc. Sydney Category Seminar 1973, Springer Lecture Notes in Math. **420** (1974), 75–103. **MR** 50/10010.

[173] S. A. Kripke, Semantic analysis of intuitionistic logic I. "Formal Systems and Recursive Functions" (ed. J. N. Crossley and M. A. Dummett), North-Holland 1965, 92–130. **MR** 34/1184.

[174] C. Kuratowski, Sur la notion d'ensemble fini. *Fund. Math.* **1** (1920), 130–131.

[175] C. Kuratowski, Sur l'opération $\bar{A}$ de l'analysis situs. Ibid. **3** (1922), 182–199.

[176] F. W. Lawvere, Functorial semantics of algebraic theories. *Proc. Nat. Acad. Sci.* **50** (1963), 869–872. **MR** 28/2143.

[177] F. E. J. Linton, Some aspects of equational categories. Proc. La Jolla conference on Categorical Algebra, Springer-Verlag 1966, 84–94. **MR** 35/233.

[178] F. E. J. Linton, Coequalizers in categories of algebras. "Seminar on Triples and Categorical Homology Theory", Springer Lecture Notes in Math. **80** (1969), 75–90. **MR** 39/5656.

[179] S. MacLane, A History of Abstract Algebra: origin, rise and decline of a movement. To appear.

[180] E. G. Manes, "Algebraic Theories". Graduate Texts in Mathematics no. 26, Springer-Verlag, 1976.

[181] A. Mostowski, An undecidable arithmetical statement. *Fund. Math.* **36** (1949), 143–164. **MR** 12, p. 2.

[182] D. S. Scott, Boolean models and nonstandard analysis. "Applications of Model Theory to Algebra, Analysis and Probability" (ed. W. A. J. Luxemburg), Holt, Rinehart and Winston 1969, 87–92. **MR** 38/4300.

[183] J. P. Serre, "Cohomologie galoisienne". Springer Lecture Notes in Math. **5** (1964). **MR** 31/4785.

[184] R. Street, Two constructions on lax functors. *Cahiers top. et géom. diff.* **XIII** (1972), 217–264. **MR** 50/436.

[185] A. Tarski, A lattice-theoretic fixpoint theorem and its applications. *Pacific J. Math.* **5** (1955), 285–309. **MR** 17, p. 574.

[186] E. J. Thiele, Über endlich axiomatisierbare Teilsysteme der Zermelo-Fraenkel' schen Mengenlehre. *Z. Math. Logik & Grundl. Math.* **14** (1968), 39–58. **MR** 36/6283.

[187] N. Yoneda, On the homology theory of modules. *J. Fac. Sci. Tokyo, Sec. I*, **7** (1954), 193–227. **MR** 16, p. 947.

Index of Definitions

Index of Notation

Symbol	Meaning	Page
ab	Theory of abelian groups	100, 259
AC	Axiom of choice	141
A_{et}	Etale topos of A	22
alg	Theory of algebraic theories	195
ann	Theory of commutative rings	21, 202
$\mathfrak{BTop}/\mathscr{E}$	2-category of bounded $\mathscr{E}$-toposes	131
bvm	Theory of boolean-valued models	322
cat	Theory of categories	47, 202
$\mathfrak{Cat}$	2-category of categories	107
$\mathfrak{cat}(\mathscr{E})$	2-category of internal categories in $\mathscr{E}$	72
$\mathfrak{Cat}(\mathscr{E})$	2-category of locally internal categories over $\mathscr{E}$	335
$\mathscr{C}(G)$	Category of continuous G-sets	288
$\mathscr{C}_f(G)$	Category of continuous finite G-sets	288
CH	Continuum hypothesis	324
$(\mathbf{C}_{\mathbb{T}}, J_{\mathbb{T}})$	Syntactic site of $\mathbb{T}$	245
C_Y	Sheaf of continuous Y-valued functions	9
$\mathscr{C}\Sigma^{-1}$	Category of fractions of $\mathscr{C}$ relative to Σ	6
ded	Theory of Dedekind real numbers	212
dofib	Theory of discrete opfibrations	209
$\mathfrak{DTop}/\mathscr{E}$	2-category of internal diagram toposes over $\mathscr{E}$	185
$\mathscr{E}^{\mathbf{C}}$	Category of internal diagrams on $\mathbf{C}$	49
$\mathscr{E}$-$\mathfrak{Cat}$	2-category of $\mathscr{E}$-categories	335
$\mathscr{E}_{\mathrm{coh}}$	Category of coherent objects of $\mathscr{E}$	236
$\mathscr{E}_{fc}$	Category of finite cardinals in $\mathscr{E}$	179
$\mathbf{E}_{\mathrm{fin}}$	Internal category of finite cardinals in $\mathscr{E}$	180
$\mathbf{E}_{\mathrm{finepi}}$	Internal category of finite cardinals and epimorphisms in $\mathscr{E}$	331

Symbol	**Meaning**	**Page**
$\mathscr{E}_{kf}$	Category of Kuratowski-finite objects in $\mathscr{E}$	296
$\mathscr{E}_{lcf}$	Category of locally constant finite objects in $\mathscr{E}$	285
end	Theory of objects-with-an-endomorphism	221
$\mathrm{Epi}(X, Y)$	Object of epimorphisms $X \twoheadrightarrow Y$	157
$\mathscr{E}_{ptr}$	Category of partially transitive objects in $\mathscr{E}$	311
eq	Equalizer map	39, 157
esp	Category of topological spaces	10
$\mathscr{E}_{tr}$	Category of transitive objects and inclusions in $\mathscr{E}$	308
$\mathscr{E}[\mathbb{T}]$	Classifying topos for $\mathbb{T}$	184, 203
$\mathscr{E}[U]$	Object classifier	117, 182
ev	Evaluation map	31
$\mathrm{ext}\,(j)$	Exterior of a topology j	102
$\mathscr{E}_{\Phi}$	Filterpower of $\mathscr{E}$ modulo Φ	320
$\mathscr{E} \models \phi$	ϕ is universally valid in $\mathscr{E}$	155
$f^{\#}, f_{\#}$	Profunctors associated with an internal functor f	61
$f^{\mathbf{C}}$	Pullback of a geometric morphism f along $\mathscr{E}^{\mathbf{C}} \longrightarrow \mathscr{E}$	114
$F \otimes_{\mathbf{C}} G$	Tensor product of profunctors	59
filt	Theory of filtered categories	202
$\mathrm{Flat}\,(\mathbf{C}^{\mathrm{op}}, \mathscr{E})$	Category of flat presheaves on $\mathbf{C}$	113
$\mathrm{Full}_{\mathscr{E}}\,(f)$	Internal full subcategory of $\mathscr{E}$ generated by f	58, 340
$\mathbb{G}_{\mathbf{C}}$	Comonad defined by an internal category $\mathbf{C}$	54
$G(\mathbf{D})$	Grothendieck category of $\mathbf{D}$	73
G-**esp**	Category of G-spaces and equivalent continuous maps	20
$\mathrm{Gl}(\Gamma, \alpha)$	Topos obtained by glueing (Γ, α)	109
$\mathbb{H}^q(A)$	qth cohomology presheaf of A	266
$H^q(\mathscr{E}; A)$	qth cohomology group of $\mathscr{E}$	261
$H^q(\mathscr{E}, X; A)$	qth cohomology group of X	261
$\check{H}^q(\mathscr{U}; A)$	qth Čech cohomology group of $\mathscr{U}$	268
$\check{H}^q(U; A)$	qth Čech cohomology group of U	270
$H^q_{\Phi}(\mathscr{E}; A)$	qth cohomology group of $\mathscr{E}$ with cosupports in Φ	294
$h_U, h(U)$	Contravariant functor represented by U	1
h^U	Covariant functor represented by U	1
IC	Implicit axiom of choice	143
$I(\mathbf{C})$	Internalization of $\mathbf{C}$	58
im	Image map	157

Symbol	Meaning	Page
int	Theory of intervals	118
int(j)	Interior of a topology j	102
Iso(X, Y)	Object of isomorphisms $X \xrightarrow{\cong} Y$	157
$\mathbf{jt}_2$	Binary Jonsson-Tarski theory	222
j_U^o, j_U^c	Open and closed topologies determined by U	94
ker	Kernel-pair map	157
kpr	Kuratowskian ordered-pair map	164
$K(X)$	Kuratowski semilattice of X	296
l	The canonical continuous functor	225
lann	Theory of local rings	202
LC	Local-internalization of **C**	338
lcn(Γ, α; X)	Category of lax cones from (Γ, α) to X	108
$L_{\mathscr{E}}$	Mitchell–Bénabou language of $\mathscr{E}$	153
lfProf	Category of left flat profunctors	210
$\varinjlim$	Object-of-components functor	51
$\varinjlim_{\mathbf{C}}$, $\varprojlim_{\mathbf{C}}$	Internal colimit and limit functors over **C**	52
$\varinjlim_f$, $\varprojlim_f$	Left and right Kan extensions along f	56
$\mathfrak{Log}$	2-category of toposes and logical functors	331
mon	Theory of monoids	190
Mono(X, Y)	Object of monomorphisms $X \rightarrowtail Y$	157
N	Natural number object	165
$\mathbb{N}^+$	Test space for sequential convergence	21
NN	Axiom of infinity	318
On	Category of ordinals	45
$[p]$	Cardinal of p	173
$\mathfrak{Prof}_{\mathscr{E}}$	Bicategory of internal profunctors in $\mathscr{E}$	62
Prof$_{\mathscr{E}}$(**C**, **D**)	Category of profunctors from **C** to **D**	59
PT	Axiom of partial transitivity	314
PX	Internal power-object of X	31, 43
$\mathscr{P}X$	External poset of subobjects of X	140
Q	Object of rationals	211
q_U	Quasi-closed topology determined by U	252

Symbol	**Meaning**	**Page**
R_c	Object of Cauchy real numbers	218
R_d	Object of Dedekind real numbers	212
$\mathscr{S}$	Category of sets	1
$\mathscr{S}_f$	Category of finite sets	25
$\uparrow$seg, $\downarrow$seg	Up-segment, down-segment	147
SG	Subobjects of 1 generate	145
$\mathrm{sh}_j(\mathscr{E})$	Category of sheaves for a topology j in $\mathscr{E}$	81
$\mathrm{Shv}(\mathbf{C}, J)$	Category of sheaves on a site $(\mathbf{C}, J)$	14
$\mathrm{Shv}(X)$	Category of sheaves on a space X	10
$\mathrm{Shv}_G(X)$	Category of G-equivariant sheaves on X	20
simpl	Theory of simplicial objects	48
slat	Theory of semilattices	296
sob	Category of sober spaces	230
spec A	Prime-spectrum of a ring A	208
$\mathrm{Spec}(\mathscr{E}, M)$	Spectrum of an $\mathbb{S}$-modelled topos	207
SS	Supports split	141
$\mathfrak{S}\mathfrak{T}\mathfrak{o}\mathfrak{p}$	2-category of spatial toposes	232
$\mathbb{T}_\mathbf{C}$	Monad defined by an internal category $\mathbf{C}$	51
$\mathbb{T}(\mathscr{E})$	Category of $\mathbb{T}$-models in $\mathscr{E}$	192, 196, 201
$\mathfrak{T}\mathfrak{o}\mathfrak{p}$	2-category of toposes and geometric morphisms	26
$\mathfrak{T}\mathfrak{o}\mathfrak{p}/\mathscr{E}$	2-category of $\mathscr{E}$-toposes	cf. 5
$\mathfrak{T}\mathfrak{o}\mathfrak{p}_N$	2-category of toposes with natural number objects	205
$\mathrm{Tors}^1(\mathscr{E}; G)$	Group of isomorphism classes of G-torsors in $\mathscr{E}$	275
TR	Axiom of transitive representatives	315
$\mathbb{T}$-$\mathfrak{T}\mathfrak{o}\mathfrak{p}$	2-category of $\mathbb{T}$-modelled toposes	205
TV	Axiom of two-valuedness	314
tvm	Theory of two-valued models	322
$\mathbb{T}^2$	Theory of morphisms of $\mathbb{T}$-models	203
wpt	Theory of well-pointed toposes	314
wtt	Theory of well-pointed, partially transitive toposes	314
$\tilde{X}$	Partial map representer for X	28
$\hat{X}$	Dense partial map representer for X	84
$\hat{X}$	Soberification of X	230
X^+	Half-sheafification of X	85

Symbol	Meaning	Page
$Y(\mathbf{C})$	Yoneda profunctor $\mathbf{C} \dashrightarrow \mathbf{C}$	61
zer	Weak Zermelo set theory	313
$\Gamma(E, p)$	Sheaf of sections of p	10
Δ	Diagonal subobject	16
δ	Classifying map of Δ	28
$\mathbf{\Delta}$	Simplicial category	48
$\Delta(S)$	Constant sheaf with stalk S	20
ν_i	Inclusion of ith factor in a coproduct	3
Ξ_D	Class of monomorphisms classified by $D \rightarrowtail \Omega$	79
Ξ_J	Class of j-bidense morphisms	91
Π_f	Right adjoint of pullback along f	36, 45
π_i	Projection on ith factor	2
ρ_V^U	Restriction map from U to V	9
$\ulcorner\sigma\urcorner$	"Name" of a subobject σ	156
Σ_f	Left adjoint of pullback along f	35
σ_U	Support functor	140
$\lvert\tau\rvert$	Interpretation of a term or formula of $L_{\mathscr{E}}$	153
ΥA	Group of units of A	198
$\lVert\phi\rVert$	Extension of a formula of $L_{\mathscr{E}}$	154
$\mathbf{\Phi}$	Class of Φ-invertible morphisms	319
$(\phi \vdash \psi)$	ϕ entails ψ	201
Ω	Subobject classifier	23
Ω_j	Closed-subobject classifier	76
$\mathbf{1}$	The degenerate topos	107
1_X	Identity morphism on X	2
$\exists f$	Internal existential quantification along f	32, 46
$\exists_f$	External existential quantification along f	144
$\forall_f$	(External) universal quantification along f	144
$\in_X$	Membership relation on X	32, 43
$\{\}$	Singleton map	27

INDEX OF NAMES

Mathematics–Logic and Problem Solving

PERPLEXING PUZZLES AND TANTALIZING TEASERS, Martin Gardner. Ninety-three riddles, mazes, illusions, tricky questions, word and picture puzzles, and other challenges offer hours of entertainment for youngsters. Filled with rib-tickling drawings. Solutions. 224pp. 5 3/8 x 8 1/2. 0-486-25637-5

MY BEST MATHEMATICAL AND LOGIC PUZZLES, Martin Gardner. The noted expert selects 70 of his favorite "short" puzzles. Includes The Returning Explorer, The Mutilated Chessboard, Scrambled Box Tops, and dozens more. Complete solutions included. 96pp. 5 3/8 x 8 1/2. 0-486-28152-3

THE LADY OR THE TIGER?: and Other Logic Puzzles, Raymond M. Smullyan. Created by a renowned puzzle master, these whimsically themed challenges involve paradoxes about probability, time, and change; metapuzzles; and self-referentiality. Nineteen chapters advance in difficulty from relatively simple to highly complex. 1982 edition. 240pp. 5 3/8 x 8 1/2. 0-486-47027-X

SATAN, CANTOR AND INFINITY: Mind-Boggling Puzzles, Raymond M. Smullyan. A renowned mathematician tells stories of knights and knaves in an entertaining look at the logical precepts behind infinity, probability, time, and change. Requires a strong background in mathematics. Complete solutions. 288pp. 5 3/8 x 8 1/2. 0-486-47036-9

THE RED BOOK OF MATHEMATICAL PROBLEMS, Kenneth S. Williams and Kenneth Hardy. Handy compilation of 100 practice problems, hints and solutions indispensable for students preparing for the William Lowell Putnam and other mathematical competitions. Preface to the First Edition. Sources. 1988 edition. 192pp. 5 3/8 x 8 1/2. 0-486-69415-1

KING ARTHUR IN SEARCH OF HIS DOG AND OTHER CURIOUS PUZZLES, Raymond M. Smullyan. This fanciful, original collection for readers of all ages features arithmetic puzzles, logic problems related to crime detection, and logic and arithmetic puzzles involving King Arthur and his Dogs of the Round Table. 160pp. 5 3/8 x 8 1/2. 0-486-47435-6

UNDECIDABLE THEORIES: Studies in Logic and the Foundation of Mathematics, Alfred Tarski in collaboration with Andrzej Mostowski and Raphael M. Robinson. This well-known book by the famed logician consists of three treatises: "A General Method in Proofs of Undecidability," "Undecidability and Essential Undecidability in Mathematics," and "Undecidability of the Elementary Theory of Groups." 1953 edition. 112pp. 5 3/8 x 8 1/2. 0-486-47703-7

LOGIC FOR MATHEMATICIANS, J. Barkley Rosser. Examination of essential topics and theorems assumes no background in logic. "Undoubtedly a major addition to the literature of mathematical logic." – *Bulletin of the American Mathematical Society.* 1978 edition. 592pp. 6 1/8 x 9 1/4. 0-486-46898-4

INTRODUCTION TO PROOF IN ABSTRACT MATHEMATICS, Andrew Wohlgemuth. This undergraduate text teaches students what constitutes an acceptable proof, and it develops their ability to do proofs of routine problems as well as those requiring creative insights. 1990 edition. 384pp. 6 1/2 x 9 1/4. 0-486-47854-8

FIRST COURSE IN MATHEMATICAL LOGIC, Patrick Suppes and Shirley Hill. Rigorous introduction is simple enough in presentation and context for wide range of students. Symbolizing sentences; logical inference; truth and validity; truth tables; terms, predicates, universal quantifiers; universal specification and laws of identity; more. 288pp. 5 3/8 x 8 1/2. 0-486-42259-3

Mathematics-Bestsellers

HANDBOOK OF MATHEMATICAL FUNCTIONS: with Formulas, Graphs, and Mathematical Tables, Edited by Milton Abramowitz and Irene A. Stegun. A classic resource for working with special functions, standard trig, and exponential logarithmic definitions and extensions, it features 29 sets of tables, some to as high as 20 places. 1046pp. 8 x 10 1/2. 0-486-61272-4

ABSTRACT AND CONCRETE CATEGORIES: The Joy of Cats, Jiri Adamek, Horst Herrlich, and George E. Strecker. This up-to-date introductory treatment employs category theory to explore the theory of structures. Its unique approach stresses concrete categories and presents a systematic view of factorization structures. Numerous examples. 1990 edition, updated 2004. 528pp. 6 1/8 x 9 1/4. 0-486-46934-4

MATHEMATICS: Its Content, Methods and Meaning, A. D. Aleksandrov, A. N. Kolmogorov, and M. A. Lavrent'ev. Major survey offers comprehensive, coherent discussions of analytic geometry, algebra, differential equations, calculus of variations, functions of a complex variable, prime numbers, linear and non-Euclidean geometry, topology, functional analysis, more. 1963 edition. 1120pp. 5 3/8 x 8 1/2. 0-486-40916-3

INTRODUCTION TO VECTORS AND TENSORS: Second Edition--Two Volumes Bound as One, Ray M. Bowen and C.-C. Wang. Convenient single-volume compilation of two texts offers both introduction and in-depth survey. Geared toward engineering and science students rather than mathematicians, it focuses on physics and engineering applications. 1976 edition. 560pp. 6 1/2 x 9 1/4. 0-486-46914-X

AN INTRODUCTION TO ORTHOGONAL POLYNOMIALS, Theodore S. Chihara. Concise introduction covers general elementary theory, including the representation theorem and distribution functions, continued fractions and chain sequences, the recurrence formula, special functions, and some specific systems. 1978 edition. 272pp. 5 3/8 x 8 1/2. 0-486-47929-3

ADVANCED MATHEMATICS FOR ENGINEERS AND SCIENTISTS, Paul DuChateau. This primary text and supplemental reference focuses on linear algebra, calculus, and ordinary differential equations. Additional topics include partial differential equations and approximation methods. Includes solved problems. 1992 edition. 400pp. 7 1/2 x 9 1/4. 0-486-47930-7

PARTIAL DIFFERENTIAL EQUATIONS FOR SCIENTISTS AND ENGINEERS, Stanley J. Farlow. Practical text shows how to formulate and solve partial differential equations. Coverage of diffusion-type problems, hyperbolic-type problems, elliptic-type problems, numerical and approximate methods. Solution guide available upon request. 1982 edition. 414pp. 6 1/8 x 9 1/4. 0-486-67620-X

VARIATIONAL PRINCIPLES AND FREE-BOUNDARY PROBLEMS, Avner Friedman. Advanced graduate-level text examines variational methods in partial differential equations and illustrates their applications to free-boundary problems. Features detailed statements of standard theory of elliptic and parabolic operators. 1982 edition. 720pp. 6 1/8 x 9 1/4. 0-486-47853-X

LINEAR ANALYSIS AND REPRESENTATION THEORY, Steven A. Gaal. Unified treatment covers topics from the theory of operators and operator algebras on Hilbert spaces; integration and representation theory for topological groups; and the theory of Lie algebras, Lie groups, and transform groups. 1973 edition. 704pp. 6 1/8 x 9 1/4. 0-486-47851-3